LIVERPOOL
JOHN MOORES UNIVERSITY
AVRIL ROBARTS LRC
TITHEBARN STR
LIVERPOOL L2 2
TEL. 0151 231 4

INSECTS IN A CHANGING ENVIRONMENT

INSECTS IN A CHANGING ENVIRONMENT

Edited by

Richard Harrington

Department of Entomology and Nematology, Rothamsted Experimental Station, Harpenden, Hertfordshire AL5 2JQ, UK

Nigel E. Stork

Department of Entomology, Natural History Museum, Cromwell Road, London, SW7 5BD, UK

**17th Symposium of the
Royal Entomological Society
7–10 September 1993
at
Rothamsted Experimental Station, Harpenden**

ACADEMIC PRESS

Harcourt Brace & Company, Publishers
London   San Diego   New York
Boston   Sydney   Tokyo   Toronto

ACADEMIC PRESS LIMITED
24/28 Oval Road, London NW1 7DX

United States Edition published by
ACADEMIC PRESS, INC.
San Diego, CA 92101

Copyright © 1995 by
The Royal Entomological Society of London

This book is printed on acid free paper

All rights reserved. No part of this book may be reproduced
or transmitted in any form or by any means, electronic or
mechanical, including photocopy, recording, or any
information storage and retrieval system without permission
in writing from the publisher

A CIP record for this book is available from the British Library

ISBN 0-12-326430-8

Typeset by Keyset Composition, Colchester
Printed in Great Britain by TJ Press (Padstow) Ltd, Padstow, Cornwall

Contents

Contributors xiii
Preface xvii

Part I. Introduction

1. The Response of Insects to Environmental Change
   J. H. LAWTON

I.	"Insects and Climate"	4
II.	Time Scales	6
III.	The Dynamics of Phytophagous Insect Populations	7
IV.	Migration and Geographic Range	12
V.	Genetic Effects	20
VI.	Concluding Remarks	22
	References	22

Part II. Changes in Climate

2. The Effects of Quaternary Climatic Changes in Insect Populations: Lessons from the Past
   G. R. COOPE

I.	Introduction	30
II.	Quaternary Entomology	31
III.	Quaternary Climatic Change	33
IV.	The Response of Insect Species to the Glacial/Interglacial Cycles	34
	References	47

3. The Hadley Centre Transient Climate Experiment
   D. A. BENNETTS

I.	Introduction	50
II.	Global Patterns of Transient Change	50
III.	Climate Change Over Western Europe	56
	References	58

4. Predicting Insect Distributions in a Changed Climate
   R. W. SUTHERST, G. F. MAYWALD AND D. B. SKARRAT

   I. Introduction 60
   II. Approaches to Comparing Climates Without Reference to
       Individual Species 61
   III. Approaches to Estimating Potential Geographical Distributions
        of Individual Species 61
   IV. Reliability and Accuracy of Data Sets Used to Estimate
       Potential Distributions 74
   V. Comparison of Approaches 74
   VI. Applications of the Above Methods to Estimating the Impacts
       of Climate Change 78
   VII. Adaptative Strategies to Changing Distributional Limits 85
   VIII. Discussion 86
   IX. Conclusions 87
       References 88

5. The Effects of Climate Change on the Agricultural
   Environment for Crop Insect Pests with Particular Reference
   to the European Corn Borer and Grain Maize
   J. PORTER

   I. Introduction 94
   II. Climate Change and Effects on the Agricultural Environment
       for Crop Insect Pests 95
   III. Climate Change and the European Corn Borer 100
       References 120

6. Aphids in a Changing Climate
   R. HARRINGTON, J. S. BALE AND G. M. TATCHELL

   I. Introduction 126
   II. Essential Aphid Biology 128
   III. The Influence of Temperature on Individual Aphids 130
   IV. The Influence of Temperature on Aphid Populations 139
   V. Conclusions 150
       References 150

7. The Effects of Climatic and Land-use Changes on the Insect
   Vectors of Human Disease
   J. LINES

   I. Introduction 158
   II. The Role of Temperature 159
   III. Potential Effects on Some Vector-borne Diseases 159

IV.	Land-use Changes	171
V.	Conclusion	172
	References	173

8. Remote Sensing and the Changing Distribution of Tsetse Flies in Africa
D. J. ROGERS

I.	Introduction	178
II.	Analysing Species' Distributions	179
III.	Presence: Mapping Tsetse Flies in Africa	181
IV.	Precedence: The Example of History	182
V.	Paradigms: The Role of Biological and Statistical Models	182
VI.	Predictions: Preparing for the Future	190
VII.	Conclusions	191
	References	192

Part III. Changes in Gas and Pollutant Levels

9. The Impact of Elevated Atmospheric CO_2 on Insect Herbivores
A. D. WATT, J. B. WHITTAKER, M. DOCHERTY, G. BROOKS, E. LINDSAY AND D. T. SALT

I.	Introduction	198
II.	Approaches to Studying the Impact of Elevated CO_2 on Insect Herbivores	198
III.	Impact of Elevated CO_2 on Insect Performance and Abundance	202
IV.	Response of Insects to Elevated CO_2 and Plant Chemistry	207
V.	CO_2, Temperature and Other Factors	213
	References	215

10. Insect Herbivores and Gaseous Air Pollutants – Current Knowledge and Predictions
V. C. BROWN

I.	Introduction	220
II.	The Gaseous Pollutants and Their Direct Effects on Plants	222
III.	Correlative Evidence for Insect/Plant/Air Pollution Interactions	225
IV.	Experimental Manipulations of Air Pollutants	228
V.	Summary of our Current Knowledge of the Effects of Air Pollutants on Insect Herbivores	242
VI.	Predictions and Future Research	243
	References	245

11. Deficiency and Excess of Essential and Non-essential Metals in Terrestrial Insects
   S. P. HOPKIN

   I. Introduction 252
   II. Natural Selection of the Elements 254
   III. Metal Pollution 259
   IV. Conclusions 264
   References 266

12. Chironomidae as Indicators of Water Quality – With a Comparison of the Chironomid Faunas of a Series of Contrasting Cumbrian Tarns
    L. C. V. PINDER AND D. J. MORLEY

    I. Introduction 272
    II. Biotic Indices and River Quality 273
    III. Chironomidae and Water Quality 274
    IV. Chironomids as Palaeoecological Indicators 277
    V. Deformities Induced by Pollution 279
    VI. The Chironomidae of Lake District Taræns 280
    VII. Discussion 287
    References 290

Part IV. Changes in Land Use

13. Southern Hemisphere Insects: Their Variety and the Environmental Pressures upon Them
    M. SAMWAYS

    I. Introduction 298
    II. The World Setting 298
    III. The Impacts 303
    IV. Conclusions and Future Perspectives 313
    V. Summary 315
    References 316

14. Species Extinctions in Insects: Ecological and Biogeographical Considerations
    N. A. MAWDSLEY AND N. E. STORK

    I. Introduction 322
    II. Insect Extinctions: Present State of Knowledge 326
    III. Extinction: Defining the Parameters 329
    IV. The External Threats Affecting Extinction in Insects 335
    V. The Intrinsic Vulnerability of Insects to Extinction 341

VI.	Extinction Rates, Community Vulnerability and Biogeography	351
VII.	Relative Extinction Rates of Invertebrates and Vertebrates: Quantifying Global Species Extinctions in Insects	354
VIII.	Extinction and Co-extinction	358
IX.	Concluding Remarks	359
	References	361

15. A World of Change: Land-use Patterns and Arthropod Communities
M. B. USHER

I.	Introduction	372
II.	The Extent of Land-use Change	373
III.	Subtle Changes within Land Uses	384
IV.	Discussion	391
	References	394

16. Insects as Indicators of Land-use Change: A European Perspective, Focusing on Moths and Ground Beetles
M. L. LUFF AND I. P. WOIWOD

I.	Introduction	400
II.	Ground Beetles and Macrolepidoptera as Indicator Groups	401
III.	Types of Land-use Change	403
IV.	Ground Beetles and Land-use Type	403
V.	Lepidoptera and Land-use Type	404
VI.	"Intensity" of Land Use	409
VII.	Monitoring the Effects of Land-use Change	410
VIII.	Habitat Scale and Pattern	413
IX.	Predicting the Effects of Land-use Change	415
X.	Conclusions	417
	References	417

Part V.   Short Communications

17. The Response of Chironomidae (Diptera) Faunas to Climate Change
S. J. BROOKS

I.	Introduction	425
II.	Chironomidae as Environmental Indicators	426
III.	Sampling Methods	427
IV.	Taphonomy	427
V.	Results	427
	References	429

18. Global Warming, Population Dynamics and Community Structure in a Model Insect Assemblage
A. J. DAVIS, L. S. JENKINSON, J. H. LAWTON, B. SHORROCKS AND S. WOOD

I.	Introduction	432
II.	Rationale	432
III.	Methods	433
IV.	Results	434
	References	438

19. Gaseous Air Pollutants – Can We Identify Critical Loads for Insects?
G. R. PORT, K. BARRETT, E. OKELLO AND A. DAVISON

I.	Introduction	442
II.	Possible Effects of Deposited Nitrogen on Insects	442
III.	Approach	444
IV.	Results	446
V.	Discussion	450
	References	452

20. Effects of Changing Land Use on Eucalypt Dieback in Australia in Relation to Insect Phytophagy and Tree Re-establishment
R. A. FARROW AND R. B. FLOYD

I.	Introduction	456
II.	Land Degradation and the Need for Trees	456
III.	Eucalypt Dieback, its Causes and Remedies	456
IV.	Selecting Insect Resistance in Trees for Re-establishment on Farmland	457
V.	Variations in Resistance to Insect Attack	458
	References	459

21. Modelling the Population of *Hydrotaea irritans* Using a Cohort-based System
J. D. AUSTIN AND J. E. HILLERTON

References	464

22. Effects of Ivermectin Residues in Cattle Dung on Dung Insect Communities Under Extensive Farming Conditions in South Africa
C. H. SCHOLTZ AND K. KRÜGER

I.	Introduction	466
II.	Methods	467
III.	Results	468
IV.	Discussion	468
	References	470

23. Monitoring the Response of Tropical Insects to Change in the Environment: Troubles with Termites
P. EGGLETON AND D. E. BIGNELL

I.	Introduction	474
II.	Troubles with Termites	476
III.	The Data So Far	491
IV.	Conclusion: General Consequences	492
	References	494

24. Potential Use of Suction Trap Collections of Aphids as Indicators of Plant Biodiversity
S. E. HALBERT, M. D. JENNINGS, C. B. COGAN, S. S. QUISENBERRY AND J. B. JOHNSON

I.	Introduction	499
II.	Idaho Suction Trap Collections Reflect Surrounding Vegetation	501
	References	503

25. Shifts in the Flight Periods of British Aphids: a Response to Climate Warming?
R. A. FLEMING AND G. M. TATCHELL

I.	Introduction	505
II.	Methods	506
III.	Results	507
IV.	Discussion	508
	References	508

26. Potential Changes in Spatial Distribution of Outbreaks of Forest Defoliators Under Climate Change
D. W. WILLIAMS AND A. M. LIEBHOLD

    References 513

Index 515

Colour plates for chapters 3, 8, 13 and 14 appear between pages 270 and 271.

Contributors

Numbers in parentheses indicate the page numbers on which the author's contributions begin

J. D. AUSTIN (461), Institute for Animal Health, Compton, Berkshire RG16 0NN, UK

J. S. BALE (125), School of Biological Sciences, University of Birmingham, Edgbaston, Birmingham B15 2TT, UK

K. BARRETT (441), Department of Agricultural and Environmental Science, The University, Newcastle upon Tyne NE1 7RU, UK

D. A. BENNETTS (49), Hadley Centre for Climate Prediction and Research, Meteorological Office, London Road, Bracknell, Berkshire RG12 2SY, UK

D. E. BIGNELL (473), School of Biological Sciences, Queen Mary and Westfield Colleges, London E1 4NS, UK

G. BROOKS (197), Institute of Terrestrial Ecology, Edinburgh Research Station, Bush Estate, Penicuik, Midlothian EH26 0QB, UK

S. J. BROOKS (425), Biodiversity Division, Department of Entomology, The Natural History Museum, Cromwell Road, London SW7 5BD, UK

V. C. BROWN (219), Department of Biology, Imperial College, Silwood Park, Ascot, Berkshire SL5 7PY, UK

C. COGAN (499), US Fish and Wildlife Service, Idaho Fish and Wildlife Research Unit, University of Idaho, Moscow, ID 83844-1136, USA

G. R. COOPE (29), Centre for Quaternary Research, Department of Geography, Royal Holloway and Bedford New College, Egham Hill, Egham, Surrey TW20 0EX, UK

A J. DAVIS (431), *Drosophila* Unit, Department of Pure and Applied Biology, The University, Leeds LS2 9JT, UK

A. DAVISON (441), Department of Agricultural and Environmental Science, The University, Newcastle upon Tyne NE1 7RU, UK

M. DOCHERTY (197), Institute of Terrestrial Ecology, Edinburgh Research Station, Bush Estate, Penicuik, Midlothian EH26 0QB, UK

P. EGGLETON (473), Biodiversity Division, Department of Entomology, The Natural History Museum, Cromwell Road, London SW7 5BD, UK

R. A. FARROW (455), Division of Entomology, CSIRO, GPO Box 1700, Canberra, ACT 2601, Australia

R. A. FLEMING (505), Forest Pest Management Institute, Forestry Canada, PO Box 490, Sault Ste. Marie, Ontario, P6A 5M7, Canada

R. B. FLOYD (455), Division of Entomology, CSIRO, GPO Box 1700, Canberra, ACT 2601, Australia

S. E. HALBERT (499), University of Idaho, College of Agriculture, Research and Extension Center, PO Box AA, Aberdeen, ID 83210-0530, USA

R. HARRINGTON (125), Department of Entomology and Nematology, Rothamsted Experimental Station, Harpenden, Hertfordshire AL5 2JQ, UK

J. E. HILLERTON (461), Institute for Animal Health, Compton, Berkshire, RG16 0NN, UK

S. P. HOPKIN (251), Department of Pure and Applied Zoology, University of Reading, PO Box 228, Reading RG6 2AJ, UK

L. S. JENKINSON (431), *Drosophila* Unit, Department of Pure and Applied Biology, The University, Leeds LS2 9JT, UK

M. JENNINGS (499), US Fish and Wildlife Service, Idaho Fish and Wildlife Research Unit, University of Idaho, Moscow, ID 83844-1136, USA

J. B. JOHNSON (499), University of Idaho, College of Agriculture, Research and Extension Center, PO Box AA, Aberdeen ID 83210-0530, USA

K. KRÜGER (465), Department of Entomology, University of Pretoria, Pretoria 0002, South Africa

J. H. LAWTON (3, 431), NERC Centre for Population Biology, Imperial College at Silwood Park, Ascot, Berkshire SL5 7PY, UK

A. M. LIEBHOLD (509), USDA Forest Services, Northeastern Forest Experiment Station, Radnor, Pennsylvania and Morgantown, West Virginia, USA

E. LINDSAY (197), Institute of Terrestrial Ecology, Edinburgh Research Station, Bush Estate, Penicuik, Midlothian EH26 0QB, UK

J. LINES (157), London School of Hygiene and Tropical Medicine, Keppel Street, London WC1E 7HT, UK

M. L. LUFF (399), Department of Agricultural and Environmental Science, The University, Newcastle upon Tyne NE1 7RU, UK

N. A. MAWDSLEY (321), Department of Entomology, The Natural History Museum, Cromwell Road, London SW7 5BD, UK and NERC Centre for Population Biology, Imperial College at Silwood Park, Ascot, Berkshire SL5 7PY, UK

G. F. MAYWALD (59), CRC for Tropical Pest Management, Gehrmann Laboratories, University of Queensland, Brisbane, Australia 4072

Contributors

D. J. MORLEY (271), Institute of Freshwater Ecology, Windermere, Cumbria, UK

E. OKELLO (441), Department of Agricultural and Environmental Science, The University, Newcastle upon Tyne NE1 7RU, UK

L. C. V. PINDER (271), NERC Institute of Freshwater Ecology, Eastern Rivers Laboratory, c/o I.T.E. Monks Wood, Abbots Ripton, Huntingdon, Cambridgeshire PE17 2LS, UK

G. R. PORT (441), Department of Agricultural and Environmental Science, The University, Newcastle upon Tyne, NE1 7RU, UK

J. PORTER (93), Climate Impacts Centre, School of Earth Sciences, Macquarie University, North Ryde 2109, NSW, Australia

S. QUISENBERRY (499), University of Idaho, College of Agriculture, Research and Extension Center, PO Box AA, Aberdeen, ID 83210-0530, USA

D. J. ROGERS (177), Department of Zoology, University of Oxford, South Parks Road, Oxford, OX1 3PS, UK

D. T. SALT (197), Institute of Terrestrial Ecology, Edinburgh Research Station, Bush Estate, Penicuik, Midlothian EH26 0QB, UK

M. J. SAMWAYS (297), Invertebrate Conservation Research Centre, Department of Zoology and Entomology, University of Natal, PO Box 375, Pietermaritzburg 3200, South Africa

C. H. SCHOLTZ (465), Department of Entomology, University of Pretoria, Pretoria 0002, South Africa

B. SHORROCKS (431), *Drosophila* Unit, Department of Pure and Applied Biology, The University, Leeds LS2 9JT, UK

D. B. SKARRATT (59), CRC for Tropical Pest Management, Gehrmann Laboratories, University of Queensland, Brisbane, Australia 4072

N. E. STORK (321), Department of Entomology, The Natural History Museum, Cromwell Road, London SW7 5BD, UK

R. W. SUTHERST (59), CRC for Tropical Pest Management, Gehrmann Laboratories, University of Queensland, Brisbane, Australia 4072

G. M. TATCHELL (125, 505), Department of Entomology, Horticulture Research International, Wellesbourne, Warwick CV35 9EF, UK

M. B. USHER (371), Scottish Natural Heritage, 2 Anderson Place, Edinburgh EH6 5NP, UK

A. D. WATT (197), Institute of Terrestrial Ecology, Edinburgh Research Station, Bush Estate, Penicuik, Midlothian EH26 0QB, UK

J. B. WHITTAKER (197), Division of Biological Sciences, Institute of Biological and Environmental Sciences, Lancaster University, Lancaster LA1 4YQ, UK

D. W. WILLIAMS (509), USDA Forest Services, Northeastern Forest Experiment Station, Radnor, Pennsylvania and Morgantown, West Virginia, USA

I. P. WOIWOD (399), Farmland Ecology Group, Department of Entomology and Nematology, Rothamsted Experimental Station, Harpenden, Hertfordshire AL5 2JQ, UK

S. WOOD (431), NERC Centre for Population Biology, Imperial College at Silwood Park, Ascot, Berkshire SL5 7PY, UK

Preface

The UN Conference on Environment and Development, held in Rio de Janeiro in 1992, highlighted international concerns about threats to the environment through anthropogenic changes in climate, levels of pollutants and land use. Such global changes impinge on all aspects of life on Earth. In terms of numbers of species and total biomass, insects dominate animal life. Their interaction with human beings is generally seen as detrimental, through direct effects on health (for example by disease transmission) or through indirect effects of competition for food resources. However, only a tiny, but clearly important, fraction of insect species cause problems to man. Many form a vital part of the ecosystems on which we depend for food and amenity. Furthermore, feedback mechanisms mean that changes in ecosystems themselves affect climate. The impacts of environmental changes on insects thus deserve much attention, and the 17th Symposium of the Royal Entomological Society was the first forum set up specifically to bring together experts working on the many aspects of this field. This volume, which summarizes the proceedings of the symposium, is divided into five parts. The introductory section sets out the magnitude of the task facing entomologists involved in unravelling the complex interactions affected by environmental change and argues the need for effort to be put into understanding a limited number of systems in detail. Subsequent chapters point out advantages and disadvantages of such an approach and cover effects of climate (Part II), gases and pollutants (Part III) and land use (Part IV). The final section (Part V) covers the shorter workshop papers presented on the final day of the symposium. It is clear that much needs to be done in order to understand the interactions between different aspects of the environment and insects, their food sources, enemies and competitors in sufficient detail to be able to superimpose models of environmental change and predict with confidence the effects on insects. We hope that this volume will form a significant section of the backdrop against which progress can be made.

It is a pleasure to acknowledge the support of the Kirby Laing Foundation and the Unilever Research and Engineering Division for their generous financial support of this symposium. The Royal Entomological Society was pleased to be able to hold the symposium at Rothamsted Experimental Station as part of the Station's 150th anniversary celebrations and we thank Professor Trevor Lewis, Director, for the provision of facilities, and members of staff for much assistance. We also thank the President, Registrar, staff and symposium committee of the Royal Entomological Society for their involvement. We are, of course, sincerely grateful to all speakers and contributors of posters, and to Trevor Lewis, John Lawton, Richard Lane, Gordon Port and Dick Vane-Wright for chairing sessions.

RICHARD HARRINGTON AND NIGEL STORK

Part I.
Introduction

1. The Response of Insects to Environmental Change
   J. H. LAWTON

1

The Response of Insects to Environmental Change

JOHN H. LAWTON

I.	"Insects and Climate"	4
II.	Time Scales	6
III.	The Dynamics of Phytophagous Insect Populations	7
	A. More than Carbon Dioxide and Temperature	8
	B. Net Effects	9
	C. Model Systems	11
IV.	Migration and Geographic Range	12
	A. Local Abundance and Size of Geographic Range	13
	B. Climate and Limits to Geographic Range	14
	C. Environmental Change and Shifts in Species' Distributions	15
	D. Rates of Spread	16
	E. Problems with Barriers in Fragmented Landscapes	19
V.	Genetic Effects	20
VI.	Concluding Remarks	22
	References	22

Abstract

The review takes as its theme a paper written by B. P. Uvarov in 1930. It makes several points. (a) On a geological time scale insects have experienced massive shifts in the Earth's climate on numerous occasions. (b) There is more to climate change than global warming – we need to consider all climatic variables. (c) Single-factor experiments on their own are potentially very misleading. (d) So too are studies on just one life-history stage. (e) There is a need to study responses of entire food chains, not single species populations, because enemies and food plants have very different climatic thresholds and responses. The chapter develops these themes, focusing on phytophagous insects. It argues the case for large-scale co-operation in the study of model systems, and for the need to focus on changes in range and distribution as well as on local population dynamic effects.

I. "INSECTS AND CLIMATE"

In August 1930, B. P. Uvarov, Senior Assistant in the Imperial Institute of Entomology in London, submitted a commissioned report on insects in relation to climate to the Dietetics Committee of the (UK) Economic Advisory Council. In preparing this report, Uvarov examined about 1300 papers and books; over 1150 of these, written in 11 languages, served as the basis for his report. More than 60 years later, it would probably be impossible for one human being to read everything that has since been published on insects in relation to climate!

Uvarov's report was subsequently published in *The Transactions of the Entomological Society of London* under the title "Insects and Climate" (Uvarov, 1931), making it a particularly appropriate starting point for this symposium. Within the paper's 247 pages Uvarov draws several insightful conclusions that are worth quoting in full; they are as relevant today as they were 60 years ago and encapsulate the problems of predicting the response of insect populations to environmental change as well as anything written more recently.

> The climate of every area on the surface of the earth has undergone a long series of great changes throughout geological history. . . . the present climate of each country is not constant, but is only a stage in the continuous evolution of climates. (p. 160)

> It would be a mistake to believe that meteorological conditions can effect only those insects which are exposed to the direct action of the atmosphere, since the conditions of the temperature, humidity, etc., in the soil, or inside a tree-trunk, or in any other habitat, are closely dependent on the intensity of solar heat, on the amount of precipitation, on the humidity and evaporating power of the air, in fact on the whole complex of phenomena covered by world climate. (p. 5)

> An insect living under natural conditions is never exposed to one isolated climatic factor, but to the continually changing combinations of several. While . . . it is necessary . . . to investigate the influence of each factor separately, it would be wrong to assume that the response of an insect to the combined action of several factors will represent merely a sum of the responses to each factor involved. Indeed, we must first satisfy ourselves whether this is true, or not, by studying experimentally all the more usual combinations of factors. (p. 80)

> ... the meteorological factors act on insects throughout their life-cycle, and the aggregated effect must be both very great and very difficult to estimate. (p. 155)
>
> Since plants depend on climatic factors for their life and metabolism in no less a degree than animals, phytophagous insects may often be affected by climate not only directly, but also through their food-plants. ... There is, for example, a certain correlation in the time of the seasonal development of an insect and its food-plant. The rate of development in both cases depends on the weather, but the latter may affect the insect and the plant in an unequal degree. As a result, the normal adjustment of the seasonal cycle of the insect to that of the plant would be destroyed ... (p. 150)
>
> (An insect) host and its parasite should be expected to have approximately the same zero of development and activities, and that they should react to any alterations in external factors by an equal increase, or decrease, in rate of metabolism. ... (However) there has already accumulated a sufficient amount of data to show this is not the case. ... Indeed the fatal limits of an insect and its parasite may be different. (p. 152)
>
> Since practically every species of insects is (*sic*) connected not with one species of parasite or predator, but with a number of them, the ultimate effect of a set of weather conditions on the pest (insect) is often very difficult to foresee (p. 155).

In a nutshell, on a geological time scale insects have seen it all before; there is more to climate change than global warming – we need to consider all climatic variables; single-factor experiments on their own are potentially very misleading; so too are studies on just one life-history stage; and we need to study responses of entire food chains, not single species populations, because enemies and food plants have very different climatic thresholds and responses. Global environmental change, of course, extends beyond climate change to include, amongst other things, the direct and indirect effects of increasing CO_2 and UV-B on plants and animals, rising levels of pollutants (e.g. SO_x and NO_x), as well as habitat destruction and modification by human beings (Vitousek, 1992). These additional variables do not negate any of Uvarov's conclusions; they reinforce them, by adding to the list of key environmental variables that need to be examined.

To keep within reasonable bounds, I am not going to say anything about pollutants, and only the briefest of remarks about habitat destruction and modification as a component of global environmental change; their effects on insects are dealt with by other contributors to the symposium. Rather, with Uvarov's insights firmly in mind, I want to review what I see as some of the main challenges and difficulties for contemporary research on the responses of insect populations to other key aspects of global environmental change, specifically rising levels of CO_2 and climate modification. Let me first preface my concerns by saying that I believe we have already made good progress with some very difficult problems; however, that should not stop us from honestly confronting the scale of the tasks ahead.

One of my concerns centres on fragmentation of effort. I believe there are major problems posed by too many different research groups studying too many different taxa, in too many different ways, with the result that we may end up not knowing enough about particular systems to say anything useful about any of them. Second, I am concerned about balance; a great deal of effort is going into research on what, for want of a better term, we can call local effects – the responses of individuals and populations to changes in temperature, CO_2, and so on. Whilst this is all well and good, I suspect that much more dramatic ecological and economic consequences will follow from large-scale changes in the distributions of insects in response to environmental change. These problems, involving geographic scales, are much harder to study under controlled conditions. But it does not mean they are not important. Nor are local abundances and geographic ranges independent phenomena, yet they are too often treated as such. Finally, we need to be extremely cautious about raising expectations by making simplistic predictions; there is a fine and difficult balance to be struck between making no predictions at all, and claiming more than is justified.

Let me now put some flesh on these deliberately provocative bones, starting with a consideration of time scales.

II. TIME SCALES

There is a considerable uncertainty about the expected timing, rate and magnitude of global climate change (Hadley Centre, 1992; Bennetts, Chapter 3, this volume). Contemporary insect populations, however, already live in a world with 25% more atmospheric CO_2 than pertained before the industrial revolution, sufficient to elicit significant changes in host plants for phytophagous insects (Bazzaz, 1990; Ayres, 1993; Siegenthaler and Sarmiento, 1993). In other words we, and insects, already live in a changing world, but probably not one in which anthropogenically derived greenhouse gases are having a detectable effect on climate – yet. However, the best available models suggest that if CO_2 levels continue to rise, we can expect quite substantial changes in the Earth's climate on a time scale of 50–100 years (Hadley Centre, 1992; Bennetts, Chapter 3, this volume). Though crude, this is quite sufficient for present purposes. It sets a time horizon for the types of phenomena worth studying. Ricklefs (1989) provides a useful summary of time scales for ecological processes setting the structure and diversity of local assemblages of species. If the global climate changes significantly within 50–100 years, Ricklefs's "rules of thumb" suggest that we can expect to see significant changes in communities of organisms, driven by changes in population processes (births, deaths and local movements) on time scales of order 1–10 generations. That is, insect population dynamics will probably track environmental changes as they happen. Competitive exclusion usually takes longer – from 10 to 100 generations, suggesting that some changes in species composition via competitive exclusion will lag behind rates of environmental change. The effects of long-distance dispersal will also be felt more slowly, Ricklefs suggests, on time spans of more than 10 but less than 1000 generations, again implying substantial lags in changes in the distributions of some insect species, but rapid, contemporary responses by others. (There is, of course, the additional problem posed by the redistribution of vegetation in response to climate change; plant migration may lag behind insects' responses, a point to which I return later.)

Even with time lags, ecological processes that change with characteristic time scales of 1–100 generations are amenable to study by normal mortals. I am therefore going to focus on local population dynamics and geographic distributions of insects. Genetic changes driven by quantitative selection on populations with time scales of roughly 7–70 generations in Ricklefs's scheme are also amenable to study, although they have not received much attention in the global change literature. I touch briefly on some genetic problems at the end of the chapter. I have, however, omitted anything more than the briefest mention of more substantial evolutionary changes (reproductive isolation, speciation and taxonomic evolution), because of the difficulties of studying processes that operate on time scales of 10^3–10^4 years, or even longer, except retrospectively by examining fossil and subfossil taxa (see Coope, Chapter 2, this volume).

We now have the framework for a discussion. Let me start by considering population processes, and in particular the impact of global change on the dynamics of phytophagous insect populations. I will then move on to consider dispersal of insects on a geographic scale.

III. THE DYNAMICS OF PHYTOPHAGOUS INSECT POPULATIONS

There are three reasons for selecting phytophagous insects for particular study in the present context. They dominate global biodiversity (Strong *et al.*, 1984; World Conservation Monitoring Centre, 1992); large numbers of people are studying their response to global environmental change; and phytophagous insects highlight particularly forcefully the difficulties of predicting the response of insect populations to global change.

A. More than Carbon Dioxide and Temperature

The two most commonly considered variables in studies of the impact of global environmental change on phytophagous insects and their host plants are CO_2 and temperature (e.g. Bazzaz, 1990; Collier et al., 1991; Cammell and Knight, 1992; Coleman and Bazzaz, 1992; Dewar and Watt, 1992; Lambers, 1993; Lincoln, 1993; Lincoln et al., 1993; Lindroth et al., 1993). But in the future real world, we also expect changes in cloud cover and rainfall (with attendant changes in humidity, evapotranspiration, etc.) as well as in UV-B (see, for example, Callaghan et al. (1992) and Ayres (1993)). Some of these changes have direct impacts on insects; others work via the host plant (by changing nutritional quality and secondary chemistry); some do both. Hence if we are to understand, and ultimately predict, changes in the local population dynamics of just one species of phytophagous insect global environmental change, we need to carry out a factorial experiment of considerable complexity (Ayres, 1993). As Uvarov reminded us 60 years ago, single factor experiments are unlikely to be very informative in these circumstances.

Unfortunately, there are also additional complexities, all of them explicitly pointed out by Uvarov; yet no one study of insects and environmental change, as far as I am aware, has taken all, or even most of them on board.

Most work on the responses of phytophagous insects to environmental change concentrates on larval (feeding) stages (e.g. Fajer et al., 1991), and often only the larger, more easily handled instars at that (e.g. Johnson and Lincoln, 1991; Lindroth et al., 1993). Whilst this is a good place to start, we will ultimately have to consider the whole life cycle, because development rates and survivorship of all other stages, as well as adult fecundity, may all be affected, if not directly, then indirectly via changes in pupal and adult body sizes and food reserves.

Nor, as Uvarov reminds us, do insect populations live in isolation. We have started to incorporate host-plant responses via changes in food nutritional quality and secondary chemistry (for brief reviews see Lambers (1993), Lincoln (1993) and Lincoln et al. (1993)). However, changes in the relative phenologies of plants and insects are likely to be just as important (Raupp and Denno, 1983; Cammell and Knight, 1992; Ayres, 1993) as changes in plant chemistry, but are receiving less

1. The Response of Insects to Environmental Change

attention. Similar remarks apply to the synchrony of enemy–victim interactions (Cammell and Knight, 1992; Landsberg and Smith, 1992; Hassell *et al.*, 1993), which may have major impacts on herbivore abundances.

Finally, virtually no insect species has sole use of a host plant. It shares it with other species of phytophagous insects, both above- and below-ground, with pathogens and with fungal mutualists. All these other species can impact positively or negatively on the target insect, usually via changes in host-plant chemistry (e.g. Karban *et al.*, 1987; Karban and Myers, 1989; Clay, 1990; Krischik *et al.*, 1991; Stovall and Clay, 1991; Masters *et al.*, 1993). All these fellow travellers, linked by their interspecific interactions, are likely to respond to environmental change and hence may further change the performance of our target insect.

In other words, the number of permutations and possible interactions rapidly becomes overwhelming. What this means is that careful, time-consuming and expensive experiments that manipulate just CO_2 and temperature, almost always on a restricted number of life-history stages, and which ignore other climatic variables, natural enemies, competitors and mutualists, as well as changes in the timing of development and nutritional status of the host plant, are unlikely to make reliable, quantitative predictions about the responses of insect populations to global change. Or, to put it bluntly, if they are used to make predictions, at best they will only be partially correct, and usually they will be wrong. We obviously have to start somewhere, and understanding the responses of phytophagous insects to rising CO_2 and temperature is an important point of departure; but it is but one small step on what will prove to be an exceptionally long, complicated and difficult journey.

This catalogue of problems is, I concede, depressingly negative; but it does no good to pretend that things are easy when they are not. There are, however, some positive ways forward, that do not involve studying everything before we can say anything. It is to these that I now want to turn.

B. Net Effects

A quantitative, rigorous means of working out the consequences of simultaneous changes in many variables is to construct a population model. However, we cannot seriously expect to be able to do this for the full gamut of environmental and biological variables embraced by global change for more than a handful of insects, at least in the foreseeable future. For reasons that I will turn to later, the current structure of the ecological research community in any case mitigates against such an approach. I also believe the resulting models run the risk of being so complex as to be impossible to validate, yielding spuriously precise answers, and not much understanding.

A sensible, tractable way forward must be to seek a middle road between too much simplicity (one- or two-factor models), and too much complexity. Qualitatively, it seems plausible that big effects swamp small ones, allowing us to guess at least the direction, if not the magnitude of the insect's net response to simultaneous changes in several environmental and biotic variables; what Jones calls "net effects" (Jones, 1991; Jones and Coleman, 1991). Using this approach, Jones and his colleagues have had some success in understanding the responses of herbivores and pathogens exploiting eastern cottonwood, *Populus deltoides*, to both ozone and shading. It remains to be seen whether similar methods help to predict net effects of global environmental change on individual insect performance, and hence ultimately upon local population dynamics.

Landsberg and Smith (1992) have recently made a similar attempt to predict outbreaks of phytophagous insects under global change, by simplying and classifying the critical variables involved. Although I disagree with some of their assumptions (e.g. there is no disruption of synchrony between the insect and its food plant), their general arguments appear sound. By considering a limited number of functional attributes, they are able to make interesting, qualitative predictions without becoming lost in a mass of detail. For example, they predict that increasing droughts and consequent water stress in plants may create outbreaks of insects specializing in young host tissues; and they point out that the impact of natural enemies on dormant life-history stages of the target insect may be particularly sensitive to climate-driven changes. All their predictions appear amenable to experimental testing.

Concentrating on net effects forces us to give more thought to identifying the critical variables, without becoming too bogged down in detail, and to explore the consequences of changes in key variables in simple "thought experiments" and population models. If nothing else, we must surely at this stage in studying impacts of global change on insects, identify single factors that typically have big effects, and distinguish them from those that are small. Changes in rainfall, for example, are likely to have at least as big an impact as rising temperatures, and yet are currently virtually ignored in the entomological literature on impacts of global change (Landsberg and Smith (1992) are a notable exception). If it turns out that the two currently most studied variables – the direct effects of temperature on rates of development and the indirect effects of CO_2 via the host plants – are the dominant effects, well and good. But I would like to see more evidence that this is, indeed, the case, based on systematic studies of other key variables, both environmental and biotic.

C. Model Systems

The task of predicting the responses of phytophagous insects to global change would be orders of magnitude easier if different species of insects, exploiting different species of plants, responded in a broadly similar manner to similar experimental manipulations. Unfortunately, abundant evidence in the general insect–plant literature suggests that different species on the same host can respond apparently idiosyncratically to experimentally induced changes in their food plant (e.g. Prestidge, 1982; MacGarvin et al., 1986; Maddox and Root, 1987; Larsson, 1989; and many others). We may ultimately understand why, but currently do not.

In the global change literature, a recent experiment by Lindroth et al. (1993) is instructive. They used 1-year-old saplings of three species of trees (*Populus tremuloides*, *Quercus rubra* and *Acer saccharum*) and penultimate instars of two species of externally feeding lepidopteran caterpillars (*Lymantria dispar* and *Malacosoma disstria*). They grew the trees for 60 days under ambient (385 ppm) or elevated (642 ppm) CO_2, and examined the performance of caterpillars feeding on the saplings over the last 10 days. Both species of caterpillars increased their feeding rates on high CO_2 *Populus*, but their growth rates declined. Their responses on *Quercus* and *Acer* differed from those on *Populus*, but in different ways. *Lymantria* grew better on high CO_2 *Quercus*, but showed no response to high CO_2 *Acer*; *Malacosoma* tended to grow less well on high CO_2 *Acer* but was unaffected by high CO_2 *Quercus*. Some, but not all these responses are predictable from the effects of CO_2 on concentrations of primary and secondary leaf metabolites, using carbon-nutrient balance theory (e.g. Tuomi et al., 1988, 1991). But the differences between caterpillar species are not well understood; nor are the reasons why carbon-nutrient balance theory sometimes makes the right predictions, and sometimes not.

Notice, incidentally, that this experiment involves only one component of environmental change (enhanced CO_2), and one life-history stage for two species of insects. Quantitative and reliable prediction is not going to be easy! The authors conclude that the "direction and magnitude of (CO_2-induced) change for both trees and insects . . . are species-specific" (p. 775). This is exactly what a large body of general literature on insect–plant interactions leads us to expect. There is a major research challenge here to see if it is possible to identify "functional groups" of phytophagous insects, that respond to changes in their food plants in ways that are consistent within groups, but differ between groups. I do not believe that species' response at this level are totally idiosyncratic, merely poorly understood.

Another barrier to progress is the relatively fragmented nature of the research effort, given the complexities of the task. If we are serious about understanding and ultimately predicting the consequences of global change for even one species of insect, we urgently need collaborative studies by teams of scientists from many laboratories, studying different aspects of the same system in continental-wide networks. There is no other way to complete the laboratory and field work, the manipulation experiments and the observations, the modelling and the synthesis, that will properly address the real complexities of the problem (Lawton, 1991), even if we focus, as I believe we must, on the net effects of variables with the biggest impacts.

A counter argument might be that since species apparently respond idiosyncratically, studying only one on a few species will overlook most of what is really interesting and challenging about the problem. My response to that is I would like to be confident that we can understand even one species properly. In any case, smaller research groups and individuals will continue to study many other species, to provide balance and richness of detail. Large-scale collaboration is not a way of working that comes easily to the ecological and entomological community, but the magnitude of the task demands it. There is no machinery to force collaboration "from the centre" across national boundaries. But there is nothing to stop groups of individuals agreeing to collaborate. International teamwork on common problems should, in a just world, improve our chances of getting funded!

IV. MIGRATION AND GEOGRAPHIC RANGE

Despite the importance of understanding local population dynamic responses to environmental change, shifts in the geographical distributions of species seem to me to pose a much greater challenge, and to present a much greater range of applied problems, not least for pest control and nature conservation. The Quaternary migrations of many species of insects in response to complex, and often rapid glacial/ interglacial climatic oscillations bear witness to the astonishing capacity of insect populations to track changing environments by moving rather than by staying put and adapting (Elias, 1991; Coope, Chapter 2, this volume; see also Davis, 1986). As Uvarov reminded us, insects have seen it all before.

One important feature of these glacial and interglacial wanderings, to which we will return later, is the labile nature of species' associations. Species that are currently found together may have had non-overlapping ranges separated by thousands of kilometres in the past; and species that occurred together five or ten thousand years ago, may no longer do so (Coope, 1978; Elias, 1991). This probably happens because no two species share exactly the same climatic limits and optima, and hence respond differently to new combinations of climatic variables generated by climate change (although other explanations are also possible (Davis, 1986)). Insects are not unique in this behaviour; labile species associations are frequent in all taxa for which we have adequate records of past and present distributions (Davis, 1986; Foster *et al.*, 1990).

A. Local Abundance and Size of Geographic Range

Interestingly, local population abundances and species' geographic ranges are not independent entities (Hanski, 1982; Brown, 1984; Lawton, 1993). Within particular taxa, species occurring over large geographic areas (i.e. those with large ranges) tend to have greater local abundances at sites where they occur than do geographically more restricted species. Examples include plants, birds, mammals, fish and variety of invertebrates from molluscs and mites to zooplankton and insects (Gaston and Lawton, 1990). There is usually considerable unexplained variation in these plots, so that an individual species can be widespread but rare everywhere, or locally common but with a small total geographic range (Rabinowitz *et al.*, 1986). Nevertheless, the general pattern appears to be robust, despite problems with definitions of geographic range (Gaston, 1991) and concerns about sampling artefacts (see Lawton (1993) for a brief discussion).

Despite its theoretical and practical interest, the pattern is not well understood. Theoretically, there are a number of ways in which a positive correlation between size of geographic range and local abundance might be generated (Hanski *et al.*, 1993; Lawton, 1993). The simplest, proposed by Brown (1984) is that species able to exploit a wide range of resources (species with "broad niches") become both widespread and locally abundant. Several metapopulation dynamic models also predict positive correlations between geographic range measured as the number (or proportion) of patches occupied and population density within patches (Gyllenberg and Hanski, 1992). Positive correlations in these models are a product of the "rescue effect" (immigration reduces the risk of local extinction), mortality during migration, and difficulties in establishing new populations.

There are two potentially important but neglected corollaries of the correlation between local abundance and size of geographic range. First, if ranges shift in response to climate change, it seems extremely unlikely that they will remain constant in size; more likely they will expand or contract, sometimes significantly so (see, for example, scenarios discussed by France (1991)). And if ranges do change in size, empirically at least we might expect concomitant changes in average local abundance that may be more than sufficient to override or mask the local population dynamic effects discussed earlier. Second, reversing this argument, big changes in local abundance driven by changes in host–plant quality, or interactions with enemies, or whatever, may result in correlated changes in range size.

It is, however, entirely unclear to me whether these speculations are reasonable, a situation that is likely to persist until the empirical correlation between average local abundance and size of geographic range is better understood. These speculations do, however, emphasize that geographic distributions and population abundances are not independent phenomena, a point that I elaborate upon in the next section.

B. Climate and Limits to Geographic Range

Climate is undoubtedly one of the key variables determining limits to species' geographic ranges (see Cammell and Knight (1992) for a valuable summary of the literature on insect pests and global change, and Sutherst, Chapter 4, this volume); however, the relative contributions of climate, habitat, enemies and competitors (to list the most obvious variables) in setting ranges have rarely been worked out for any organism, including insects.

One reason why the correlation between size of geographic range and local abundance is invariably "noisy" is that population densities vary across species' ranges. Very crudely, densities tend to be greatest near the centre of a species' range, and decline towards boundaries (Hengeveld and Haeck, 1982; Brown, 1984; Wiens, 1989). Brown (1984) argues that two phenomena are involved. First, species tend to inhabit a progressively smaller proportion of local patches towards the edge of their range and second, average population densities within occupied patches also decline. In practice, textures of distribution and abundance are often more complex than a gradual decline from the centre to the edge of the range, with multimodal patterns of abundance being common and perhaps even the norm (Taylor and Taylor, 1979; Brown, 1984; Rogers and Randolph, 1986; Root, 1988; Wiens, 1989).

Because average abundances vary across species' ranges, it follows that one or more of the key demographic rates (birth, death, immigration and emigration) also change across the range, in response to changes in environment, resources, and interactions with other species; that is, a species' population dynamics must be very different near the centre compared with the edge of its range (Richards, 1961; Huffaker and Messanger, 1964). Important, recently documented examples of markedly different population dynamics across species ranges are provided by studies on insects, at both local (Randall, 1982) and geographic scales (Rogers and Randolph, 1986; see also Rogers, Chapter 8, this volume) and mammals (Caughley *et al.*, 1988). At some point close to the range boundary, rates of population increase from low densities (r) must on average be zero. Beyond the point where r = zero, "sink populations" (Pulliam, 1988) with negative average r may be sustained by immigration from "source populations" deeper within the geographic range, where overall population performance (but not necessarily density) is higher. *In extremis* some populations may persist only because of immigration from the core (Harrison *et al.*, 1988; Harrison, 1991).

Drawing these arguments together, a sound theoretical framework now exists to describe, in general terms, what can happen to population processes across a species' geographic range. In practice, the details of what actually happens have not been worked out fully for *any* species of insect, or for that matter, any kind of organism.

C. Environmental Change and Shifts in Species' Distributions

I have spent some time discussing these general problems because we need to understand them in order to properly predict the impacts of environmental change upon species' distributions. As has already been noted, widespread deleterious changes in the environment that lead to a general decline in population abundance via an increasing death rate or falling birth rate, should result in overall contraction of a species' range, even in the absence of habitat destruction. If the original range had a single, well-defined centre, we expect range contraction towards that core; if there were originally multiple modes, we expect range contraction *and* fragmentation into former "hot spots". Conversely, overall improvements in the environment will lead to range expansion from former, isolated strongholds. Notice, however, that improvements or declines in "environmental quality" may involve direct and indirect effects of the numerous variables outlined in Section III. In other words, predicting changes in range is just as difficult as predicting changes in local abundance. Indeed, by now it should be obvious that they are not separate problems; they are different facets of the same problem.

But, one could argue, will it not at least be reasonably easy to predict changes in the limits to species ranges, if populations move to higher latitudes as the Earth's climate warms? No, for several reasons.

Although it is relatively straightforward, and interesting, to correlate present distributions with several climatic (and other) variables (see Sutherst et al., Chapter 4, this volume), it is not straightforward to predict future distributions of organisms using these statistical models, in combination with predictions from general circulation models (GCMs) of future climatic zones. At best, such predictions rest on a number of untested assumptions, and ignore some major difficulties. Amongst these difficulties is the fact that we expect several components of climate to change and, as Uvarov reminded us, insect responses may be sensitive to significant interactions between these components. There may be no exact analogues of present climates in the future. To the extent that this is true, simply extrapolating from present distributions will be in error.

Second, extrapolation confuses realized and fundamental niches (Pacala and Hurtt, 1993). Present distributions are not solely (or even mainly) limited by climate, but also by interactions with other organisms – food plants, competitors, enemies. Following Uvarov, and as we have already observed, environmental change will alter both timing and rates of interactions with these other organisms. It will also lead to the complete uncoupling of some associations, and the establishment of some entirely new ones, as species respond idiosyncratically to a new world order. Again, to the extent that these biological interactions matter in determining geographic ranges, simple, statistical extrapolation from present distributions and climatic variables are likely to be in error.

It is, of course, easier to point out these difficulties than it is to do anything about them. One approach, discussed by Davis et al. (Chapter 18, this volume) is to create experimental communities which allow us to test whether interspecific interactions do lead to misleading predictions about changes in distribution, based on statistical models which ignore the interactions. However, laboratory systems are not a substitute for field studies, and in the end we are going to have to pay much more attention to biotic as well as to environmental constraints on the limits to species' distributions.

D. Rates of Spread

For the majority of species of insects, potential migration rates are likely to be sufficiently fast to track changing environments more or less as they happen. Interestingly, Uvarov at one point in his paper (p. 124), documents several examples of rapid changes in the distributions of insects in response to unusual weather conditions. The main constraints for most insects will be much slower rates of movement by vegetation (Davis, 1986). We can say this with some confidence, because reasonably good data now exist on migration rates for both insects and plants, albeit in different contexts.

For plants, the best information comes from postglacial range expansions. Maximum rates of spread for plant taxa in Europe, calculated by Huntley and Birks (1983) are up to 2 km yr^{-1}. They are matched by northward migration rates of 2 km yr^{-1} for white spruce (Ritchie and MacDonald, 1986), regarded by botanists as unusually rapid (e.g. Pastor, 1993). They are nearly an order of magnitude faster than more typical rates of invasion by other tree species in eastern North America during the early postglacial (average 0.3 km yr^{-1}, range $0.1-0.4 \text{ km yr}^{-1}$; Davis (1981)) and by Scots pine in Britain ($0.4-0.8 \text{ km yr}^{-1}$; Gear and Huntley (1991)). Although such calculations require cautious interpretation (Bennett, 1986), even the fastest rates are slow compared with observed rates of spread of many invading insects.

New data on rates of spread for 25 species of Lepidoptera that have invaded Britain during roughly the last 100 years, derived from numerous published records by D. J. L. Agassiz (pers. comm.) are shown in Fig. 1; most rates exceed 2 km yr^{-1}. Two recent invaders with leaf-mining caterpillars (*Phyllonorycter leucographella* and *P. platani*) have been subject to particularly detailed study (Nash *et al.*, unpublished) and are currently spreading at 10.3 and 8.6 km yr^{-1}. A tephritid fly *Urophora cardui* is currently expanding north in Germany at 2.1 km yr^{-1} (Eber and Brandl, 1994). Other invading insects have moved even faster – up to roughly 100 km yr^{-1} for *Pieris rapae* (Andow *et al.*, 1990), and between 20 and 70 km yr^{-1} for several recently invading Hemiptera and Lepidoptera in Hungary (Kozár and Dávid, 1986).

We also need to remember that a small minority of insects, including a number of important agricultural pests (e.g. rice planthoppers, *Nilaparvata lugens* and *Sogatella furcifera*; Wada *et al.*, 1987) migrate huge distances every year from a limited number of over-wintering refugia, to occupy virtually all suitable habitats whilst conditions are favourable.

Although we do not currently understand why invading insects spread at rates that vary by several orders of magnitude (the variation appears to have nothing to do with the ability to fly, or lack of it, body size, wing loading or type of host plant), if these rates are in any sense representative of the potential for insects to spread under global change, the rate-limiting step will be the migration of vegetation, not the migration of insects. Most insects have the potential to spread extremely quickly, and to effectively track climatic changes as they happen.

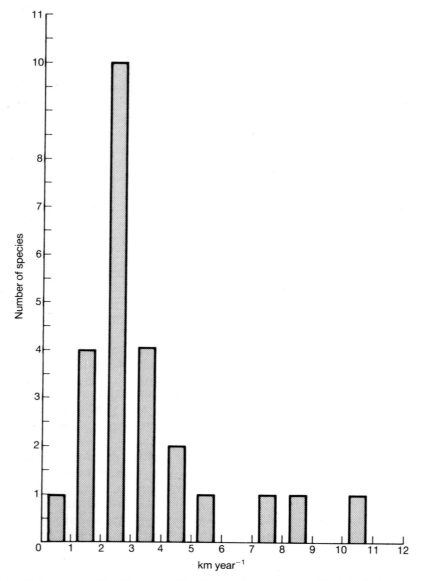

Fig. 1. Rates of spread for 25 species of Lepidoptera that have invaded Britain during roughly the last 100 years, derived from numerous published records by D. J. L. Agassiz (pers. comm.)

E. Problems with Barriers in Fragmented Landscapes

One major problem that may constrain movements by many species of insects with specialized requirements, however, is habitat fragmentation. Examples that spring to mind in Europe include species associated with ancient forests and dead wood, calcareous grassland, and heathland. These, and many other habitats have suffered major fragmentation and destruction by human activities, and now exist in tiny, isolated remnants in an inhospitable agricultural and urbanized landscape. Hence, although insects (and most other taxa) have seen it all before, those species that survived and recolonized large parts of the globe after the last glaciation, did so across a relatively intact landscape.

Detailed studies on the ability of species to cross unfavourable habitats are few, but such data as there are, are not encouraging for species of high conservation interest. Some of the best recent work has been carried out on British butterflies by C. D. Thomas and colleagues (Thomas *et al.*, 1992; Thomas and Jones, 1993). Suitable habitat patches isolated by as little as 1–10 km of unfavourable ground may remain uncolonized for long periods of time, and gaps greater than 10 km may not be crossed by some species (e.g. *Hesperia comma*) in 100 years. Such species are unlikely to survive climatic change on isolated habitat patches unaided, even if these are currently protected as nature reserves.

Whilst these data are discouraging, it is entirely unclear what proportion of, say, the European entomofauna they represent. Many natural postglacial habitats must have been fragmented and isolated by deep forest, albeit in much larger blocks than extant remnants; *Phragmites* reed beds come to mind. Many insects are naturally also extremely mobile (e.g. host-alternating aphids). Indeed, there is no evidence that present distributions of many British taxa (insects, birds, other animals and plants) – specifically patterns of species' replacement or turnover along geographic transects (beta-diversity) – are constrained by powers of dispersal (Harrison *et al.*, 1992). Within distances of a few hundred kilometres, on continents in the temperate zone, current patterns of species distribution appear not to be shaped primarily by isolation (Harrison, 1993). Given predicted scales of habitat change under global environmental change over the next 100 years, this is a less gloomy scenario than is often painted. But it is not grounds for complacency. Climate change is unlikely to make the task of insect conservation any easier! (See Peters and Darling (1985) for further discussions on nature conservation and global climate change).

V. GENETIC EFFECTS

One of the most remarkable things about the Quaternary wanderings of many insects, documented by Coope, Elias and their colleagues, is the fact that they appear to have taken place without the smallest indication of morphological evolution in the migrating species. In Coope's (1978) own words: "There is no evidence of any morphological evolution during the last half million years at least. . . . For the past few glacial/interglacial cycles . . . almost all our fossil insects match their modern counterparts with extraordinary exactness even down to the intimate internal sclerites of the genitalia". On this evidence (which is now abundant), significant morphological evolution by insects in response to climate change seems extremely unlikely on time scales of 10^2–10^3 years. Physiological or behavioural evolution are possibilities, but unproven.

This does not, however, rule out measurable genetic changes in insect populations on time scales of a few 100 years; paradoxically, these may be in species that track changing environments most rapidly.

Over the last 300–400 years, the cynipid gallwasp *Andricus quercuscalicis* has spread north and west across Europe, following the introduction of one of its two principal hosts, the Turkey oak *Quercus cerris*. The spread of *A. quercuscalicis* mimics the kind of rapid range expansion that we expect many insects to show under global environmental change. Spread appears to have taken place in a series of "founder-flush" events, with populations established by small numbers of colonizing individuals that are subsamples of the full range of genotypes present in the original, core range. New populations then act as sources for further colonization. Repeated many times, this pattern of spread has resulted in a dramatic, linear loss of genetic variability with distance from the original range in south-eastern Europe (Stone and Sunnucks, 1993). The invasion has resulted in substantial reduction in allelic variation (including alleles with a frequency as high as 40% in the native range) (Fig. 2) and loss of heterozygosity.

Loss of genetic variability in the invaded range of colonizing species has been shown in a wide variety of taxa (see Stone and Sunnucks (1993) for a summary); but the loss of variability seen in *A. quercuscalicis* is among the most dramatic yet recorded. An important contributory factor may be the relatively fragmented and scattered nature of suitable patches of introduced *Q. cerris*, which increases the isolation of new *A. quercuscalicis* populations. We can only speculate whether global environmental change will generate analogous distributions of host plants and insects. Much depends upon how plant populations themselves respond, and migrate across fragmented landscapes.

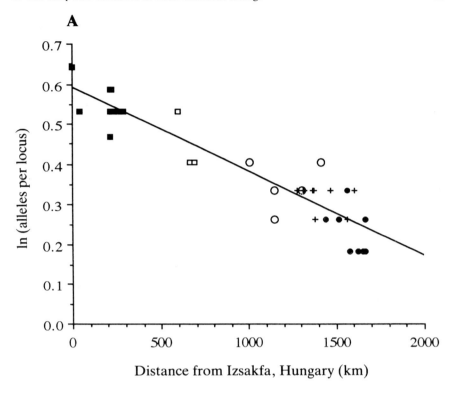

Fig. 2. The relationship between distance from Izsafka, Hungary (in the core of the original European distribution) and allelic diversity for *Andricus quercuscalicis*, a gall wasp that has spread rapidly across Europe (Stone and Sunnucks, 1993). (■) Hungary, Slovenia and Austria; (□) Germany and Italy; (○) France, Belgium and the Netherlands; (+) northern and eastern Britain; (●) southern and western Britain.

Nor is it clear how loss of genetic variability might influence the ability of insect populations to cope with, and adapt to, the changing climatic regimes, photoperiods, new combinations of food plants, competitors, enemies and so on, that global change and migration imply. Simberloff (1986) has reviewed some of the evidence for genetic effects (specifically inbreeding depression) on the establishment of invading insects, and concludes, tentatively, that they may play a part in reducing the performance of colonists. The problem deserves much more attention, building on the extensive experience of insect population geneticists, particularly with *Drosophila*.

VI. CONCLUDING REMARKS

As I have pointed out several times in this review, it is easier to identify problems than it is to solve them. In many, indeed most cases, I have no better idea than anybody else how best to move forward. But this does not mean that we should ignore the problems. Nor does it mean that ecologists and entomologists are stupid, or that most of past and present research on impacts of climate change is a waste of time. They are not, and it isn't. It means, simply, that the problems are very hard, and will require a great deal of effort and time to solve. Whether we have that time remains to be seen. It is over 60 years since Uvarov brought together what was then known about insects and climate; in another 60 years the world will be a very different place. This gives an unusual urgency to the problems addressed in this symposium, and presents us with an unusual responsibility. Part of that responsibility is for us to honestly confront the scale and magnitude of the task and to act accordingly.

Acknowledgements

I am extremely grateful to David Agassiz for permission to reproduce Fig. 1, and to Graham Stone for making his data available prior to publication. I thank Charles Godfray, Bill Kunin and Shahid Naeem for helpful comments on the manuscript. Preparation of this review was supported by the core grant to the NERC Centre for Population Biology.

REFERENCES

Andow, D. A., Kareiva, P. M., Levin, S. A. and Okubo, A. (1990). Spread of invading organisms. *Landscape Ecol.* **4**, 177–188.
Ayres, M. P. (1993). Plant defense, herbivory, and climate change. *In* "Biotic Interactions and Global Change" (P. M. Kareiva, J. G. Kingsolver and R. B. Huey, eds), pp. 75–94. Sinauer, Sunderland, MA.
Bazzaz, F. A. (1990). The response of natural ecosystems to the rising global CO_2 levels. *Ann. Rev. Ecol. Syst.* **21**, 167–196.
Bennett, K. D. (1986). The rate of spread and population increase of forest trees during the postglacial. *Phil. Trans. Roy. Soc. Lond., B* **314**, 523–531.
Brown, J. H. (1984). On the relationship between abundance and distribution of species. *Am. Nat.* **124**, 255–279.
Callaghan, T. V., Sonesson, M. and Somme, L. (1992). Responses of terrestrial plants and invertebrates to environmental change at high latitudes. *Phil. Trans. Roy. Soc. Lond., B* **338**, 279–288.
Cammell, M. E. and Knight, J. D. (1992). Effects of climatic change on the population dynamics of crop pests. *Adv. Ecol. Res.* **22**, 117–162.
Caughley, G., Grice, D., Barker, R. and Brown, B. (1988). The edge of range. *J. Anim. Ecol.* **57**, 771–785.

Clay, K. (1990). Fungal endophytes of grasses. *Ann. Rev. Ecol. Syst.* **21**, 275–297.
Coleman, J. S. and Bazzaz, F. A. (1992). Effects of CO_2 and temperature on growth and resource use of co-occurring C_3 and C_4 annuals. *Ecology* **73**, 1244–1259.
Collier, R. H., Finch, S., Phelps, K. and Thompson, A. R. (1991). Possible impact of global warming on cabbage root fly (*Delia radicum*) activity in the UK.*Ann. Appl. Biol.* **118**, 261–271.
Coope, G. R. (1978). Constancy of insect species versus inconstancy of Quaternary environments. *In* "Diversity of Insect Faunas" (L. A. Mound and N. Waloff, eds), pp. 176–187. *Symposia of the Royal Entomological Society of London* **9**. Blackwell Scientific, Oxford.
Davis, M. B. (1981). Quaternary history and the stability of forest communities. *In* "Forest Succession. Concepts and Application" (D. C. West, H. H. Shugart and D. B. Botkin, eds), pp. 132–153. Springer-Verlag, New York.
Davis, M. B. (1986). Climatic instability, time lags, and community disequilibrium. *In* "Community Ecology" (J. Diamond and T. J. Case, eds), pp. 269–284. Harper and Row, New York.
Dewar, R. C. and Watt, A. D. (1992). Predicted changes in the synchrony of larval emergence and budburst under climatic warming. *Oecologia* **89**, 557–559.
Eber, S. and Brandl, R. (1994). Ecological and genetic spatial patterns of *Urophora cardui* (Diptera: Tephritidae) as evidence for population structure and biogeographical processes. *J. Anim. Ecol.* **63**.
Elias, S. A. (1991). Insects and climate change. *BioScience* **41**, 552–559.
Fajer, E. D., Bowers, M. D. and Bazzaz, F. A. (1991). The effects of enriched CO_2 atmospheres on the buckeye butterfly, *Junonia coenia*. *Ecology* **72**, 751–754.
Foster, D. R., Schoonmaker, P. K. and Pickett, S. T. A. (1990). Insights from paleoecology to community ecology. *TREE* **5**, 119–122.
France, R. L. (1991). Empirical methodology for predicting changes in species range extension and richness through climate warming. *Int. J. Biomet.* **34**, 211–216.
Gaston, K. J. (1991). How large is a species' geographic range? *Oikos* **61**, 434–438.
Gaston, K. J. and Lawton, J. H. (1990). Effects of scale and habitat on the relationship between regional distribution and local abundance. *Oikos* **58**, 329–335.
Gear, A. J. and Huntley, B. (1991). Rapid changes in the range limits of Scots pine 4000 years ago. *Science* **251**, 544–547.
Gyllenberg, M. and Hanski, I. (1992). Single-species metapopulation dynamics: a structured model. *Theoret. Pop. Biol.* **42**, 35–66.
Hadley Centre (1992). "The Hadley Centre Transient Climate Change Experiment". Meteorological Office, Bracknell.
Hanski, I. (1982). Dynamics of regional distribution: the core and satellite hypothesis. *Oikos* **38**, 210–221.
Hanski, I., Kouki, J. and Halkka, A. (1993). Three explanations of the positive relationship between distribution and abundance of species. *In* "Species Diversity in Ecological Communities: Historical and Geographical Perspectives" (R. E. Ricklefs and D. Schulter, eds), pp. 108–116. University of Chicago Press, Chicago.
Harrison, S. (1991). Local extinction in a metapopulation context: an empirical evaluation. *Biol. J. Linn. Soc.* **42**, 73–88.
Harrison, S. (1993). Species diversity, spatial scale, and global change. *In* "Biotic Interactions and Global Change" (P. M. Kareiva, J. G. Kingsolver and R. B. Huey, eds), pp. 388–401. Sinauer, Sunderland, MA.
Harrison, S., Murphy, D. D. and Ehrlich, P. R. (1988). Distribution of the bay checkerspot butterfly, *Euphydryas editha bayensis*: evidence for a metapopulation model. *Am. Nat.* **132**, 360–382.

Harrison, S., Ross, S. J. and Lawton, J. H. (1992). Beta diversity on geographic gradients in Britain. *J. Anim. Ecol.* **61**, 151–158.

Hassell, M. P., Godfray, H. C. J. and Comins, H. N. (1993). Effects of global change on the dynamics of insect host-parasitoid interactions. *In* "Biotic Interactions and Global Change" (P. M. Kareiva, J. G. Kingsolver and R. B. Huey, eds), pp. 403–423. Sinauer, Sunderland, MA.

Hengeveld, R. and Haeck, J. (1982). The distribution of abundance. I. Measurements. *J. Biogeogr.* **9**, 303–306.

Huffaker, C. B. and Messanger, P. S. (1964). The concept and significance of natural control. *In* "Biological Control of Insect Pests and Weeds" (P. De Bach, ed.), pp. 74–117. Chapman & Hall, London.

Huntley, B. and Birks, H. J. B. (1983). "An Atlas of Past and Present Pollen maps for Europe: 0–13,000 Years Ago". Cambridge University Press, Cambridge.

Johnson, R. H. and Lincoln, D. E. (1991). Sagebrush carbon allocation patterns and grasshopper nutrition: the influence of CO_2 enrichment and soil mineral limitation. *Oecologia* **87**, 127–134.

Jones, C. G. (1991). Interactions among insects, plants, and microorganisms: a net effects perspective on insect performance. *In* "Microbial Mediation of Plant–Herbivore Interactions" (P. Barbosa, V. A. Krischik and C. G. Jones, eds), pp. 7–35. Wiley, New York.

Jones, C. G. and Coleman, J. S. (1991). Plant stress and insect herbivory: towards an integrated perspective. *In* "Response of Plants to Multiple Stresses" (H. A. Mooney, W. E. Winner and E. J. Pell, eds), pp. 249–279. Academic Press, San Diego.

Karban, R. and Myers, J. H. (1989). Induced plant responses to herbivory. *Ann. Rev. Ecol. Syst.* **20**, 331–348.

Karban, R., Adamchak, R. and Schnathorst, W. C. (1987). Induced resistance and interspecific competition between spider mites and vascular wilt fungus. *Science* **235**, 678–680.

Kozár, F. and Dávid, A. N. (1986). The unexpected northward migration of some species of insects in Central Europe and the climatic changes. *Anz. Schändlingskde., Pflanzenschutz, Umweltschutz* **58**, 90–94.

Krischik, V. A., Goth, R. W. and Barbosa, P. (1991). Generalised plant defense: effects on multiple species. *Oecologia* **85**, 562–571.

Lambers, H. (1993). Rising CO_2, secondary plant metabolism, plant–herbivore interactions and litter decomposition. *Vegetatio* **104/105**, 263–217.

Landsberg, J. and Smith, M. S. (1992). A functional scheme for predicting the outbreak potential of herbivorous insects under global atmospheric change. *Aust. J. Bot.* **40**, 565–577.

Larsson, S. (1989). Stressful times for the plant stress-insect performance hypothesis. *Oikos* **56**, 277–283.

Lawton, J. H. (1991). Ecology as she is done, and could be done. *Oikos* **61**, 289–290.

Lawton, J. H. (1993). Range, population abundance and conservation. *TREE* **8**, 409–413.

Lincoln, D. E. (1993). The influence of plant carbon dioxide and nutrient supply on susceptibility to insect herbivores. *Vegetatio* **104/105**, 273–280.

Lincoln, D. E., Fajer, E. D. and Johnson, R. H. (1993). Plant–insect herbivore interactions in elevated CO_2 environments. *TREE* **8**, 64–68.

Lindroth, R. L., Kinney, K. K. and Platz, C. L. (1993). Responses of deciduous trees to elevated atmospheric CO_2: productivity, phytochemistry, and insect performance. *Ecology* **74**, 763–777.

MacGarvin, M., Lawton, J. H. and Heads, P. A. (1986). The herbivorous insect communities of open and woodland bracken: observations, experiments and habitat manipulations. *Oikos* **47**, 135–148.

Maddox, G. D. and Root, R. B. (1987). Resistance to 16 diverse species of herbivorous insects within a population of goldenrod, *Solidago altissima* – genetic variation and heretability. *Oecologia* **72**, 8–14.

Masters, G. J., Brown, V. K. and Gange, A. C. (1993). Plant mediated interactions between above- and below-ground insect herbivores. *Oikos* **66**, 148–151.

Nash, D. R., Agassiz, D. J. L., Godfray, H. C. J. and Lawton, J. H. (in press). The pattern of spread of invading species: two leaf-mining moths colonising Great Britain. *J. Anim. Ecol.*

Pacala, S. W. and Hurtt, G. C. (1993). Terrestrial vegetation and climate change: integrating models and experiments. *In* "Biotic Interactions and Global Change" (P. M. Kareiva, J. G. Kingsolver and R. B. Huey, eds), pp. 57–74. Sinauer, Sunderland, MA.

Pastor, J. (1993). Northward march of spruce. *Nature* **361**, 208–209.

Peters, R. L. and Darling, J. D. S. (1985). The greenhouse effect and nature reserves. *BioScience* **35**, 707–717.

Prestidge, R. A. (1982). Instar duration, adult consumption, oviposition and nitrogen utilization efficiencies of leafhoppers feeding on different quality food (Auchenorrhycha: Homoptera). *Ecol. Entomol.* **7**, 91–101.

Pulliam, H. R. (1988). Sources, sinks, and population regulation. *Am. Nat.* **132**, 652–661.

Rabinowitz, D., Cairns, S. and Dillon, T. (1986). Seven forms of rarity and their frequency in the flora of the British Isles. *In* "Conservation Biology. The Science of Scarcity and Diversity" (M. E. Soulé, ed.), pp. 182–204. Sinauer, Sunderland, MA.

Randall, M. G. M. (1982). The dynamics of an insect population throughout its altitudinal distribution: *Coleophora alticolella* (Lepidoptera) in northern England. *J. Anim. Ecol.* **51**, 993–1016.

Raupp, M. J. and Denno, R. F. (1983). Leaf age as a predictor of herbivore distribution and abundance. *In* "Variable Plants and Herbivores in Natural and Managed Systems" (R. F. Denno and M. S. McClure, eds), pp. 91–124. Academic Press, New York.

Richards, O. W. (1961). The theoretical and practical study of natural insect populations. *Ann. Rev. Entomol.* **6**, 147–162.

Ricklefs, R. E. (1989). Speciation and diversity: the integration of local and regional processes. *In* "Speciation and its Consequences" (D. Otte and J. A. Endler, eds), pp. 599–622. Sinauer, Sunderland, MA.

Ritchie, J. M. and MacDonald, G. M. (1986). The patterns of post-glacial spread of white spruce. *J. Biogeogr.* **13**, 527–540.

Rogers, D. J. and Randolph, S. E. (1986). Distribution and abundance of tsetse flies (*Glossina* spp.). *J. Anim. Ecol.* **55**, 1007–1025.

Root, T. (1988). "Atlas of Wintering North American Birds". University of Chicago Press, Chicago.

Siegenthaler, U. and Sarmiento, J. L. (1993). Atmospheric carbon dioxide and the ocean. *Nature* **365**, 119–125.

Simberloff, D. (1986). Introduced insects: A biogeographic and systematic perspective. *In* "Ecology of Biological Invasions of North America and Hawaii" (H. A. Mooney and J. A. Drake, eds), pp. 3–26. Springer-Verlag, New York.

Stone, G. N. and Sunnucks, P. (1993). Genetic consequences of an invasion through a patchy environment – the cynipid gallwasp *Andricus quercuscalicis* (Hymenoptera: Cynipidae). *Molec. Ecol.* **2**, 251–268.

Stovall, M. E. and Clay, K. (1991). Adverse effects on fall armyworm feeding on fungus-free leaves of fungus-infected plants. *Ecol. Entomol.* **16**, 519–523.

Strong, D. R., Lawton, J. H. and Southwood, T. R. E. (1984). "Insects on Plants. Community Patterns and Mechanisms". Blackwell Scientific, Oxford.

Taylor, R. A. J. and Taylor, L. R. (1979). A behavioural model for the evolution of spatial dynamics. *In* "Population Dynamics" (R. M. Anderson, B. D. Turner and L. R. Taylor, eds), pp. 1–27. *20th Symposium of the British Ecological Society*. Blackwell Scientific, Oxford.

Thomas, C. D. and Jones, T. M. (1993). Partial recovery of a skipper butterfly (*Hesperia comma*) from population refuges: lessons for conservation in a fragmented landscape. *J. Anim. Ecol.* **62**, 472–481.

Thomas, C. D., Thomas, J. A. and Warren, M. S. (1992). Distribution of occupied and vacant butterfly habitats in fragmented landscapes. *Oecologia* **92**, 563–567.

Tuomi, J., Niemelä, P., Chapin, F. S., III, Bryant, J. P. and Sirén, S. (1988). Defensive responses of trees in relation to their carbon/nutrient balance. *In* "Mechanisms of Woody Plant Defenses Against Insects. Search for Pattern" (W. J. Mattson, J. Levieux and C. Bernard-Dagan, eds), pp. 57–72. Springer-Verlag, New York.

Tuomi, J., Fagerström, T. and Niemelä, P. (1991). Carbon allocation, phenotypic plasticity, and induced defenses. *In* "Phytochemical Induction by Herbivores" (D. W. Tallamy and M. J. Raupp, eds), pp. 85–104, Wiley-Interscience, New York.

Uvarov, B. P. (1931). Insects and climate. *Trans. Ent. Soc. London* **79**, 1–247.

Vitousek, P. M. (1992). Global environmental change: An introduction. *Ann. Rev. Ecol. Syst.* **23**, 1–14.

Wada, T., Seino, H., Ogawa, Y. and Nakasuga, T. (1987). Evidence of autumn migration in the rice planthoppers, *Nilaparvata lugens* and *Sogatella furcifera*: analysis of light trap catches and associated weather patterns. *Ecol. Entomol.* **12**, 321–330.

Wiens, J. A. (1989). "The Ecology of Bird Communities. Foundations and Patterns", Vol. 1. Cambridge University Press, Cambridge.

World Conservation Monitoring Centre (1992). "Global Biodiversity: Status of the Earth's Living Resources". Chapman & Hall, London.

Part II.
Changes in Climate

2. The Effects of Quaternary Climatic Changes on Insect Populations: Lessons from the Past
   G. R. COOPE
3. The Hadley Centre Transient Climate Experiment
   D. A. BENNETTS
4. Predicting Insect Distribution in a Changed Climate
   R. W. SUTHERST, G. F. MAYWALD and D. B. SKARRAT
5. The Effect of Climate Change on the Agricultural Environment for Crop Insect Pests with Particular Reference to the European Corn Borer and Grain Maize
   J. PORTER
6. Aphids in a Changing Climate
   R. HARRINGTON, J. S. BALE AND G. M. TATCHELL
7. The Effects of Climate and Land-use Changes on the Insect Vectors of Human Disease
   J. LINES
8. Remote Sensing and the Changing Distribution of Tsetse Flies in Africa
   D. ROGERS

2

The Effects of Quaternary Climatic Changes on Insect Populations: Lessons from the Past

G. R. COOPE

I. Introduction	30
II. Quaternary Entomology	31
III. Quaternary Climatic Change	33
IV. The Response of Insect Species to the Glacial/Interglacial cycles	34
A. The Insect Fauna of the Last Interglacial	35
B. The Insect Fauna of the Last Glaciation	36
C. Examples of Changes in the Geographical Ranges of Species	36
D. Summary	45
V. Prospect	46
References	47

Abstract

Insect remains are extremely abundant in almost all terrestrial and freshwater deposits that accumulated during the Quaternary period (the last 2.4 million years). These remains provide the only objective information on the precursors of living insect faunas. They show that there was a remarkable degree of specific constancy during this period in spite of (or because of) the frequent glacial/interglacial climatic oscillations that occurred at this time. This fossil evidence shows that the response of thermally sensitive species to these climatic changes was to alter their geographical ranges rather than to evolve to fit the ever-changing environmental conditions. Alterations of the geographic ranges of many species can be shown to have taken place on an enormous scale even during the latest glacial/interglacial cycle. Thus by tracking acceptable environments from place to place as the climatic changes dictated, the actual environment in which a species lived remained relatively constant in spite of the frequency and intensity of Quaternary climatic changes. Species extinctions on a global scale during this period seem to have been remarkably uncommon. If the ability to track acceptable environments is an essential ingredient in the survival strategy of insect species when facing the climatic changes of the recent past, it may likewise be an essential option for them if they are to survive any humanly induced climatic changes of their immediate future.

I. INTRODUCTION

In the last few decades the investigation of Quaternary subfossil insects has burgeoned into a subject in its own right; a vigorous hybrid born of the union of entomology and palaeontology. Although, in the past these subfossils have been paid modest attention, it has been small in comparison to that lavished on the vertebrates and on the plants. However, insect remains are extremely abundant in many near-surface deposits such as those that accumulated in bogs, marshes or at the bottom of lakes and small ponds. These remains provide the only objective information on the precursors of the living insect faunas. They are the representatives of the ancestors of the species that make up the insect communities of the present day. We thus have direct access to the ways in which species, and populations of species, responded to the great climatic oscillations of the Quaternary period.

There is still today a firmly held belief that insects are undergoing rapid evolution at the present time, and that the successive glacial and interglacial climatic fluctuations were the cause of widespread extinctions. According to this view, there must have been a rapid turnover of insect species in the geologically recent past and this rate can be extrapolated to include the ongoing evolution of insect species at the present day.

There are many reasons for this very immediate and dynamic view of the evolutionary process. The enormous diversity of insect species, their short generation time, their high level of heterozygosity, their numerous scattered subpopulations, their apparent close adaptation to local environmental conditions and their small size, all suggest that the evolutionary clock should tick faster for insects than it does for other organisms. With these characteristics they would seem to be well placed to adapt very rapidly to any changes in their circumstances. Furthermore, early workers on Quaternary fossils frequently credited them with new species names, often implying that they were either ancestral but distinct from living forms or else representatives of totally extinct species. These views of the fossil record tended to reinforce the belief that there had been rapid evolution and widespread extinctions during the Quaternary period.

The fossils themselves, however, tell quite a different story. They show a remarkable degree of species constancy, even as far back as Miocene times. From deposits in Alaska that date from over 5 million years ago, fossil insect assemblages have been discovered that include species that have exactly the same morphologies as modern forms, right down to the intimate structures of their male genitalia (Matthews, 1974, 1976). Even some of today's sibling species were present in those remote times with exactly the same subtle similarities and differences that distinguish them at the present day.

Throughout the whole of the Quaternary period, namely during the last 2.4 million years, there is scarcely any fossil evidence of morphological change in insect species. At the present time, Quaternary fossils of well over 2000 species have been recognized from Britain alone, but none shows any morphological differences when compared with present-day equivalent species. Furthermore, because fossil assemblages are largely made up of the same species as those that constitute present-day communities, there is no reason to believe that they have changed their environmental requirements in the intervening time. It is a general rule that, for the most part, species kept the same company in the past as they do today. In other words, it would appear that their physiology must have remained just as constant as their morphology. To some entomologists unfamiliar with the Quaternary fossil record, this long-term species constancy may seem so unexpected as to be almost unbelievable. Yet it is a fact of life with which we have to live.

The reasons for the discrepancy between our evolutionary expectations based on studies of living animals and the evidence provided by the late Cenozoic fossils are intriguing but, in the context of the present symposium, they must not detain us here. We are concerned only with the basic problem of how insect species responded to the enormous environmental changes of the past couple of million years, and how they managed to do so without finding it necessary to evolve into something else. The current programme of investigation into Quaternary fossil insects goes some way towards a solution to this curious enigma.

II. QUATERNARY ENTOMOLOGY

The occurrence and mode of preservation of Quaternary fossil insects has been well documented in the past few decades (see Buckland and Coope, 1991 for a review and extensive bibliography). There are, however, some significant aspects of these fossils and their interpretation that should be emphasized here. Of paramount importance is the fact that they are not mineralized as are the fossils of classical palaeontology, but consist of their original skeletal chitin, a material that would appear to be almost indestructible provided that the deposit in which it occurs has remained waterlogged, and preferably anaerobic, since the time when it was laid down. Occasionally, whole bodies are found interbedded in peat layers but, more usually, the fossils are found as disarticulated skeletal elements. This means that keys to species identification are of only limited value and most determinations have to be done by direct comparison with well-identified modern material. There are, however, hidden advantages in this procedure in that it is often possible to see important features that would be concealed by adjacent body parts in the fully articulated animal.

Because, in this particular case, both neontologists and palaeontologists base their taxonomies on exoskeletal characters, it is possible to use the same sorts of morphological features to differentiate between our fossils as those employed to differentiate between living insect species. We can utilize the same taxonomic system for both. Here at least there can be no argument that palaeontologist and neontologist use such different concepts of species that they are not comparing like with like. The status of our fossil species is precisely the same as that of the living species with which they are equated.

Most of the inferences that are drawn from Quaternary fossils have been based on evidence from the Coleoptera. It must be emphasized, however, that there is nothing peculiar about the beetles that make them in any way exceptional – they are simply so robust that they make very good fossils. The interpretations made from this beetle evidence may well be applicable to other groups of insect that have not been blessed with such a comprehensive fossil background. Furthermore the present-day species of beetles display characteristics that have encouraged many biologists to see in them evidence of ongoing speciation and rapid evolutionary change. Their fossil record therefore provides a valuable test-bed for a variety of evolutionary, ecological and biogeographical hypotheses.

Fossils of other orders of insect are also commonly found in Quaternary deposits, many of which would repay greater specialist attention. Diptera are abundant as heads, or rather distorted abdomens. The larval heads of chironomids are frequent in lake deposits and have been used to interpret various parameters of past lake environments (Walker, 1987; and see Brooks, Chapter 17, this volume). Larval sclerites of Trichoptera are also common in similar sediments and have been used in palaeolimnological investigations (Wilkinson, 1984; Williams, 1988). Hemiptera are also fairly abundant and often identifiable even to the extent of preserving intriguing colour variants. Innumerable characteristic pieces of Hymenoptera look very promising but, particularly in the case of the Parasitica, their present-day taxonomy seems barely adequate yet to permit useful identifications to be made. Remains of Odonata are remarkably rare being restricted to isolated segments and the distinctive middles of the heads of damsel flies. All of these "non-beetle" fossils can be matched exactly with modern specimens even if we can not be sure of their specific identity. It would thus seem that they, too, show a similar degree of evolutionary stability as do the Coleoptera.

Unfortunately no remains of Lepidoptera have yet been found in Quaternary deposits, apart from scattered caterpillar jaws. Presumably this is because the frailty of their skeletons leaves little opportunity for their survival as fossils.

There are a number of important gaps in the current state of our knowledge of Quaternary insects. One of the most significant of these is the fact that nothing is known of equatorial fossil insect faunas. It is always possible that species stability and evolutionary rates were different at latitudes well away from those that suffered the climatic contrasts of the glacial cycles but without the fossils we have no objective evidence. There are so many treasured hypotheses that could be tested by an investigation into tropical Quaternary insect fossils but, so far, nobody has taken up this difficult challenge. For those who might like to tackle an almost impossible task, there are plenty of tropical lake sediments available in many countries and also high altitude peat bogs where fossil insects might well have been preserved. Likewise, there have been no investigations, as far as I know, of the geologically recent fossil insect faunas from oceanic islands. What a contribution such a study might contribute to our understanding of island biogeography or speciation rates!

III. QUATERNARY CLIMATIC CHANGE

With the advent of deep coring in both the ocean depths and the polar ice caps, a great deal has recently been discovered about the numbers and intensities of major climatic fluctuations. The ocean core drilling programme has shown that there were many more climatic oscillations in the Quaternary period of glacial/interglacial status than could possibly have been inferred from the terrestrial stratigraphic records alone (see Ruddiman and Raymo, 1988, for a review of the last 3 million years).

Hints of this complexity can be gleaned from loess studies in eastern Europe (Kukla, 1970) and in central China (Kukla, 1987). Deep drilling of the ice caps has increased the resolving power of this system immensely, though it is limited to the last two glacial/interglacial cycles and is thus not able to extend the record as far back in time as the ocean bottom cores. The ice cores show that the latest two glacials, and their intervening interglacial, were much more climatically complex than could have been inferred from the ocean cores. This is partially because the mixing time of the ocean limits resolution to changes that take place during periods in excess of 1000 years; however, the ice cores have another outstanding advantage in that they permit annual snow layer counting, introducing an extraordinary degree of precision into their timing. They are able to show, for instance, that climatic changes take place with great rapidity; from glacial to interglacial within a matter of decades (Dansgaard et al., 1989). It is interesting to note that similar rates of climatic change were inferred long ago from fossil insect assemblages from western Britain (Coope and Brophy, 1972).

It is inevitable that these numerous, intense and sudden climatic changes during the last 2.4 million years must have had drastic effects on the insect faunas of the times.

IV. THE RESPONSE OF INSECT SPECIES TO THE GLACIAL/INTERGLACIAL CYCLES

At first sight it might seem paradoxical that insect species should show such extreme stability in the face of the extraordinary instability of the Quaternary climate. However, it has been argued that it was the climatic inconstancy itself that was responsible to a large extent for the maintenance of species constancy (Coope, 1970, 1978). This is because species responded to these larger-scale climatic changes, not by Darwinian evolution, but by tracking acceptable conditions from place to place as the climatic zones moved back and forth across the continents. The precise driving mechanisms for these movements may have varied from species to species but this does not alter the fact that adequate movement *was* achieved regardless of any specific idiosyncrasies. In fact, if too much attention is paid to the details whereby the movements of particular species were achieved, it could easily obscure our appreciation of the general picture. Thus, provided that the species had adequate mobility and there was sufficient land surface available over which to move, the environment in which it lived remained effectively the same in spite of the climatic changes. From any individual species' point of view, it was

2. The Effects of Quaternary Climatic Changes on Insect Populations

merely the geography that had changed. There was thus no reason to evolve.

Furthermore, since populations were continuously being subdivided and rejoined in their forced marches across the continents, genetic as well as physical isolation must also have been repeatedly broken down and the gene pools were kept well stirred. Evolution, under these conditions, must have been almost impossible. Even within the limited time of the Last Interglacial (Eemian, Ipswichian) and the Last Glaciation (Weichselian, Wurm, Devensian), namely the last 120 000 years, there have been spectactular changes in the British insect fauna. In this chapter examples will be drawn from this period only, because it is the most thoroughly investigated so far. However, there is accumulating evidence that similar changes in our insect faunas accompanied each of the previous climatic oscillations back to at least three quarters of a million years ago, and almost certainly to the very start of the Quaternary period.

A. The Insect Fauna of the Last Interglacial

One of the most spectacular interglacial insect faunas so far found in Britain was discovered in the foundation excavations for Uganda House in Trafalgar Square, London. There can be little doubt that this fauna dates from the thermal maximum of the Last (Eemian) Interglacial. The insect assemblage included a large number of exotic scarabaeid beetles (not surprising in view of the numbers of bones of elephant, hippopotamus and lion that were found in the same deposit). Amongst these, the predominant species were *Caccobius schreberi* L., *Oniticellus fulvus* (Goeze) and *Onthophagus furcatus* (F.), all of which have present-day ranges that do not reach as far north as the British Isles. The most abundant species of *Onthophagus* was clearly related to the widespread *O. fracticornis* Prey but was much smaller and had a number of features that made it identical to *O. massai* Bar., a species which today is endemic to the island of Sicily (Coope, 1990). This will not be the only case of a so-called endemic species being found as a fossil well away from the area where it happens to be living today (see p. 39).

The Trafalgar Square insect assemblage also included a rare Mediterranean bark beetle, *Scolytus koenigi* Sch. which lives on various species of *Acer*. It is interesting to note, therefore, that leaves of *Acer monspesulanum* have been recorded from the same interglacial. The whole fossil insect assemblage from this site leaves little room to doubt that the climate at this time was considerably warmer than it is in Britain today.

B. The Insect Fauna of the Last Glaciation

The insect fauna of Britain during the predominantly cold period that followed the Last Interglacial was very different from the one just described from the Trafalgar Square deposits. All of the southern species had disappeared by this time and they were replaced by assemblages that were dominated by species with arctic or cold continental distributions at the present day. Such is the diversity of these faunas (for instance, Coope, 1968; Briggs et al., 1985), that only some selected highlights can be given here. From time to time the climate ameliorated for short, temperate interludes (called interstadials) during which cold-adapted insect species disappeared from our assemblages and were replaced by relatively thermophilous ones (Coope, 1959; Coope and Angus 1975).

In order to illustrate the profound changes in the geographical ranges of insect species as they responded to these climatic events, attention will be concentrated on two selected time intervals. Firstly, between about 40 000 and 25 000 radiocarbon years ago (the latter part of the Upton Warren Interstadial Complex), an episode of cold continental climate brought into Britain many exotic species that live today far from these islands. This was followed by the period of maximum ice extension when glaciers covered most of the northern half of Britain. At this time much of this country was reduced to conditions resembling polar desert and few, if any, insects were able to survive. The second selected period is from the closing phase of the glaciation, between about 13 000 and 10 000 radiocarbon years ago, when the ice sheets had largely retreated from Britain. During this period the climate was very unstable becoming rapidly temperate at 13 000 years ago (the Lateglacial Interstadial) before reverting more gradually to arctic conditions once again (the Loch Lomond Stadial partly equivalent in time to the Younger Dryas period of continental authors). A second episode of sudden and intense climatic warming took place at about 10 000 years ago at the start of the present interglacial period (Atkinson et al., 1987).

C. Examples of Changes in the Geographical Ranges of Species

In the examples that follow species will be considered, as far as possible, in their taxonomic order.

The distinctive species, *Carabus maeander* Fisch., has been found in two sites in Britain (Coope, 1962; Briggs et al., 1985). Today this is an eastern Siberian and North American species that lives no nearer to Britain than the Lena River valley in central Siberia.

2. The Effects of Quaternary Climatic Changes on Insect Populations

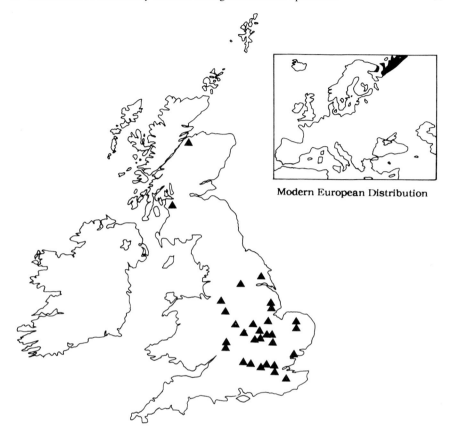

Fig. 1. Fossil sites in the British Isles for *Diacheila polita* Fald. ▲ Devensian (Weichselian) Glacial and Interstadial sites. There are no Lateglacial records from the British Isles.

Diacheila polita Fald. (Fig. 1) was a widespread species in Britain throughout much of the colder episodes of the Upton Warren Interstadial Complex. It also occurred as a fossil at many sites on the adjacent continent, that broadly date from the same period (e.g. Lindroth, 1948; Helle *et al.*, 1981). Today it is one of the characteristic species of the tundra of northern Eurasia and Alaska reaching as far west in arctic Russia as the Kola Peninsula. After the retreat of the ice it did manage to return to Britain (Walker *et al.*, 1993) but did not survive the great climatic warming at 13 000 years ago.

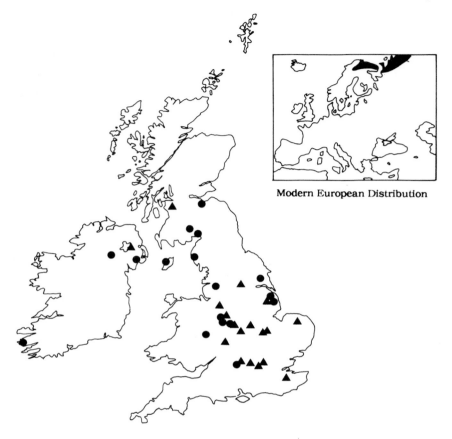

Fig. 2. Fossil sites in the British Isles for *Diacheila arctica* Fald. ▲ Devensian (Weichselian) Glacial and Interstadial sites. ● Devensian (Weichselian) Lateglacial cold sites.

Diacheila arctica Gyll. (Fig. 2) is today another rare northern species with a similar geographical range to that of *D. polita*. It was never as common as *D. polita* during glacial times in Britain but it became widespread and common during the colder parts of the Lateglacial episode. There are no fossil records of this species from the postglacial period, suggesting that it did not survive the sudden climatic warming at the beginning of the present interglacial.

Amongst the dytiscid water beetles, the exclusively northern species *Colymbetes dolabratus* (Payk.) seems to have been particularly wide-

2. The Effects of Quaternary Climatic Changes on Insect Populations

spread in Britain during all cold climatic phases. It apparently did not survive the postglacial climatic warming at 10 000 years ago.

Fossils of the hydrophilid genus *Helophorus* include a relatively large number of species that do not now occur in Britain, such as *H. aspericollis* Angus, *H. jacutus* Popp. and *H. mongoliensis* Angus, all of which are today wholly eastern Asiatic species. *H. obscurellus* Popp. is chiefly Asiatic in its distribution today, occurring both on the northern tundras and the high cold steppes of central Asia. It extends westwards only as far as Kanin in arctic Russia. The northern and eastern species *H. sibiricus* Motsch. was also abundant in Britain during the colder parts of the Upton Warren Interstadial. Both these latter species recolonized Britain after the retreat of the ice sheets but failed to survive the climatic amelioration at the start of the postglacial.

Deposits at Tattershall in Lincolnshire, date from a very short but temperate phase at the beginning of the Upton Warren Interstadial, namely about 43 000 years ago. These have yielded a large and complex fossil insect of mostly temperate species (Girling, 1974), amongst which was a form of *Ochthebius* almost identical to *O. dilatatus* Steph. but with quite distinct male genitalia. Robert Angus (1993) has recently shown that the fossil is *O. figueroi* Garr. Vall. & Regil., a Spanish "endemic" recently described from the headwaters of the river Ebro (see p. 41 for another Spanish "endemic" from the same site).

The Staphylinidae from the cold episodes include a considerable number of exotic species and only a selection will be given here. The Omaliinae are particularly abundant in all of these "cold" faunas. One of the most unexpected of these is *Holoboreaphilus nordenskioeldi* Makl. whose fossil occurrences in Britain, mirrors closely that of *Diacheila polita*. Today this is an almost circumpolar species in the high north extending from Alaska westwards as far as the Kanin peninsula in arctic Russia. It managed to recolonize Britain after the retreat of the ice (Coope, 1982; Walker *et al.*, 1993) but apparently failed to survive the abrupt climatic warming at 13 000 years ago at the start of the Lateglacial Interstadial warm period. Other exotic omaliines include *Pycnoglypta lurida* Gyll., *Olophrum boreale* Payk., *Acidota quadrata* Zett. and *Boreaphilus henningianus* Sahlb. (Fig. 3), all of which returned to Britain in the Lateglacial period but which have seemingly died out here since. However, some of these species are so inconspicuous that it would not be at all surprising if some of them were found one day living at high altitudes somewhere in the north of these islands.

Another unexpected fossil occurrence in Britain has been *Anotylus* (= *Oxytelus*) *gibbulus* Epp., which today is almost wholly confined to the Caucasus mountains. (There is a single, somewhat questionable,

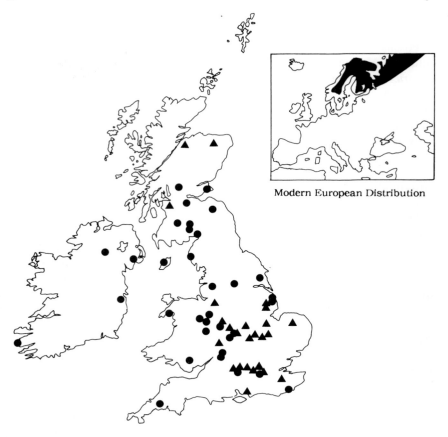

Fig. 3. Fossil sites in the British Isles for *Boreaphilus henningianus*. J. Sahlb. ▲ Devensian (Weichselian) Glacial and Interstadial sites. ● Devensian (Weichselian) Lateglacial cold sites.

occurrence from near Vladivostok; see Hammond *et al.*, 1979.) This species has been found very rarely in deposits of the Last Interglacial in England and as an uncommon species in the Upton Warren Interstadial. It has not yet been found in any Lateglacial deposits. It is interesting to note, therefore, that *A. gibbulus* was the most abundant staphylinid species in Britain during an earlier interglacial which probably was the one that immediately preceded the Eemian Interglacial discussed above.

The most abundant medium-sized dung beetle throughout the whole of the Upton Warren Interstadial Complex, and usually occurring in deposits wherever there are bones of mammoth, woolly rhinoceros and bison, was *Aphodius holdereri* Reitt. represented by almost all skeletal parts including the male genitalia. Today this species is restricted to the high plateau of Tibet and adjacent parts of western China (Coope, 1973). It was often associated with large numbers of another central Asiatic species, *Aphodius jacobsoni* Kosh. (Briggs et al., 1985). Here again the numerous skeletal elements were associated with the characteristic male genitalia.

During the temperate phase of the Upton Warren Interstadial Complex, one of the most abundant dung beetles at Tattershall in Lincolnshire was the Spanish "endemic" species *Aphodius bonvouloiri* Har. (Coope and Angus, 1975). It was associated with numerous relatively thermophilous species of Coleoptera but none of the Asiatic species mentioned above.

The exclusively moss-eating byrrhid beetles are common constituents of many of our glacial fossil assemblages. *Simplocaria metallica* Sturm and *Curimopsis cyclolepidia*. Munst. are both high latitude species (*C. cyclolepidia* also occurs in central Asia) that appear to be rare today throughout the whole of their ranges. *S. metallica*, in particular, was very abundant and widespread in Britain during the colder parts of the Upton Warren Interstadial and also in the colder parts of the Lateglacial period. Thus, rarity at the present day must not be viewed as an intrinsic property of a particular species but an indication that really suitable habitats may not be available to it anywhere at the present time. They are merely eking out an existence in suboptimal conditions. On the evidence of the fossil record, future circumstances may well alter in their favour once again, with the advent of the next glacial period: that is, of course, provided that some ecological catastrophe does not intervene in the meantime to exterminate them altogether.

Ladybirds are one of the most popular groups of beetles but unfortunately do not preserve well as fossils. In spite of this, several exotic species have been found in deposits dating the Last Glaciation. *Ceratomegilla ulkie* Crotch was discovered in an organic lens interbedded within river gravels at Dorchester on Thames in Oxfordshire, dating from about 40 000 years ago (Briggs et al., 1985) and I have recently recognized this species in a fossil assemblage belonging to an early phase of the Last Glaciation at Shropham, Norfolk. This species is today a widespread species in northern Canada and in Asia as far west as North Kazakhstan, the Altai Mountains. Mongolia and western China (M. Majerus in lit. 2/11/93).

A second unusual species of ladybird, *Hippodamia arctica* Sch. was widespread in Britain all through the colder part of the Upton Warren Interstadial Complex. Today, this is a species of the high north of Europe and Asia. This species was one of the early colonizers of the British Isles after the retreat of the ice but it disappeared from our records for a short time during the thermal maximum of the Lateglacial Interstadial. Its fossils then reappeared as the climate subsequently cooled down and it persisted throughout the period of arctic climate of the Loch Lomond Stadial. It does not seem to have survived in Britain after the second of the sudden and intense episodes of climatic warming at the start of the postglacial period. The way in which *Hippodamia arctica* moved back and forth in response to the dictates of the changing Lateglacial climate is an excellent illustration of the way insect species in general reacted to such climatic fluctuations.

Amongst the fossils of phytophagous species from the Upton Warren Interstadial, no doubt the most exciting was the discovery at Brandon near Coventry of *Chrysolina septentrionalis* Men. (Coope, 1968). At the present day, this is an exclusively arctic Siberian and Alaskan species.

Fossils of other orders of insect show that they also responded to the large-scale climatic changes in the same manner, though they have been less well investigated. For example, amongst the Hemiptera Heteroptera, two high northern saldid species, *Chiloxanthus stellatus* Curtis and *Calacanthia tribomi* (J. Sahlb.) were discovered at the Brandon site near Coventry (Coope, 1968) dating from the cold phase of the Upton Warren Interstadial Complex. Here again they responded to climatic changes by altering their geographical ranges rather than by eevolving to fit the new conditions.

It is perhaps appropriate here to make a general comment upon some of the "endemic" species demonstrated in the previous paragraphs, to have occurred as fossils well outside their present-day geographical ranges. In the absence of this fossil record, such restricted distributions might have been construed as indicating an area of origin for the species. In the light of the fossil evidence presented here, these restricted ranges must be viewed rather differently; namely as the latest phase in a complex pattern of ever-shifting geographical distributions that owe more to environmental history than they do to their place of evolutionary origin.

Finally, and rather unexpectedly, during the brief but intense thermal maximum of the Lateglacial Interstadial, between 13 000 and 12 200 years ago, temperatures in Britain rose very suddenly, reaching values rather warmer than those of the present day. A number of relatively southern species took advantage of this short interlude to extend their ranges well to the north even of their present-day distributional limits. Figures 4–6 show examples of three such species. Of these, only *Bembidion octomaculatum* Goez (Fig. 4) is still a British species but is a great rarity holding onto a precarious foothold in the southeast of England apparently being

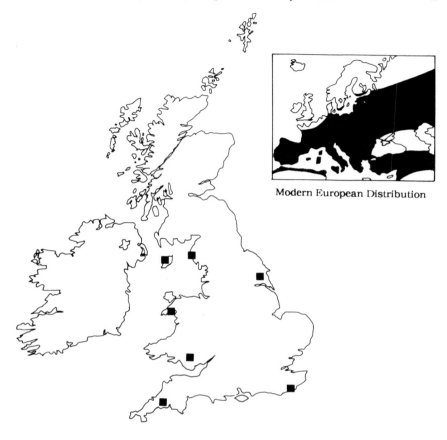

Fig. 4. Fossil sites in the British Isles for *Bembidion octomaculatum* Goeze. ■ Devensian (Weichselian) Lateglacial temperate sites.

unable to establish long-term colonies here. At the present day it regularly reaches southern Scandinavia but again does not seem to be able to establish a regular breeding population there either (Lindroth, 1985). During the temperate part of the Lateglacial Interstadial, however, this species seems to have had thriving populations as far north as the Isle of Man (Coope, 1971) and St Bees on the Cumbrian coast (Coope and Joachim, 1980) and to have been widespread over much of England from Bodmin Moor in Cornwall to Folkestone in Kent.

Bembidion grisvardi Dew. and/or *B. ibericum* Pioc. (Fig. 5) are both southern European species that are very difficult to distinguish from one

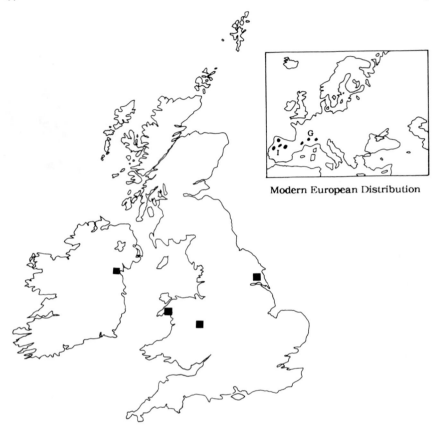

Fig. 5. Fossil sites in the British Isles for *Bembidion grisvardi* Dew. or *ibericum* Pioc. ■ Devensian (Weichselian) Lateglacial temperate sites.

another even when complete specimens are available. They cannot be distinguished on the basis of the fossil evidence even though both pronotum and elytra have been found. They indicate that either or both species were present in Britain during the temperate part of the Lateglacial Interstadial.

Asaphidion cyanicorne Pand. is another species in this group that reached as far north as southern Scotland at this time (Fig. 6). They are included here to show how rapidly insect species can move in response to sudden climatic changes. The complete replacement of a thoroughly arctic assemblage of beetles by one dominated by relative thermophiles, took place so quickly that it was impossible to measure using any

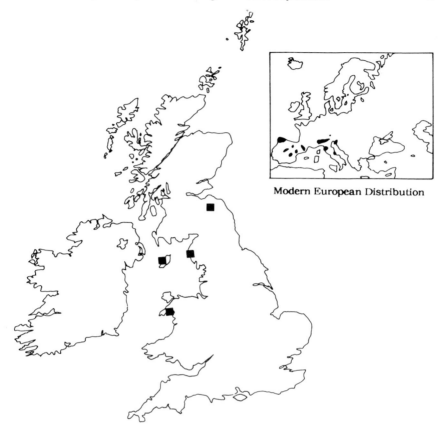

Fig. 6. Fossil sites in the British Isles for *Asaphidion cyanicorne* Pand. ■ Devensian (Weichselian) Lateglacial temperate sites.

available geochronological technique. It would seem that the change from fully glacial to fully interglacial climatic conditions was completed in *less* than a century (Coope and Brophy, 1972). The speed of these climatic changes was such that it is difficult to see how species could have adjusted to them by any process of Darwinian selection.

D. Summary

These selected highlights from the fossil insect record of the last glacial/interglacial cycle illustrate that large-scale changes in the geographic ranges of species took place right across the taxonomic spectrum of the Coleoptera. There are hints too that other orders of insect, such as the Hemiptera, adopted the same stratagem for avoiding the problems of climatic change. Thus, even within this relatively short time span, some species readily altered their distributions, by several thousand kilometres, and changed latitude with apparent ease. A dynamic picture emerges of alterations of arctic/continental and temperate faunas crossing and recrossing the British Isles in orderly processions of species at the dictate of a climate that oscillated back and forth between glacial and interglacial modes.

V. PROSPECT

The long-term stability of insect species shown by the Quaternary fossil record should not delude us into believing that they are invulnerable to environmental change. The conditions of the glacial/interglacial cycles to which both species and communities have become adapted over the last few million years, operated in a free world in which the opportunity existed of moving whenever the need arose. These idyllic conditions no longer prevail. Human activities such as the drainage of wetlands and the clearing of forests have not only had a destructive effect on the habitats themselves, but serve to isolate the surviving fragments in a sea of hostility. Commercial enterprises such as intensive agriculture or forest monoculture exacerbate still further this isolation, so that even the most sensitively sited nature reserve will become, under any changing climatic regime, an inescapable prison, as isolated as an oceanic island.

If population movement is an essential component of the survival strategy of a species during times of climatic change, and if some aspect of global warming now seems likely, our nature reserves should be planned accordingly. For flightless species at least, provision ought to be made for corridors of acceptable habitat to link up major reserves which in their turn must be large enough to permit adequate environmental diversity. Blighted property adjacent to motorways or even railways might be pressed into service in this context and areas liable to flooding alongside river courses could similarly be adapted as linear reserves, without encroaching upon valuable agricultural land. If such a programme seems impracticable in the harsh political and economic climate of the present day (and we can always hope for a global warming here too) then a deliberate plan to translocate species may be necessary as a poor, but perhaps the only, substitute for natural dispersion. If we do not provide opportunities for species to adjust their geographical ranges in the face of climatic changes as they have done so effectively in the past, it is difficult to see how any other than some pest species will be able to survive the environmental changes that we unwittingly may impose upon them in the future.

REFERENCES

Angus, R. B. (1993). Spanish "endemic" *Ochthebius* as a British Pleistocene fossil. *Latissimus* **2**, 24–25.
Atkinson, T. C., Briffa, K. R. and Coope, G. R. (1987). Seasonal temperatures in Britain during the past 22 000 years, reconstructed using beetle remains. *Nature* **325**, 587–592.
Briggs, D. J., Coope, G. R. and Gilbertson, D. D. (1985). The Chronology and Environmental Framework of Early Man in the Upper Thames Valley. *British Archaeological Reports* **135**, 1–176.
Buckland, P. C. and Coope, G. R. (1991). "A Bibliography and Literature Review of Quaternary Entomology". J. R. Collins Publications, Department of Archaeology & Prehistory, University of Sheffield, pp. 1–85.
Coope, G. R. (1959). A Late Pleistocene insect fauna from Chelford Cheshire. *Proc. R. Soc. Lond. B* **151**, 70–86.
Coope, G. R. (1962). A Pleistocene insect fauna with arctic affinities from Fladbury, Worcestershire. *Quart. J. Geol. Soc. Lond.* **118**, 103–123.
Coope, G. R. (1968). An insect fauna from Mid-Weichselian deposits at Brandon, Warwickshire. *Phil. Trans. R. Soc. Lond. B* **254**, 425–456.
Coope, G. R. (1970). Interpretations of Quaternary insect fossils. *Ann. Rev. Entomol.* **15**, 97–120.
Coope, G. R. (1971). The fossil Coleoptera from Glen Ballyre and its bearing upon the interpretation of the Lateglacial environment. *In* "A Field Guide to the Isle of Man" (G. P. S. Thomas, ed.), pp. 13–15. Quaternary Research Association, Liverpool.
Coope, G. R. (1973). Tibetan species of Dung Beetle from later-Pleistocene deposits in England. *Nature* **245**, 335–336.
Coope, G. R. (1978). Constancy of insect species versus inconstancy of the Quaternary environment. *In* "Diversity of Insect Faunas" (L. A. Mound and N. Waloff, eds), pp. 176–187. *Symposia of the Royal Entomological Society* **9**, Blackwell, Oxford.
Coope, G. R. (1982). Coleoptera from two Late Devensian sites in the lower Colne valley, West London, England. *Quat. Newsl.* **38**, 1–6.
Coope, G. R. (1990). The invasion of Northern Europe during the Pleistocene by Mediterranean species of Coleoptera. *In* "Biological Invasions in Europe and the Mediterranean Basin" (F. di Castri, A. J. Hansen and M. Debussche, eds), pp. 203–215. Kluwer, Dordrecht.
Coope, G. R. and Angus, R. B. (1975). An ecological study of a temperate interlude in the middle of the last glaciation, based on fossil Coleoptera from Isleworth, Middlesex. *J. Anim. Ecol.* **44**, 365–391.
Coope, G. R. and Brophy, J. A. (1972). Late Glacial environmental changes indicated by a coleopteran succession from North Wales. *Boreas* **1**, 97–142.
Coope, G. R. and Joachim, M. J. (1980). Lateglacial environmental changes interpreted from fossil Coleoptera from St Bees, Cumbria, N.W. England. *In* "Studies in the Lateglacial of North West Europe" (J. J. Lowe, J. M. Gray and E. J. Robinson, eds), pp. 55–68. Pergamon, Oxford.
Dansgaard, W., White, J. W. C. and Johnsen, S. J. (1989). The abrupt termination of the Younger Dryas event. *Nature* **339**, 532–533.

Girling, M. A. (1974). Evidence from Lincolnshire of the age and intensity of the mid-Devensian temperate episode. *Nature* **250**, 270.

Hammond, P. M., Morgan, A. and Morgan, A. V. (1979). On the *gibbulus* group of *Anotylus*, and the fossil occurrences of *Anotylus gibbulus* (Staphylinidae). *Syst. Ent.* **4**, 215–221.

Helle, M., Sonstegaard, E., Coope, G. R. and Rye, N. (1981). Early Weichselian peat at Brumunddal, southwestern Norway. *Boreas* **10**, 369–379.

Kukla, G. J. (1970). Correlations between loeses and deep sea sediments. *Geologiska Föreningens i Stockholm Förhandlingar* **92**, 148–180.

Kukla, G. J. (1987). Loess stratigraphy in central China. *Quat. Sci. Rev.* **6**, 307–374.

Lindroth, C. H. (1948). Interglacial insect remains from Sweden. *Årsbok Sveriges geologiska undersökning* **C42**, 1–29.

Lindroth, C. H. (1985). "The Carabidae (Coleoptera) of Fennoscandia and Denmark", pp. 1–225. *Fauna Entomologica Scandinavica* **15** (1). E. J. Brill, Leiden.

Matthews, J. V. Jr (1974). A preliminary list of insect fossils from the Beaufort Formation, Meighen Island, District of Franklin. *Geological Survey of Canada, Papers*, **74-1A**, 203–206.

Matthews, J. V. Jr (1976). Insect fossils from the Beaufort Formation: geological and biological significance. *Geological Survey of Canada, Papers* **76-1B**, 217–227.

Ruddiman, W. F. and Raymo, M. E. (1988). Northern Hemisphere climatic regimes during the past 3 Ma: possible tectonic connections. *Phil. Trans. R. Soc. Lond.* B **318**, 411–430.

Walker, I. R. (1987). Chironomidae (Diptera) in Palaeoecology. *Quat. Sci. Rev.*, **6**, 29–40.

Walker, M. J. C., Coope, G. R. and Lowe, J. J. (1993). The Devensian (Weichselian) Lateglacial Palaeoenvironmental Record from Gransmoor, East Yorkshire, England. *Quat. Sci. Rev.* **12**, 659–680.

Wilkinson, B. J. (1984). Interpretation of past environments from sub-fossil Caddis larvae. *In* "Proc. 4th Int. Symposium on Trichoptera" (J. C. Morse, ed.), pp. 447–452. *Series Entomologicae* **30**. W. Junk, The Hague.

Williams, N. E. (1988). The use of caddisflies (Trichoptera) in palaeoecology. *Palaeogeography, Palaeoclimatology, Palaeoecology* **62**, 493–500.

3

The Hadley Centre Transient Climate Experiment

D. A. BENNETTS

    I. Introduction 50
   II. Global Patterns of Transient Change 50
      A. The Magnitude of Transient Change 50
      B. The Effects of Man-made Aerosols 53
      C. The Pattern of Transient Change 53
      D. Sea Level Rises 54
  III. Climate Change Over Western Europe 56
      A. Temperature 56
      B. Precipitation 56
      C. Soil Moisture 57
      D. Storminess 57
      E. Hadley Centre Predictions of Climate Change for Western Europe 57
      F. Present-Day General Circulation Model Limitations 58
      References 58

Abstract

The results of the first transient, climate change experiment undertaken at the Hadley Centre are presented. The experiment sought to determine the climate response, over a 75-year period, expected to arise from a scenario in which atmospheric CO_2 concentrations increased by 1% per year (compound). Under this scenario, CO_2 would double in about 70 years. Key conclusions are that there is likely to be asymmetry in the warming of the two hemispheres with the northern hemisphere warming faster, and that little surface warming is expected in some oceanic regions before about 2050. There is general support for the assessment provided in the IPCC Report (1990) and Supplement (1992). Changes in temperature, precipitation, soil moisture and storminess expected over western Europe by the year 2030 are discussed.

I. INTRODUCTION

In 1990 the Intergovernmental Panel on Climate Change (IPCC) presented an assessment of the information relevant to climate change issues (Houghton et al., 1990). One of the main questions addressed was "How much do we expect the climate to change and how quickly?". In formulating an answer to this some assumptions on the future rate of emission of the man-made greenhouse gases needed to be made and this led to the development of a number of "scenarios". The central one was the "Business-as-Usual" scenario in which it was broadly assumed that future emissions would continue to follow present-day trends.

In their 1992 supplement (Houghton et al., 1992), IPCC re-assessed their earlier conclusions in the light of recent research. During the intervening 2 years there had been continued improvement of the general circulation models (GCMs) and a few research groups, including the Hadley Centre, had carried out "transient" experiments with "coupled" Atmosphere–Ocean GCMs (AOGCMs).

This chapter describes the Hadley Centre results and gives an assessment of the rates of change of climatic conditions that are implied for the first half of the twenty-first century.

II. GLOBAL PATTERNS OF TRANSIENT CHANGE

A. The Magnitude of Transient Change

The northern hemisphere warms faster than the southern hemisphere partly because of the stronger climate "feedback" mechanisms over land areas and partly because the southern hemisphere, being largely covered by oceans, has a higher thermal inertia than the northern hemisphere. Although this result was anticipated by Houghton et al. (1990), the present experiment confirms that prediction and adds points of detail.

Figure 1 shows the response of the annually averaged, global-mean, surface air temperature (Murphy, 1992). Initially the rate of change is slow, with a mean global warming of only 0.1°C per decade, and almost no warming in the southern hemisphere. Subsequently the rate of change increases, the global mean rate of increase reaching 0.3°C per decade at the time of CO_2 doubling (within the IPCC estimated range of 0.15–0.4°C per decade).

The above figures are put into perspective through comparison with the natural, interannual variability of global mean temperature in which variations of up to 0.3°C often occur within a decade. Consequently, over

3. The Hadley Centre Transient Climate Experiment

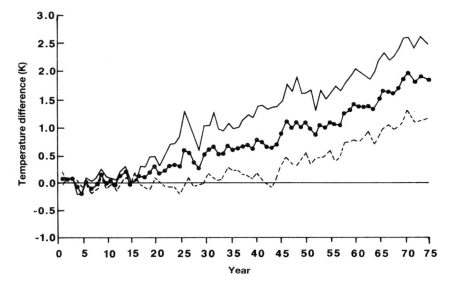

Fig. 1. The annually averaged mean surface temperature response (●—●) globally, (———) for the northern hemisphere and (–––) for the southern hemisphere.

one or two decades, natural variability may mask any change expected from increased greenhouse gases. Only over several decades can the anthropogenic effects (at present-day emission rates) be expected to impose a noticeable trend on the natural variability.

Care must be taken in interpreting the results shown in Fig. 1. During the first two to three decades little warming appears to take place and it is believed that this is partly a consequence of experimental design. The phenomenon is also present in at least one other transient experiment and has become known as the "cold start" problem. It probably arises because the real climate has been subject to a steadily increasing forcing due to greenhouse gases for well over a century whereas in the experiment the model climate was brought to (near) equilibrium and then subjected to a sudden change in the rate of forcing at year 0.

The effects of the cold start make it difficult to ascribe calendar years to the simulated years of the experiment. However, through the use of a simple energy-balance model calibrated against the GCM, this can be partly overcome and estimates of climate change, derived using such a model, are shown against calendar years in Fig. 2.

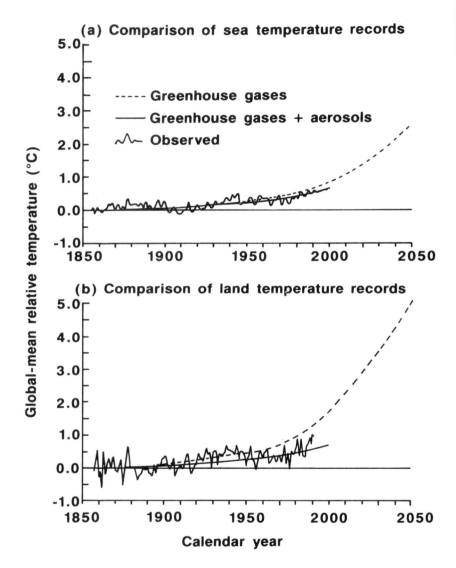

Fig. 2. The simulated increase in the annual mean surface air temperature due to observed historical increases in greenhouse gases for 1850–1990, and a scenario in which CO_2 increases by 1% per annum (compound) during the period 1990–2050 for (a) sea areas, (b) land areas. Shown also is an assessment of the possible impact of aerosols.

3. The Hadley Centre Transient Climate Experiment

B. The Effects of Man-made Aerosols

It has been suggested (Houghton *et al.*, 1992) that sulphate aerosols could offset some 25% of the radiative forcing due to man-made emissions of greenhouse gases. The Hadley Centre experiment did not include the effects of sulphate aerosols directly, but their possible effects on global warming can be investigated using the energy-balance model discussed earlier.

Charlson *et al.* (1991) have estimated the radiative effects of sulphate aerosols. There are large uncertainties but, to illustrate the possible effects in terms of the global temperature response, the low-range value given by Charlson *et al.* was used as input to the energy-balance model.

As expected, the effects are larger over land areas, close to the sources of aerosols, as is evident in the larger displacement of the solid curves (greenhouse gas + aerosols) from the dashed curves (greenhouse gases alone) in Fig. 2(b) compared to that over sea areas (Fig. 2(a)).

For both the land and sea areas there is a marked improvement in the match between model and observations when the effects of aerosols are included, although this must remain a very tentative result until there are both better observations of aerosol concentrations and their effects can be represented more realistically in the models. The present results, derived from simpler models, are instructive and suggestive, but do not provide the necessary, rigorous treatment to give confidence in the predictions. However, there is confidence that the sign of the change is correct with aerosols having a net cooling effect.

C. The Pattern of Transient Change

One of the main reasons for undertaking transient experiments is to determine the pattern and rate of onset of climate change over the next few decades. Until recently such information has not been available directly since previous GCM experiments were only capable of determining the very long-term ("equilibrium") response. Transient changes were then inferred. Plate 1 shows the results for both an "equilibrium" experiment and the current transient experiment; the similarities and differences can be readily seen.

In comparing the two predictions it is important to note that, because the thermal inertia of the oceans is now represented, the transient experiment is not at equilibrium at the time of effective CO_2 doubling. Averaged over years 66–75, the decade centred on the time of CO_2

doubling, the global mean warming in the transient run is 1.7°C above the starting point of the experiment (see Fig. 1); this compares with an (estimated) equilibrium response of 2.7°C. The temperature rise in the transient experiment is some 60% of the equilibrium response – the remaining 40% is, in the words used in the IPCC report, "committed" and will be realized in subsequent years as the deep ocean warming becomes uniformly distributed.

After discounting this difference in the absolute magnitude of the response, other features can be identified. Of immediate note is the fact that while the warming over the Arctic region remains generally similar, the southern hemisphere has marked differences, particularly in the Southern Ocean and Antarctica. Averaged over the southern hemisphere there remains a warming trend, but it is much weaker than in the equilibrium experiment and in many parts of the Southern Ocean there is an absence of warming, surface temperatures being much the same as today. A similar lack of warming in surface temperatures is also evident over the Northwest Atlantic, to the south of Greenland.

These changes in the climate response result directly from the inclusion of a full-depth model of the ocean and the consequent ability of the GCM to simulate both the thermal inertia and the effect of changes in the ocean circulations.

D. Sea Level Rises

A number of factors contribute to changes in the volume of water in the oceans. Of these the most important are:

(1) the (vertical) temperature structure of the oceans (since that directly influences the volume occupied by the water);
(2) the changes in the amount of ice held in the major ice sheets on Antarctica and Greenland;
(3) the melting of mountain glaciers; and
(4) the changes in the water storage on land, primarily in snow, soil moisture and ground water.

Floating sea ice is not a factor since any increase or decrease in extent has no effect on sea level.

The first and last of these can be assessed directly from the current numerical experiment. The second and third are taken from estimates produced from the IPCC report. The individual contributions and total

3. The Hadley Centre Transient Climate Experiment

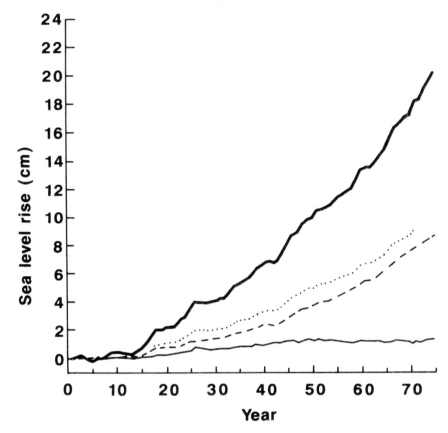

Fig. 3. Sea level rise calculated from the transient experiment. ——, Total; ·····, glacier melt; – – – –, thermal expansion; ———, water budget (computed, excluding glacier melt).

rise are shown in Fig. 3. However, neither the dynamics of the permanent ice sheets nor the mountain glaciers are represented explicitly in the GCM and appropriate data for those aspects have been obtained from Houghton et al. (1990). The mean, global rate of increase at the time of doubling of CO_2 is approximately 4 cm per decade.

For comparison, the IPCC report estimates the change to be from 2 to 7 cm per decade at the time of CO_2 doubling for the (IPCC, 1990) Business-as-Usual scenario. The present estimate falls within that range.

III. CLIMATE CHANGE OVER WESTERN EUROPE

Detailed study of the results illustrated in Plate 1, together with the results of similar transient experiments carried out at other international climate prediction centres, allow the creation of a scenario of the expected climate change over western Europe for the early part of the twenty-first century (typically 2030).

A. Temperature

Warming is expected to be a minimum near the Atlantic coast (0.2°C per decade – 0.75°C by 2030) rising to about 0.35°C per decade in eastern Europe. In particular there is expected to be a decrease in the intensity of very cold spells in winters and an increased number of hot summers. Confidence in this prediction is fairly high in the UK as it lies between an ocean to the west, which will warm only slowly, and land to the east which will warm much more rapidly. The largest warming will therefore occur with winds from the east, which will mitigate the cold spells associated with easterly winds in winter, and enhance the warm spells associated with easterlies in summer.

Because of the frequent occurrence of near freezing temperature in winter in the UK, the expected warming is likely to produce a marked decrease (of order 25%) in the occurrence of frosts.

For comparison there are observed interannual variations of up to 1°C in the annual mean surface air temperature over western Europe in some decades. Consequently, over a period of one or two decades, natural variability could mask any warming due to increasing greenhouse gases.

B. Precipitation

Most experiments suggest that, in winter, northern and central Europe are likely to experience an increase in precipitation, a few regions experiencing up to 40% more but most having an increase of about 20%. The UK will probably experience a similar trend and, for the same reason as the reduction in frosts, the frequency of snowfall is expected to be markedly reduced. The area in Europe where decreases are most probable is the Iberian Peninsula.

In summer the south of Europe is expected to become drier but there is no firm guidance on the sign of change elsewhere; in the Hadley Centre model there is reduced precipitation from the Mediterranean up to the latitude of northern UK.

The models also indicate a change in the character of the rainfall with convective precipitation (showers) becoming more frequent. It is therefore likely that, even in areas where the precipitation increases, there will be a decrease in the number of rain days, and an increase in the amount of rainfall on a "rain" day.

The interannual variability in seasonal rainfall over western Europe is typically some 30–40%, rising occasionally to in excess of 60%. A few decades of data will therefore be necessary to separate the anticipated trend due to global warming from the natural variability.

C. Soil Moisture

The models provide little indication on the sign of changes of soil moisture during winter for most of Europe; the implication is that there is likely to be little change from present-day values. The exception is the Iberian Peninsula where there will probably be a drying.

Most models predict that southern Europe will become drier in summer but there is considerable disagreement between the models as to how far north the drier area will extend; the Hadley Centre model suggests that it would extend to cover much of the UK.

D. Storminess

In the Hadley Centre experiment there is a slight indication that the storm tracks in the North Atlantic will extend eastwards as the climate changes. This could indicate a small increase in the number of storms crossing northern UK. However, there is poor agreement amongst the different models (Houghton *et al.*, 1992) and more work is required to establish the veracity of this prediction.

E. Hadley Centre Predictions of Climate Change for Western Europe

The details of the changes predicted by the Hadley Centre model are illustrated in Plate 2. There is broad consensus with the results of the other models but, as would be expected, there are differences in detail. Clearly care must be taken in interpreting the results on this scale – note the grid-point resolution of the GCM which is clearly visible in Plate 2, soil moisture content. Each square represents 1 grid box.

F. Present-day General Circulation Model Limitations

All present-day GCM results must be treated with caution. The models contain many approximations, and hence uncertainties which, through a lack of understanding of the underlying physical and chemical processes of the climate system, cannot yet be quantified or removed. Particular care must be taken when using the results in connection with regional (continental scale) aspects of climate change.

Acknowledgements

The work described in this report was carried out at the Hadley Centre for Climate Prediction and Research under the Department of the Environment Climate Prediction Programme (contract PECD 7/12/37).

REFERENCES

Charlson, R. J., Langer, J., Rodhe, H., Leovy, C. B. and Warren, S. (1991). Perturbation of the Northern hemisphere radiative balance by backscattering from anthropogenic sulfate aerosols. *Tellus* **43A-B**, 152–163.

Houghton, J. T., Jenkins, G. J. and Ephraums, J. J. (1990). "Climate Change, The IPCC Scientific Assessment". Cambridge University Press, Cambridge.

Houghton, J. T., Callandar, B. A. and Varney, S. K. (1992). "Climate Change, 1992, The Supplementary Report to the IPCC Scientific Assessment". Cambridge University Press, Cambridge.

Murphy, J. M. (1992). "A Prediction of the Transient Response of Climate". *Climate Research Technical Note* **32**. Hadley Centre, Bracknell.

4

Predicting Insect Distributions in a Changed Climate

R. W. SUTHERST, G. F. MAYWALD AND
D. B. SKARRATT

I. Introduction	60
II. Approaches to Comparing Climates Without Reference to Individual Species	61
III. Approaches to Estimating Potential Geographical Distributions of Individual Species	61
A. Climograms	65
B. BIOCLIM	65
C. Multivariate Statistics	65
D. Physiological Data and Population Dynamics Models	66
E. CLIMEX	67
IV. Reliability and Accuracy of Data Sets Used to Estimate Potential Distributions	74
V. Comparison of Approaches	74
VI. Applications of the Above Methods to Estimating the Impacts of Climate Change	78
A. Climate Change Scenarios	78
B. Impact Studies	78
C. Nature of Impacts	79
VII. Adaptative Strategies to Changing Distributional Limits	85
VIII. Discussion	86
IX. Conclusions	87
References	88

Abstract

Climate plays a dominant role in determining the distribution and abundance of insects, either through its direct effects on population processes or through its effects on host plants or animals. Projected climate change is likely to be associated with change not only the average temperatures and moisture availability, but also atmospheric composition and the variability of climates in different regions.

The possible mechanisms by which climate change might impact on insects and hence on their geographical distributions are considered. The wide range of different philosophies and analytical techniques used to describe the geographical distributions of insects in relation to climate is reviewed. Finally, some thoughts are presented on possible future directions.

I. INTRODUCTION

The abundance of a species depends on its physical environment, including the climate, as well as its own biological characteristics and the influence of other organisms (Andrewartha and Birch, 1954). Geographical distributions of species are the consequence of their population attributes and capacity to tolerate climatic conditions during the unfavourable season each year. Exceptions occur when the species is limited by host availability or interactions with other species. Ideally, explanation and prediction of the geographical distribution of a species will be based on a deductive approach, using physiological data on demographic parameters and field biology, linked using mechanistic models. In practice, that has not yet been achieved for even the most intensively studied species, while the need is for predictive methods suitable for applications to large numbers of rare species, pests and species beneficial to humans. It is therefore necessary to fall back to different degrees on inductive approaches, which rely on inferring climatic requirements from observed distributions (Meats, 1981), and to accept levels of precision and uncertainty that are not necessarily intellectually satisfying. Experience has shown that valid and useful conclusions can often be made about the limits of the potential geographical distribution of species based on inference. This has been possible even when no information is available on the biological mechanisms involved or on the population dynamics of that species (Cook, 1929; Booth, 1988; Sutherst *et al.*, 1989; Lindenmayer *et al.*, 1991). The degree of confidence in such estimates increases with knowledge of the species' demographic parameters and its biology in the field.

The limitations of both deductive and inductive approaches have led to the development of a variety of methods to estimate the nature and duration of climatic conditions which limit species' distributions. In this chapter the different approaches are reviewed and compared and then the use is described of the CLIMEX model (Sutherst and Maywald, 1985; Maywald and Sutherst, 1989, 1991; Sutherst *et al.*, 1991) to address issues of climate change.

The climate is made up of diurnal and annual cycles of temperature, moisture, light and wind. In the context of climate change, atmospheric composition, principally carbon dioxide (CO_2) also needs to be considered. The dynamics of each variable depends on the geographical zone, latitude and altitude. The key variables are maximum and minimum temperatures, rainfall and evaporation, day-length and their seasonal patterns. The role of climate alone is discussed but that should not be taken to imply that other factors are unimportant.

II. APPROACHES TO COMPARING CLIMATES WITHOUT REFERENCE TO INDIVIDUAL SPECIES

There have been numerous attempts to classify climatic regions, with questionable benefits. The division of the world into ecological zones has been one of the most widely used tools for matching the climates of different geographical areas (Papadakis, 1966; Walter, 1985). It is helpful in identifying regions with similar plant communities but becomes less useful when the requirements of individual species are considered. The climates at specific locations have also been described in an agronomic context by Walter and Leith (1960), using graphs to portray annual temperature and moisture balance regimens. Meteorological data from specific locations have also been compared using the "Match Climates" facility in the CLIMEX model. CLIMEX compares the climates of different locations (Fig. 1) using a more sensitive algorithm than statistical correlation techniques (Maywald and Sutherst, 1991). The "Match Index, MX" is calculated using four component indices, describing the similarity of maximum temperature I_{tmax}, minimum temperature, I_{tmin}, total annual rainfall, I_{rtot}, and the seasonal rainfall pattern, I_{rpat}, as follows:

$$MX = \sqrt{(((I_{tmax} + I_{tmin})/2) \times I_{rtot} \times I_{rpat})}$$

In the case of Cambridge (UK) and Cooma (Australia) the respective values of the Match Index and the four components are: 0.74, 0.61, 0.88, 0.99, 0.74 indicating a close similarity between rainfall in particular with less similar maximum temperatures. Indices comparing Cambridge with other Australian towns are given in Table 1.

III. APPROACHES TO ESTIMATING POTENTIAL GEOGRAPHICAL DISTRIBUTIONS OF INDIVIDUAL SPECIES

The size of the geographical range of species is related to local abundance in favourable parts of the habitat (Brown, 1984; Gaston and Lawton,

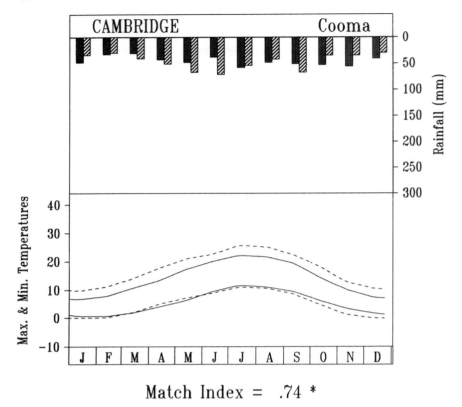

Fig. 1. Comparison of meteorological data from Cambridge, UK, with the most similar location in the CLIMEX Australian database, Cooma, using the "Match Index" facility. The solid line and dark shading refer to Cambridge, the broken line and hatching refer to Cooma; the * indicates that the data from the latter location has been displaced by 6 months to take account of the difference in hemispheres.

Table 1. Comparison of the climates of some capital cities of Australia with Cambridge UK, using the CLIMEX "Match Index" facility (see text for explanation)

City	Match Index	Maximum temperature	Minimum temperature	Rainfall (annual total)	Rainfall (seasonal pattern)
Hobart	0.72	0.64	0.71	0.97	0.79
Canberra	0.69	0.46	0.85	0.90	0.81
Melbourne	0.58	0.44	0.50	0.86	0.82
Adelaide	0.46	0.35	0.43	0.86	0.62
Sydney	0.38	0.31	0.33	0.54	0.85

1990). Species with wide ranges tend to be more locally abundant. Further, individual species respond to climate independently so that the response of communities of species is the sum of the individual responses rather than of a linked web (Hengeveld, 1990).

Kingsolver (1989) emphasized two roles that weather can play in the population ecology of species: firstly, as a limiting factor that determines the relative importance of various biotic determinants of population dynamics and, secondly, as an influence on physiological rate processes that mediate interspecific interactions. These effects will in turn affect the distribution of the species.

Carter and Prince (1988) used the suprapopulation concept – "a population of populations which go extinct locally and recolonise" – to explain the population dynamics that limit the geographical distribution of species. In their thesis, the limits are determined, not by the physiological limits of an individual, but by the combined limits at the individual, population and suprapopulation levels. They used the standard epidemic model:

$$dy/dt = \beta xy - \lambda y$$

where x is the number of susceptible sites, y is the number of infective sites, β is the infection rate and λ is the removal rate.

In order for a species to persist at a given site the individual must be able to complete its life cycle and reproduce; the population birth rates must exceed the death rates and, at the suprapopulation level, the colonization rates must exceed the extinction rates. While the model is very general, and is therefore of more value in explaining mechanisms limiting distributions than in predicting outcomes in specific situations, it may explain the failure of deduction-based attempts to explain geographical distributions on the basis of physiological limits of individuals. When applied to insect populations, the model parameters would usually reflect high numbers of susceptible sites, high infection rates and probably high removal rates compared with plants.

There has been a long history of attempts to develop methods with which to estimate the potential geographical range of species and plant communities in relation to climate. Cook (1929), Messenger (1959), Sutherst and Maywald (1985), Nix (1986), Caughley *et al.* (1987), Maywald and Sutherst (1989, 1991), Busby (1991), Lindenmayer *et al.* (1991) and Rogers and Williams (1993) based their approaches primarily on the use of inference from observed geographical distributions of species in the field, supported in some cases with observations on the biology of the species. Messenger and Fitters (1954), Rogers (1979),

Rogers and Randolph (1986), Meats (1981, 1989a,b) and Caughley et al. (1988) started from the premise that it was possible to use a deductive, experimental approach to predict the limits of distributions, with gaps in knowledge being filled by inferring boundaries from field observations. Given the deficiencies inherent in both the inductive and deductive approaches, it is desirable to derive an integrated method that is rigorous, practical in most real-life situations, and able to provide some insight into the biological responses of a species to climate. Meats (1989a) gave a well-balanced discussion of these problems.

The main difficulties with the inductive approach relate firstly to the uncertain role of climate, as opposed to other factors, in determining the distribution of some species. Secondly, the extent of heterogeneity of environmental conditions in the area already occupied by the species determines the range of parameter values available for use in extrapolation to different areas (Sutherst and Maywald, 1985). This limitation was recognized by Dobesberger and MacDonald (1993) in relation to the use of discriminant analysis and is illustrated by the restricted, though still locally useful, area of application of the findings of Rogers and Williams (1993) and Rogers and Randolph (1993). Thirdly, there is inevitably a degree of curve fitting with the various models and this limits their explanatory capacity.

On the other hand the deductive approach, based on defining demographic parameters from physiological experiments, relies on exhaustively detailed data collection to define maximum development and survival times under combinations of fluctuating conditions. Unfortunately, estimation of the limits of tolerance of climatic conditions by an organism is notoriously difficult and unreliable. On the one hand there are great practical difficulties in measuring maximum survival rates under representative, experimental conditions. Classical, deductive experimentation concentrates on providing reliable estimates of the mean value of dependent variables, with much higher errors always being associated with the extreme high or low values. It is these maximum values that are needed to estimate the limits of tolerance of a population under field conditions. The problem is exacerbated by the need for the results to be applicable under a wide range of climates, with heterogeneity of local environments making such applications even more difficult. Climatic conditions vary in intensity and duration, not only seasonally, but daily, and those patterns vary with every geographical location. As a result, success has yet to be achieved using the deductive approach *per se* for even the most intensively studied species of insects.

In this chapter the current availability of predictive tools is reviewed and suggestions made for the development of new tools.

A. Climograms

Cook (1929) used climographs based on population densities to relate pest numbers to climate in North America. "Normal", "Occasional" and "Possible" outbreak zones were defined – based on the frequency of outbreaks – and related to monthly temperature and rainfall at each site. Such climographs or climograms have been used commonly since, with a variant development by Rogers (1979) and Rogers and Randolph (1986), based on mortality rates rather than abundance. They combined mortality climograms of three species of tsetse fly (*Glossina* spp.), with different responses to temperature, with a fecundity plane for *Glossina morsitans*, based on experimental conditions. The authors then related the different geographical distributions of the three species to temperature and saturation deficit. They nevertheless had to infer the limiting climatic conditions for the flies from the atmospheric temperatures and saturation deficits at the observed limit of distribution in the field in west Africa.

B. BIOCLIM

BIOCLIM (Busby, 1991) is a software package which produces a specific "climate profile" or "environmental envelope" for a species based on means and ranges (diurnal, seasonal, quarterly, annual) of climate using up to 24 input meteorological variables. The choice of these variables has been refined in the recent version of BIOCLIM to include only the 16 variables considered to have the greatest biological significance (Busby, 1991). The core of BIOCLIM is a high resolution "climate surface", used to interpolate temperature and precipitation between meteorological stations for any given grid point. Such surfaces need to be developed from topographical maps for the area of interest and already exist with different resolutions for most countries (IIASA, Laxenburg; Zedx Inc. Boalsburg, Pennsylvania; Hutchinson, 1988). BIOCLIM-based approaches have been used very successfully in conservation biology and forestry in particular (Nix, 1986; Caughley *et al.*, 1987; Booth, 1988; Kohlmann *et al.*, 1988; Lindenmayer *et al.*, 1991).

C. Multivariate Statistics

Multivariate statistical techniques can be used to select predictive variables and classify them into two or more classes. A number of such techniques have been used to describe climatic limits to the distributions of various species of plants, mammals and insects. They are applicable to any set of data, regardless of whether the predictive variables are derived from transformed or raw meteorological data, modelling results or non-climatic variables such as soils, vegetation or satellite imagery.

Caughley et al. (1987) analysed the climatic requirements of three species of kangaroos (*Macropus* spp.) in Australia. They used interpolated meteorological data with a grid size of one degree, 12 climatic variables common to the BIOCLIM model and discriminant functions to characterize and compare the requirements of each species. The approach enabled them to define climatic envelopes, in terms of mean annual temperature and precipitation and their seasonal variation across Australia. The authors were then able to draw broad conclusions about the limiting effects of climate on the different species. Discriminant function analysis has been used often by entomologists to discriminate between areas of varying risk, for example Kohlmann et al. (1988) related distributional limits of chromosomal taxa in the Australian grasshopper *Caledia captiva* (F.) to climate, while Berryman and Stark (1985) forecast forest pest outbreaks and Pearson and Meyer (1990) forecast the risk of infestation of the blueberry maggot, *Rhagoletis mendax*. Discriminant function analysis was combined with geographical information systems (GIS) by Dobesberger and MacDonald (1993) to forecast the potential occurrence of pest infestations, using *R. mendax* as an example. Rogers and Williams (1993) used discriminant function analysis, and satellite imagery as an indication of vegetation cover in a GIS to predict areas suitable for tsetse fly (*Glossina* spp.) (Rogers, Chapter 8, this volume) and the livestock tick *Rhipicephalus appendiculatus* in parts of Africa. Diekmann (1988) used a discriminant function to predict the potential global distribution of the pathogen Ascochyta blight on chickpea based on temperature and rainfall data at different phases of the crop's growth.

D. Physiological Data and Population Dynamics Models

Meats (1989a) attempted to estimate the distribution limits of the Queensland fruit fly *Dacus tryoni* in relation to climate using laboratory data, but had to resort to major, overriding assumptions derived from field correlations between the limits and climatic indicators. This recourse to field and meteorological conditions illustrates the difficulty inherent in all extrapolation from laboratory conditions to the field in that it is difficult to translate a laboratory temperature or humidity reading into a meteorological measure. The second constraint is the requirement to define, in biological terms from laboratory data, the point at which the

4. Predicting Insect Distributions in a Changed Climate

balance of births and deaths in a population result in the species failing to persist beyond the boundary of its distribution in any direction.

Population dynamic models are the tool with most potential for estimating impacts of climate on insect populations and hence, in theory, on geographical distributions. They can describe abundance and phenology in a wide range of environments as well as the mechanisms involved in the species' response to particular climatic conditions. Unfortunately, there are very few models which can quantify abundance sufficiently accurately to allow reliable projections of the effects of climate on distributions. Nevertheless, phenological models can provide valuable insights into the effects of climate on specific population attributes such as over-wintering success, diapause mechanisms, number of generations and timing of events.

Baker (1991) described a day-degree model which derived the number of generations that are possible in a given environment. The model was developed as a component of a general risk assessment procedure for quarantine insects and, as such, it provides one indicator of the potential of the species to cause economic losses. In relation to distributions, the question remains as to the minimum number of generations that are essential for survival.

In the present state of the world's scientific environment it is doubtful whether the data required to develop realistic model descriptions of many species will be able to be collected in the foreseeable future. There is therefore a valuable role for inferential methods for estimating population parameters from field observations on insect development, fecundity and phenology. Dallwitz and Higgins (1978) described a statistical approach for estimating development rates as a function of ambient temperature from observations on insect development under field conditions. More recently Hudes and Shoemaker (1988) described an inferential method for modelling insect phenology. Such pragmatic approaches offer the prospect of rapid progress when resource constraints would otherwise stifle progress (Worner, 1991).

E. CLIMEX

CLIMEX is a dynamic climate-matching model that is based on an integrated inductive and deductive approach as appropriate to the particular situation. It is designed to describe and interpret the climatic requirements of a species or other biological entity with a minimum of data on biological processes. The model relies largely on inferring the species' requirements from its observed geographical distribution, supported as far as possible by field observations on the phenology of populations. CLIMEX attempts to emulate the process by which species so successfully integrate the effects of climate over time, and to reflect the results of their response in the form of a geographical distribution. Experimental data on biological processes – when available – help to refine, explain and increase confidence in CLIMEX projections.

The CLIMEX approach is based on the observations of Southwood (1977) and Sutherst and Maywald (1985) that the year can be divided into a period suitable for population growth and another period during which the species has to survive to be present at the start of the following growth season. Any method which purports to describe the suitability of the climate of a given environment for a species must take account of both the potential for population growth during the favourable season and the risk of extinction during the unfavourable season. A description of the phenology of populations of a species is also vital for any interpretation of field situations.

The CLIMEX approach is consistent with the theory that the numbers of a species are related to environmental gradients, with greater abundance at the core of their distributions and declining densities towards the margins (Krebs, 1978). Caughley *et al.* (1988) referred to it as the "ramp" theory.

1. CLIMEX, seasonal phenology and relative abundance

Conditions that favour population growth are described in CLIMEX by a temperature (TI) and moisture (MI) -based, weekly "Population Growth Index, GI_W" ($GI_W = TI \times MI$) scaled between 0 and 1. The weekly GI_W values are summed and averaged to give an annual GI_A, scaled between 0 and 100, describing the overall potential for population growth of the target species in the given environment. The GI_W describes the seasonal suitability of climate for the development, fecundity and survival of a species. It is assumed that there is a high correlation between the effects of temperature on development rates and fecundity. When averaged over one generation, the GI_W is analogous to a scaled *finite rate of increase* which combines fecundity and mortality rates. Both measures suffer from their failure to incorporate the effects of temperature on development rates of multivoltine species. This deficiency becomes evident when attempting to integrate the values over periods much longer than one generation, when multiplicative effects become apparent and the relationship between the annual total population size and GI_A or finite rate of increase is not linear for a variety of reasons including negative feedback effects. The weekly GI_W is used to infer the period of seasonal activity of the species without reference to life-cycle stages or generations, although the latter is easily computed if the thermal requirements of the species are known in terms of day-degrees per generation. The annual GI_A gives an indication of the likely relative abundance of a species across its range as determined by climate with all other factors being equal.

Estimation of the temperature and moisture responses of a species, in order to derive parameter values for the GI, relies on a degree of approximation where biological data are not available. Identification of the limiting effects automatically provides a pointer to those conditions that favour population growth. Estimates are possible by examining the seasonal phenology of populations in the field. One of the most sensitive parameters is the temperature below which no population growth or development is possible; that is, the development threshold. Fortunately, the scope for error is limited by first identifying the region inhabited by the species. Often the development rates of cool temperate species respond when temperatures exceed 5–8°C, temperate–subtropical species around 8–12°C and tropical species around 12–15°C (e.g. Meats, 1989b). Similar groupings have been described for plants by Nix (1981).

2. CLIMEX and geographical distributions

Conditions that prevent a species from surviving the unfavourable season also limit its geographical distribution. They are described in CLIMEX by four "Stress Indices" based on extremes of temperature or moisture. Stress is accumulated linearly above or below a threshold value each week, so each stress function has two parameters, a threshold and a rate. In the case of extreme temperatures, two different effects can be modelled. In the first, cold stress (CS) is accumulated when the minimum temperature is below a threshold indicating the action of lethal low temperatures. In the second, cold stress is a function of an inadequate number of day-degrees to sustain metabolism. The limiting factor needs to be identified for each species and that is often possible by comparing CLIMEX results with field distributions. In practice, the measures are often correlated which probably reflects the importance of both the length and severity of the winter in limiting many species.

The weekly stress values are accumulated during the year, using an exponential function to reflect the diminishing ability of an organism to withstand a given week's stress as the period of stress accumulation lengthens. Provision is also made in CLIMEX for limiting effects being dependent on interactions between temperature and moisture, for example, Hot–Wet. These are rarely needed but have been essential on occasions to explain distributions of species such as the European wasp

Vespula germanica (Spradbery and Maywald, 1992). Care is needed to take irrigation practices into account when moisture is limiting a species in a modified environment and CLIMEX has a facility to enable that to be done.

Stress is a function of time multiplied by the amount by which the temperature or moisture departs from the threshold; for example, a fruit fly may survive for 3 days at 5°C or 10 days at 8°C. CLIMEX describes stress with a linear function, accumulating stress above or below a threshold value, at a rate that is proportional to the extent to which the condition lies outside the threshold. Thresholds are mathematical conveniences rather than meaningful biological parameters. As in any such linear relationship, the threshold and rate parameters are highly correlated. In CLIMEX, high values of the rate parameters result in abrupt boundaries with steep gradients being predicted and increased sensitivity to threshold values.

The sensitivity of each CLIMEX parameter enables it to be fitted to distributional data visually with limited margin for error. Small changes in the rate parameters result in large changes in predicted distributions. A sensitivity analysis of the threshold and rate parameters for Cold Stress for the Colorado beetle in North America is shown in Table 2.

The four stress indices together enable the limiting factors to be described in any geographical direction within a species' range and so are global rather than local in nature. When the stress values are combined with the GI_A, they describe the overall suitability of the location for propagation and persistence of the species, in the form of an "Ecoclimatic index, EI" scaled between 0 and 100. The Ecoclimatic Index is defined as follows:

$$EI = GI_A \times SI \times SX,$$

where the annual GI, termed $GI_A = 100 \, (\Sigma_{I=1}^{52} GI_w)/52$, where GI_W is the weekly Growth Index, $SI = (1 - CS/100)(1 - DS/100)(1 - HS/100)(1 - WS/100)$ and CS, DS, HS and WS represent cold, dry, hot and wet stress respectively; and SX is the product of interaction terms involving a temperature and a moisture stress; for example, Hot–Wet.

The EI value is intended to present the overall conclusions in a simple format, suitable for policy makers. Interpretation of the index is best done by examining its component parts, for example the EI for Colorado beetle, *Leptinotarsa decemlineata* in Beijing is 13, which is derived from the following annual components: $GI_A = 19$; Cold Stress (CS) = 16; Dry Stress (DS) = 18; Heat Stress (HS) = 0; and Wet Stress (WS) = 0. These six values describe the suitability of Beijing for the species in readily interpretable and meaningful biological terms.

Table 2. Sensitivity of CLIMEX Cold Stress (CS) indices, based on either minimum temperatures (TMIN) or day-degrees (DD), and illustrated by values for Colorado potato beetle around its northern limits in North America. TTCS and DTCS are the temperature (T) and day-degree (D) threshold (T) values for cold stress (CS) respectively, while THCS and DHCS are the corresponding rate (H) parameters which determine the rate of accumulation of cold stress when the threshold values are triggered

TMIN	TTCS (°C) THCS	-5 .00013	-5 .00026	-5 .00052	1 .00013	1 .00026	1 .00052	5 .00013	5 .00026	5 .00052
Chicago		1	3	5	13	26	51	29	62	124
Fargo		21	43	94	52	103	222	88	176	353
Winnipeg		45	91	181	81	173	347	126	253	507

DD	DTCS (DD) DHCS	1 0.001	10 0.0001	10 0.001	20 0.0001	30 0.00001	30 0.00005
Chicago		24	32	321	75	12	65
Fargo		31	37	369	90	14	76
Winnipeg		38	46	460	110	17	92

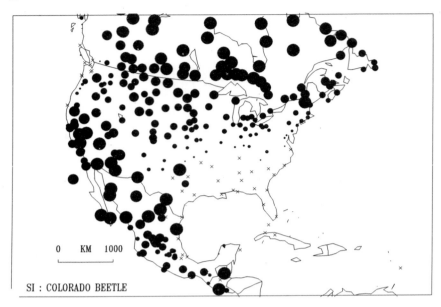

Fig. 2. Distribution of total stress for Colorado beetle, *Leptinotarsa decemlineata*, in North America, estimated using the CLIMEX model. The amount of stress is proportional to the size of the circles. Crosses indicate Stress = 0.

Maximum abundance is related in CLIMEX terms to two measures. The first is the absence of stress effects which limit the size of the population that survives through the unfavourable season. This is illustrated in Fig. 2 for the Colorado beetle in North America. The local abundance in relation to climate is also dependent on how suitable the conditions are for reproduction and development, as expressed by the CLIMEX GI_A, Fig. 3. Thus there is a degree of concordance between the two indices, with the potential for population growth in the favourable season decreasing towards the edge of the distribution as the severity of the unfavourable season reaches limiting levels. The reduction in population growth occurs through a combination of less favourable conditions and the resulting reduction in the length of the season suitable for growth. CLIMEX offers the opportunity to explore the relationship between species' distributions and climatic gradients in a way not previously possible.

Given that the distribution of a species is limited by climate, CLIMEX can describe the response of that species, even when parameters are fitted visually. The rigour that can be applied using the CLIMEX approach is

4. Predicting Insect Distributions in a Changed Climate 73

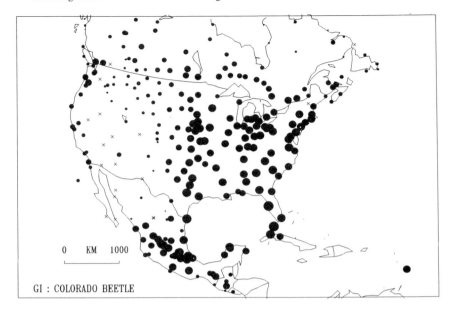

Fig. 3. Distribution of CLIMEX annual population Growth Indices (GI_A) for Colorado beetle, *Leptinotarsa decemlineata*, in North America. The value of GI_A is proportional to the size the circles. Crosses indicate $GI_A = 0$.

far greater than is evident at first sight because the database linking observed distributions with meteorological data can be deceptively rich. It can cover a wide range of parameter values and vast number of combinations and permutations of climatic conditions. Exploitation of this spatial source of variability has hardly been touched in the ecological literature on insect population dynamics. Of course, in situations where the original species' range is confined to an area with a limited range of climates, it is likely that limiting factors will remain undefined in some directions. For example, Europe does not include any sites with a hot–wet climate, so the response of a European species to such conditions must remain undefined until experience is gained from an introduction to such an environment. This limitation illustrates one of the weaknesses of relying on any method based on inference alone, but often inference is the only available option, apart from direct climate matching applied to selected months of the year if necessary.

CLIMEX has been used on numerous occasions to predict areas that have climates suitable for the establishment of exotic species (e.g. Worner, 1988; Sutherst et al., 1989). The model has also been combined with GIS (Perry et al., 1991; Mayer et al., 1992) and with an expert system (Sutherst et al., 1991) to take non-climatic limiting factors, such as availability of suitable vegetation and host animals, and management actions or real time events into account.

3. CLIMEX and annual changes in geographical distributions

Norval and Perry (1990) used CLIMEX to investigate annual changes in the distribution of the tick *R. appendiculatus* in Zimbabwe. They showed that the changes were related to rainfall and the associated reduction in maximum temperatures.

IV. RELIABILITY AND ACCURACY OF DATA SETS USED TO ESTIMATE POTENTIAL DISTRIBUTIONS

While extensive effort has been made to ensure the integrity of meteorological data and to develop techniques to define accurately the relationship between a given distribution limit and meteorological or other variables, similar effort or validation techniques for checking the validity of species distribution and abundance records are rarely reported. In one such study, Lindenmayer et al. (1991) plotted cumulative frequency distributions of output from the BIOCLIM package to detect discontinuities indicating the likelihood of geocoding errors in the distribution data of the possum, *Gymnobelideus leadbeateri*.

A checklist to define the relative reliability of each type of data set used in any assessment of species distributions, seasonal phenology and relative abundance is shown in Table 3.

V. COMPARISON OF APPROACHES

The attributes of different approaches to defining potential geographical distributions of species and the relative ability of each approach to address the questions related to different population attributes are compared in Table 4.

The main features that distinguish the different approaches are whether they rely on inductive or deductive processes and whether they are static or dynamic. The strengths and weaknesses of the inductive versus deductive approaches have been discussed above. It is evident that a purely deductive approach is likely to be beyond the resources of most scientists and is equally vulnerable to the unknown influence of non-

Table 3. Considerations in rating the relative reliability of data sets used to define potential distributions of species

Data type	Attribute
Meteorological	Long-term averages: duration and variances
	Parameter range within fitted species distribution
	Annual data for successive years
Species distribution	Present/Abundance index/absent
	Hosts and interacting species present
Physiological	All critical life stages identified
	Constant/variable temperature, moisture and day-length
	Means, variance, extreme values of development and survival
Population	Seasonal phenology
	Relative abundance
	Mortality and fecundity
	Movements

physiological limiting effects, such as epidemiological processes operating in the field. To date such approaches have had to be combined with major inferential inputs in order to match field observations. On the other hand inductive methods have had numerous and often spectacular successes as well as some more limited outcomes due to inadequate consideration of the non-climatic influences or of the limited range of climatic parameters within the area used to fit model parameter values. Integrated approaches offer the best prospect of reliable results.

Two differences in philosophy underlie the use of climograms, multivariate statistics or BIOCLIM on one hand and CLIMEX on the other. The former methods approach the problem by comparing patterns of annual meteorological data, while CLIMEX approaches it from the perspective of how a species experiences the climate week by week during the year. Climographs emphasize the abruptness of the limits of distributions, while CLIMEX describes the gradual decline in the densities of populations at the extremes of their distributions. Discriminant function analysis describes transition zones between two states and so falls in between the former two approaches. In practice, one of the greatest differences between the different approaches is the temporal resolution of the meteorological data. The statistical methods often use data as coarse as annual extreme values while CLIMEX uses weekly values.

Nix (1987) argued that the effort devoted to problems of definition and classification of agroecological regions has been misdirected. A stronger focus on measurement of primary attribute data, such as water balance, and storage within a spatially referenced system (geographic information system) will be more rewarding. This argument is as relevant to

Table 4. Attributes of methods used to describe geographical distributions of species. Ratings of performance are given with + signs on an increasing scale from + to +++

Method \ Attribute	Precision	Dynamic (D) Static (S)	Inductive	Deductive	Combined inductive/ deductive	Seasonal pattern of activity	Population density
CLIMOGRAMS	+	S	+			+	
BIOCLIM	+–+++	S	+				
Multivariate statistics	+–+++	S	+				
Physiological	+–++	S		+			
Physiological and population data	++	S		++	+		
Epidemiological analytical models	+	D		+			
CLIMEX	+–+++	D	+		+	++	++
Mechanistic population models	++	D		++		+++	+++
Inferential/statistical population models	+–+++	D	+	++	+	+	+

entomology as it is to agronomy, because – as pointed out by Hengeveld (1990) – communities are made up of dynamic mixtures of species, each of which responds in its own way to climate. Primary attribute data enable dynamic models such as CLIMEX to be used to describe the temporal dynamics of a species' response to climate in any new environment, so providing greater interpretation of information than is possible using descriptive tools such as multivariate statistics or agroclimatic zonation.

There has been a great deal of emphasis on the accuracy of techniques used to describe insect distributions in relation to climate, but much greater errors are likely to arise from inadequacies in the data on the distribution of the species concerned, or misinterpretation of mechanisms limiting distributions, such as barriers to migration and, rarely, hybrid zones, or inappropriate use of models to extrapolate beyond the range of reliable parameter values. This latter danger exists both in the spatial dimension with geographical data sets and in projections of the impacts of climate change following estimation of parameter values using data with limited ranges (Rogers and Randolph, 1993). Until all data sets and model functions used in an analysis are rated on reliability and domain of applicability, there is a danger that results from both statistical and simulation modelling studies will be interpreted as having a greater reliability than is warranted from the weakest data set or model function involved.

Across the globe there is a vast range of permutations of climatic conditions to be taken into account when developing systems for predicting the potential geographical distributions and relative abundance of a large number of species of agricultural, environmental or public health interest. The logistics demand reliance on computer-based data processing techniques. In the context of climate change, predictive systems need to be able to address hypothetical shifts in climate as well as the emergence of new climates in some regions. In addition, both the direct and indirect effects of increases in carbon dioxide concentrations on insects need to be considered. Given the constraints in funding, the task of addressing such a global problem can only be achieved if generic tools are developed to handle the logistics generated by the number of target species, locations of interest and evolving climatic scenarios. As knowledge of the potential geographical distribution *per se* is of limited value, some indication is also desirable of the potential, relative abundance in relation to climate at any given point within the projected distribution. When such tools have been developed it will be necessary to link the results of species-specific analyses to the species' interaction with its hosts, predators and parasitoids, and other factors including those resulting from human intervention.

From these considerations it is apparent that in predicting the potential range of a species there will often be higher levels of uncertainty and lower levels of precision of the results than may be the norm in experimental or some population dynamics studies. The most promising approach combines inference from observed geographical distributions and field observations on populations on one hand with deductive, experimental studies into the biological mechanisms that limit survival in unfavourable seasons combined with modelling on the other hand. The amount of information available will vary greatly with the species concerned. There is a need to note the relative precision and reliability of the data available from different sources – meteorology, geographical distributions, field observations on populations and experimental studies (Table 3) – when attributing confidence levels to results for a given species. Hence the outcomes of different investigations will vary widely in sophistication, balance and uncertainty and more effort needs to be put into making such attributes of any study more explicit to users.

VI. APPLICATIONS OF THE ABOVE METHODS TO ESTIMATING THE IMPACTS OF CLIMATE CHANGE

A. Climate Change Scenarios

There is a great deal of uncertainty surrounding the likely changes in regional climates under the influence of the enhanced greenhouse effect. Results from different general circulation models (GCMs) vary substantially and reliable regional predictions await further enhancement of the GCMs to take account of topography, ocean–atmosphere interactions and the effects of cloud cover of different types. In the face of these uncertainties, tools for estimating impacts of climate change need to be built in such a way that they can accommodate a wide range of climate scenarios at the regional level. In the meantime, the use of the CLIMEX model is illustrated using a global scenario which envisages a rise of $0.1°C$ per degree of latitude with an increase of 20% in summer rainfall and a decrease of 10% in winter rainfall. Many other scenarios have been proposed for different regions.

B. Impact Studies

Potential impacts of climate change on insect population dynamics and geographical distributions have been reviewed by Anon. (1989), Prestidge and Pottinger (1990), Cammell and Knight (1991) and Porter et al. (1991). The latter listed the following effects of temperature on insects: limiting geographical ranges; over-wintering; population growth rates; number of generations per annum; length of growing season; crop–pest synchronization; interspecific interactions; dispersal and migration, and availability of host plants and refugia. Sutherst (1991a,b) outlined an approach suitable for defining impacts of climate change on agricultural pests while Sutherst (1990) explored some impacts expected in the Australasian region and Sutherst (1993) reviewed possible impacts on public health. Sutherst et al. (1989), Porter et al. (1991) and Spradbery and Maywald (1992) considered impacts on particular key pest species at the global and regional level. Dennis and Shreeve (1991) assessed the vulnerability of species of British butterflies to climate change, with emphasis on the size of their geographical range, the abundance within the range, host plant and habitat associations and dispersal ability. Landsberg and Stafford Smith (1992) described a scheme for analysing insect populations in terms of functional attributes relevant to global atmospheric change.

C. Nature of Impacts

1. Length of growing season and the associated spatial consequences

The length of the growing season determines whether univoltine species can complete their life cycle in 1 year and how many generations multivoltine species can complete.

(a) Univoltine species
At higher latitudes many plant species are unable to complete their life cycle and produce seed (Porter, Chapter 5, this volume). Similarly, some univoltine species of arthropods are unable to complete development to ensure that the appropriate life stage is available to withstand the unfavourable conditions during the winter. Sutherst (1990) gave an example in which the southern limit of the distribution of the livestock tick *Haemaphysalis longicornis* in New Zealand is determined by the inadequate length of the suitable season on the South Island. As shown in Fig. 4 the effect of global warming in Europe would be to increase greatly the risk from this species of Far Eastern origin. Such a result contrasts with the likely impact of such scenarios on the life cycle of the ixodid tick of major public health and livestock importance, *Ixodes ricinus*, in Europe. This latter species is able to over-winter in different stages and to

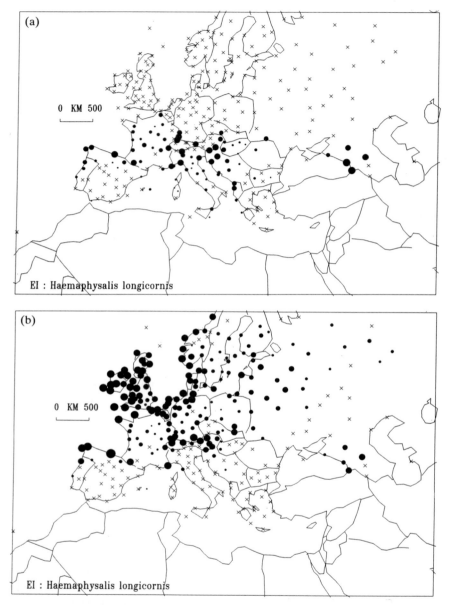

Fig. 4. The potential distribution of the livestock tick, *Haemaphysalis longicornis*, in Europe as indicated by the CLIMEX Ecoclimatic Index, EI, (a) current climate, (b) Greenhouse scenario (see text). The value of EI is proportional to the size of the circles. Crosses indicate EI = 0.

4. Predicting Insect Distributions in a Changed Climate 81

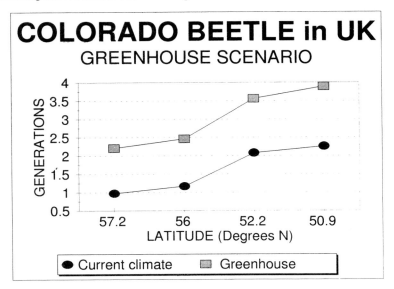

Fig. 5. Number of generations of the Colorado beetle possible at different latitudes in the UK.

complete its life cycle in 1, 2 or 3 years in different environments, so the effect of global warming will be to accelerate the life cycle in northern areas and to displace it from southern areas that become too hot.

(b) Multivoltine species
The likely impacts of changes in the length of the season vary according to the nature of the life cycle of the species concerned. Global warming will have the effect of increasing the number of generations that summer-active species can complete each year. For example, using CLIMEX, the climate change scenario above is estimated to increase the numbers of generations of the Colorado beetle significantly in different parts of the UK, as shown in Fig. 5.

The impact that such warming will have on the abundance and hence distribution of a species will depend on its life cycle characteristics and on its ability to disperse rapidly during the season of population growth. Impacts may not be significant on species which already have numerous generations each year and reach equilibrium population densities during the year as a result of negative feedback mechanisms. In such cases, further generations may simply have an additive effect. In contrast, species which exhibit near exponential population growth, associated with

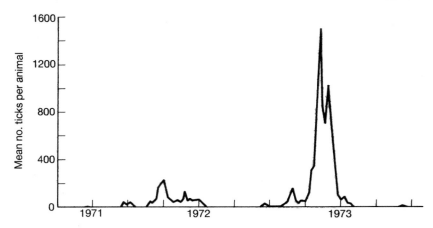

Fig. 6. Populations of the cattle tick, *Boophilus microplus*, in southern Queensland during two years with average differences in summer temperature of 1.6°C. After Sutherst (1983).

cumulative effects from the growth of two to four generations from a small over-wintering base each year, may respond very dramatically to global warming in some areas. Data on the cattle tick, *Boophilus microplus*, near the southern limit of its range in Australia (Fig. 6, after Sutherst, 1983) illustrate how an increase of only 1.6°C in the average summer temperature resulted in a major explosion of the population. This was associated with the hatching of a large cohort of third generation eggs which fail to hatch in normal seasons due to the onset of low temperatures in autumn. Such outbreaks are associated with epidemics of tick-borne diseases that result from the intermittent transmission of the pathogens to hosts which fail to become infected while they are young with natural immunity to the disease.

2. Change in severity of limiting factors

While growing conditions determine the opportunities for species to exploit their environments, limiting factors determine whether the species is able to survive in those environments during the unfavourable season each year. Climate change will affect the incidence and severity of limiting factors as well as the favourableness of growing conditions. Two different types of limiting effects are described above. The first is the occurrence of short periods of extreme conditions which are lethal to the species concerned. This is the most commonly quoted limiting effect. Sutherst *et al.* (1991) illustrated the effect that an increase in temperature distribution of the Colorado beetle in northern Europe ... d to be limited by prolonged low temperatures. There is ..., which is probably as common, and that is the oc... of conditions which are not immediately lethal, but which are outs...e the range that supports the species' development. While the first type of stress is related to extreme minimum temperatures, for example, the latter is related to the long duration of low maximum temperatures in winter that are inadequate to enable the species to maintain its metabolism. The southern limit of the tick, *B. microplus*, in Australia is an example of such an effect (Sutherst, 1990).

A third limiting effect related to temperature is the specific requirement of some species, such as the gipsy moth, *Lymantria dispar* L., for particular regimens of cold for induction, maintenance and termination of diapause (Tauber et al., 1986). Such requirements need to be described in such a way that they reflect the climatic limits to the species' distribution. The complex biology underlying diapause control poses real challenges to the production of a generic approach to describing the effect of diapause on geographical distributions using models such as CLIMEX.

3. Incidence of extreme events

As pointed out by Wigley (1985), a small shift in the mean of a frequency distribution of temperature or rainfall is expected to have a disproportionate effect on the frequency of extreme events. For example, a 20% increase in the average rainfall at Lawes in southern Queensland would result in more than doubling the frequency of rainfall events over 350 mm (Fig. 7). Intermittent occurrence of extreme climatic events such as heatwaves, severe winters, floods and droughts or sequences of seasons with above or below average temperature or rainfall can have major effects on arthropod abundance and distributions as illustrated by the example of the cattle tick above (Fig. 6). When the species concerned is highly mobile it can invade territory far beyond its normally accepted limits. Such changes in distributions, as illustrated by the examples of Williams et al. (1985), Murray and Nix (1987) and Norval and Perry (1990) are likely to increase in frequency with global warming. There is therefore a need to pay more attention to the dynamic nature of species' distributions and to collect more accurate data on the fluctuations in the distributions from year to year in order to relate them to climate.

4. Implications of latitude under climate change

Day-length is an important element of the climate experienced by insects and its variation with season and latitude can have major consequences

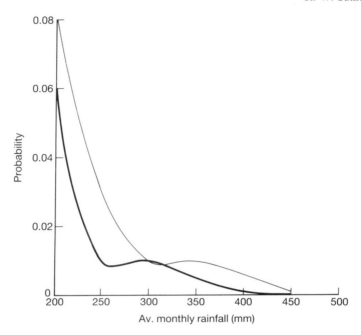

Fig. 7. Effect of a 20% increase in average rainfall on the frequency of extremely high rainfall events in southern Queensland. ———, Historic; ———, climate change scenario.

for insects that are affected by geographical shifts in climatic patterns. Seasonal changes in day-length with latitude form complex patterns that require adaptation to different amplitudes and rates of change. Insects have adapted to such changes by developing highly plastic responses to day-length under the control of polygenic inheritance (Tauber et al., 1986). As a result they are able to respond readily to changes in the interactions between temperature and day-length that arise from abnormal seasons or from movements into new latitudes. Hence it is likely that such a change in climate will not be a constraint to the spread of insects. Evidence for historical movements of species between different latitudes (Coope, Chapter 2, this volume) supports this view.

5. Link to host crops and animals

An isolated focus on the direct impacts of climate change on insects can lead to oversights of the concurrent effects that the changes may have on host plants and other animals. As many insect species are host specific and limited in their ranges indirectly by climate through its effect on the host plant, it is essential to take a holistic view of the insect–host system and the other components of an insect's environment. As a first step, it is necessary to make projections of the impact of climate change on the potential change in the geographical distribution of the host plant. As plant phenology affects insect feeding, it needs to be taken into account, ideally by coupling crop–pest models. These will need to have a sufficiently short time step to account for shifts in timing of temperature and moisture optima of plants for specific functions such as flowering.

The enhancement of crop growth under the long day-lengths experienced at high latitudes in a warmer climate may be a disadvantage to some insect populations. The rate of plant maturation may be increased by an amount in excess of that experienced by the insect population as a result of increases in temperature alone, thus effectively turning the plant into a "short season" variety. Alternatively, some plants may simply grow larger providing a greater food supply.

Collier *et al.* (1991) gained useful insights into the likely impact of different climate change scenarios on the synchronization of populations of the cabbage root fly (*Delia radicum*) in the UK using a phenological model. Such changes are likely to be reflected in changed geographical distributions.

6. Interactions with other insect species

Insect parasitoids and predators, and other competitors, will all be affected by climate change and the dynamics of the interactions will be mediated by climate (Kingsolver, 1989). Such changes can influence the distributions of species through their effect on the fitness of the populations towards the edge of the species' range.

VII. ADAPTATIVE STRATEGIES TO CHANGING DISTRIBUTIONAL LIMITS

In the face of global climate change and the altered risks posed by pests of agriculture, public health and the natural environment, the world community needs to re-evaluate its relaxed attitude to quarantine barriers. As many of the world's most damaging insect pests are tropical or subtropical, they pose new risks if allowed to become established in temperate zones where they may be perceived as not being of immediate concern.

In agriculture it is not feasible to design effective adaptive strategies to the spread of insect pests without taking into account the impacts of cropping systems and likely changes in crop species grown in the new climate experienced in each area (Porter, Chapter 5, this volume). Direct extrapolation from current homologous situations may often be adequate to enable effective adaptation. On the other hand the interaction of climate, soil types, latitude effects and altered markets on each crop will require holistic studies that will demand close integration of pest and crop science with links to economists. Comprehensive global databases of species distributions, relative abundance and biological attributes, as pioneered by CAB International, will be required. Models, such as CLIMEX, linked to databases will enable the information contained in the databases to be extracted, enhanced and interpreted for management purposes.

VIII. DISCUSSION

The original climograms of Cook (1929) and others provided useful working tools which are still valid today. Their deficiencies lie in their lack of integration of variables into readily interpretable outputs for comparison with other years or places.

Multivariate analytical techniques, such as discriminant analysis, provide a pragmatic, descriptive approach to defining the climatic and other variables which limit the distributions of organisms. The apparent simplicity is offset by the limited insights that the approach can provide into the factors limiting the distribution of the species. Austin (1985) concluded that multivariate statistical approaches, applied to gross meteorological variables in order to describe distributions, are a retrograde step in the progress to developing more informative procedures. They are best restricted to fitting models to data chosen with some understanding of their biological function, rather than for selecting variables from large numbers chosen without thought for their biological relevance.

Murray and Nix (1987) used the GROWEST model to relate large-scale changes in the distribution of the biting midge *Culicoides brevitarsis* in New South Wales, Australia to annual differences in rainfall. GROWEST uses the same Growth Index concept that is incorporated into CLIMEX, but lacks the stress indices. As a result, an arbitrary winter cut-off point had to be assumed.

The CLIMEX approach to describing effects of climate differs greatly from traditional process-based population models in relying largely on inference to estimate parameter values. In this regard it is similar to the approach of Hudes and Shoemaker (1988) and fills a role where detailed models are not feasible. It also complements process-based models by deriving parameter values that result from the integration of influences of all climatic variables. It is rarely possible to reach such a holistic view with a process model due to the difficulties of identifying and then measuring extreme values of every significant variable.

One of the most useful functions of CLIMEX is to enable users to explore different hypotheses regarding the factors limiting the distribution of a species. Limitations in interpretation of currently available results arise from the need to approximate by defining one set of growth conditions for the whole life cycle. This is evident with insect species which undergo obligate diapause development and it also affects estimates of the suitability of different seasons for phenological phases in the growth of plants. This latter restriction can be important when overlaying host plant and insect responses to climate where the synchronization of insect attack and particular plant growth phases is important (Dewar and Watt, 1992).

IX. CONCLUSIONS

(1) Approaches which address geographical distributions *per se* are too limiting. Distributions are a consequence of the population dynamics of the species so the seasonal phenology and relative abundance of a species in different locations need to be considered.
(2) Purely descriptive methods are too limiting. The need is for more interpretation of the mechanisms that limit distributions. For example, diapause requirements are important limiting factors on geographical distributions of temperate species and they need to be taken into account when extrapolating to new environments.
(3) There is a need to take more account of non-climatic variables limiting distributions. This can be done using geographic information systems and/or expert systems with data from a variety of sources including satellite imagery and high resolution soil, host plant and animal maps.
(4) More sensitive indicators are needed of phenology, abundance and the number of generations a year.
(5) Generic approaches must be incorporated into user-friendly software to handle logistics of describing and comparing scenarios.
(6) A generic population model is needed, with flexible requirements for inputs and outputs ranging from the CLIMEX-type level to full life-cycle descriptions.

(7) Inferential methods are needed to by-pass experimental approaches to estimating development rates and phenology of species as well as their geographical distributions. The capacity to use inferential approaches to investigate the limiting effect of climate on each life-cycle stage of univoltine species of both animals and plants will increase understanding of the mechanisms which limit species.
(8) Coupling of crop and insect models, as well as overlaying of their climatic requirements on geographical platforms, is needed to develop holistic impact assessments.
(9) There is a need to document the significance of extreme climatic events in determining geographical distributions. Do species permanently occupy new and favourable local habitats colonized during extreme climatic events, and how frequently do they adapt genetically to the new environments?

The above list illustrates some of the great opportunities ahead for ecologists to contribute to a better understanding and more effective management of insects in a changed climate.

Acknowledgements

We acknowledge the statistical advice of Mrs A. Bourne.

REFERENCES

Andrewartha, H. G. and Birch, L. C. (1954). "The Distribution and Abundance of Animals". Chicago University Press, Chicago.

Anon. (1989). "The Potential Effects of Global Climate Change in the United States". Executive summary, Report to Congress, Office of Policy, Planning and Evaluation. United States Environmental Protection Agency, Washington DC.

Austin, M. P. (1985). Continuum concept, ordination methods, and niche theory. *Ann. Rev. Ecol. Syst.* **16**, 39–61.

Baker, C. R. B. (1991). The validation and use of a life-cycle simulation model for risk assessment of insect pests. *EPPO Bulletin* **21**, 615–622.

Berryman, A. A. and Stark, R. W. (1985). Assessing the risk of forest insect outbreaks. *Z. ang. ent.* **99**, 199–208.

Booth, T. H. (1988). Which wattle where?: selecting Australian acacias for fuelwood plantations. *Plants Today* **1**, 85–90.

Brown, J. H. (1984). On the relationship between abundance and distribution of species. *Am. Nat.* **124**, 255–279.

Busby, J. R. (1991). BIOCLIM – a bioclimate analysis and prediction system. *Plant Protection Quarterly* **6**, 8–9.

Cammell, M. E. and Knight, J. D. (1991). Effects of climate change on the population dynamics of crop pests. *Adv. Ecol. Res.* **22**, 117–162.

Carter, R. N. and Prince, S. D. (1988). Distribution limits from a demographic viewpoint. *In* "Plant Population Ecology" (A. J. Davy, M. J. Hutchings and A. R. Watkinson, eds), pp. 165–184. Blackwell, Oxford.

Caughley, G., Short, J., Grigg, G. C. and Nix, H. (1987). Kangaroos and climate: analysis of distribution. *J. Anim. Ecol.* **56**: 751–761.

Caughley, G., Grice, D., Barker, R. and Brown, B. (1988). The edge of the range. *J. Anim. Ecol.* **57**, 771–785.

Collier, R. H., Finch, S., Phelps, K. and Thompson, A. R. (1991). Possible impact of global warming on cabbage root fly (*Delia radicum*) activity in the UK. *Ann. Appl. Biol.* **118**, 261–271.

Cook, W. C. (1929). A bioclimatic zonation for studying the economic distribution of injurious insects. *Ecology* **10**, 282–293.

Dallwitz, M. J. and Higgins, J. P. (1978). "User's Guide to DEVAR. A Computer Program for Estimating Development Rate as a Function of Temperature". *Division of Entomology Report* **2**, CSIRO, Australia.

Dennis, R. L. H. and Shreeve, T. G. (1991). Climatic change and the British butterfly fauna: opportunities and constraints. *Biol. Conserv.* **55**, 1–16.

Dewar, R. C. and Watt, A. D. (1992). Predicted changes in the synchrony of larval emergence and budburst under climatic warming. *Oecologia* **89**, 557–559.

Diekmann, M. (1988). Use of climatic parameters to predict the global distribution of Ascochyta blight on chickpea. *Plant Disease* **76**, 409–412.

Dobesberger, E. J. and MacDonald, K. B. (1993). An application of geographic information systems and discriminant analysis to forecast the potential occurrence of pest infestation: an example using blueberry maggot (Diptera: Tephritidae). *In* "Proc. Symp. Agrometeorology and Plant Protection" (D. Rijks and B. Hopper, eds). *NAPPO Bull.* **9**.

Gaston, K. J. and Lawton, J. H. (1990). Effects of scale and habitat on the relationship between regional distribution and local abundance. *Oikos* **58**, 329–335.

Hengeveld, R. (1990). Theories on species responses to variable climates. *In* "Landscape – Ecological Impact of Climatic Change" (M. M. Boer and R. S. de Groot, eds), pp. 249 IOS Press, Amsterdam.

Hudes, E. S. and Shoemaker, C. A. (1988). Inferential method for modeling insect phenology and its application to the spruce budworm (Lepidoptera: Tortricidae). *Environ. Entomol.* **17**, 97–108.

Hutchinson, M. F. (1988). A new objective method for spatial interpolation of meteorological variables from irregular networks applied to the estimation of monthly mean solar radiation, temperature, precipitation and windrun. *In* "Proceedings of the United Nations University Workshop on Need for Climatic and Hydrologic Data in the Agriculture of South-East Asia, Canberra, 12–15 December 1983" (E. A. Fitzpatrick, ed.). United Nations University, Tokyo.

Kingsolver, J. G. (1989). Weather and the population dynamics of insects: integrating physiological and population ecology. *Physiol. Zool.* **62**, 314–334.

Kohlmann, B., Nix, H. and Shaw, D. D. (1988). Environmental predictions and distributional limits of chromosomal taxa in the Australian grasshopper *Caledia captiva* (F.). *Oecologia* **75**, 483–493.

Krebs, C. J. (1978). "Ecology: the Experimental Analysis of Distribution and Abundance", 2nd edn. Harper and Row, New York.

Landsberg, J. and Stafford Smith, M. (1992). A functional scheme for predicting the outbreak potential of herbivorous insects under global atmospheric change. *Aust. J. Bot.* **40**, 565–577.

Lindenmayer, D. B., Nix, H. A., McMahon, J. P., Hutchinson, M. F. and Tanton, M. T. (1991). The conservation of Leadbeater's possum, *Gymnobelideus leadbeateri* (McCoy): a case study of the use of bioclimatic modelling. *J. Biogeogr.* **18**, 371–383.

Mayer, D. G., Atzeni, M. G. and Butler, D. G. (1992). Adaptation of CLIMEX for spatial screwworm fly population dynamics. *Math. Comp. Sim.* **33**, 445–450.

Maywald, G. F. and Sutherst, R. W. (1989). CLIMEX: recent developments in a computer program for comparing climates in ecology. "Proceedings Simulation Society of Australia/IMACS Eighth Biennial Conference Canberra, 25–27 September 1989" pp. 134–140. Central Printery, Australian National University.

Maywald, G. F. and Sutherst, R. W. (1991). "User's Guide to CLIMEX a Computer Program for Comparing Climates in Ecology", 2nd edn. *Division of Entomology Report* **48**. CSIRO, Australia.

Meats, A. (1981). The bioclimatic potential of the Queensland fruit fly, *Dacus tryoni*, in Australia. *Proc. Ecol. Soc. Aust.* **11**, 151–161.

Meats, A. (1989a). Bioclimatic potential. *In* "World Crop Pests. Fruit Flies: their Biology, Natural Enemies and Control" (A. S. Robinson and G. Hooper, eds), pp. 241–52. Elsevier, Amsterdam.

Meats, A. (1989b). Abiotic mortality factors – temperature. *In* "World Crop Pests. Fruit Flies: their Biology, Natural Enemies and Control". (A. S. Robinson and G. Hooper, eds), pp. 229–240. Elsevier, Amsterdam.

Messenger, P. S. (1959). Bioclimatic studies with insects. *Ann. Rev. Entomol.* **4**, 183–206.

Messenger, P. S. and Fitters, N. E. (1954). Bioclimatic studies of three species of fruit flies in Hawaii. *J. Econ. Entomol.* **47**, 756–765.

Murray, M. D. and Nix, H. A. (1987). Southern limits of distribution and abundance of the biting midge *Culicoides brevitarsis* Kieffer (Diptera: Ceratopogonidae) in south eastern Australia. *Aust. J. Zool.* **35**, 575–585.

Nix, H. A. (1981). Simplified simulation models based on specified minimum data sets: the CROPEVAL concept. *In* "Application of Remote Sensing to Agricultural Production Forecasting" (A. Berg, ed.), pp. 151–169. A A. Balkema, Rotterdam.

Nix, H. A. (1986). A biogeographic analysis of Australian elapid snakes. *In* "Atlas of Australian Elapid Snakes" (R. Longmore, ed.), pp. 4–15. Bureau of Flora and Fauna, Canberra, ACT.

Nix, H. (1987). The role of crop modelling, minimum data sets and geographic information systems in the transfer of agricultural technology. *In* "Agricultural Environments" (A. H. Bunting, ed.), pp. 113–117. CAB International, Wallingford.

Norval, R. A. I. and Perry, B. D. (1990). Introduction, spread and subsequent disappearance of the brown ear-tick, *Rhipicephalus appendiculatus*, from the southern lowveld of Zimbabwe. *Exp. Appl. Acarol.* **9**, 103–111.

Papadakis, J. (1966). "Climates of the World and their Agricultural Potentialities". Avola, Cordoba.

Pearson, G. A. and Meyer, J. R. (1990). Discriminant models for predicting risk of blueberry maggot (Diptera: Tephritidae) infestations in southeastern North Carolina. *J. Econ. Entomol.* **83**, 56–532.

Perry, B. D., Kruska, R., Lessard, P., Norval, R. A. I. and Kundert, K. (1991). Estimating the distribution and abundance of *Rhipicephalus appendiculatus* in Africa. *Prev. Vet. Med.* **11**, 261–268.

Porter, J. H., Parry, M. L. and Carter, T. R. (1991). The potential effects of climatic change on agricultural insect pests. *Agric. For. Meteorol.* **57**, 221–240.

Prestidge, R. A. and Pottinger, R. P. (1990). "The Impact of Climate Change on Pests, Diseases, Weeds and Beneficial Organisms Present in New Zealand Agricultural and

Horticultural Systems". MAF Technology, Ruakura Agricultural Centre, Hamilton, New Zealand.

Rogers, D. J. (1979). Tsetse population dynamics and distribution: a new analytical approach. *J. Anim. Ecol.* **48**, 825–849.

Rogers, D. J. and Randolph, S. E. (1986). Distribution and abundance of tsetse flies (*Glossina* spp.). *J. Anim. Ecol.* **55**, 1007–1025.

Rogers, D. J. and Randolph, S. E. (1993). Distribution of tsetse and ticks in Africa: past, present and future. *Parasitol. Today* **9**, 266–271.

Rogers, D. J. and Williams, B. G. (1993). Monitoring trypanosomiasis in space and time. *Parasitology* **106**, S77–S92.

Southwood, T. R. E. (1977). Habitat, the templet for ecological strategies? *J. Anim. Ecol.*, **46**, 337–365.

Spradbery, J. P. and Maywald, G. F. (1992). The distribution of the European or German wasp, *Vespula germanica* (F.) (Hymenoptera: Vespidae), in Australia: past, present and future. *Aust. J. Zool.* **40**, 495–510.

Sutherst, R. W. (1983). Variation in the numbers of the cattle tick, *Boophilus microplus* (Canestrini), in a moist habitat made marginal by low temperatures. *J. Aust. Ent. Soc.* **22**, 1–5.

Sutherst, R. W. (1990). Impact of climate change on pests and diseases in Australasia. *Search* **21**, 230–232.

Sutherst, R. W. (1991a). Pest risk analysis and the greenhouse effect. *Rev. Agric. Entomol.* **79**, 1177–1187.

Sutherst, R. W. (1991b). Predicting the survival of immigrant insect pests in new environments. *Crop Protection* **10**, 331–333.

Sutherst, R. W. (1993). Arthropods as disease vectors in a changing environment. In "Environmental Change and Human Health" (J. V. Lake, G. R. Bock and K. Ackrill, eds), pp. 124–145. *Ciba Foundation Symposium* **175**. Wiley, Chichester.

Sutherst, R. W. and Maywald, G. F. (1985). A computerised system for matching climates in ecology. *Agric. Ecosystems and Environ.* **13**, 281–99.

Sutherst, R. W., Spradbery, P. and Maywald, G. F. (1989). The potential geographical distribution of the Old World screw-worm fly, *Chrysomya bezziana*. *Med. Vet. Entomol.* **3**, 273–280.

Sutherst, R. W., Maywald, G. F. and Bottomley, W. (1991). From CLIMEX to PESKY, a generic expert system for pest risk assessment. *EPPO Bulletin* **21**, 595–608.

Tauber, M. J., Tauber, C. A. and Masaki, S. (1986). "Seasonal Adaptations of Insects". Oxford University Press, New York.

Walter, H. (1985). "Vegetation of the Earth, and Ecological Systems of the Geo-Biosphere". Springer, New York.

Walter, H. and Leith, H. (1960). "Klimadiagramm – Weltatlas". VEB Gustav Fischer Verlag, Jena.

Wigley, T. M. L. (1985). Impact of extreme events. *Nature* **316**, 106–107.

Williams, J. D., Sutherst, R. W., Maywald, G. F. and Petherbridge, C. T. (1985). The southward spread of buffalo fly (*Haematobia irritans exigua*) in Eastern Australia and its survival through a severe winter. *Aust. Vet. J.* **62**, 367–369.

Worner, S. P. (1988). Ecoclimatic assessment of potential establishment of exotic pests. *J. Econ. Entomol.* **81**, 973–83.

Worner, S. (1991). Use of models in applied entomology: the need for perspective. *Environ. Entomol.* **20**, 768–773.

5

The Effects of Climate Change on the Agricultural Environment for Crop Insect Pests with Particular Reference to the European Corn Borer and Grain Maize

J. PORTER

I. Introduction	94
II. Climate Change and Effects on the Agricultural Environment for Crop Insect Pests	95
A. Effects of Increased CO_2	95
B. Direct Effects of Climate Change on Crops	96
C. Likely Adjustments in Agriculture	98
III. Climate Change and the European Corn Borer	100
A. The European Corn Borer	100
B. Mapping Potential Distribution and Flight Activity	101
C. Projected Changes in Climate	106
D. Impacts of Climate Change on ECB	107
E. Implications of Climate Change for the ECB	112
F. Refining the Approach	118
References	120

Abstract

The European corn borer (*Ostrinia nubilalis*) is a major pest of grain maize in Europe. This chapter examines the sensitivity of the European corn borer and grain maize to changes in temperature. The distribution of the European corn borer and grain maize has been mapped on the basis of their thermal (degree-day) requirements under the present (1951–1980) temperatures. In all regions where, on the basis of temperature alone, grain maize can potentially be cultivated, there is the potential for at least one generation of the European corn borer to develop. Projections of possible changes in temperature from three general circulation models have been used to investigate the impacts of anthropogenically induced climate change on the European corn borer and grain maize. The results indicate that substantial shifts in the potential distribution of both the pest and its host crop could occur. Furthermore, in some areas an additional generation of the European corn borer is indicated to occur. The rate at which such shifts in distribution could occur will depend on the rate of climate change and thus the level of greenhouse gas emissions. One of the general circulation models used in this study gives estimates of transient, time-dependent changes in temperature induced by different emissions levels. It seems possible that the European corn borer would be able to adapt to all but the most extreme rates of climate change projected to occur. Degree-day requirements have also been used to estimate the timing of peak flight activity of the adults under current and possible future climates. The results indicate that flight activity of the European corn borer could occur several weeks earlier than at present as a result of the higher temperatures. Changes in distribution and the phenology of the European corn borer could have a significant impact on the cultivation of grain maize in Europe. Some of the implications of climate change for the European corn borer and consequently for grain maize production are discussed.

I. INTRODUCTION

This chapter is divided into two sections. The first examines the general effects of climate change on the agricultural environment for crop insect pests. It draws heavily on the assessment of the impacts of climate change on agriculture made by the Intergovernmental Panel on Climate Change (IPCC) (Parry *et al.*, 1990) and the book on climate change and world agriculture by Parry (1990). However, the interpretation of how climate-change induced changes in agriculture may affect the agricultural environment for insect pests rests solely with the author of this chapter.

The second part examines in more detail how change in climate could affect the European corn borer (ECB) (*Ostrinia nubilalis* Hbn.), a major pest of grain maize. The effects of climate change on both the ECB and its host are examined. The distribution of the ECB and grain maize in Europe has been mapped on the basis of their thermal (degree-day) requirements under the current climate and a range of scenarios of possible future climates. In addition, degree-day requirements have been used to examine the effects of climate change on the timing of peak flight activity of adult ECBs.

II. CLIMATE CHANGE AND EFFECTS ON THE AGRICULTURAL ENVIRONMENT FOR CROP INSECT PESTS

There are four broad types of effects of climate change on agriculture that are likely to have an impact on the agricultural environment for insects: firstly, the direct physiological effects on crops resulting from the increase in the level of ambient carbon dioxide (CO_2) in the atmosphere; secondly, the direct effects of changes in climate on the crops themselves; thirdly, the effects of climate change on crop potential and finally, the changes in agricultural management likely to occur in response to the effects of climate change. The possible impacts of such effects for insect pests are now considered.

A. Effects of Increased CO_2

The direct effects of increased CO_2 on insects are covered by Watt *et al.*, Chapter 9, this volume (also Nicolas and Sillans, 1989) and only a brief summary is included here. As CO_2 is taken up by all crop plants through the process of photosynthesis, any increase in the atmospheric concentration of CO_2 is likely to increase photosynthetic rates and therefore growth. Changes in the concentration of CO_2 are also likely to bring about changes in the quality of plant species which may in turn affect insect pests feeding on them. In the absence of any physiological adaptation by crops, increased levels of atmospheric carbon are likely to result in greater carbon : nitrogen ratios in plant tissues, which may stimulate greater feeding activity in some insects (e.g. see Lincoln *et al.*, 1986).

An additional effect of increased CO_2 is its influence on water use by plants. Under higher levels of CO_2, water use efficiency tends to increase as a result of closure of the stomata and consequently a reduction in transpiration (Parry, 1990). This would be beneficial for crop productivity in semi-arid areas where moisture is currently limiting. However, it may not be of benefit to insect pests because reduced transpiration would tend to result in higher leaf surface temperatures and consequently a reduction in relative humidity which could make conditions less favourable for the development of some species.

B. Direct Effects of Climate Change on Crops

1. Changes in temperature

The effects of climate change on agricultural crops will depend on the magnitude of the change in climate and prevailing climatic and environmental conditions. In regions where agricultural production is currently limited by temperature, one of the most important effects of an increase in temperature would be to extend the crop growing season and reduce the growing period required by crops for maturation (Parry, 1990). This could be important for some insect species depending on how it alters pest–crop synchrony. Elsewhere any increase in temperature could lead to temperature-induced crop stress. Temperature-induced stress can cause changes in plant physiology resulting in changes in the levels of chemical or morphological defences and/or nutritional quality of the host (Tingey and Singh, 1980, cited in Benedict and Hatfield, 1988). As a result of these changes, insects feeding on the plants would have altered growth, development, reproduction, survival and/or behaviour. Similar effects can occur as a result of water-induced plant stress. Indeed, it is often difficult to distinguish the effects of temperature and water-induced stress on insect herbivores (Benedict and Hatfield, 1988).

2. Changes in available moisture

In many areas agricultural productivity is related to the amount of water available for crop growth. Even with no change in precipitation levels, changes in temperature would have an effect on moisture availability and therefore crop growth (Parry, 1990). Higher temperatures would lead to an increase in evapotranspiration which could result in water-stress in non-irrigated crops particularly if there was no compensatory increase in precipitation (e.g. Kenny and Harrison, 1992). Similarly a reduction in precipitation could result in water stress for some crops in some areas. However, there are great uncertainties surrounding the possible future changes in precipitation levels, particularly at the regional level and so it is difficult to make any assessment of changes in moisture availability for crops. An additional factor for consideration is that the increase in water use efficiency associated with higher levels of CO_2 could offset any shortage of available moisture.

A reduction in moisture availability could affect insect pests in two ways. Firstly, as a result of changes to the microclimate within the crop canopy. Temporary water stress could affect leaf temperature and humidity as a result of changes in stomatal behaviour. On the other hand long-term water stress can produce permanent changes in the crop canopy. Either could alter the suitability of the environment for insects. The second type of effect is associated with changes in food quality. When plants are stressed by certain changes in weather, particularly lack of moisture, they can become better sources of food for insects due to increased levels of nitrogen. It has been suggested that the young of some polyphagous insects, feeding from a range of host plants, suffer from a shortage of nitrogenous food (White, 1984). They adapt to this inadequate environment by producing increased numbers of offspring. When weather conditions stress plants and stimulate an increase in nitrogen, it is much more likely that young insects will survive, thereby leading rapidly to a population explosion and hence a pest outbreak. However, not all insects are limited by low levels of nitrogen and increased nitrogen may have a deleterious effect on some either through direct effects or an increase in plant defensive chemicals (Waring and Cobb, 1992). In contrast, Rhoades (1985) suggested that drought-induced stress could decrease the plant's defensive system which could increase host-plant suitability for some insects. It seems as though changes in host-plant suitability in relation to water stress can be beneficial, detrimental or result in no change and the overall effect would depend on the situation under consideration (Holtzer et al., 1988). An increase in precipitation could be beneficial for some insects if it resulted in a more favourable relative humidity and did not adversely affect temperature within the canopy. However, increased humidity could also encourage the development of pathogenic micro-organisms within insects (Ferro, 1987).

3. Changes in crop potential

Climate change is likely to result in changes in crop potential; that is, the suitability of an area for crop growth. This could occur either as a result of the direct effects of changes in climate on crop potential or because of the impacts of climate change on other important environmental factors such as soils, water supply, diseases, weeds and, of course, insect pests. The interaction between climate and agriculture means that any change in climate is likely to result in a spatial shift in climatic resources for agriculture. This lengthening of the growing season would result in a poleward and altitudinal shift in the thermal limits to agriculture (see the example of grain maize below). Similarly, changes in precipitation or the availability of moisture will alter the limits of crop suitability. Changes in potential yields in the core food-producing areas would also alter overall crop potential and would probably have a greater impact on overall food production than shifts in the potential limits at the boundaries of current agricultural regions (Parry, 1990). Changes in soil fertility and erosion, water availability or the occurrence of insect pests, diseases or weeds could all alter crop potential. Some of the effects of climate change on insect pests have been considered elsewhere (see, e.g., EPA, 1989; Porter et al., 1991; Sutherst, 1991; Pimental et al., 1992). Any change in agricultural potential, and consequently profitability, could result in changes in crop types and cropping patterns (see below).

C. Likely Adjustments in Agriculture

A number of management responses to the impacts of climate change on agriculture are expected to occur as the impacts are perceived. Such changes are likely to alter the agricultural environment for insect pests.

1. Changes in crop type

In areas where temperature is currently limiting and climate change results in a substantial increase in the temperature during the growing season it is likely that there will be a shift towards crops or crop varieties which have higher thermal requirements and could take advantage of the warmer growing season (Yoshino et al., 1988). In areas where higher temperatures could lead to higher rates of evapotranspiration and subsequently reduced available moisture, or where rainfall is reduced, it would be more useful to switch to drought-resistant crops. The introduction of new crops or cultivars could have serious implications for existing crops by providing alternative hosts and "green bridges" (temporary hosts or over-wintering sites to bridge the "temporal gap" between hosts) for pests. New crops could provide new over-wintering sites for pests thereby maintaining local population levels (Crawford et al., 1989). To illustrate, the current expansion of grassland to combat soil erosion in North America is expected to create new alternative hosts for the Russian wheat aphid (*Diuraphis noxia*) which is a major pest of wheat and barley (Kindler and Springer, 1989).

2. Changes in cropping patterns

A switching of crop types implies a change in crop location and land use. Such changes will depend not only on the effect of climate change on agricultural potential and the comparative advantage of land, but also on economics and changes in price that are largely determined by changes in potential in other areas (Parry, 1990). The broad-scale changes in crop suitability and locations imply a poleward shift of present-day agricultural zones (Parry et al., 1990). Spatial shifts in the distribution of crops would have implications for insect pests because the distribution of the pests is strongly dependent on the availability of a suitable host. Thus, changes in cropping patterns and the introduction of different crops may open up new areas for pest invasion. Shifts in the limits of potential for grain maize cultivation in Europe and the implications of this for the ECB are examined in more detail below. Furthermore, the spatial and temporal occurrence of both host crops and non-hosts has a great influence on a pest's natural enemy complex and its importance in preventing pest outbreaks (Ferro, 1987). As such, changes in the natural enemy complex could be expected to occur as a result of the combination of changes in cropping patterns and changes in the distribution of natural vegetation.

3. Changes in irrigation

In some areas it is likely that there will be a substantial increase in the need for irrigation to counteract the effects of moisture loss through increased evapotranspiration. For example, in eastern England a 2°C increase in mean temperatures with no change in precipitation would lead to a significant increase in the demand for irrigation (Rowntree et al., 1989). The use of irrigation can, depending on the type of irrigation system used, lead to higher humidity levels which could increase the longevity of some pest species. However, as already mentioned, increased humidity could also encourage the development of pathogenic microorganisms.

4. Changes in planting and harvesting dates

It is likely that climatic changes would result in changes in the planting date of crops to take advantage of the earlier start to the growing season or to avoid some of the adverse effects of drought stress later in the season which may occur in some regions (Parry et al., 1989; Carter et al., 1991a). This, combined with the faster rate of development of crops, would enable earlier harvesting. Changed planting dates would alter pest–crop synchrony and could make conditions more or less suitable for the insect pest depending on whether or not it enabled the pest to invade during vulnerable stages in the crop's life cycle. If crops are planted so that the crop phenology is not in synchrony with the insect's phenology then the rate of insect colonization is likely to be retarded (Ferro, 1987). Similarly, if the crop is harvested before insects that have a diapausing stage enter dispause, then the over-wintering population would tend to be reduced and the rate of colonization is likely to be lower the following spring (Ferro, 1987). The situation will be made more complicated by the effects of temperature on the rate of development of both insects and crops which would also affect pest–crop synchrony (see below in relation to the ECB).

5. Changes in pest control strategies

Changes in crop or cultivar types combined with changes in the crop–pest complex (including insects, diseases and weeds) associated with climate change are likely to result in changes in pest-control strategies which may have a significant impact on some insects. For example, the use of insecticides for pest control is generally associated with insect mortality. However, in some circumstances insecticides and other pesticides (fungicides, herbicides, etc.) can have a beneficial effect on insects by inhibiting or stimulating feeding, growth and reproduction, and an indirect effect as a result of destruction of natural enemies or through host-plant mediated effects (Heinrichs, 1988). Some of the implications of climate change for control of the ECB are discussed in more detail below.

III. CLIMATE CHANGE AND THE EUROPEAN CORN BORER

A. The European Corn Borer

The ECB is a phytophagous lepidopteran of the family Pyralidae, and is of agricultural significance over much of the northern hemisphere. It is probably native to southern Europe, but is now widely distributed in the northern hemisphere between latitudes of about 10°N to 58°N in Europe, Asia, northern Africa and North America (Beck, 1989). The distribution and density of the ECB is determined by a complex combination of availability of a suitable host plant, agricultural practices and climate. The ECB thrives on a large number of host plants, but, from an agricultural standpoint, maize is by far the most important. For the ECB the availability of maize, in terms of acreage, makes it the most important host in Europe. The ECB can be found on all types of maize (grain, silage or forage). However, ECB survival in silage maize is low because the whole crop is harvested thereby removing most of the crop material suitable for ECB over-wintering sites. With grain maize only the corn cobs are harvested and so survival is much higher. In areas such as southern England and southern Sweden, where very little grain maize is grown commercially, populations generally remain at low levels and are found mainly on wild hosts such as mugwort (*Artemisia vulgaris*) or hops (*Humulus* spp.) (Borg, 1949; Beirne, 1952). In northern Europe the climate is less favourable for ECB development than it is further south. For example, in the UK the current climate is marginal for ECB development and it is generally only an immigrant visitor, although it may breed in southern England in particularly favourable years (Carter, 1984). It may be the less favourable climate or the predominance of silage maize, rather than grain maize, or the combination of both factors that means the ECB is currently not a major pest in the UK.

B. Mapping Potential Distribution and Flight Activity

1. The bioclimatic model

As temperature plays such an important role in determining the rate of development of insects, a simple bioclimatic index known as Effective Temperature Sum (ETS) or degree-days is often used to examine phenological development (e.g. AliNazee, 1976; Butts and McEwen, 1981). ETS represents the accumulation of temperature above a critical threshold throughout a fixed time period, usually calculated on a daily basis. Thus, for an annual summation based on daily mean temperatures:

$$\text{ETS} = \sum_{i=1}^{365} \delta i (T_i - T_b)$$

where:

$$\delta i = 1 \text{ for } T_i > T_b$$

$$\delta i = 1 \text{ for } T_i > T_b$$

T_i is the mean temperature on day i and T_b is the threshold or base temperature.

A specific level of ETS is required before each phenological stage can be reached, so in this way the timing of different stages of development can be determined. ETS can also be used to map zones of insect distribution (e.g. Scriber and Hainzee, 1987). ETS is widely used to predict the phenological development of the ECB, particularly with respect to adult emergence and flight activity in order to time chemical control measures effectively (Stengal, 1982; Despins and Roberts, 1984). The time of peak adult flight or when 50% of adults have flown is generally taken as the critical time for control applications.

The use of ETS in examining ECB development is not without drawbacks. There are obvious dangers in the oversimplification of using a single numerical value to characterize the climate over a whole season. Such single values give an indication of average conditions, but show nothing of climatic variability, the rate of change of climate conditions within this variability or the occurrence of extreme events. For many insects, including the ECB, a steady rate of warming or cooling is crucial as it allows acclimatization to the change in temperature, whereas sudden extremes of temperature do not and can therefore result in high mortality. Use of ETS gives no indication of such events. Furthermore, the method of ETS calculation assumes a linear relationship between temperature and insect (or crop) development, yet it is known that the relationship between growth rate and temperature is not directly proportional near the upper and lower threshold temperatures (Baker, 1980). Also, the threshold temperature is generally assumed to remain constant throughout the development period, yet this can vary depending on the stage of development (e.g. Matteson and Decker, 1965). The actual ETS requirements are therefore different to the theoretical straight line requirements. However, one study suggested that this is immaterial when ETS is used for forecasting the development of the ECB in the field because the amount of development taking place within this range of temperature is insignificant (Matteson and Decker, 1965). A further criticism of the use of ETS is that it does not take into account many of the other climatic factors influencing growth and development, such as precipitation, and windspeed. However, despite these criticisms, the system has been employed successfully for many organisms in a wide range of environments. In this study ETS has been used to examine flight activity as an indication of the timing of ECB development and also to map the potential distribution of the ECB in Europe.

(a) ETS and the ECB

ETS requirements for ECB development vary with environmental conditions and due to genetic adaptation of individual populations (Beck and Apple, 1961; Ohnesorge and Reh, 1987). However, in this study it is assumed that a single multivoltine population, referred to as the representative population, having the same ETS requirement exists across the whole of Europe. An ETS value of 726 degree-days $\geq 10°C$ is required for each generation of the representative population to develop. This value was obtained by extrapolating information on ETS requirements for peak flight activity of ECB populations from a number of locations across North America (Porter, 1991). It is assumed that the ECB can develop in all areas where sufficient temperature accumulates to enable this ETS value of 726 degree-days to be met. It is possible to delimit the potential distribution of successive generations of the population based on these thermal limits. The timing of flights of the first generation of the representative population was calculated by determining the date on which the ETS required for peak flight (372 degree-days $\geqslant 10°C$) was reached.

(b) ETS and grain maize
As temperature is often a limiting factor for the production of maize, ETS is also commonly used to determine zones of potential suitability for the crop in both Europe and North America (see, e.g., Chapman and Brown, 1966; Hough, 1978). It is therefore possible to map the potential distribution of the ECB and its host plant. A study by Carter et al. (1991a,b) examined the effects of climate change on the potential distribution of grain maize within Europe. An ETS value of 850 degree-days $\geqslant 10°C$ was used to delimit the potential distribution of grain maize. The work outlined here for the ECB refers to the study by Carter et al. when considering the impact of climate change on grain maize.

2. The European Mapping System

In order to examine the impacts of climate change on the ECB it is necessary to map the potential distribution and flight activity under current climatic conditions, then perturb the climate to simulate a change in climate and remap the distribution and flight activity. To do this a geographical information system known as EMSYS – the European Mapping System – has been used. EMSYS is based on a 0.5° latitude by 1.0° longitude grid for Europe (35°N to 72°N by 12°W to 42°E). Mean temperatures for the period 1951–1980 were interpolated to the EMSYS grid and provide the baseline or current climate of the region (Carter et al., 1991a). Maps produced using the baseline climate can then be used as a reference point for assessing the impact of climate change on ECB distribution and development.

3. Potential distribution and flight activity under the current climate

(a) Potential distribution
Figure 1 shows the long-term potential distribution of the ECB in Europe under baseline (1951–1980) climatic conditions. The potential distribution

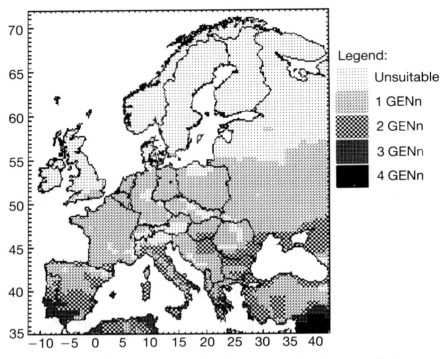

Fig. 1. Potential number of generations of the European corn borer based on effective temperature sum for the baseline climate (1951–1980).

delimits the maximum geographical range based on climate alone. However, a number of factors, including the lack of a suitable host, may prevent the ECB from occupying its entire potential range and thus the actual distribution may be more restricted. These calculated limits of potential distribution are broadly consistent with observed limits obtained from existing maps and information from published literature (CAB, 1971; Carter, 1984; references in Porter, 1991). The potential distribution of grain maize (based on ETS) lies approximately 100–200 km to the south of the limits of ECB distribution extending from the north-west coast of France across northern central Europe and the Commonwealth of Independent States along a latitude of about 52°N (Carter *et al.*, 1991a; Porter *et al.*, 1991). The actual distribution of grain maize will, of course, be influenced by many other environmental, socioeconomic and political factors. Nevertheless, on the basis of ETS alone, wherever there is the potential for grain maize cultivation there is also the potential for at least one generation of the ECB to develop. However, near the northern limits

5. Climate Change and the European Corn Borer

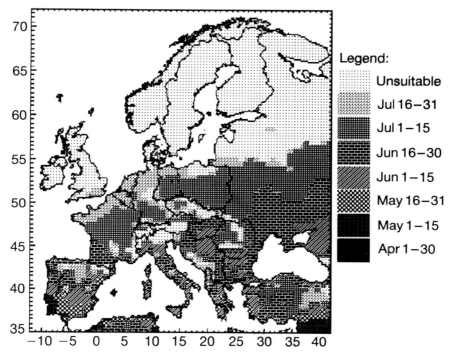

Fig. 2. Estimated dates of peak flight activity of the first generation of European corn borer under the present climate, based on ETS using current (1951–1980) temperatures.

to ECB distribution temperatures are insufficient to enable grain maize cultivation.

(b) Peak flight activity
Given the range of climatic conditions found within Europe, estimated peak flight activity extends over several months for the region as a whole (Fig. 2). Peak flight activity in southern Spain, Italy, Greece and Turkey is estimated to occur in May, or even earlier in some climatically favourable coastal regions. Further north in southern England, the Benelux countries, north Germany and in some of the high altitude areas of southern Europe, due to the cooler climate and therefore slower rate of development, peak flight activity is indicated to occur in July. The calculated timing of adult flights shown in Fig. 2 compares well with the information available on observed flight activity in the region (e.g. Anglade, 1975; Donlika, 1975; Cordillot, 1989).

C. Projected Changes in Climate

General circulation models (GCMs) are considered to be the most credible way of estimating changes in climate. However, it is important to remember that the results from GCMs are merely projections and not predictions of climate change. Details and descriptions of various GCMs can be found elsewhere in this volume (Bennetts, Chapter 3) and in Bach (1988); Mitchell *et al.* (1990) and references therein. Two types of simulation experiment have been conducted using GCMs to estimate future climate. These are equilibrium-response experiments and transient-response experiments. The equilibrium-response experiments generally consider the climatic effects of a sudden doubling of carbon dioxide and this is known as a $2 \times CO_2$ scenario. Transient-response experiments consider gradual increases in greenhouse gas emission levels, include the effect of the oceans in delaying the response of the climate to increasing emission levels and estimate the rate as well as the magnitude of climate change. Relatively few transient experiments have been made (see IPCC, 1992 for details and Bennetts Chapter 3). The effects of a $2 \times CO_2$ scenario on the potential distribution of the ECB has been examined elsewhere (Porter *et al.*, 1991) and will not be considered here. Instead this chapter uses the results from two transient experiments conducted with the GCM from the Goddard Institute of Space Studies (GISS).

The GISS experiments were among the first transient-response experiments to be conducted and use a very simple ocean model. Climate changes were estimated using three scenarios (A, B and C) of changing greenhouse gas concentrations (only two of which are used here) to provide an indication of how the changes in climate depend upon the rate of greenhouse gas emissions (Hansen *et al.*, 1988). Scenario A is considered to be the "worst case" scenario and assumes growth rates of emissions will continue to rise as they did in the 1970s and 1980s, representing an increase of 1.5% per annum. Scenario C assumes drastic, and unrealistic, reductions in growth rates of emissions between 1990 and 2020 such that the climate system is stabilized after 2000 and there are no further greenhouse gas induced changes in climate. Temperature changes estimated by scenarios A and C are shown in Table 1.

It is useful to compare the GISS scenarios with the estimates made by the Intergovernmental Panel on Climate Change (IPCC) using a range of transient GCMs. The IPCC estimates are for an increase in global mean annual surface temperature of between 0.7 and 1.5°C above the current 1990 level, with a best estimate of 1.1°C by 2030, rising to between 1.6 and 3.5°C by 2070, with a best estimate of 2.4°C (Houghton *et al.*, 1990). The equivalent estimates made by GISS scenario A are for an increase of

Table 1. Temperature increases for Europe for the periods 1990s, 2020s and 2050s estimated by GISS scenarios A and C. These figures are relative to 1958 – the beginnning of the experiments. Estimates for the 2050s are not available for scenario C because the experiment finishes at 2039

	1990s	2020s	2050s
Scenario A	0.5–1.5°C	1.5–2.5°C	3.0–5.0°C
Scenario C	0.0–1.0°C	0.0–1.5°C	

about 1.6°C by 2030 and in excess of 4.0°C by 2070 (extrapolated using a diagram in Hansen et al. (1988), scenario A ends at 2062). GISS scenario C estimates a temperature increase of only 0.4°C by 2030. Thus, it can be seen that, in terms of changes in the mean global surface temperature, GISS scenario A is on the high side of the IPCC estimates, while GISS scenario C with its drastic emissions reductions is unrealistically low. However, these scenarios were designed to cover a range of possibilities for future climatic warming and, as such, provide useful information for examining the likely range of impacts of climate changes. Thus, following Carter et al. (1991b) GISS scenarios A and C have been used in this study to delimit the upper and lower bounds of projections of future global climate changes. It is important to emphasize that there is a large element of uncertainty attached to regional estimates of temperature change from GCMs, and these results should be interpreted accordingly.

In order to develop climate change scenarios for Europe the estimates of temperature changes for each GISS GCM grid box were interpolated to the EMSYS grid using a simple linear interpolation method and then added to the baseline (1951–1980) temperatures at each EMSYS grid box (Carter et al., 1991a). One of the major difficulties with assessing the effects of climate change on insect pests is the problem of differences in spatial scale. In the main, insects are influenced by local or perhaps regional climatic (and other) conditions. However, the coarse resolution of GCM grids means that estimates of possible changes in local climatic conditions are unavailable as yet. Even at the regional level very few estimates are currently available. The GISS GCM has a horizontal grid resolution of around 700 km by 1000 km (7.8° latitude by 10° longitude) (Hansen et al., 1988).

D. Impacts of Climate Change on ECB

In order to examine the effects of climate change on the ECB, the baseline climate was adjusted using the temperature changes estimated by

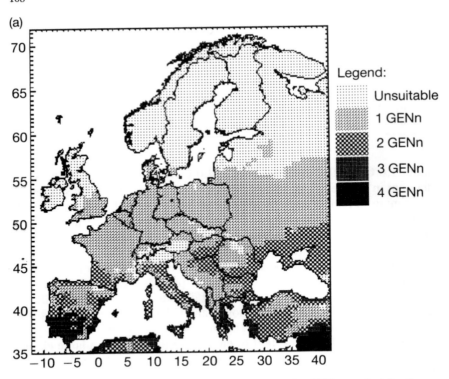

Fig. 3a. Potential distribution of the European corn borer under GISS scenario A for the 1990s.

the GISS scenarios and the potential distribution and flight activity were then mapped by calculating ETS for the 1990s, 2020s and 2050s. In regions where temperature is the primary factor governing the length of the development season any increase in temperature is likely to result in an extension of the development season. Furthermore, as temperature is such an important factor governing the rate of insect development, higher temperatures would lead to an increase in the rate of development. The impacts of climate change on ECB distribution and development will now be examined.

1. Impacts on potential distribution

Figures 3 and 4 indicate that the combination of an extension of the development season and faster rates of development would lead to a northward expansion in the limits of potential distribution of the ECB.

5. Climate Change and the European Corn Borer

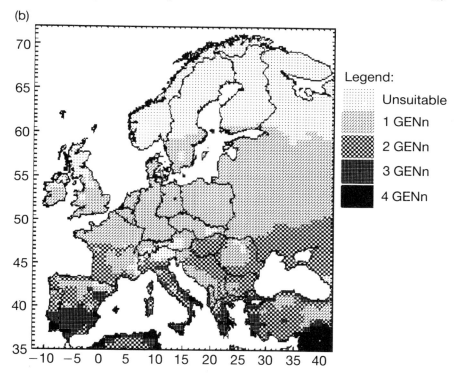

Fig. 3b. Potential distribution of the European corn borer under GISS scenario A for 2020s.

Based on ETS requirements alone, the temperature changes projected by the GISS scenario A result in a northward extension in potential ECB distribution from the 1990s limit of 60–555 km by the 2020s to 165–1055 km by the 2050s. This implies a northward shift in potential ECB limits that would accelerate from 20–185 km per decade between the 1990s and 2020s to 55–350 km per decade between the 2020s and 2050s. This reflects the exponential rate of greenhouse gas emissions, and consequently temperature change, under GISS scenario A. The lower temperature changes projected under GISS scenario C indicate a small extension in the limit of ECB distribution, with a northward shift of 0–500 km (0–165 km per decade) between the 1990s and 2020s. To take advantage of the favourable climatic conditions the ECB would require the presence of a suitable host. Work by Carter *et al.* (1991a,b) indicates that a northward expansion in the limits for potential grain maize cultivation is likely to occur as a result of the temperature changes

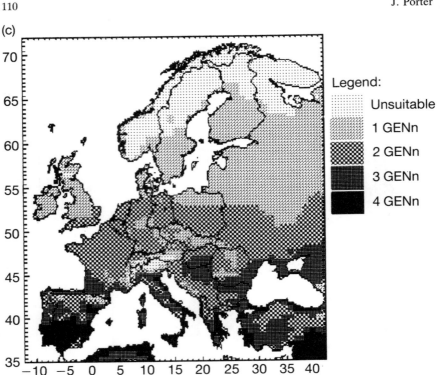

Fig. 3c. Potential distribution of the European corn borer under GISS scenario A for the 2050s.

estimated by the GISS scenarios. The rate of shift for potential grain maize cultivation would accelerate from around 150 km per decade between the 1990s and 2020s to 240 km per decade for the 2020s to 2050s (Carter *et al.*, 1991a). The higher thermal requirements of grain maize mean that it responds more slowly to the temperature increases. As under the current climate there are areas at the northern limits of its distribution where conditions would be less favourable for the ECB due to the absence of its main host plant. Due to the fact that the above calculations have been made using gridded data and represent discrete multiples of EMSYS grid box dimensions they should be regarded merely as approximations of the rate of spatial shift. Nevertheless, they do give some indication of the effects of possible greenhouse gas mitigation policies on the rate and extent of impact on ECB and grain maize distributions.

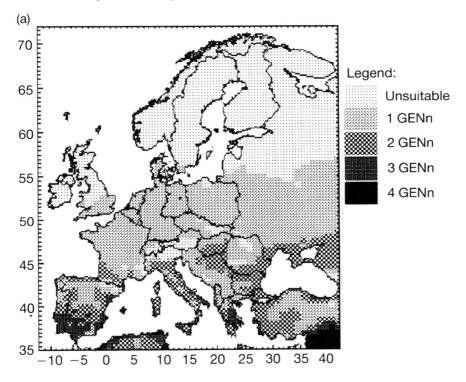

Fig. 4a. Potential distribution of the European corn borer under GISS scenario C for the 1990s.

2. Impacts on phenology

To examine the influence of climate change on flight activity the date of peak flight under the GISS scenarios was calculated and then compared to the baseline. Owing to the combination of an earlier start to the development season and faster development rates, under the higher temperatures flight activity of the first generation is estimated to occur several weeks earlier than at present. Figure 5 indicates that under GISS scenario A by the 1990s flights could occur 0–15 days earlier, by the 2020s flight activity could be between 5 and 25 days earlier and this could increase to up to 45 days earlier by the 2050s. With the temperature changes projected by GISS scenario C flight activity is estimated to occur 0–15 days earlier in both the 1990s and the 2020s (Fig. 6). It must be remembered that little confidence can be put into regional temperature

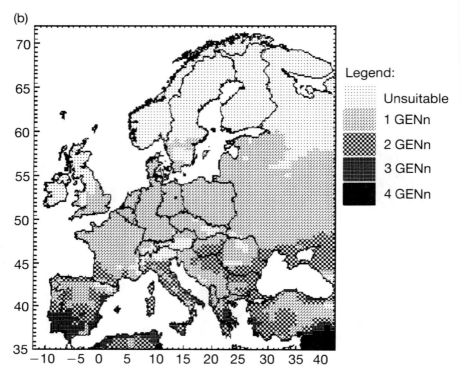

Fig. 4b. Potential distribution of the European corn borer under GISS scenario C for the 2020s.

changes projected from such a low resolution GCM as the GISS model; however, the results shown in Figs 5 and 6 give an indication of the magnitude of effect that could occur due to the climate changes estimated by the GISS scenarios.

E. Implications of Climate Change for the ECB

It seems likely that anthropogenically–induced climate change could provide opportunities for the ECB to become established in parts of northern Europe where it is currently not found. The temperature increases estimated by GISS scenarios A and C suggest that the limits for ECB distribution may move northwards at a rate of between 0 and 350 km per decade depending on the region and scenario under consid-

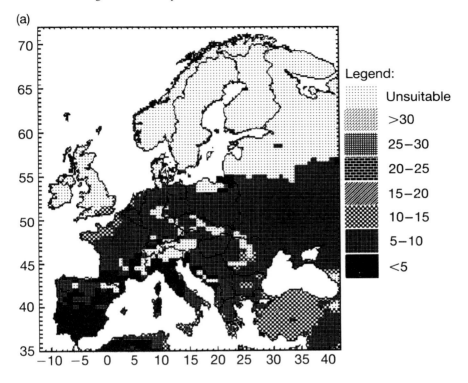

Fig. 5a. Difference in timing of peak flight activity of the first generation of the European corn borer under GISS scenario A for the 1990s (number of days earlier than baseline).

eration. In Switzerland the distribution of the ECB is currently expanding southwards through the Rhine Valley and Rhine tributaries at a yearly expansion rate estimated at 3–10 km, sustained by the yearly increase in the area of grain maize production (Cordillot, 1989). Furthermore, in Ohio (USA) between 1921 and 1938 the ECB spread at a rate of about 19–24 km year^{-1} (Neiswander, 1962). Thus, it would appear from comparison of estimates of possible future spread with known rates of migration that, except under the highest rates of temperature change projected under GISS scenario A, the ECB would have the ability to migrate provided that it is able to find a suitable host plant. The work by Carter *et al.* (1991a,b), discussed above, suggests increased opportunities for grain maize cultivation in northern Europe. Thus, in areas that are currently unsuitable for ECB development, more favourable climatic

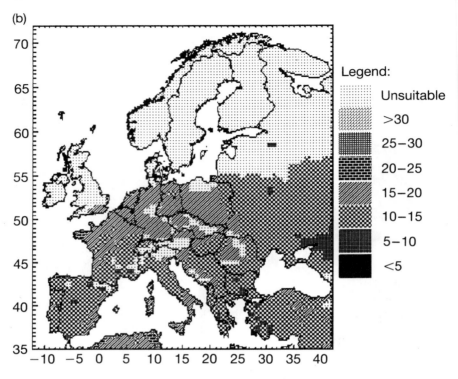

Fig. 5b. Difference in timing of peak flight activity of the first generation of the European corn borer under GISS scenario A for the 2020s (number of days earlier than baseline).

conditions and the introduction of grain maize cultivation may encourage ECB invasion. However, in southern Europe, where maize cultivation already depends heavily on irrigation (Bignon, 1990), there may be decreasing potential for cultivation due to increased drought stress (Parry et al., 1989; Kenny and Harrison, 1992). Thus, in contrast to the increasing potential in northern Europe, conditions could become less favourable for the ECB in southern Europe due to the absence of grain maize.

An additional factor influencing ECB populations would be the presence of natural enemies and changes in interspecific interactions with these enemies. Populations of natural enemies would also be affected by changes in climate. If they respond with a proportionately greater increase in population than the ECB, then they are likely to exert greater control. However, natural enemies may not move as rapidly as the ECB into new areas and thus ECB population levels and consequently pest

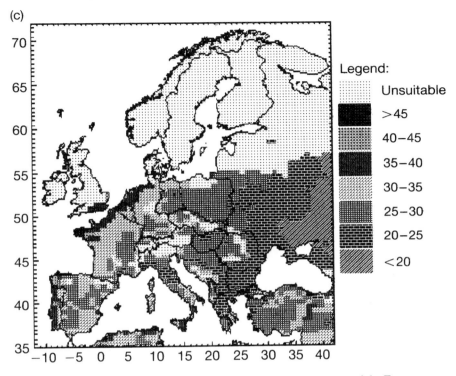

Fig. 5c. Difference in timing of peak flight activity of the first generation of the European corn borer under GISS scenario A for the 2050s (number of days earlier than baseline).

status may increase. For example, in the 1970s the level of grain maize cultivation in northern France increased as a result of the introduction of early maturing cultivars. Favourable climatic conditions and the absence of effective natural enemies enabled ECB populations to reach far higher levels there than they had in southern France resulting in high yield losses (Leclant, 1977). Changes in interspecific interactions could have a significant effect on the use of biological control (see below).

In areas where grain maize is already cultivated the greatest problems arising from the effects of climatic changes are likely to stem from the faster rates of development and, in some areas, the increased number of generations of the ECB developing each year. The effect of changes in the development rate of the ECB depends on how the timing of flight or, more importantly, oviposition and larval development following flight, relate to the development of the crop. Changes in flight activity would have serious implications for grain maize yields if flight activity, and

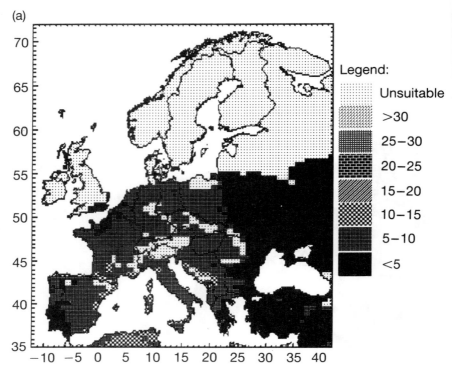

Fig. 6a. Difference in timing of peak flight activity of the first generation of the European corn borer under GISS scenario C for the 1990s (number of days earlier than baseline).

consequently larval development, were to occur when the crop was still in its early, more vulnerable, stages of development. With no change in the planting date of maize, climate changes would be likely to result in earlier infestation of the crop. However, earlier planting is likely to be one of the management responses to climate change (see above) and so earlier flight activity may not be so critical. Faster rates of development would also result in shorter development times. Comparison of Figs 1, 3 and 4 suggests that, as the limits for ECB distribution move northwards, there would be opportunities for additional generations to develop further south. Given the temperature changes projected by GISS scenario A, by the 2050s an additional generation of the ECB could occur in many regions where it is currently known to be present (cf. Figs 1 and 3(c)). This could have serious implications for grain maize production as additional generations are likely to result in higher population levels and may increase damage and crop loss.

5. Climate Change and the European Corn Borer

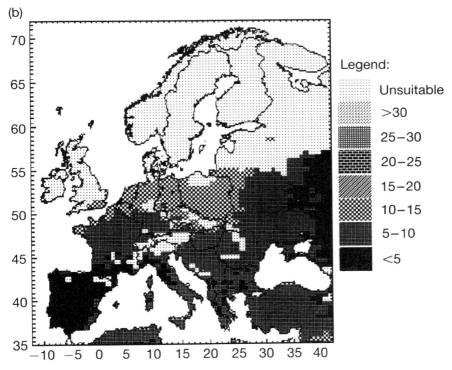

Fig. 6b. Difference in timing of peak flight activity of the first generation of the European corn borer under GISS scenario C for the 2020s (number of days earlier than baseline).

Many of the management problems likely to stem from the effects of changes in climate on insect pests will be associated with the use of control measures. An earlier start to ECB development and more generations produced per year is likely to lead to both earlier and increased use of control because control measures are generally used against each generation of the ECB (Bigler and Brunetti, 1986; Hudon et al., 1989). Furthermore, as a result of the direct effects of changes in climate on insecticides, additional changes in the frequency and level of applications may be required in order for effective control to be maintained. Under changed climatic conditions insecticide deposition (the amount and pattern of insecticide on a leaf surface) and persistence are likely to change. To illustrate, under the higher temperatures found in tropical regions the majority of even the most persistent insecticides may be lost from the target surface in a few days (Hill, 1987). In contrast, insecticides used against the ECB in France were estimated to last for approximately 20 days (Leclant, 1977). Furthermore, higher rainfall intensity tends to wash insecticide residues from leaf surfaces at a much greater rate than more gentle rainfall. There are suggestions that, due to the higher temperatures, rainfall in parts of Europe may increasingly be derived from heavy convective rainstorms resulting in increases in intense rainfall events (Mitchell et al., 1990). Such changes would potentially exacerbate the problem of insecticide loss. Additional applications of insecticides may therefore be required to counteract this short-lived effectiveness as well as to control additional generations of the ECB. However, the present knowledge of yield loss and insecticide spray thresholds used to calculate control requirements may be inadequate under a changed climate and it is not possible to extrapolate current models of insecticide efficacy with any certainty (Treharne, 1989). Nevertheless, increased use of control measures is likely to result in an increase in the cost of maize production particularly where more expensive chemical or biological control methods are used. An additional implication resulting from increased insecticide applications is that excessive use may encourage selection for ECB resistance to the effects of insecticides.

Biological control may become more or less effective in the future depending on the level of climate change and its effect on the natural enemies of the ECB. Biological control of the ECB using organisms such as the egg parasite *Trichogramma maidis* has proved successful in a number of countries in Europe (e.g. Bigler and Brunetti, 1986; Robert, 1987). As with chemical control, timing of application is critical and it is recommended that the parasites be released at the time of adult flight to coincide with egg laying and the start of each new generation. Changes in climate will have differential effects on the population dynamics of the ECB and the predators, parasites and pathogens used to control it. However, at present, it is unclear how changes in climate would affect the interaction between the ECB and its natural enemies. Climate changes that would be favourable for the ECB may not, of course, be favourable for its natural enemies and could therefore alter their effectiveness as control agents.

F. Refining the Approach

Examination of the implications of climate change for the ECB using an approach such as that outlined in this chapter, particularly with respect to changes in flight activity, is open to criticism on the grounds of giving too much precision to impacts at the local level, especially given the uncertainties surrounding GCM-derived projections of climate change. Furthermore, this study assumes that temperature is the most important climatic variable governing ECB development and hence distribution. However, a number of studies have shown that other climatic factors such as rainfall, windspeed and relative humidity can have a significant effect on ECB survival, behaviour and the timing of activities such as emergence and flight (e.g. Sappington and Showers, 1983; Beck, 1989). Furthermore, it has been suggested that climate change may disrupt the diapause strategies of many insects as the linkages between temperature or moisture regimes and day-length are altered (Sutherst, 1991). The effect of day-length on ECB development and its influence on diapause initiation are not included in this assessment, principally because the temperature–photoperiod interactions involved are complex and not well understood. However, some discussion of the influence of diapause on the possible expansion of ECB distribution has been made by Porter (1991). Despite their absence in this study, it is recognized that factors such as diapause and the influence of other climatic variables should be taken into account when assessing the impacts of climate change for the ECB. Ultimately, use of an integrated simulation model comprising an ECB development submodel (see, e.g., Loewer *et al.*, 1973) and a maize development submodel would enable a more detailed examination of the possible impacts of climate change on both the ECB and maize. Such information could then be used to begin assessing possible changes in the profitability of maize cultivation and the economic effects of climate change for maize growers. However, such a model would require a detailed understanding of the relationships between the ECB and grain maize and other external factors influencing the interactions between them. Although there have been many studies on ECB biology since it was first identified as an important pest, very little is known about its population dynamics or its dispersal and oviposition strategies. Information about both population dynamics and the mechanisms of ECB dispersal under present climatic conditions is required in order to begin a full assessment of how the ECB may respond to future changes in climate. In addition, such an assessment would require future estimates of regional or even local scale changes in a number of climatic variables. Currently such information is available to only a limited degree. Nevertheless, the potential consequences of climate changes for the ECB, particularly its possible impact on grain maize cultivation, are considerable and so it is important to begin assessing its response to climate change.

Acknowledgements

This work was originally conducted at the University of Birmingham and was supported, in part, by the Commission of European Communities (Contract No. EY 4C.0017 UK). I thank T. R. Carter (Finnish Meteorological Institute) and M. L. Parry (University College London) for their contribution to this work, F. Bigler (Swiss Federal Research Station for Agronomy, Zurich) for providing information about the ECB and P. Burns and S. Harding (Sydney) for their understanding regarding the production of this chapter.

REFERENCES

AliNazee, M. T. (1976). Thermal unit requirements for determining adult emergence of the western cherry fruit fly (*Diptera: Tephritidae*) in the Williamette Valley of Oregon. *Environ. Entomol.* **5**, 397–402.

Anglade, P. (1975). Corn pest management in Western Europe exemplified by French systems. *In* "Report on the International Project on *Ostrinia nubilalis*, Phase II Results" (B. Donlika, ed.), pp. 30–33. Centre of Information of Ministry of Agriculture and Food, Budapest, Hungary.

Bach, W. (1988). Development of climatic scenarios: A. From general circulation models. *In* "The Impact of Climatic Variations on Agriculture. Vol. 1: Assessments in Cool Temperate and Cold Regions" (M. L. Parry, T. R. Carter and N. T. Konijn, eds), pp. 125–157. Kluwer, Dordrecht.

Baker, C. R. B. (1980). Some problems in using meteorological data to forecast the timing of insect life cycles. *Bull. OEPP/EPPO Bull.* **10**, 83–91.

Beck, S. D. (1989). Developmental and seasonal biology of *Ostrinia nubilalis*. *In* "Biology and Population Dynamics of Invertebrate Crop Pests" (G. E. Russell, ed.), pp. 45–82. Intercept, Dover.

Beck, S. D. and Apple, J. W. (1961). Effects of temperature and photoperiod on voltinism of geographic populations of the European corn borer, *Pyrausta nubilalis*. *J. Econ. Entomol.* **54**, 550–558.

Beirne, B. P. (1952). "British Pyralid and Plume Moths". F. Warne, London.

Benedict, J. H. and Hatfield, J. L. (1988). Influence of temperature-induced stress on host plant suitability to insects. *In* "Plant Stress–Insect Interactions" (E. A. Heinrichs, ed.), pp. 139–166. Wiley, New York.

Bigler, F. and Brunetti, R. (1986). Biological control of *Ostrinia nubilalis* Hbn. by *Trichogramma maidis* Pint. et Voeg. on corn for seed production in southern Switzerland. *J. Appl. Entomol.* **102**, 303–308.

Bignon, J. (1990). "Agrometeorologie et Physiologie du Maïs Grain dans la Communauté Européenne". Commission des Communautés Européenes, Luxembourg. (In French and English.)

Borg, Å (1949). The overwintering of *Pyrausta nubilalis* in Skane. *Ent. Tidskr.* **70** (4), 270–271.

Butts, R. A. and McEwen, F. L. (1981) Seasonal populations of the diamondback moth: *Plutella xylostella* (*Lepidoptera: Plutellidae*) in relation to day-degree accumulation. *Canad. Entomol.* **113**, 127–131.

CAB (1971). "Distribution Maps of Insect Pests, Series A (Agricultural), No. 11 (revised)". Commonwealth Institute of Entomology, London.

Carter, D. J. (1984). "Pest Lepidoptera of Europe with Special Reference to the British Isles". Junk, Dordrecht.

Carter, T. R., Parry, M. L. and Porter, J. H. (1991a). Climatic change and future agroclimatic potential in Europe. *Int. J. Clim.* **11**, 251–269.

Carter, T. R., Porter, J. H. and Parry, M. L. (1991b). Climatic warming and crop potential in Europe: prospects and uncertainties. *Global Environmental Change* **1**, 291–312.

Chapman, L. J. and Brown, D. M. (1966). The climates of Canada for agriculture. "The Canadian Land Inventory Report No. 3". Canada Dept of Forestry and Rural Development, Ottawa.

Cordillot, F. P. (1989). "Dispersal Flight and Oviposition Strategies of the European Corn Borer, *Ostrinia nubilalis* Hbn., (Lepidoptera: Pyralidae)" PhD thesis, Basel University, Switzerland.

Crawford, J. W., Duncan, J. M., Ellis, R. P., Griffiths, B. S., Hillman, J. R., MacKerron, D. K. L., Marshall, B., Ritz, K., Robinson, D., Wheatley, R. E., Woodford, J. A. T. and Young, I. M. (1989). "Global Warming. The Implications for Agriculture and Priorities for Research". Scottish Crop Research Institute, Invergowrie.

Despins, J. L. and Roberts, J. E. (1984). Phenology of adult European corn borer in Virginia. *J. Econ. Entomol.* **77**, 588–590.

Donlika, B. (1975). Pest control on the Danube Plain illustrated by Hungarian examples. *In* "Report on the International Project on *Ostrinia nubilalis*, Phase II Results" (B. Donlika, ed.), pp. 35–41. Centre of Information of Ministry of Agriculture and Food, Budapest, Hungary.

EPA (1989). "The Potential Effects of Global Climate Change on the United States, Vol. 2: National Studies". Review of the Report to Congress, US Environmental Protection Agency, Washington.

Ferro, D. N. (1987). Insect pest outbreaks in agroecosystems. *In* "Insect Outbreaks" (P. Barbosa and J. C. Schultz, eds), pp. 195–216. Academic Press, San Diego.

Hansen, J., Fung, I., Lacis, A., Rind, D., Lebedeff, S., Ruedy, R., Russell, G. and Stone, P. (1988). Global climate changes as forecast by the GISS 3-D model. *J. Geophys. Res.*, **93**, 9341–9364.

Heinrichs, E. A. (1988). Global food production and plant stress. *In* "Plant Stress–Insect Interactions" (E. A. Heinrichs, ed.), pp. 1–34. Wiley, New York.

Hill, D. S. (1987). "Agricultural Insect Pests of Temperate Regions and their Control". Cambridge University Press, Cambridge.

Holtzer, T. O., Archer, T. L. and Norman, J. M. (1988). Host plant suitability in relation to water stress. *In* "Plant Stress–Insect Interactions" (E. A. Heinrichs, ed.), pp. 111–138. Wiley, New York.

Hough, M. N. (1978). Mapping areas of Britain suitable for maize on the basis of Ontario Units. *ADAS Q. Rev.*, **31**, 217–221.

Houghton, J. T., Jenkins, G. J. and Ephraums, J. J. (eds) (1990). "Climate Change: The IPCC Scientific Assessment". Cambridge University Press, Cambridge.

Hudon, M., LeRoux, E. J. and Harcourt, D. G. (1989). Seventy years of European corn borer (*Ostrinia nubilalis*) in North America. *In* "Biology and Population Dynamics of Invertebrate Crop Pests" (G. E. Russell, ed.), pp. 1–44. Intercept, Dover.

IPCC (1992). "Climate Change: The IPCC 1990 and 1992 Assessments. IPCC First Assessment Report Overview and Policymakers Summaries and 1992 IPCC Supplement". World Meteorological Organization and United Nations Environment Program, Geneva and Nairobi.

Kenny, G. J. and Harrison, P. A. (1992). Thermal and moisture limits of grain maize in Europe: model testing and sensitivity to climate change. *Clim. Res.* **2**, 113–129.

Kindler, S. D. and Springer, T. L. (1989). Alternate hosts of Russian wheat aphid (*Homoptera: Aphididae*). *J. Econ. Entomol.* **82**, 1358–1362.

Leclant, F. (1977). Pest control methods for maize in France. *Ann. Appl. Biol.* **87**, 270–276.

Lincoln, D. E., Couvet, D. and Sionit, N. (1986). Response of an insect herbivore to host plants grown in carbon dioxide enriched atmospheres. *Oecologia* **69**, 556–560.

Loewer, O. J., Jr, Huber, R. T., Barrett, J. R. Jr, and Peart, R. M. (1973). Simulation of the effects of weather on an insect population. *Simulation* **22**, 113–118.

Matteson, J. W. and Decker, G. C. (1965). Development of the European corn borer at controlled and variable temperatures. *J. Econ. Entomol.* **58**, 344–349.

Mitchell, J. F. B., Manabe, S., Tokioka, T. and Meleshko, V. (1990). Equilibrium climate change. *In* "Climate Change: The IPCC Scientific Assessment" (J. T. Houghton, G. J. Jenkins and J. J. Ephraums, eds), pp. 131–172. Cambridge University Press, Cambridge.

Neiswander, C. R. (1962). An adventure in adaptation: the European corn borer, *Ostrinia nubilalis* (Hubn). *Ohio Agric. Expt. Stn Res. Bull.*, **916**, 35 pp.

Nicolas, G. and Sillans, D. (1989). Immediate and latent effects of carbon dioxide on insects. *Ann. Rev. Ent.* **34**, 97–116.

Ohnesorge, B. and Reh, P. (1987). Untersuchungen zur populationsdynamik des maizsunsters *Ostrinia nubilalis* Hbn (*Lepidoptera: Pyralidae*) in Baden-Wurttemberg I. populationsstruktur, apparenz verteilung im habitat. *J. Appl. Ent.* **103**, 288–304 [in German with English summary].

Parry, M. L. (1990). "Climate Change and World Agriculture". Earthscan, London.

Parry, M. L., Carter, T. R. and Porter, J. H. (1989). The greenhouse effect and the future of UK agriculture. *J. R. Agric. Soc.* **150**, 120–131.

Parry, M. L. with Duinker, P. N., Morison, J. I. L., Porter, J. H., Reilley J. and Wright, L. J. (1990). The Potential Impacts of Climate Change: Impacts on Agriculture and Forestry. *In* "Climate Change, The IPCC Impacts Assessment" (W. J. McG. Tegart, G. W. Sheldon and D. C. Griffiths, eds), pp. 2.i–2.45. Australian Government Publishing Service, Canberra.

Pimental, D., Brown, N., Vecchio, F., La Capra, V., Hausman, S., Lee, O., Diaz, A., Williams, J., Cooper, S. and Newburger, E. (1992). Ethical issues concerning potential global climate change on food production. *Journal of Agricultural and Environ. Eth.* **5**, 113–146.

Porter, J. H. (1991). "Some Implications of Climatic Change for the European Corn Borer in Europe". PhD thesis, University of Birmingham.

Porter, J. H., Parry, M. L. and Carter, T. R. (1991). The potential effects of climatic change on agricultural insect pests. *Agric. For. Met.* **57**, 221–240.

Rhoades, D. F. (1985). Offensive-defense interactions between herbivores and plants: their relevance in herbivore population dynamics and ecological theory. *Am. Nat.* **125**, 205–238.

Robert, Y. (1987). Forecasting annual crop pests: advantages and limitations. *In* "Rational Pesticide Use" (K. J. Brent and R. K. Atkin, eds) Proceedings of the Ninth Long Ashton Symposium, pp. 269–283. Cambridge Univeristy Press, Cambridge.

Rowntree, P. R., Callender, B. A. and Cochrane, J. (1989). Modelling climate change and some potential effects on agriculture in the UK. *J. R. Agric. Soc.* **150**, 153–170.

Sappington, T. W. and Showers, W. B. (1983). Effects of precipitation and wind on populations of adult European corn borers (*Lepidoptera: Pyralidae*). *Environ. Entomol.* **12**, 1193–1196.

Scriber, J. M. and Hainzee, J. H. (1987). Geographic invasion and abundance as facilitated by differential host plant utilization abilities. *In* "Insect Outbreaks" (P. Barbosa and J. C. Schultz, eds), pp. 469–504. Academic Press, San Diego.

Stengal, M. (1982). Essai de mise au point de la prevision des degats pour la lutte contre la pyrale du mais (*Ostrinia nubilalis*) en Alsace (Est de la France). *Entomophaga*, **27**, 105–114 (in French).

Sutherst, R. W. (1991). Pest risk analysis and the greenhouse effect. *Rev. Agric. Entomol.* **79**, 1177–1187.

Treharne, K. (1989). The implications of the 'greenhouse effect' for fertilisers and agrochemicals. *In* "The Greenhouse Effect and UK Agriculture" (R. M. Bennett, ed.), CAS Paper 19, 67–78. Centre for Agricultural Strategy, Reading.

Waring, G. L. and Cobb, N. S. (1992). The impact of plant stress on herbivore population dynamics. *In* "Insect–Plant Interactions Vol. IV" (E. Bernays, ed.), pp. 167–240. CRC Press, Boca Raton, FL.

White, T. C. R. (1984). The abundance of invertebrate herbivores in relation to the availability of nitrogen in stressed food plants. *Oecologia* **63**, 90–105.

Yoshino, M. M., Horie, T., Seino, H., Tsujii, H., Uchijima, T. and Uchijima, Z. (1988). The effects of climatic variations on agriculture in Japan. *In* "The Impact of Climatic Variations on Agriculture. Vol. 1: Assessments in Cool Temperate and Cold Regions" (M. L. Parry, T. R. Carter and N. T. Konijn, eds), pp. 725–863. Kluwer, Dordrecht.

6

Aphids in a Changing Climate

R. HARRINGTON, J. S. BALE AND G. M. TATCHELL

I. Introduction	126
II. Essential Aphid Biology	128
III. The Influence of Temperature on Individual Aphids	130
A. Development and Fecundity	130
B. Mortality	133
C. Sublethal Effects	134
D. Movement	135
E. Morph Determination	136
F. Implications of a Warmer Climate	137
IV. The Influence of Temperature on Aphid Populations	139
A. Life-cycle Strategies	139
B. Phenology	144
C. Abundance	149
V. Conclusions	150
References	150

Abstract

Aphids are major pests of agriculture, silviculture and horticulture and are hence important targets for studies of the implications of a changing environment. Their short generation times and immense capacity for increase give them the potential for rapid responses to such changes, but the complexity and variability in their life cycles make them more of a challenge than many other insects.

The Rothamsted Insect Survey has used suction traps since 1965 to monitor aphids, and these data have been analysed to detect trends in aphid phenology over time and in relation to weather, particularly temperature in winter. A brief description of this system is given.

Aspects of aphid biology that are influenced by temperature are described. Examples of optimum and threshold temperatures for development, fecundity, movement and morph determination are presented, and the lethal and sublethal effects of low temperatures discussed. This information provides the necessary background against which to consider the effects of changing climate on individual aphids and populations.

Aphids which are anholocyclic (remain parthenogenetic throughout the winter) in a given area fly earlier after mild winters and in greater numbers in the spring migration. For *Myzus persicae*, an increase in mean temperature in January and February of 1°C advances the timing of the spring migration by about 2 weeks. Holocyclic aphids that overwinter as diapausing eggs are more cold hardy than active stages, and such species are less affected by low winter temperatures. In *Rhopalosiphum padi*, a species that is capable of both strategies, the proportion of anholocyclic individuals in autumn is related to temperature in the previous winter.

Phenologies of two cereal aphid species, *Sitobion avenae* and *Metopolophium dirhodum*, follow a south–north trend: specific events occur earlier further south but with separate components each side of latitude 54°N. Below this latitude the trend is correlated with winter temperature. Phenological trends with time have been shown for five species: specific events occur earlier in more recent years.

The fundatrix of the high arctic aphid *Acyrthosiphum svalbardicum* produces sexual morphs and a few viviparae, which themselves produce entirely sexual morphs. However, these later sexual morphs can only mature and mate in exceptionally warm seasons. Manipulative experiments mimicking a warmer climate have shown that the species can produce 11 times as many eggs under such a situation.

In northern temperate zones, predicted increases in temperature are likely to exacerbate aphid damage, and there are many examples of serious aphid-borne virus problems in crops following mild winters. However, much remains to be done to understand how the effects of temperature on aphids interact with the influence of other climate variables and their effects on the aphids' competitors, natural enemies and host plants. The relatively simple arctic system may be a very good place to start.

I. INTRODUCTION

Current models of expected climate change have been summarized elsewhere in this volume (see Bennetts, Chapter 3). Whilst it is necessary to be cautious over the interpretation of these models, there is a need to assess their implications for potential effects on insects. Aphids are

6. Aphids in a Changing Climate

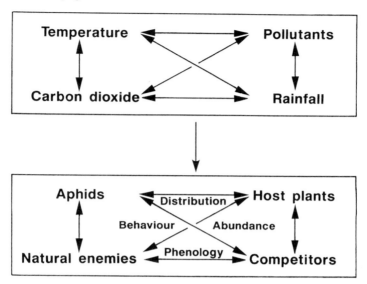

Fig. 1. Potential interactions affecting the influence of climate change on aphids.

important pests of agriculture, silviculture and horticulture, and are an obvious subject for such studies. They offer many unique opportunities for studying interactions with climate change, but also present several difficulties. Their short generation times and immense capacity for reproduction give them the potential for rapid and dramatic responses to change, enabling observation and experimentation on responses to short-term variation in weather to be extrapolated to the longer term effects of climate change. However, the complexity and variation in their life cycles make assessment of their possible responses more of a challenge than for insects with simpler life cycles.

This chapter begins with a brief review of features of aphid biology that are influenced by temperature. This is the only climate variable to be considered here in detail and particular attention is given to predicted changes over the next half century. Changes in other climatic variables and in carbon dioxide levels (see Watt *et al.*, Chapter 9), pollutants (see Brown, Chapter 10) and land use (see Luff and Woiwod, Chapter 16) may all have direct or indirect effects on aphid populations and are considered elsewhere in this volume. All are likely to interact in complex ways to confound predictions made by consideration of a single variable. Also, the effects of changing climatic and related variables on aphid host plants, natural enemies and competitors will influence effects on aphids themselves (Fig. 1). Such is the complexity of the interactions that it could be a long time before they are understood fully and modelled in anything other than an empirical fashion. At present, there is still much to be learnt about the effects of single components of the environment on individual taxa and it is these simple interactions that are considered here, firstly by looking at the effects of temperature on individual aphids and then by considering how these effects may translate into changes in the population dynamics of aphids and the consequences for selected applied problems.

Discussion is confined largely to the northern temperate zone for which most data are available, and to the subfamily Aphidinae (classification according to Heie, 1980) which includes a large proportion of the pest species in this region. The most significant temperature changes are expected to occur in winter (see Bennetts, Chapter 3), and therefore discussion is focused mainly on this season and on the potential effects of the temperature changes predicted by the Hadley Centre model.

II. ESSENTIAL APHID BIOLOGY

Eastop (1977) lists 3742 species of aphid worldwide. They have become adapted to a wide range of climatic conditions and occur in each of the major zoogeographic regions, although the greatest number of species is found in the northern temperate zone (Dixon, 1987; Dixon et al., 1987). They have evolved a variety of life-cycle strategies with the common feature of parthenogenetic viviparity for at least part of the year (Fig. 2). Embryogenesis can be initiated in the grandparent and very shortly after the parent becomes adult it can begin to reproduce. Generation times can be as little as a week under optimal conditions and within a week of the onset of reproduction around 50 offspring may be born. The capacity for rapid population increase is therefore immense, contributing to the pest status of many aphids. Short generation times enable a rapid response to environmental conditions: for example, by the production of apterous (wingless) or alate (winged) forms, depending on the nutritional status of the host plant or the degree of crowding of aphids (Kawada, 1987).

Most aphids have an annual sexual phase which is usually initiated by a threshold night length that ensures that the oviparae (sexual females) will normally reach maturity before leaf fall or die back of the host plant on which the egg is produced. The alternation of a period of parthenogenetic reproduction with a sexual phase is a primitive feature of aphids, and those that do this are termed holocyclic. No species is known to have lost secondarily the parthenogenetic phase of the life cycle and only a few

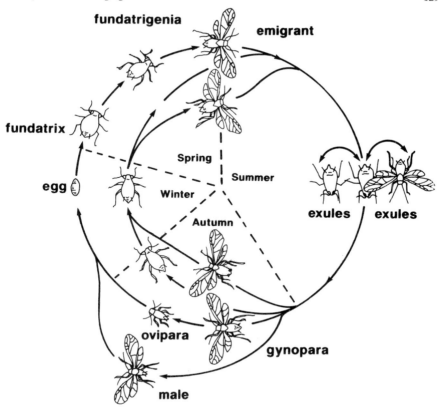

Fig. 2. Typical life-cycle options of an aphid with holocylic heteroecious clones (outer ring) as well as anholocyclic clones (inner ring). Adapted from Dixon (1973).

have lost entirely the sexual phase to become exclusively anholocyclic (Blackman and Eastop, 1984). Many species, particularly amongst the agricultural pests, have some anholocyclic clones which continue to reproduce parthenogenetically throughout the year. Some (androcyclic) clones produce males and parthenogenetic females but no oviparae, and hence contribute to the sexual phase but also continue in the asexual phase through the winter (Blackman, 1971, 1972; Simon *et al.*, 1991). Any given clone will always respond in the same way to a given set of environmental conditions, although different clones within a species may respond differently or at different intensities of a given cue, allowing regional adaptation and local variation in morph determination (MacKay, 1989; Mittler and Wilhoit, 1990).

There are two distinct strategies amongst holocyclic aphids, but many variations on each (Lampel, 1968). Mating and egg production usually occur on a limited range of closely related woody angiosperms, with which the species has primitive associations (Moran, 1992). Such plant species are termed the primary host. Many species produce all of their parthenogenetic generations on the primary host and are termed monoecious. Other species, especially amongst the Aphidinae, migrate in early summer to a completely different group of host plants. These (usually) herbaceous plants are the secondary host plants and include many agricultural crops. Aphids that alternate between a primary and a secondary host plant are termed heteroecious. In autumn, alatae (termed gynoparae) of heteroecious species return to the primary host and produce the oviparae. Alate males are usually produced on the secondary host, by the same parents that produce the gynoparae, and return to the primary host to mate. Some species have lost their association with the primary host and are monoecious on what was the secondary host. In such cases the males need not be alate and are often produced by the same parents that produce the oviparae.

III. THE INFLUENCE OF TEMPERATURE ON INDIVIDUAL APHIDS

Temperature is undoubtedly the main abiotic factor controlling aphid development, fecundity, movement and, to some extent, morph determination. In this section, the influence of temperature on these features is considered in relation to individual aphids to provide a basis for later discussion of effects on populations. Temperature responses differ between species, between clones within a species, between morphs within a clone and, for any of these groupings, between regions. Here some specific examples of temperatures that elicit given responses are illustrated and the consequences of climate change considered.

A. Development and Fecundity

Development does not occur below a lower threshold temperature. Studies on 16 aphid species are summarized by Honek and Kocourek (1990). In most cases, thresholds were calculated by extrapolation from development rates at higher temperatures, and the extrapolations were not all done in the same way, making comparisons difficult. The lowest developmental threshold estimated was 0.5°C for *Drepanosiphum platanoidis* (Schrank) (Wellings, 1981), although Zhou et al. (1989) and Carter *et al.* (1982) give −2.2°C and −3.6°C for *Metopolophium dirhodum* (Walker) and *Sitobion avenae* (Fabricius) respectively. The highest threshold recorded was 7.1°C for *Brevicoryne brassicae* (L.) (Campbell *et al.*, 1974). Within *Acyrthosiphon pisum* (Harris), reported thresholds range between 2.3 and 6.3°C (Kilian and Nielson, 1971; Campbell and Mackauer, 1975; Lamb *et al.*, 1987).

Upper developmental limits of 25°C for *Rhopalosiphum padi* (L.) (Elliott and Kieckhefer, 1989) and *S. avenae* (Kieckhefer *et al.*, 1989) and between 25 and 30°C for *Myzus persicae* (Sulzer) and *Macrosiphum euphorbiae* (Thomas) (Barlow, 1962) have been reported.

Most studies of aphid developmental parameters have been carried out at constant temperatures. However, differences occur between developmental rates at fluctuating temperatures and at constant temperatures representing the mean of the fluctuations (Liu and Meng, 1989; Michels and Behle, 1989). The differences are particularly noticeable at temperatures around the developmental threshold (Fig. 3). For example, at a mean temperature precisely equal to the lower developmental threshold, no development will occur. However, at a temperature which fluctuates around this mean, development will occur during periods at which the temperature is above the threshold. The converse is true at the upper limit of development. Under winter field conditions, the mean temperature at which no development occurs will be well below the lower developmental threshold. At Rothamsted, for example, when considering all dates over the 10 years from 1981 to 1990 on which mean screen temperature fell below 0°C, the maximum screen temperature was on average 3°C higher.

Most studies have shown aphid development rates and fecundity to be greatest at between 20 and 25°C (e.g. Kenten, 1955; Barlow, 1962; Dean, 1974; Liu and Hughes, 1987). When and wherever current maximum temperatures do not reach these levels, the direct effect of a warmer climate on individual aphids will be to accelerate development and fecundity. When and where the maximum currently exceeds the temperature optimum, the direct effect of warmer conditions will depend on the balance between the detrimental effect of supra-optimal temperatures and beneficial effect of increased suboptimal temperatures. In the UK, periods with temperatures greater than 20°C are usually short, and a warmer climate would be expected to raise development rates and fecundity.

The number of day-degrees above the lower developmental threshold required to complete development from birth to the onset of reproduction ranges from 93 for *Schizaphis graminum* (Rondani) (Walgenbach *et al.*, 1988) to 250 for *D. platanoidis* and *Drepanosiphum acerinum*

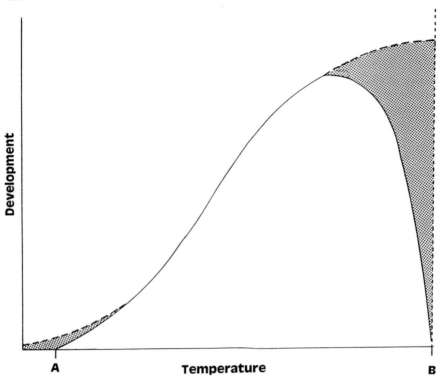

Fig. 3. Model depicting relationships between temperature and rate of development under constant (solid line) and fluctuating (dotted line) temperatures. A, lower developmental threshold at constant temperature; B, upper developmental threshold at constant temperature. Shaded areas represent potential extra development at fluctuating compared to constant temperatures.

(Walker) (Wellings, 1981) and is dependent on nutritional and other environmental conditions, as well as temperature. For *A. pisum* different studies have shown development times to range from 99 to 147 day-degrees (Honek and Kocourek, 1990). Stinner *et al.* (1974) suggested that a sigmoid function relating temperature to development is more accurate than the linear day-degree relationship. At Rothamsted, a mean of 2142 day-degrees above a threshold of 4°C was accumulated each year from 1966 to 1990 (range 1982–2456 day-degrees above 4°C). Assuming a requirement of 120 day-degrees per generation, this would allow 18 generations per year in an anholocyclic aphid. A warming of an average of 2°C would increase this to 23 generations per year.

B. Mortality

It is generally recognized that aphid eggs are more cold hardy than the active forms. There have been few direct assessments of the cold hardiness of aphid eggs and such studies have been based mainly on measurements of supercooling points (i.e. the temperature at which the egg freezes). This temperature is very consistent in over-wintering aphid eggs at −35 to −40°C across a range of species (Somme, 1964, 1969; Parry, 1979; James and Luff, 1982). Whilst there is increasing evidence that supercooling points are not good indicators of cold hardiness in aphids because the over-wintering stages die before they freeze (Bale, 1987, 1991), it is clear from the distribution of aphids in relation to winter temperatures that their eggs are able to survive at temperatures close to the limit of their supercooling.

There have been more extensive studies on the cold tolerance of active stages of anholocyclic clones. Low supercooling points (−23 to −21°C) were recorded for adults and nymphs of anholocyclic clones of *M. persicae* (O'Doherty and Bale, 1985) and *S. avenae* (Knight and Bale, 1986). Comparison of these laboratory-determined supercooling points with temperatures at which marked decreases were observed in overwintering populations of *M. persicae* (Harrington and Cheng, 1984) and *S. avenae* (Knight et al., 1986) suggested that the aphids were killed by exposure to subzero temperatures as much as 10–15°C above their freezing temperature. This was borne out in subsequent laboratory studies (Bale et al., 1988) and confirmed in winter field experiments (Clough et al., 1990). Work in Canada on *Diuraphis noxia* (Mordvilko) has found a similar pattern of results with supercooling points of the active stages between −24.9 and −26.8°C but a steady decline in winter field populations at temperatures between 0 and −10°C (Butts, 1992).

Because of the large discrepancy between the freezing temperature and lethal low temperature in the active stages, Bale (1987) has recommended the use of the LT_{50} (where T = temperature) as a means of comparing levels of cold hardiness between aphid species, age groups and populations subject to different environmental conditions. Using this approach, Clough et al. (1990) found that in *M. persicae* first-born nymphs were more cold hardy than nymphs born later in the reproductive sequence, nymphs were consistently more cold hardy than adults and nymphs and adults were progressively more cold hardy when reared at lower temperatures. Furthermore there was no significant difference in the LT_{50} of aphids cooled in isolation or in feeding contact with their host plants, indicating that the mortality arises from a direct effect of low temperature on the aphid rather than through any plant-mediated process.

There was a close agreement between the LT_{50} values calculated for *M. persicae* in the laboratory and the temperatures at which 50% of nymphs and adults were killed in a winter field experiment (Clough et al., 1990). However, it is important to bear in mind that the number of individuals killed at any potentially lethal low temperature will depend on how long the insects have been exposed to that temperature (Bale, 1987). The determination of aphid LT_{50} values is usually based on very brief exposures (1 min) at a series of decreasing temperatures, thus reflecting the lethal effect of temperature but not the interaction with time. Recent data suggest that differences in cold tolerance between aphid species and clones may be more apparent using an LT_{50} approach where T = time. As an example, first instar nymphs from clones of *M. persicae* and *R. padi* reared at 20°C had an identical LT_{50} (temperature) of −15.3°C, but when maintained at a constant 0°C the LT_{50} (time) values were 107.3 and 41.5 h respectively (Howling, Bale and Harrington, unpublished data). Given that the cold hardiness of anholocyclic aphids is relatively limited compared to most other overwintering insects (exposures of 3–6 h at −15°C kill 100% of the population) and that temperatures between −10 and −15°C occur infrequently in UK winters (and in some winters not at all), it seems that the lethal effect of low temperature (and species and clonal differences in cold tolerance) may be more apparent after longer exposures at less severe temperatures (0 to −10°C) than with brief exposures at lower temperatures.

C. Sublethal Effects

By definition, the LT_{50} describes the temperature for a fixed exposure period or time at a constant exposure temperature at which 50% of survivors are thereafter capable of development and reproduction. Criteria for assessing survival after cold exposure range from the ability to move appendages (legs and antennae) to co-ordinated walking behaviour, but rarely include assessments of growth, development and reproduction (Bale, 1987; Leather et al., 1993). Intuitively, it is perhaps unwise to assume that a temperature or exposure period that kills 50% of a sample population should have no effects on the 50% of survivors.

In a series of experiments in which first instar or adult *R. padi* reared at 20°C were exposed to −5 and −7.5°C for 1 and 6 h, the sublethal effects of cold exposure were assessed in the surviving aphids. Data presented in Table 1 comparing the −7.5°C for 6 h treatment with a control population indicate the profound effects of this sublethal damage (Hutchinson and Bale, 1994). In all assessments made (rate of development, mean daily

Table 1. Effects of sublethal exposure to low temperatures on first instar nymphs and adults of *Rhopalosiphum padi*

	Aphids cooled to −7.5°C for 6 h			
Assessment	First instar	Adult	Control	
Development time to adult (days)	10.3 ± 0.14	—	6.6 ± 0.14	$P<0.001$
Mean daily reproduction	4.22 ± 0.26	5.87 ± 0.19	7.58 ± 0.14	$P<0.001$
Reproductive life (days)	7.91 ± 0.89	5.58 ± 0.45	11.9 ± 0.30	$P<0.001$
Total fecundity	32.30 ± 3.41	31.88 ± 2.40	89.36 ± 1.28	$P<0.001$
Adult longevity	8.04 ± 0.91	5.58 ± 0.45	17.44 ± 1.14	$P<0.001$

reproduction, length of reproductive life, total fecundity and adult longevity), the sublethal effects increased from exposure at −5 to −7.5°C and from 1 to 6 h. Also, 50% of nymphs born on the first day of the reproductive life of aphids cooled as first instars or newly moulted adults to −7.5°C for 6 h died within 3 days of birth. A similar pattern of results has been obtained with *S. avenae* (Parish and Bale, 1993) in which development, reproduction and longevity were all reduced after sublethal exposure of nymphs and adults to subzero temperatures (−5 and −10°C for 1 and 6 h), although prior acclimation at 10°C reduced the scale of these effects. With *S. avenae* all nymphs born to adults surviving exposure to −5°C for 6 h died within 48 h of birth.

These two studies provide clear evidence that cold stress arising from only a single exposure at temperatures that would occur in most UK winters can have a marked effect on the key processes that influence aphid population dynamics – development, reproduction and longevity.

D. Movement

1. Flight

Aphids will not take off or fly at temperatures below a critical level. Dry and Taylor (1970) studied six species and found none to take off at temperatures below 14°C. The authors were surprised at the high threshold temperatures required and suspected that rearing conditions might have influenced their results. However, the figures agree closely with those of Walters and Dixon (1984) who noticed the absence of *S. avenae* in suction trap samples on days with maximum temperatures below 11°C, with temperatures of 16°C being needed to trap this species on 50% of the days considered. Wiktelius (1981), using a similar method, described seasonally variable flight thresholds for *R. padi* that were highest in spring (16–17°C), lowest in autumn (9–10°C) and intermediate in summer (13–14°C), and proposed that these thresholds were associated with different seasonal forms. Robert and Rouzé-Jouan (1976a) found that the temperature threshold for flight of *R. padi* varied not only with season but also with location, with higher thresholds further south in France, suggesting adaptation to local conditions.

Many factors complicate the interpretation of flight thresholds. For example, most species take off at certain times of day and not at other times, even if the temperature is suitable. If temperature is suitable for flight, other conditions, such as wind speed and light intensity, may prevent take off (see review by Robert, 1987a). Aphids do not necessarily fly even if all conditions for flight are suitable: they fly to locate suitable sites for feeding and reproduction, and once located may settle irrespective of conditions.

2. Walking

During winter, anholocyclic aphids may need to move to locate new feeding sites as existing ones deteriorate. For example, *M. persicae* feeding on *Brassica oleracea* L. prefers senescing to mature or young leaves. As the winter progresses, older leaves die and fall from the plant. If movement is prevented for extended periods, aphids become trapped on fallen leaves, reducing their chance of locating a new food source (Harrington and Cheng, 1984; Harrington and Taylor, 1990). Little information is available on thresholds for movement of apterous aphids, but Smith (1981) found that *S. avenae* was capable of movement at 3°C. However, at −1°C first instars could not move, and at −4°C all aphids were immobile. Aitchison (1978) found some aphids in Manitoba to be active at temperatures as low as −5.5°C.

E. Morph Determination

Environmental and plant conditions interact with genotype to determine whether aphids are alate or apterous and whether they remain parthenogenetic or switch to the production of sexual morphs. The exposure of first instar *S. avenae* nymphs to −5°C for as little as 1 hour causes a significant reduction in the proportion of aphids that become alate (Parish and Bale, 1990). Suppression of alatae in winter might be advantageous as low temperature would inhibit flight and apterae are more fecund. At low temperatures, flight is not possible and any movement to new feeding sites would have to be by walking. High temperatures reduce alate production in *Aphis craccivora* Koch (Johnson, 1966), *Aphis fabae* Scop. (Tsitsipis and Mittler, 1976), *B. brassicae* (Lamb and White, 1966), *Megoura viciae* (Buckton) (Lees, 1959), *S. avenae* (Dedryver *et al.*, 1990) and *A. pisum* (Kenten, 1955). Whilst high temperatures may encourage movement, and thus contact between aphids, and increase crowding by accelerating development and reproduction, the expected increase in alatae does not occur above upper limiting temperatures that differ between 11 and 30°C for different species studied.

The critical night length for production of sexual forms increases with increasing temperature (see review by Kawada, 1987), hence high temperatures in autumn tend to delay the production of sexual forms (Fig. 4). Above 25°C the photoperiodic response can disappear completely so that no sexual morphs are produced, even in long nights (Lees, 1959). In contrast, the sexual morphs of *Myzocallis kuricola* (Matsumura) are produced in response only to low temperature (Shibata, 1952), but such examples are rare.

F. Implications of a Warmer Climate

The short, temperature-dependent generation times of aphids make them particularly sensitive to warmer conditions. In temperate regions, where aphids are most frequently living at temperatures below their developmental optima, higher temperatures will increase potential developmental and reproductive rates.

In warmer winters, individuals from anholocyclic clones will spend less time below the developmental threshold, will experience less sublethal stress, will be inhibited less often from moving and are less likely to be killed directly by low temperature. It is possible that higher aphid survival will put more pressure on the host plants and that there will be stronger stimuli to trigger production of alatae. Threshold temperatures for flying would be reached earlier in the year and more often.

The suppression of alate production by low temperature may be less marked, leading to a greater number of alatae available to fly in spring. However, at high temperatures, a reduced tendency for wing formation could lead to a lowering of migration potential in mid summer at a time when it might be beneficial for aphids to escape from dense populations on hosts of poor nutritional quality. Sexual morph production may be delayed at higher autumn temperatures, benefiting holocyclic clones by an extended period of parthenogenetic development at times of low risk

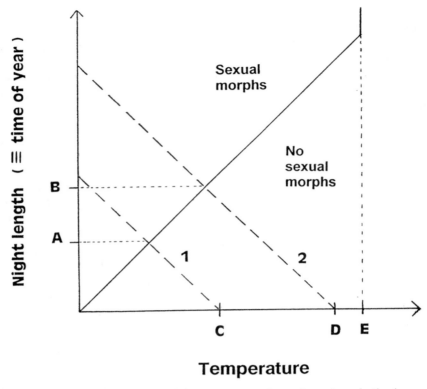

Fig. 4. Interaction between photoperiod, temperature and sexual morph production in a hypothetical aphid under current (dashed line 1) and warmer (dashed line 2) conditions. The solid line represents the relationship between critical night length and temperature; that is, sexual morphs are induced to the left of the line. If the temperature increases from C to D, the time at which sexual morph induction occurs will be delayed by the time it takes night length to change from A to B. E is the temperature above which sexual morph induction will be inhibited regardless of night length.

of mortality. However, this would be detrimental if there was no parallel delay in the time of leaf fall from deciduous primary hosts.

Thus, in general, the direct effects of a warmer climate on individual aphids in temperate regions are likely to be beneficial in terms of individual performance, with important consequences for population dynamics and their pest status. These consequences will now be considered in relation to life-cycle strategy options, seasonal phenology and abundance.

IV. THE INFLUENCE OF TEMPERATURE ON APHID POPULATIONS

A. Life-cycle strategies

Whether a species is monoecious or heteroecious has been determined over evolutionary time, and climate will not change these relationships over the few decades considered here. This section describes the effect of temperature on the proportion of holocyclic to anholocyclic individuals in species capable of both strategies. This ratio is likely to be very sensitive to changes in climate.

Two factors largely determine the ratio of holocyclic to anholocyclic individuals in such species: temperature and host plant distribution. Even the most cold-tolerant species show high mortality at temperatures around $-10°C$, and in areas where winter temperatures frequently fall to this level and below, anholocycly is not possible. In warmer areas, anholocycly occurs, but where the primary host is common or in species that are secondarily monoecious, holocycly may still persist (Fig. 5).

Life-cycle variation in *M. persicae* around the world depends on the interaction between climate and genotype (Blackman, 1974). The potential for sexual reproduction may be retained throughout the range of the

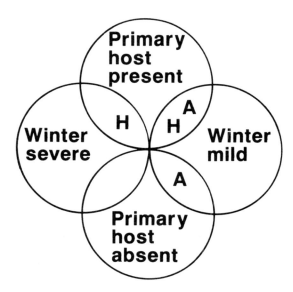

Fig. 5. Factors affecting balance of holocycly and anholocycly in heteroecious aphid species. H, holocycly possible; A, anholocycly possible.

species, but be suppressed in some areas due to the absence of photoperiodic cues necessary to trigger it, or to high temperatures which inhibit the photoperiodic response (Fig. 4).

In Britain, the bird cherry aphid *R. padi* is an important vector of barley yellow dwarf viruses (BYDV) in autumn-sown cereals. The epidemiology of this disease may alter as a consequence of the effect of climate change on the life cycle strategy of the vector. In autumn, three migrant forms may occur: gynoparae and males are in search of the primary host *Prunus padus* L. where the sexual cycle is completed, while exules (which are born on and remain on secondary host plants) from anholocyclic clones may colonize and infect winter sown cereal crops with BYDV. Data obtained by trapping alate female *R. padi* alive in suction traps in autumn and then giving them a choice of *P. padus* or barley on which to reproduce (Tatchell et al., 1988), have shown that the proportion of gynoparae to exules differs between years at Rothamsted. These differences are associated with the temperature the previous winter, fewer exules occurring following more severe winters. Furthermore, these preliminary data suggest that there is a threshold around 20 day-degrees accumulated below 0°C between December and February. Fewer day-degrees below zero than this threshold result in a significant proportion of individuals from anholocyclic clones in the following autumn (Fig. 6). This relationship probably occurs because severe winters kill the active forms from anholocyclic clones more readily than the relatively cold-hardy eggs from holocyclic clones. In samples collected from 1.5 m suction traps over three autumns (1984–1986), a greater proportion of exules was found (using the method of Plumb (1971)) at Starcross in south-west England than at Rothamsted in eastern England (Fig. 7). At Starcross, the mean temperature during the previous (1983–1985) December to February periods was 5.1°C (24 day-degrees below zero experienced) whereas at Rothamsted it was 2.8°C (71 day-degrees below zero experienced) indicating that the proportion of the population that is anholocyclic may differ between regions in association with winter temperature. There is little difference in the abundance of the primary host plant in the two areas. The difference of 2.3°C is similar to the increase expected at a single site by the year 2050, again indicating the potential effect of climate change on life-cycle strategy.

BYDV is currently a greater problem in autumn-sown than spring-sown cereals in Britain. The later in autumn the crop is sown, the fewer aphids will colonize it. In Britain, *S. avenae* is also an important vector. Autumn flights of *R. padi* are usually much greater than those of *S. avenae*, and *R. padi* is thought to be more important as a source of primary crop infection. *S. avenae* is more cold tolerant (Hand, 1980) and is thought to

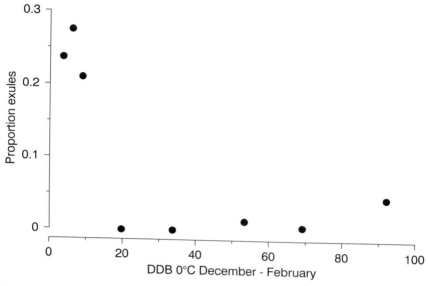

Fig. 6. Proportion of female *R. padi* that were from anholocyclic clones in samples from 12.2 m suction trap at Rothamsted in autumns (from 1st October) 1986–1993 in relation to temperature (day degrees below zero) the previous winter (December to February).

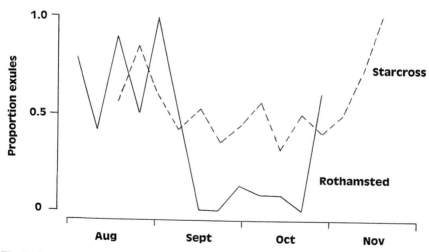

Fig. 7. Proportion of female *R. padi* that were from anholocyclic clones in samples from a 1.5 m suction trap in autumns 1984–1986 at Rothamsted (solid line) and Starcross (dotted line).

survive and move more readily in winter than *R. padi* and to play an important role in secondary spread of virus. Extension of the autumn flight period of anholocyclic clones in warmer conditions and greater movement and survival of vectors in winter will increase BYDV risk. There is also some evidence that warmer winters are likely to reduce the duration of the latent period after inoculation during which an infected plant is unable to act as a source of infection (Tatchell and Wang, unpublished data).

Gynoparae and males of *R. padi* can also transmit BYDV (Halbert *et al.*, 1992; Foster *et al.*, 1993; Tatchell and Taylor, unpublished data) and may play a role in primary infection if confined overnight or in bad weather on a cereal crop whilst moving from an infected source to the primary host. Delay in initiation of sexual morph production in warmer conditions may therefore extend the duration of the risk of primary BYDV infection by these aphids.

Extension of the range over which anholocyclic clones can survive will extend the area where the effects of the virus are of economic importance. However, virus incidence depends on the number of individual viruliferous aphids feeding and moving in the crop through winter, not on the proportion of individuals in the population that are anholocyclic. The relationship between winter temperature and the proportion of anholocyclic individuals in the following autumn is thus of interest when assessing the effect of climate change on life-cycle strategy, but it will only affect virus spread if it affects actual numbers entering the crop. This is more likely to depend on complex interactions involving weather throughout the summer which will influence aphids directly and also indirectly through effects on crop (and other grasses) growth (Robert and Rouzé-Jouan, 1976a; A'Brook, 1981) and natural enemy success. Competition between holocyclic and anholocyclic clones through the summer might also be important, and as yet nothing is known about these effects.

In spring-sown crops, there is much evidence for a link between winter temperature and virus incidence. For example, following the mild winter of 1988/89 many spring-sown cereals were damaged by BYDV. Mild winters in the mid-1970s were followed by serious virus infection in Scottish seed potato crops and sugar beet (Turl, 1980, 1983; Woodford *et al.*, 1983; Harrington *et al.*, 1989). Frequently, immature crops are more susceptible to infection than mature crops and the earlier infection occurs, the greater the economic loss (Doodson and Saunders, 1970; Smith and Hallsworth, 1990). As many species that are largely anholocyclic fly earlier after mild winters this can lead to an upsurge in virus incidence. Virus problems will also be exacerbated in vegetatively

6. Aphids in a Changing Climate 143

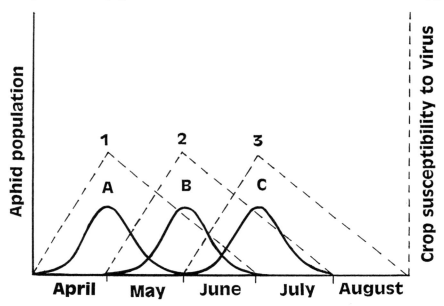

Fig. 8. Theoretical interaction of crop and aphid phenology in different climates. Curved lines represent the phenology of the spring migration of the aphid after A, mild; B, average; and C, cold winters. Dashed lines represent phenologies of crop susceptibility to virus after 1, early; 2, average; and 3, late planting dates.

produced crops because of the increasing reservoir of background infection with virus in seed stocks. The extent of increases in virus incidence in a warmer climate will depend partly on the extent to which crop phenologies change in relation to those of the aphid. For example, in the East Midlands and East Anglia, maincrop potatoes and sugar beet are normally planted in late April and emerge in mid May. Earliest plantings are in late March (emerging mid April) and latest plantings around late May (emerging mid June). Younger potato crops are more likely to acquire virus and subsequent translocation is more rapid (Beemster, 1972; Sigvald, 1985). The earliest that *M. persicae* has been recorded in suction traps in the area is late March. Early first records are usually followed by larger numbers in the spring migration (Harrington *et al.*, 1992a). Thus in years of early aphid flight, early-emerging crops are likely to be available to receive all but the very first aphids (Fig. 8). Crops emerging later, but at an average time, may emerge before the latter part of the spring migration, and populations of apterae may develop and spread virus. Late-emerging crops may escape the spring migration altogether. In years when aphids fly at an average time, early-sown crops may emerge in time for the start of the aphid migration while they are still susceptible to virus infection, but the peak of the migration may come too late to cause widespread infection. Crops with average sowing dates may be at most risk, and late sowings may suffer considerable secondary spread. In a late aphid year, late-emerging crops will be most at risk, and early-sown crops may escape aphid infestations altogether. In the case of viruses with many vector species, especially the non-persistently transmitted viruses, planning a planting strategy based on such interactions may be virtually impossible!

An interesting exception to the general finding that sexual morph production, where genetically possible, is environmentally controlled is found in the high arctic aphid *Acyrthosiphon svalbardicum* Heikinheimo, which feeds on mountain avens (*Dryas octopetala* L.) in a growing season with continuous daylight. In this species, the fundatrix gives birth directly to both sexual morphs, together with a small number of viviparae, and this occurs in experimental regimes regardless of photoperiod and temperature (Strathdee *et al.*, 1993a). The viviparae produce entirely sexual morphs which can only reach maturity and mate in warm years (Strathdee *et al.*, 1993b). In a field manipulation experiment that simulated a global warming with an average rise in temperature of 3°C, 11 times as many eggs were produced as in a control population at the natural temperature (Strathdee *et al.*, 1993b). The production of sexual forms in the generation after the fundatrix ensures the aphids' survival from year to year, but the potential extra generation makes the aphid well placed to take advantage of warmer conditions (Fig. 9).

B. Phenology

The annual pattern of morph production varies greatly between aphid species and between clones. In many holocyclic, heteroecious clones, there are three main periods of alate activity: one in spring, as aphids move from a primary to a secondary host plant, a second in summer as crowding and poor nutrition induce the production of alate exules, and a third in autumn as gynoparae and males return to a primary host (Taylor, 1985; Robert and Rouzé-Jouan, 1976b; Robert, 1987b). The long-term data from the Rothamsted Insect Survey suction trap network (Taylor, 1986; Tatchell, 1991) enable the occurrence of specific events in the life cycle, such as the time of peak spring migration, to be examined in relation to climatic data. Relationships between aphid abundance during specific periods and climatic variables can also be investigated.

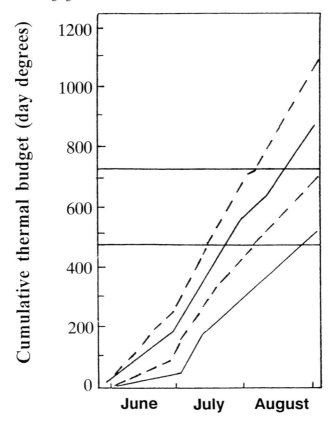

Fig. 9. Estimated ranges of accumulated day degrees C above zero that would have been experienced by *A. svalbardicum* in Ny Alesund, Sweden, each year from 1969 to 1992 under present conditions (solid lines) and if the average climate had been 2°C warmer (dotted lines). Horizontal lines represent the minimum thermal budget required for the first (lower line) and second (upper line) sexual generations to produce eggs. After Strathdee *et al.* (1993b).

The appearance of the first specimen of a given species in the suction trap samples indicates that the number of that species flying has reached a threshold level for detection by the system. A strong negative correlation has been identified between winter temperature prior to the event and the time of the first record in suction trap samples for a number of aphid species, particularly those that are largely anholocyclic in the area of the trap (Turl, 1980; A'Brook, 1983; Walters and Dewar, 1986; Harrington *et al.*, 1990). The fit with spring temperature is generally less good

(Harrington *et al.*, 1990). It is not surprising that winter temperatures are correlated more strongly with the time of first record of aphid species which are largely anholocyclic than those which are largely holocyclic, in view of the enhanced cold hardiness of diapausing eggs compared to active forms, and the ability of active forms to continue development and reproduction when temperature is sufficiently high. It does, however, seem surprising that winter climate is apparently so much more important than that in spring in determining early season activity of anholocyclic aphids. In view of the reproductive potential of aphids, it might be expected that a warm spring would encourage rapid development and reproduction and thus compensate for all but the coldest winters, when virtually all active forms are killed. The sublethal effects of low temperature on aphid individuals described earlier may explain the apparent failure of aphids to recover quickly in warm springs from the effects of a severe winter.

For holocyclic species, relationships between occurrence of the first alate in trap samples and winter temperature are generally weak (Harrington *et al.*, 1990). For some holocyclic species, early season activity is related more closely to spring temperatures (Thomas *et al.*, 1983), probably through an effect on aphid development immediately after egg-hatch.

Using observed relationships between the times of the first suction trap records, or the abundance of certain aphid species, and winter weather, it is possible to deduce the likely effect of warmer winters on early season aphid populations (Fig. 10) (Bale *et al.*, 1992; Woiwod and Harrington, 1994). For *M. persicae*, each 1°C rise in the January–February mean temperature advances the onset of the spring migration by about 2 weeks. Thus a temperature increase of 3°C will bring forward the average date of the first record by approximately 6 weeks. This is within the range of current experience. Future extremes will fall outside this range and the likelihood of error associated with linear extrapolation of predictions will be greater.

For *M. persicae* the linear relationships for different sites examined throughout England and France between the date on which the first aphid is trapped in spring, and winter temperature, are not significantly different from parallel (Harrington *et al.*, 1992b). In other words, for any given change in winter temperature, the change in date of first record will be the same number of days at all sites (Fig. 11). However, the relationships are displaced on the time axis in association with latitude such that, for any given winter temperature, the first aphid will tend to appear later further north. Thus a latitude-associated component other than temperature, possibly photoperiod, is also influencing the timing of

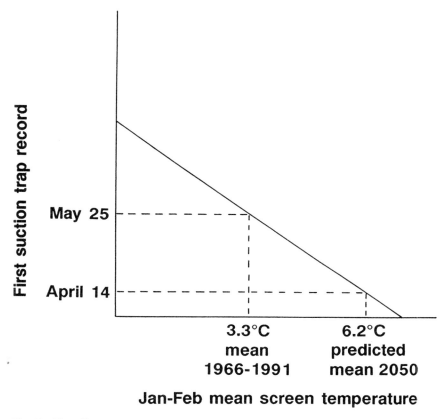

Fig. 10. The effect of predicted warmer winters on date of first suction trap record of *M. persicae* at Rothamsted.

migrations. Robert (1987b) showed that the phenologies of *R. padi* and *M. persicae* become earlier from northern Scotland to southern France.

Examination of the flight phenologies of two aphid pests of cereals, as recorded in 18 suction traps in Britain between 1975 and 1984, has revealed differences between species that may be associated with aphid life-cycle strategy, but with an overall north–south trend (Clark *et al.*, 1992). *S. avenae* is largely anholocyclic on Gramineae in southern Britain (Hand, 1989) with increasing proportions of monoecious, holocyclic clones further north (Walters and Dewar, 1986). In contrast, *M. dirhodum* is predominantly holocyclic in Britain, alternating between its primary hosts (*Rosa* spp.) and secondary graminaceous hosts. Detailed

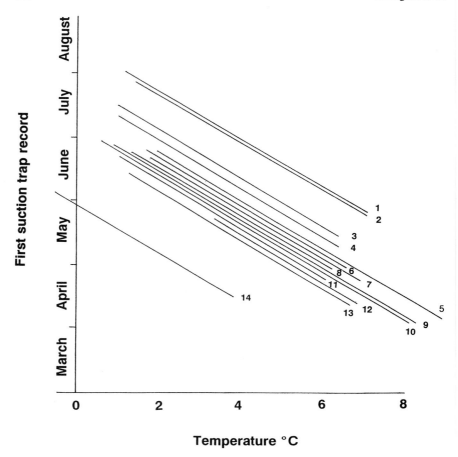

Fig. 11. Date of first record of *M. persicae* in relation to mean screen temperature in January and February at a range of suction trap sites. The relationships are significant (at least $P < 0.05$) at all sites shown. Latitude of sites: (1) 55.2° Newcastle; (2) 55.5° Ayr; (3) 56.5° Dundee; (4) 55.9° East Craigs; (5) 50.6° Starcross; (6) 52.1° Hereford; (7) 51.2° Wye; (8) 52.3° Broom's Barn; (9) 50.3° Arras; (10) 52.9° Kirton; (11) 51.8° Rothamsted; (12) 48.1° Rennes; (13) 51.7° Writtle; (14) 48.1° Colmar.

study of the overall pattern and duration of flight activity, together with the specific dates on which the first aphid, peak activity, and 25, 50 and 75% of the migration occurred, showed that both *S. avenae* and *M. dirhodum* followed a general north–south trend with specific events occurring earlier further south. However, there were separate components on either side of latitude 54°N. To the south the trend line was from south south west to north north east, while to the north the line was from south south east to north north west. Below 54°N, the trend was associated with winter temperature for both species; above 54°N there was no significant relationship with temperature, possibly because the temperature gradient along the northern trend line was much less than that for the south, making it less easy to identify associations.

The data from the Rothamsted suction trap network have been used to examine phenological trends with time to see if detectable trends of response to climate change are already evident (see Fleming and Tatchell, Chapter 25). Five common, widely distributed species with a well-defined spring flight period were examined. Data from eight traps, all with at least 20 years of data, were used and the date of the 5th, 10th, 25th, 50th and 75th centile of the spring migration identified for each year. The centile dates were regressed on year such that a negative relationship indicated phenologies becoming earlier with time. The number of significant relationships was no greater than would be expected by chance, but a highly significant number of the slopes were negative. These results are consistent with the hypothesis that phenologies are becoming earlier, and whilst this does not yet prove an association with temperature, there is a significant relationship ($P < 0.001$) between increasing temperature in the northern hemisphere and time elapsed since 1964 (Jones and Wigley, 1990).

Flight phenologies of *R. padi* in France have been classified according to the seasonal pattern of alatae caught in suction traps (Hullé *et al.*, 1994). Seven patterns were identified according to the timing and size of the main period of seasonal activity. Where the aphid is almost exclusively anholocyclic, a relatively large and early spring migration and a small autumn migration occurred. Where it is holocyclic, there was always a large autumn migration. At sites where the proportion of holocyclic to anholocyclic clones varied, patterns were influenced particularly by April and September temperatures. Large autumn migrations only occurred if temperatures at these times were less than average.

C. Abundance

Winter temperature is also correlated with the abundance of anholocyclic aphids leaving overwintering sites later in the year (Harrington *et al.*, 1990). There is a significant relationship between the abundance of *M. persicae* in trap samples up to the beginning of July and temperature the previous winter (Harrington *et al.*, 1992b). If abundance is accumulated beyond this date, relationships are less significant as summer migrants are included.

V. CONCLUSIONS

It is true that little has been done even to begin to unravel experimentally the effects of the many and complex interactions between aphids and climate (Fig. 1). If progress is to be made it has been suggested that collaboration on a standard system is necessary (see Lawton, Chapter 1, this volume; Robert, 1987b), but, even then, the interactions may be too numerous and complex for a mechanistic approach to provide valuable insights within an acceptable time. However, such interactions may be less complex in the arctic where there may be excellent opportunities for productive study. There is value in the empirical approaches made possible by long data sets as these account, by default, for all current interactions even though they may not be understood fully. They at least enable sensible suggestions to be made about the potential implications of climate change and areas of work where targeted collaboration between entomologists, plant scientists and mathematical modellers could put these suggestions on to a more mechanistic footing. The danger of diverting limited resources into detailed studies of many aspects of a single system is that it would fail to take account of the great variation in response between even closely related species, a feature for which the insects are renowned and a key feature likely to influence their future abundance and distribution.

Acknowledgements

The authors are grateful to Dr Y. Robert, Dr M. Hullé, Dr C. Dedryver and Mrs L. A. D. Turl for valuable comments on the manuscript.

REFERENCES

A'Brook, J. (1983). Forecasting the incidence of aphids using weather data. *EPPO Bull.* **13**, 229–233.

A'Brook, J. (1981). Some observations in west Wales on the relationships between numbers of alate aphids and weather. *Ann. Appl. Biol.* **97**, 11–15.

Aitchison, C. W. (1978). Notes on low temperature and winter activity of Homoptera in Manitoba. *Manitoba Entomol.* **12**, 135–137.

Bale, J. S. (1987). Insect cold hardiness: freezing and supercooling – an ecophysiological perspective. *J. Insect Physiol.* **33**, 899–908.

Bale, J. S. (1991). Insects at low temperature: a predictable relationship? *Functional Ecol.* **5**, 291–298.

Bale, J. S., Harrington, R. and Clough, M. S. (1988). Low temperature mortality of the peach–potato aphid *Myzus persicae*. *Ecol. Entomol.* **13**, 121–129.

Bale, J. S., Harrington, R. and Howling, G. G. (1992). Aphids and winter weather I. Aphids and climate change. In "Proceedings 4th European Congress of Entomology, Godollo, Hungary, 1-6 September, 1991" (L. Zombori and L. Peregovits, eds). pp. 139-143. Hungarian Natural History Museum, Budapest.

Barlow, C. A. (1962). The influence of temperature on the growth of experimental populations of *Myzus persicae* (Sulzer) and *Macrosiphum euphorbiae* (Thomas) (Aphididae). *Can. J. Zool.* **40**, 145-156.

Beemster, A. B. R. (1972). Virus translocation in potato plants and mature-plant resistance. In "Viruses of Potatoes and Seed Potato Production" (J. A. de Bokx, ed.), pp. 144-151. PUDOC, Wageningen.

Blackman, R. L. (1971). Variation in the photoperiodic response within natural populations of *Myzus persicae* (Sulz.). *Bull. Entomol. Res.* **60**, 533-546.

Blackman, R. L. (1972). The inheritance of life-cycle differences in *Myzus persicae* (Sulz.) (Hem., Aphididae). *Bull. Entomol. Res.* **62**, 281-294.

Blackman, R. L. (1974). Life-cycle variation of *Myzus persicae* (Sulz.) (Hom., Aphididae) in different parts of the world, in relation to genotype and environment. *Bull. Entomol. Res.* **63**, 595-607.

Blackman, R. L. and Eastop, V. F. (1984). "Aphids on the World's Crops: An Identification and Information Guide". John Wiley, Chichester.

Butts, R. A. (1992). Cold hardiness and its relationship to overwintering of the Russian Wheat Aphid (Homoptera: Aphididae) in southern Alberta. *J. Econ. Entomol.* **85**, 1140-1145.

Campbell, A. and Mackauer, M. (1975). Thermal constants for development of the pea aphid (Homoptera: Aphididae) and some of its parasites. *Can. Entomol.* **107**, 419-423.

Campbell, A., Frazer, B. D., Gilbert, N., Gutierrez, A. P. and Mackauer, M. (1974). Temperature requirements of some aphids and their parasites. *J. Appl. Ecol.* **11**, 431-438.

Carter, N., Dixon, A. F. G. and Rabbinge, R. (1982) "Cereal Aphid Populations: Biology, Simulation and Prediction". PUDOC, Wageningen.

Clark, S. J., Tatchell, G. M., Perry, J. N. and Woiwod, I. P. (1992). Comparative phenologies of two migrant cereal aphid species. *J. Appl. Ecol.* **29**, 571-580.

Clough, M. S., Bale, J. S. and Harrington, R. (1990). Differential cold-hardiness in adults and nymphs of the peach potato aphid *Myzus persicae*. *Ann. Appl. Biol.* **116**, 1-9.

Dean, G. J. (1974). Effect of temperature on the cereal aphids *Metopolophium dirhodum* (Wlk), *Rhopalosiphum padi* (L.) and *Sitobion avenae* (F.). *Bull. Entomol. Res.* **63**, 401-409.

Dedryver, C. A., Bos, C. and Wegorek, P. (1990). Deux exemples de variabilité du polymorphisme chez *Sitobion avenae* F. (Homoptera, Aphididae). In "Régulation des Cycles Saisonniers Chez les Invertebrés" (P. Ferron, J. Missonnier and B. Mauchamp, eds), pp. 133-136. *Les Colloques de l'I.N.R.A.* **52**.

Dixon, A. F. G. (1973). "The Biology of Aphids". Edward Arnold, London.

Dixon, A. F. G. (1987). The way of life of aphids: host specificity, speciation and distribution. In "Aphids: their Biology, Natural Enemies and Control. Vol. A" (A. K. Minks and P. Harrewijn, eds), pp. 197-207. Elsevier, Amsterdam.

Dixon, A. F. G., Kindlmann, P., Leps, J. and Holman, J. (1987). Why are there so few species of aphids, especially in the tropics? *Am. Nat.* **129**, 580-592.

Doodson, J. K. and Saunders, P. J. W. (1970). Some effects of barley yellow dwarf virus on spring and winter cereals in field trials. *Ann. Appl. Biol.* **66**, 361-374.

Dry, W. W. and Taylor, L. R. (1970). Light and temperature thresholds for take-off by aphids. *J. Anim. Ecol.* **39**, 493-504.

Eastop, V. F. (1977). Worldwide importance of aphids as vectors. In "Aphids as Virus Vectors" (K. F. Harris and K. Maramorosch, eds), pp. 3-62. Academic Press, New York.

Elliott, N. C. and Kieckhefer, R. W. (1989). Effects of constant and fluctuating temperatures on immature development and age-specific life tables of *Rhopalosiphum padi* (L.) (Homoptera: Aphididae). *Can. Entomol.* **121**, 131–140.

Foster, G. N., Holmes, S. J. and Bone, S. F. (1993). Ten years' experience of infectivity indexing as a method of predicting the risk of barley yellow dwarf virus outbreaks in autumn-sown cereals in the west of Scotland. *Proceedings Crop Protection in Northern Britain* **1993**, 97–101.

Halbert, S. E., Connelly, B. J., Bishop, G. W. and Blackmer, J. L. (1992). Transmission of barley yellow dwarf virus by field collected aphids (Homoptera: Aphididae) and their relative importance in barley yellow dwarf epidemiology in southwestern Idaho. *Ann. Appl. Biol.* **121**, 105–121.

Hand, S. C. (1980). Overwintering of cereal aphids. *IOBC, WPRS Bull.* **3**, 59–61.

Hand, S. C. (1989). The overwintering of cereal aphids on Gramineae in southern England, 1977–1980. *Ann. Appl. Biol.* **115**, 17–29.

Harrington, R. and Cheng Xia-Nian (1984). Winter mortality, development and reproduction in a field population of *Myzus persicae* (Sulz.) in England. *Bull. Entomol. Res.* **74**, 633–640.

Harrington, R. and Taylor, L. R. (1990). Migration for survival: fine scale population redistribution in an aphid *Myzus persicae*. *J. Anim. Ecol.* **59**, 1177–1193.

Harrington, R., Dewar, A. M. and George, B. (1989). Forecasting the incidence of virus yellows in sugar beet in England. *Ann. Appl. Biol.* **114**, 459–469.

Harrington, R., Tatchell, G. M. and Bale, J. S. (1990). Weather, life cycle strategy and spring populations of aphids. *Acta Phytopathologica et Entomologica Hungarica* **78**, 121–129.

Harrington, R., Dewar, A. M., Howling, G. G. and Bale, J. S. (1992a). The value of statistical models in aphid forecasting. *Proceedings Brighton Crop Protection Conference – Pests and Diseases – 1992*, 965–972.

Harrington, R., Hullé, M., Pickup, J. and Bale, J. (1992b). Forecasting the need for early season aphid control: geographical variation in the relationship between winter temperature and early season flight activity of *Myzus persicae*. *In* "Monitoring and Forecasting to Improve Crop and Environment Protection". Association of Applied Biologists, Warwick.

Heie, O. E. (1980). "The Aphidoidea (Hemiptera) of Fennoscandia and Denmark". *Faun. Entomol. Scand.* **9**, 236 pp.

Honek, A. and Kocourek, F. (1990). Temperature and development time in insects: a general relationship between thermal constants. *Zool. Jb. Syst.* **117**, 401–439.

Hullé, M., Coquio, S. and Laperche, V. (1994). Patterns in flight phenology of a migrant cereal aphid species and implications in pest forecasting. *J. Appl. Ecol.* **31**, 49–58.

Hutchinson, L. A. and Bale, J. S. (1994). Effects of sub-lethal cold stress on the aphid *Rhopalosiphum padi*. *J. Appl. Ecol.* **31**, 102–108.

James, B. D. and Luff, M. L. (1982). Cold-hardiness and development of eggs of *Rhopalosiphum insertum*. *Ecol. Entomol.* **7**, 277–282.

Johnson, B. (1966). Wing polymorphism in aphids IV. The effect of temperature and photoperiod. *Ent. Exp. Appl.* **9**, 301–313.

Jones, P. D. and Wigley, T. M. L. (1990). Global warming trends. *Sci. Am.* **263**, 66–73.

Kawada, K. (1987). Polymorphism and morph determination. *In* "Aphids: their Biology, Natural Enemies and Control. Vol. A" (A. K. Minks and P. Harrewijn, eds), pp. 255–268. Elsevier, Amsterdam.

Kenten, J. (1955). The effect of photoperiod and temperature on reproduction in *Acyrthosiphon pisum* (Harris) and on the forms produced. *Bull. Entomol. Res.* **46**, 599–624.

Kieckhefer, R. W., Elliott, N. C. and Walgenbach, D. D. (1989). Effects of constant and fluctuating temperatures on developmental rates and demographic statistics of the English Grain Aphid (Homoptera: Aphididae). *Ann. Entomol. Soc. Am.* **82**, 701–706.

Kilian, L. and Nielson, M. W. (1971). Differential effects of temperature on the biological activity of four biotypes of the pea aphid. *J. Econ. Entomol.* **64**, 153–155.

Knight, J. D. and Bale, J. S. (1986). Cold hardiness and overwintering of the grain aphid *Sitobion avenae*. *Ecol. Entomol.* **11**, 189–197.

Knight, J. D., Bale, J. S., Franks, F., Mathias, S. and Baust, J. G. (1986). Insect cold hardiness: supercooling points and pre-freeze mortality. *Cryoletters* **7**, 194–203.

Lamb, K. P. and White, D. (1966). Effect of temperature, starvation and crowding on production of alate young by the cabbage aphid (*Brevicoryne brassicae*). *Ent. Exp. Appl.* **9**, 179–184.

Lamb, R. J., MacKay, P. A. and Gerber, G. H. (1987). Are development and growth of pea aphids, *Acyrthosiphon pisum*, in North America adapted to local temperatures? *Oecologia* **72**, 170–177.

Lampel, G. (1968). "Die Biologie des Blattlaus-Generationswechsels." Gustav Fischer, Jena.

Leather, S. R., Walters, K. F. A. and Bale, J. S. (1993). "The Ecology of Insect Overwintering." Cambridge University Press, Cambridge.

Lees, A. D. (1959) The role of photoperiod and temperature in the determination of parthenogenetic and sexual forms in the aphid *Megoura viciae* Buckton. I. The influence of these factors on apterous virginoparae and their progeny. *J. Insect Physiol.* **3**, 92–117.

Liu Shu-Sheng and Hughes, R. D. (1987). The influence of temperature and photoperiod on the development, survival and reproduction of the sowthistle aphid, *Hyperomyzus lactucae*. *Ent. Exp. Appl.* **43**, 31–38.

Liu Shu-Sheng and Meng Xueduo (1989). The change pattern of development rates under constant and variable temperatures in *Myzus persicae* and *Lipaphis erysimi*. *Acta Ecologica Sinica* **9**, 182–190.

MacKay, P. A. (1989). Clonal variation in sexual morph production in *Acyrthosiphon pisum* (Homoptera: Aphididae). *Environ. Entomol.* **18**, 558–562.

Michels, G. J. and Behle, R. W. (1989). Influence of temperature on reproduction, development, and intrinsic rate of increase of Russian Wheat Aphid, Greenbug, and Bird Cherry-Oat Aphid (Homoptera: Aphididae). *J. Econ. Entomol.* **82**, 439–444.

Mittler, T. E. and Wilhoit, L. (1990). Sexual morph production by two regional biotypes of *Myzus persicae* in relation to photoperiod. *Environ. Entomol.* **19**, 139–143.

Moran, N. (1992). The evolution of aphid life cycles. *Ann. Rev. Entomol.* **37**, 321–348.

O'Doherty, R. and Bale, J. S. (1985). Factors affecting the cold hardiness of the peach potato aphid *Myzus persicae*. *Ann. Appl. Biol.* **106**, 219–228.

Parish, W. E. G. and Bale, J. S. (1990). Effects of short term exposure to low temperature on wing development in the grain aphid *Sitobion avenae* (F.) (Hem., Aphididae). *J. Appl. Entomol.* **109**, 175–181.

Parish, W. E. G. and Bale, J. S. (1993). Effects of brief exposures to low temperature on the development, longevity and fecundity of the grain aphid *Sitobion avenae* (Hemiptera: Aphididae). *Ann. Appl. Biol.* **122**, 9–21.

Parry, W. H. (1979). Factors affecting low temperature survival of *Cinara pilicornis* eggs on Sitka Spruce. *Int. J. Biometeor.* **23**, 185–193.

Plumb, R. T. (1971). The control of insect-transmitted viruses of cereals. *Proc. 6th Brit. Ins. & Fung. Conf.* **1**, 307–313.

Robert, Y. (1987a). Dispersion and migration. *In* "Aphids, their Biology, Natural Enemies and Control, Vol. A" (A. K. Minks and P. Harrewijn, eds); pp. 299–313. Elsevier, Amsterdam.

Robert, Y. (1987b). Aphid vector monitoring in Europe. In "Current Topics in Vector Research, Volume 3" (K. F. Harris, ed.), pp. 81–129. Springer-Verlag, New York.

Robert, Y. and Rouzé-Jouan, J. (1976a). Neuf ans de piégeage de pucerons des céréales: *Acyrthosiphon (Metopolophium) dirhodum* Wlk., *A. (M.) festucae* Wlk., *Macrosiphum (Sitobion) avenae* F., *M. (S.) fragariae* Wlk. et *Rhopalosiphum padi* L. en Bretagne. *Rev. de Zool. Agric. et de Path. Veg.* **75**, 67–80.

Robert, Y. and Rouzé-Jouan, J. (1976b). Activité saisonnière de vol des pucerons [Hom. Aphididae] dans l'ouest de la France. Résultats de neuf années de piégeage (1967–1975). *Ann. Soc. Ent. Fr.* (N.S.) **12**, 671–690.

Shibata, B. (1952). "Ecological Studies of Plant Lice. VII. With Reference to Experimental Production of Sexual Females by Constant Low Temperature". Utsunomiya University College of Agriculture, Series B2, 1–8 (in Japanese with English abstract).

Sigvald, R. (1985). Mature-plant resistance of potato plants against potato virus Y° (PVY°). *Potato Res.* **28**, 135–143.

Simon, J. C., Blackman, R. L. and Le Gallic, J. F. (1991). Local variability in the life cycle of the bird cherry-oat aphid, *Rhopalosiphum padi* (Homoptera: Aphididae) in western France. *Bull. Entomol. Res.* **81**, 315–322.

Smith, H. G. and Hallsworth, P. B. (1990). The effects of yellowing viruses on yield of sugar beet in field trials, 1985 and 1987. *Ann. Appl. Biol.* **16**, 503–511.

Smith, R. K. (1981). "Studies on the Ecology of Cereal Aphids and Prospects for Integrated Control." PhD thesis, University of London.

Somme, L. (1964) Effects of glycerol on cold hardiness in insects. *Can. J. Zool.* **42**, 89–101.

Somme, L. (1969). Mannitol and glycerol in overwintering aphid eggs. *Norw. J. Entomol.* **16**, 107–111.

Stinner, R. E., Gutierrez, A. P. and Butler, G. D. (1974). An algorithm for temperature-dependent growth rate simulation. *Can. Ent.* **106**, 519–524.

Strathdee, A. T., Bale, J. S., Block, W. C., Webb, N. R., Hodkinson, I. D. and Coulson, S. J. (1993a). Extreme adaptive life-cycle in a high arctic aphid, *Acyrthosiphon svalbardicum*. *Ecol. Entomol.* **18**, 254–258.

Strathdee, A. T., Bale, J. S., Block, W. C., Coulson, S. J., Hodkinson, I. D. and Webb, N. R. (1993b). Effects of temperature elevation on a field population of *Acyrthosiphon svalbardicum* (Hemiptera: Aphididae) on Spitsbergen. *Oecologia* **96**, 457–465.

Tatchell, G. M. (1991). Monitoring and forecasting aphid problems. In "Aphid–Plant Interactions: Populations to Molecules" (D. C. Peters, J. A. Webster and C. S. Chlouber, eds) pp. 215–231. USDA/Agricultural Research Service, Oklahoma.

Tatchell, G. M., Plumb, R. T. and Carter, N. (1988). Migration of alate morphs of the bird cherry aphid (*Rhopalosiphum padi*) and implications for the epidemiology of barley yellow dwarf virus. *Ann. Appl. Biol.* **112**, 1–11.

Taylor, L. R. (1985). An international standard for the synoptic monitoring and dynamic mapping of migrant pest aphid populations. In "The Movement and Dispersal of Agriculturally Important Biotic Agents" (D. R. MacKenzie, C. S. Barfield, G. G. Kennedy, R. D. Berger and D. J. Taranto, eds) pp. 337–418. Claitor's Publishing Division, Baton Rouge.

Taylor, L. R. (1986). Synoptic dynamics, migration and the Rothamsted Insect Survey. *J. Anim. Ecol.* **55**, 1–38.

Thomas, G. G., Goldwin, G. K. and Tatchell, G. M. (1983). Associations between weather factors and the spring migration of the damson-hop aphid, *Phorodon humuli*. *Ann. Appl. Biol.* **102**, 7–17.

Tsitsipis, J. A. and Mittler, T. E. (1976). Influence of temperature on the production of parthenogenetic and sexual females by *Aphis fabae* under short-day conditions. *Ent. Exp. Appl.* **19**, 179–188.

Turl, L. A. D. (1980). An approach to forecasting the incidence of potato and cereal aphids in Scotland. *EPPO Bull.* **10**, 135–141.
Turl, L. A. D. (1983). The effect of winter weather on the survival of aphid populations on weeds in Scotland. *EPPO Bull.* **13**, 139–143.
Walgenbach, D. D., Elliott, N. C. and Kieckhefer, R. W. (1988). Constant and fluctuating temperature effects on developmental rates and life statistics of the greenbug (Homoptera: Aphididae). *J. Econ. Entomol.* **81**, 501–507.
Walters, K. F. A. and Dewar, A. M. (1986). Overwintering strategy and the timing of the spring migration of the cereal aphids *Sitobion avenae* and *Sitobion fragariae*. *J. Appl. Ecol.* **23**, 905–915.
Walters, K. F. A. and Dixon, A. F. G. (1984). The effect of temperature and wind on the flight activity of cereal aphids. *Ann. Appl. Biol.* **104**, 17–26.
Wellings, P. W. (1981). The effect of temperature on the growth and reproduction of two closely related aphid species on sycamore. *Ecol. Entomol.* **6**, 209–214.
Wiktelius, S. (1981). Diurnal flight periodicities and temperature thresholds for flight for different migrant forms of *Rhopalosiphum padi* L. (Hom., Aphididae). *Z. Ang. Ent.* **92**, 449–457.
Woiwod, I. P. and Harrington, R. (1994). Flying in the face of change – the Rothamsted Insect Survey. *In* "Long-term Research in Agricultural and Ecological Sciences" (R. A. Leigh and A. E. Johnston, eds). CAB International, Wallingford.
Woodford, J. A. T., Harrison, B. D., Aveyard, C. S. and Gordon, S. C. (1983). Insecticidal control of aphids and the spread of the potato leafroll virus in potato crops in eastern Scotland. *Ann. Appl. Biol.* **103**, 117–130.
Zhou, X-L., Carter, N. and Mumford, J. (1989). A simulation model describing the population dynamics and damage potential of the rose grain aphid, *Metopolophium dirhodum* (Walker) (Hemiptera: Aphididae), in the UK. *Bull. Entomol. Res.* **79**, 373–380.

7

The Effects of Climatic and Land-Use Changes on Insect Vectors of Human Disease

J. LINES

I.	Introduction	158
II.	The Role of Temperature	159
III.	Potential Effects on Some Vector-borne Diseases	159
	A. Dengue	159
	B. Japanese Encephalitis (JE)	162
	C. Tick- and Mite-borne Diseases	163
	D. Chagas' Disease, African Trypanosomiasis and Leishmaniasis	163
	E. Lymphatic Filariasis (Elephantiasis)	164
	F. Onchocerciasis	165
	G. Malaria	165
IV.	Land-use Changes	171
V.	Conclusion	172
	References	173

Abstract

An increase in global temperatures may alter the distribution of an arthropod-borne pathogen either by affecting the distribution of the vector, or by accelerating the development of the pathogen within the vector. In those cases where the vector is relatively long-lived, in comparison with the period of development of the parasite, the main question is how the distribution of the vector will be changed. Temperature changes may also affect the parasite directly, in particular by reducing the time taken to mature in the vector. This will be particularly important with short-lived vectors such as mosquitoes. In the case of malaria, the effects of climatic warming may already have been felt in the east African highlands, and large areas of central Asia may also be vulnerable to a dramatic resurgence. However, in most places, man-made environmental changes at the local scale – including urbanization, deforestation, and agricultural and water resource developments – are as much or more of a threat, in terms of vector-borne disease, as global climate change.

I. INTRODUCTION

Environmental changes that result from global and local processes have the same kinds of effects on vectors of human disease as on other insects. Both distribution and abundance are likely to change, in ways that are not easy to predict. What we really need to know, though, are the effects on the distribution and abundance of the parasites. This chapter looks at some of the evidence concerning the role of temperature as a limiting factor on some of the major vector-borne diseases. In Chapter 8, this volume, Rogers examines the case of one vector for which detailed predictions relating to distributional change have been made, by a statistical model derived and validated with existing data. For most vectors, though, we are still a long way from being able to make such detailed predictions. Examples of previous reviews of this question have dealt with it along with other effects on health (Shope, 1991; Cook, 1992), or have focused on relatively well-understood disease transmission processes in specific examples (Bradley, 1993a). Little attempt seems to have been made to review systematically, for each of the major pathogens, whether and when the direct effects of temperature on the pathogen itself are likely to be more important than indirect effects on the vector. Such a broad review is a daunting task, and this chapter represents only my own first steps, following those of Sutherst (1993), in trying to build up an overall view. In doing so, comparisons are made between global and local man-made processes – in other words, between what we can speculate about the effects of climate change and what we can already see happening locally as a consequence of changes in land use.

The aim here, therefore, is for generalization. This inevitably involves some neglect of the complexities and anomalies that may be very important locally. Estimates of the amount and timing of the temperature rises to be expected over the next few decades are imprecise. From the predictions made by the International Panel on Climate Change (Houghton et al., 1990), Bradley (1993a) took a rise of 2°C by 2025 as a representative figure, and we will follow this example. As a very broad average, an increase of 2°C can be thought of as roughly equivalent to 180 km in latitude or 300 m in altitude (Peters, 1993).

II. THE ROLE OF TEMPERATURE

Temperature changes associated with global warming could affect vector-borne pathogens in three ways. The least direct, but perhaps most conspicuous, is by changing entire habitats and communities of organisms. If entire communities shift together, as a consequence of global warming, vector and parasite may well shift along with the rest. Unfortunately, communities may well be disrupted, and in this case it is very difficult to predict what will happen. New associations may appear. Of course, temperature also has direct effects on the vector, and may be more or less important as a factor limiting distribution and local abundance. For the pathogen, the most direct effects of temperature are those that affect the parasite itself. Some vector-borne diseases are more sensitive in this way than others. Temperature controls the time it takes for the parasite to mature in the vector, that is, the time an infected vector takes to become infectious. This is usually of the order of a few days to a couple of weeks in favourable conditions. If this period is relatively long compared to the median life span of the vector, transmission is likely to be especially sensitive to temperature. Table 1 shows that such temperature-sensitive transmission systems comprise mainly the mosquito-borne diseases. The following section deals more or less briefly with a selection of these diseases in turn, ending up with the most important, malaria.

III. POTENTIAL EFFECTS ON SOME VECTOR-BORNE DISEASES

A. Dengue

Dengue is the most widespread and important of the human arboviral diseases. There is no vaccine, and its severe manifestations – dengue haemorrhagic fever and dengue shock syndrome – can have a high case fatality rate in the absence of intensive supportive treatment. It is widely distributed throughout south and southeast Asia, the western Pacific, tropical Africa, Latin America and the Caribbean.

Dengue is almost entirely associated with man-made environments because its main vector, *Aedes aegypti*, breeds almost entirely in

Table 1. Major arthropod-borne diseases and their vectors. A vector with a "long" life span is one in which median life expectancy exceeds the period of maturation of the parasite; that is, a majority of individuals that become infected live long enough to become infectious

Disease	Vector	Life span of vector compared to incubation period of pathogen
Dengue	*Aedes* mosquitoes	Short
Japanese encephalitis	*Culex* mosquitoes	Short
Relapsing fever, Lyme disease, etc.	Ticks	Long
Chagas' disease	Reduviid bugs	Long
African trypanosomiasis	Tsetse flies	Long
Leishmaniasis	Sandflies	Fairly long
Malaria	*Anopheles* mosquitoes	Short
Lymphatic filariasis	Mosquitoes	Short
Onchocerciasis	Blackflies	Fairly short

man-made containers. These are typically small rather than large and clean rather than polluted, such as flower vases, discarded tyres, and water storage drums and tanks. It is, therefore, essentially an urban disease, although in places like Thailand it also spreads into rural villages where large pots are traditionally used for water storage. Its container-breeding habits, and the ability of its eggs to resist desiccation, have enabled *A. aegypti* spread throughout the tropics, often following the international trade in used tyres. It has not, however, penetrated far into the temperate zones.

There is another container-breeding vector of dengue, *Aedes albopictus*, which is currently spreading around the world in the same way as *A. aegypti* did previously. Unlike its predecessor, however, this species does have temperate-adapted strains which, in China for example, can transmit dengue far to the north of the limits of *A. aegypti* distribution (Fan et al., 1989). *A. albopictus* has now reached Italy (Dalla Pozza and Majori, 1992) and is already firmly established on the eastern seaboard of the USA, again well to the north of *A. aegypti*'s limits (Rai, 1991).

Dengue is already spreading at an alarming rate (WHO, 1993), perhaps more rapidly than any other tropical disease, driven mainly by the ever-accelerating process of urbanization and assisted by human move-

7. Effects of Climatic and Land-use Changes on Insect Disease Vectors

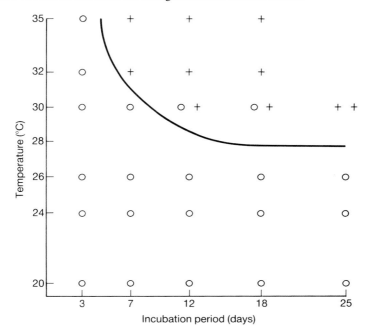

Fig. 1. The relationship between temperature and dengue transmission at intervals following an infective bloodmeal. Crosses represent successful and circles unsuccessful transmission. From Watts *et al.* (1987) with permission.

ment. Global warming is likely to exacerbate this, but how much, and in what way? We know (Watts *et al.*, 1987) that the relationship between temperature and the extrinsic incubation period of dengue is highly non-linear (Fig. 1). At the lower end of the scale, the virus does not develop at all in the mosquito, or at least the process takes so long that no infected females are likely to survive long enough to transmit. Just above this point, however, even small increases in temperature greatly increase the rate of development. At the higher end of the range, on the other hand, temperature ceases to be such an important limiting factor.

A 2°C rise is therefore unlikely to have a major effect on the local distribution of dengue in the tropics. Away from the equator, on the other hand, it may greatly affect the course and intensity of epidemics, by lengthening the summer period during which the vector is abundant and transmission can easily occur. A 2°C rise would also extend the altitude at which transmission is likely to occur. In general, however, the distribution and severity of dengue outbreaks is not well correlated either with altitude or, at least within the tropics, with latitude; other factors, mostly to do with people and their domestic habits, are more important.

Where global warming could be decisive is at the latitudinal limits of the disease's current distribution (Nawrocki and Hawley, 1987). Dengue has an animal reservoir only in the rain forests of southeast Asia; elsewhere it is an anthroponosis. Moreover, infected people are infectious for only a short period. The virus therefore relies for winter survival on either continuing low-level transmission or on transovarial (mother to offspring) transmission through the overwintering stages of the mosquito. We must therefore consider the possibility that less severe winters and warmer springs, coupled with the northerly spread of *A. albopictus*, could make dengue transmission in Europe and the USA a real hazard. Is the domestic environment in rich countries so different that these mosquitoes will fail to find sufficient breeding sites and access to human hosts? We do not know.

B. Japanese Encephalitis (JE)

This is another arboviral disease in which only a few of the people infected get severe symptoms, but since these include brain damage and are often fatal, JE is a disease greatly to be feared. It is transmitted by ricefield-breeding mosquitoes of the genus *Culex*, and is widely distributed in the rice-growing areas of south, east and southeast Asia. It is a zoonosis particularly associated with birds and pigs, which act as maintenance and amplifying hosts, and without which JE is not usually a problem.

At a constant 32°C an infective bloodmeal with JE virus can lead to salivary gland infection in 3–5 days; at 26°C, the same process takes 21 days (Leake and Johnson, 1987). Thus, JE development is sensitive to temperature over a higher range than dengue or malaria. This leads to a rather paradoxical situation. The typical pattern in equatorial countries such as Malaysia, Indonesia and the Philippines, with fairly constant rainfall and moderately high temperature, is one of sporadic low-level transmission all year round. Further from the equator, in India, Laos and China, for example, summer temperatures are more extreme and coincide with the rainy season peak in vector abundance. This leads to a short and well-defined season of epidemic transmission which can be so intense that the overall annual incidence exceeds that in the equatorial zone. In Nagasaki, Japan, incidence rates during summer epidemics between 1950 and 1969 were correlated positively with temperature and negatively with rainfall (Burke and Leake, 1988).

Thus, it seems unlikely that JE will spread away from its classical "ricefield with birds and pigs" ecological setting. Within these areas, it is probable that global warming will cause more intense and perhaps more prolonged transmission, although this may be modified by indirect effects on the timing of rain and of rice cultivation.

C. Tick- and Mite-borne Diseases

Tick- and mite-borne infections, including Lyme disease, tick-typhus and scrub-typhus, are very patchily distributed and tend to be confined within particular ecological habitats or settings. Because of ticks' relatively long life span compared to the incubation period of the pathogens they carry, the effects of climate change on the epidemiology of tick-borne disease are likely to reflect changes in the distribution and abundance of the vector. For this reason, considerable effort has been made to model the distribution of ticks of veterinary importance. It seems that in at least one case, abiotic factors (including temperature) are a primary limiting factor. Randolph (1993) has analysed detailed data on the distribution and seasonal abundance of the tick *Rhipicephalus appendiculatus* in five southern African locations, and related them to climatic- and satellite-derived data. She found that night-time minimum temperature was the most important environmental variable, the questing larvae needing a low saturation deficit to replenish moisture lost during the day. She concluded that night-time moisture availability is likely to be a critical factor throughout the range of this species. However, the association between seasonal temperatures and satellite indices is positive in some places and negative in others, so that it may not be possible to derive a universally consistent relationship between tick numbers, temperatures and satellite data.

There have been significant achievements in the use of climatic and satellite data to model the distribution of ticks and tsetses. This progress has clearly depended on the existence of extensive and detailed records of vector abundance, from a variety of locations and covering relatively long periods of time. However, for vectors of exclusively medical importance, there are currently very few data sets of this kind.

D. Chagas' Disease, African Trypanosomiasis and Leishmaniasis

The trypanosomiases are, like the tick-borne diseases, cases where the extrinsic cycle of the parasite is short compared to the life span of the vector, and direct effects of temperature on the parasites are therefore expected to be rather less important than effects on the vector. Animal reservoir hosts are necessary to maintain the parasite, but a variety of animals can fulfil this role, and as a very general rule, it is rare to find both vector and reservoir present without the parasite being there too.

The case of African trypanosomiasis is considered elsewhere in this volume (Rogers, Chapter 8). Chagas' disease, or American trypanosomiasis, is transmitted by triatomine bugs and causes chronic degenerative diseases of the heart and intestine. One might expect global warming to lead to a shift in its distribution, northwards from Mexico into the USA, or southwards into the southern cone of the continent. However, it is very clear that Chagas' disease is a disease of poverty, even in the centre of its distribution. Even if the vectors do spread to more northerly latitudes, Chagas' disease transmission is likely to be constrained by the north–south gradient in the availability of poor housing and suitable reservoir hosts. In the USA, there have been three cases of Chagas' disease attributed to local transmission, compared to two acquired through blood transfusion (Moncayo, 1992); it is my guess that any future increase is just as likely to come from the latter route as the former.

The leishmaniases, both cutaneous and visceral, have such complex epidemiological and ecological patterns that general predictions about the effects of global warming are probably not useful. It is not easy to infer from the existing disease distributions when, where and how temperature is a limiting factor. What is required is a detailed examination of each of the important combinations of sandfly vector and reservoir host, and this will not be attempted here.

E. Lymphatic Filariasis (Elephantiasis)

The filarial nematode *Wuchereria bancrofti* is transmitted by a variety of mosquitoes in Asia, Africa, the western Pacific and the east coast of Latin America. However its distribution is not well correlated with that of suitable vector species, and in this it is a useful example of how climatic factors other than temperature can modify transmission. The L3 infective larva does not enter its human host by being injected along with the mosquito saliva. Rather it apparently falls or breaks out of the mosquito mouthparts in a droplet of haemocoel, and must then find its way into the hole in the skin left behind by the proboscis of the mosquito (Lindsay *et al.*, 1984). It is therefore not surprising that within the tropics the distribution of this parasite is closely associated with humidity, which presumably makes it easier or allows more time for the larva to find its destination. Thus, predicting the effects of global warming must take into account the likely changes in humidity as well as temperature.

W. bancrofti is transmitted in both urban and rural environments. Its urban vector is *Culex quinquefasciatus*, and this species, by virtue of its ability to breed in highly polluted water, is now the most abundant human-biting mosquito in most tropical cities. In some places, people suffer hundreds of bites a night. Thus, while global warming is likely to cause an increase in the distribution and intensity of filariasis transmission, the influence of continued urbanization will almost certainly be equally or more important.

F. Onchocerciasis

The nematode *Onchocerca volvulus* causes skin and eye disease in west and east Africa and some parts of Central and South America. Its blackfly vector has a life span which is neither much longer nor much shorter than the extrinsic period of the parasite. However, closely related *Onchocerca* species are widely transmitted between cattle by other blackflies in the temperate zone, and it therefore appears that development can continue at least in summer temperatures outside the tropics. Instead, the distribution of *O. volvulus* seems to depend simply on the presence of competent human-biting vectors. In west Africa and Sudan, these breed in the white water rapids of large rivers, in east Africa they are confined to the rivers and streams of hill forests, and in Latin America to forest rivulets. Altitude, at least in Africa, appears to be no obstacle, except on the highest mountains.

In west Africa, the forest zone and savannah zones have their own vectors and associated parasites. Depending on what happens to the vegetation zones in the region, we might expect the boundary between these to be moved some a few hundred kilometres northwards by global warming. Replacement of the savannah by the forest form would almost certainly be a good thing, as the latter is less pathogenic than the former.

Local man-made environmental changes can also be important. In west Africa and Sudan, for example, the building of dams, large or small, tends to create ideal breeding sites in the fast flows of the spillway.

G. Malaria

Malaria is by far the most important insect-borne disease, killing more than 1 million people every year, most of them African children. The

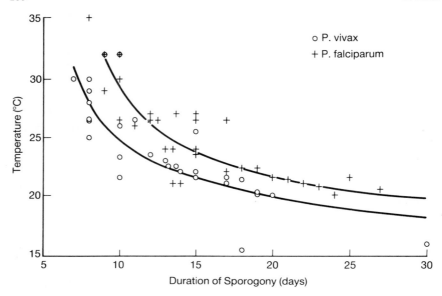

Fig. 2. The relationship between temperature and the duration of the sporogonic cycle in two species of malaria, *Plasmodium falciparum* and *P. vivax*. Redrawn from Macdonald (1952).

importance of temperature as a factor influencing the intensity of malaria transmission is relatively well studied. As with the other mosquito-borne diseases, the length of the extrinsic cycle is very sensitive to temperature, the shape of the curve (Fig. 2) bearing a striking resemblance to that of an arbovirus disease seen in Fig. 1. Again, it seems that low temperatures are limiting, and the rate of development increases rapidly with temperature at the lower end of the scale and more slowly at the higher end. At any given temperature, *Plasmodium vivax* develops faster than *P. falciparum*, and this is presumably one reason why the former extends further into temperate zones than the latter.

One might expect the probability that the parasite establishes itself in the mosquito to be temperature dependent. Lines *et al.* (1991) estimated the probability that a bloodmeal results in mosquito infection in natural populations of *Anopheles gambiae s.l.* in a holoendemic area of Tanzania. They measured the sporozoite rates in successive gonotrophic age groups, and these implied that a constant proportion of females was becoming infected in every cycle. The data were compared with a similar data set, collected by the same techniques, from 25 years earlier, in the early

1960s. The comparison suggested that the probability of infection had increased 2.5-fold, from about 8 to about 21%, in the intervening years. Temperature records from a nearby town indicated that over this period there had been a more or less steady increase of about 0.5°C. At first sight, it seemed that this warming could have been responsible for the increase in the mosquito infection probability, but this explanation was ruled out by a more careful analysis. This showed a greater proportion of the mosquitoes in the earlier data set had been collected during the rather cooler months of the year, so that in fact there was little difference between the two in terms of "mosquito-degrees". After considering a variety of alternative explanations, it was concluded that drug use had caused the increase in the infection probability in two ways. Firstly, during the earlier study, human infectiousness had probably been suppressed by a level of chloroquine use in the community; since then drug use has, if anything, increased still further, but widespread resistance may have released the parasite from its suppressive effects. Secondly, drug use may have selected for more infectious parasite strains, adapted to a higher rate of turnover between human hosts. It therefore seems that the main effect of temperature, at least in these conditions, is not on the probability of parasite development in the vector, but on its duration.

Bradley (1993a) has combined the information in Fig. 2 with Macdonald's expression for the "basic case reproduction rate" (R_0) of malaria. This is the number of secondary cases arising in a fully susceptible population from a single primary case, after one cycle through the mosquito. If R_0 is greater than 1, the infection will spread exponentially; if R_0 is less than 1, the disease will die out.

R_0 is given by the expression

$$R_0 = \frac{ma^2 bp^n}{-r\log_e p}$$

where m is the number of female mosquitoes per person; a is the frequency with which each female bites a human; b is the probability that an infectious bite will lead to human infection; p is the proportion of mosquitoes that survive through 1 day: r is the recovery rate in people; and n is the duration of extrinsic cycle.

Bradley (1993a) derived an expression relating n to temperature in the manner shown in Fig. 2, and made reasonable assumptions about the other parameters, taking them to be unaffected by temperature change. He then analysed the effect on R_0 of a 2°C rise in temperature from a range of different starting temperatures.

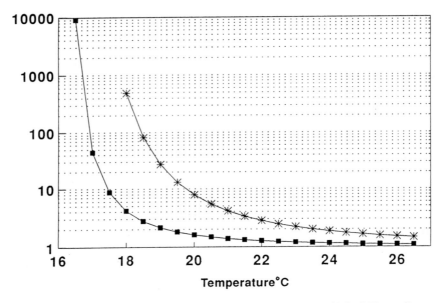

Fig. 3. The effect of temperature on the basic case reproduction rate (R_0) of *Plasmodium falciparum*, for two vectors with moderate or high rates of daily survivorship ($p = 0.8$ (—*—) or 0.95 (—■—)), supposing an increase in temperature from t to $t + 2°C$. The effect is expressed as the ratio between the value of R_0 at $t + 2$ and that at t. (Reproduced from Bradley (1993a) with permission from the CIBA Foundation.

The results are shown in Fig. 3. At low temperatures, a 2°C rise makes a big difference, increasing the basic case reproduction rate dramatically. This is the part of the graph where parasite development in the vector has more or less stopped at the lower temperature, and the effect of the 2°C rise is to permit it just to begin. This creates very large ratios, but does not mean that R_0 will be greater than 1, even at the higher temperature. More realistic situations are seen above about 20°C. The effect then is still substantial, especially with a relatively short-lived vector, and could bring R_0 to above 1 in previously non-malarious areas. At still higher temperatures, above 24°C for example, the effect is much smaller: no more than two-fold, in fact. This is minor compared, say, to the variation in mosquito density between neighbouring villages (Magesa *et al.*, 1991; Smith *et al.*, 1993).

This analysis is broadly consistent with the proposition mentioned earlier, that a 2°C change is equivalent to 300 m in altitude or 180 km in latitude. It stresses that, as with dengue, the most conspicuous effects of global warming are likely to be seen at the edge of the parasite's distribution, where altitude or geography make temperature a key factor. In thoroughly malarious areas, things are likely to get a little worse, but in such places other factors become more important, and temperature recedes into the background. Such factors include human immunity, and in hot dry areas, vector density.

It must be stressed that these calculations are based on measures of the extrinsic cycle at constant temperature, and the effect of diurnal variations may be very large. So far, moreover, we have taken no account of the effects of global warming on rainfall, which perhaps surprisingly is expected to increase marginally in most places, although in a more variable and patchy way than temperature. This will certainly affect the availability of breeding sites, which seems to be the main limiting factor on most tropical mosquitoes. The outcome will depend on whether the local vector population tends (as most do) to peak in the rains, or whether it is one of the few (such as *Anopheles culicifacies* in parts of Sri Lanka) with a dry season peak.

It has also been suggested that increased humidity is likely to lead to increased vector survival. In fact, vector survival is generally inferred from the age-structure of the mosquito population, and changes in survival rates are therefore hard to separate from changes in recruitment. For this reason there are little reliable data on how survival is affected by humidity. However, Lindsay *et al.* (1991) observed that in irrigated rice areas of west Africa, dry season peaks of malaria mosquitoes may exceed those of the rainy season in density, but are not accompanied by the expected peak in malaria. They suggested that this could be due to the very high temperatures, which might kill the parasite inside the mosquito, or could reduce mosquito survival. Evidence in favour of the latter explanation was provided by Ijumba *et al.* (1990), in a ricefield area of Kenya (see Table 2). They also found high densities when the ricefields were flooded during the dry season, and showed that female mosquitoes caught at this time had extraordinarily low parous rates, and consisted almost entirely of very young individuals. It is nevertheless clear that we still have very little idea of how humidity affects mosquito survival.

A third important entomological unknown is whether increased temperature will shorten not only the extrinsic cycle of the parasite but also the gonotrophic cycle of the vector. This seems likely, and it has been suggested (Bradley, 1993a) that this will increase the frequency of feeding on man and so the rate of transmission. However, we do not really know the causes of adult mortality in mosquitos, or where in the cycle – feeding, egg development, or oviposition – it tends to occur. If mortality occurs mainly during feeding or oviposition, then it may well be roughly constant per cycle, however long the cycle takes. The expression

Table 2. *Anopheles arabiensis*: parous rates and sporozoite rates by season in Karime village, Mwea–Tebere irrigation scheme, Kenya. From Ijumba *et al.* (1990)

	Short rains Oct.–Nov.	Dry season Feb.	Long rains May–June
Parous rate	14%	24%	61%
$n =$	791	120	639
Sporozoite rate	0.05%	0%	1.2%
$n =$	4054	54	4204

given earlier for the basic case reproduction rate could then be recast in time units of "per cycle" rather than "per day". It would not be unreasonable to assume that temperature has proportional effects on the extrinsic and the gonotrophic cycles, and if so, the entire expression becomes independent of temperature.

Can we put these observations together to make predictions about the likely effects of global warming? The most dramatic consequences are likely to be seen in places like the highlands of east Africa, Madagascar, Nepal and Papua New Guinea, where temperature is clearly a limiting factor. Signs that this has already started may be seen in the analysis by Loevinsohn (1994) of climatic data and malaria incidence in Rwanda. Since 1987, which was unusually warm and wet, case incidence rates have increased by more than three-fold in some districts and two-fold in the country as a whole. Loevinsohn analysed records at one clinic, at which a consistent and high level of service had been offered over a long period, and found that variation in temperature and rainfall could explain 80% of the variety in monthly malaria incidence. The greatest increases in incidence occurred in the high altitude parts of the clinic's catchment area, which is also where monthly incidence rates were most closely related to temperature rather than rainfall. Interestingly, minimum temperature was a better predictor of incidence than either mean or maximum temperature. All this confirms that malaria in east African highlands is constrained by cold, and is extremely sensitive to small changes in temperature.

Another region of particular concern is the vast area of the former USSR from which malaria was driven during the eradication era (Molineaux, 1988). Warmer springs and hotter summers will, unless there is a decline in rainfall, greatly increase the probability of resurgence. The hazard will be particularly acute where political disturbances interfere with surveillance and promote large-scale population movement.

It seems unlikely that malaria will return to western Europe, except perhaps as small self-limiting outbreaks arising from imported cases or from "airport malaria", when an infected mosquito arrives as a stowaway on an incoming aircraft (Bradley, 1993a). Many, if not all, of the factors that drove malaria from Europe in the first place are still there, and if transmission started it could no doubt be stopped again. Moreover, in the case of *P. falciparum*, the European vectors are not susceptible to tropical strains of the parasite; it seems that the European strain of *P. falciparum* is extinct (Molineaux, 1988).

IV. LAND-USE CHANGES

Against these prospects, we must set the changes that we can see happening now as a result of land use changes – mainly through water resource development and urbanization. These have been reviewed elsewhere (IRRI, 1988; Service, 1989; Knudsen and Sloof, 1992; Bradley, 1993b; Sutherst, 1993; Walsh *et al.*, 1993; Lines *et al.*, 1994), and only a few illustrative points are drawn here.

Rice irrigation may not always increase malaria, but there are many places in Africa where it does (e.g. Coosemans, 1985). Indeed in every continent, man-made environmental changes – mainly water-resource development, deforestation and urbanization – are having the most profound effects on malaria. The effects of these are complex, and dependent on the breeding sites and habitat preferences of the local vector. Thus, ricefields are the main breeding sites for malaria mosquitoes in rice-growing areas of Africa and China but not in southeast Asia. The greatest effects may be seen from the interactions between local development processes like rice projects and global warming – for example in the highlands of Madagascar.

Malaria in most of southeast Asia is associated with the forest, and deforestation leads to a temporary surge in malaria, as people move to the forest, followed by a greater fall, as the highly efficient shade-loving vector *Anopheles dirus* is replaced by the rather less efficient *Anopheles minimus*. Although it is excluded by cold from the highest peaks, malaria in this geographic region is a feature of hill country rather than the plains, because that is where the forest remains. The vectors that breed in the hill streams and forest puddles are much more efficient than the plains species. This illustrates how other ecological factors can override the effect of temperature. In Africa, the opposite situation prevails. The efficient vectors are lowland sun-loving savannah species, and malaria is negatively associated with altitude. In northern Tanzania, deforestation has recently led to the spread of malaria upwards into the hilltop villages of the Usambara mountains (Matola et al., 1987). This was attributed partly to the creation of sunlit breeding sites, but also to the great increase in temperature (from 12.8–15.6°C to 17.7–20.8°C) that accompanied forest clearing. This is considerably larger than the 0.5–1.0°C increase seen in highland Rwanda. Clearly, local processes can be just as important as global ones.

Urbanization, the third major category of land-use change, is generally inimical to malaria, because *Anopheles* mosquitoes with one or two exceptions cannot breed in containers or in highly polluted water. The main exception is *A. stephensi*, one of the most important vectors of the Indian subcontinent. It has adapted to breeding in overhead water tanks, with the result that in many places, there is the unusual situation of transmission in town being just as intense as in the surrounding countryside. *Aedes aegypti* and *Culex quinquefasciatus*, on the other hand, are well adapted to urban conditions throughout the tropics, and are responsible for the transmission in urban areas of dengue and filariasis respectively. There is little doubt that the process of increasing urbanization will continue for a long time to come. For the inhabitants of tropical cities, the risk of dengue and filariasis transmission is likely to be changed much more by the local environmental consequences of continued urbanization than by climate changes at the global or regional level.

V. CONCLUSION

Global climate change is indeed likely to have important consequences for the distribution of vector-borne diseases. In some cases, the effects on disease could be predicted if we could predict the effects on the vectors. This can be expected for those diseases where the life span of the vector is long compared to the extrinsic cycle of the pathogen. With short-lived vectors (which in effect means the mosquitoes) a further degree of complication can be expected, due to the temperature dependence of the pathogen's rate of development in the vector. We might expect the transmission of vector-borne disease to spread to places that are presently too cold, and to increase in intensity in places that are already warm enough. In fact, the former seems likely to be more important than the latter. This is because of the non-linearity of the relationship between temperature and incubation period, seen in several vector–pathogen systems.

Thus at the current limits of distribution of vectors and of the diseases they transmit, the effects of global warming may be more significant than those of local man-made environmental changes. Elsewhere the effects of such land-use changes – deforestation, irrigation, urbanization – are already having profound effects on the distribution and severity of vector-borne disease, effects that are likely to be much greater than those of a modest rise of a few degrees in temperature. The largest effects of all may arise from the interactions between these global and local processes.

Acknowledgements

I am grateful to Colin Leake and Chris Curtis for helpful suggestions, and to the Overseas Development Administration for their support of the Tropical Disease Control and Urban Health Programmes at the London School of Hygiene and Tropical Medicine.

REFERENCES

Bradley, D. J. (1993a). Human tropical diseases in a changing environment. In "Environmental Change and Human Health" (J. V. Lake, G. R. Bock and K. Ackrill, eds), pp. 146–170. *Ciba Foundation Symposium* **175**. Wiley, Chichester.

Bradley, D. J. (1993b). Environment and health and problems of developing countries. In "Environmental Change and Human Health" (J. V. Lake, G. R. Bock and K. Ackrill, eds), pp. 234–246. *Ciba Foundation Symposium* **175**. Wiley, Chichester.

Burke, D. S. and Leake, C. J. (1988). Japanese Encephalitis. In "The Arboviruses: Epidemiology and Ecology Vol. 3" (T. Monath, ed.), pp. 63–92. CRC, Boca Raton, FL.

Cook, G. C. (1992). Effect of global warming on the distribution of parasitic and other infectious diseases. *J. R. Soc. Med.* **85**, 688–691.

Coosemans, M. H. (1985). Comparaison de l'endémie malarienne dans une zone de riziculture et dans une zone de culture de coton dans la pleine de la Ruzizi, Burundi. *Ann. Soc. Belge Méd. Trop.* **65** (Suppl. 2), 187–200.

Dalla Pozza, G. D. and Majori, G. (1992). First record of *Aedes albopictus* establishment in Italy. *J. Am. Mosq. Cont. Assoc.* **8**, 318–320.

Fan Wufang, Yu Shurong and Cosgriff, D. M. (1989). The re-emergence of dengue in China. *Rev. Infect. Dis.* **2**, S847–S853.

Houghton, J. T., Jenkins, G. J. and Ephraums, J. J. (1990). "Climate Change, the IPCC Scientific Assessment". Cambridge University Press, Cambridge.

Ijumba, J. N., Mwangi, R. W. and Beier, J. C. (1990). Malaria transmission potential of *Anopheles* mosquitoes in the Mwea–Tebere irrigation scheme, Kenya. *Med. Vet. Ent.* **4**, 425–432.

IRRI (1988). "Vector-borne Disease Control in Humans through Rice Agroecosystem Management". International Rice Research Institute, Los Baños, the Philippines.

Knudsen, A. B. and Sloof, R. (1992). Vector-borne disease problems in rapid urbanization: new approaches to vector control. *Bull. Wld Hlth Org.* **70**, 1–6.

Leake, C. J. and Johnson, R. T. (1987). The pathogenesis of Japanese encephalitis virus in *Culex tritaeniorhynchus* mosquitoes. *Trans. Roy. Soc. Trop. Med. Hyg.* **81**, 681–685.

Levinsohn, M. E. (1994). Climatic warming and increased malaria incidence in Rwanda. *Lancet* **343**, 714–718.

Lindsay, S. W., Denham, D. A. and McGreevy, P. B. (1984). The effect of humidity on the transmission of *Brugia pahangi* infective larvae to mammalian hosts. *Trans. Roy. Soc. Trop. Med. Hyg.* **8**, 19–22.

Lindsay, S. W., Wilkins, H. A., Zieler, H. A., Daly, R. J., Petrarca, V. and Byass, P. (1991). Ability of *Anopheles gambiae* mosquitoes to transmit malaria during the dry and wet season in an area of irrigated rice cultivation in The Gambia. *J. Trop. Med. Hyg.* **94**, 313–324.

Lines, J. D., Wilkes, T. J. and Lyimo, E. O. (1991). Human malaria infectiousness measured by age-specific sporozoite rates in *Anopheles gambiae* in Tanzania. *Parasitology* **102**, 167–177.

Lines, J., Harpham, T., Leake, C. and Schofield, C. (1994). Trends, Priorities, and Policy Directions in the Control of Vector-borne Diseases in Urban Environments. *Health Policy & Planning*, in press.

Macdonald, G. (1952). The analysis of the sporozoite rate. *Trop. Dis. Bull.* **49**, 569–585.

Magesa, S. M., Wilkes, T. J., Mnzava, A. E. P., Njunwa, K. J., Myamba, J., Kivuyo, M. V. P., Hill, N., Lines, J. D. and Curtis, C. F. (1991). Trial of pyrethroid impregnated bednets in an area of Tanzania holoendemic for malaria. Part II. Effects on the malaria vector population. *Acta Trop.* **49**, 97–108.

Matola, Y. G., White, G. B. and Magayuka, S. A. (1987). The changed pattern of malaria endemicity and transmission at Amani in the eastern Usambara mountains, north-eastern Tanzania. *J. Trop. Med. Hyg.* **90**, 127–134.

Molineaux, L. (1988). The epidemiology of human malaria as an explanation of its distribution, including some implications for its control. *In* "Malaria: Principles and Practice of Malariology (W. H. Wernsdorfer and I. McGregor, eds), pp. 913–998. Churchill Livingstone, Edinburgh.

Moncayo, A. (1992). Chagas' disease: Epidemiology and prospects for interruption of transmission in the Americas. *World Health Stat. Quart.* **45**, 276–279.

Nawrocki, S. J. and Hawley, W. A. (1987). Estimation of the northern limits of the distribution of *Aedes albopictus* in North America. *J. Am. Mosq. Cont. Assoc.* **3**, 314–317.

Peters, R. L. (1993). Conservation of biological diversity in the face of climate change. *In* "Global Warming and Biological Diversity" (R. L. Peters and T. E. Lovejoy, eds), pp. 15–30. Yale University Press, New Haven.

Rai, K. S. (1991). *Aedes albopictus* in the Americas. *Ann. Rev. Entomol.* **36**, 459–484.

Randolph, S. E. (1993). Climate, satellite imagery and the seasonal abundance of the tick *Rhipicephalus appendiculatus* in southern Africa: a new perspective. *Med. Vet. Entomol.* **7**, 243–258.

Service, M. W. (1989). Urbanisation: a hot-bed of vector-borne diseases. *In* "Demography and Vector-borne Diseases" (M. W. Service, ed.), CRC Press, Boca Raton, FL.

Shope, R. (1991). Global climate change and infectious diseases. *Environ. Health Perspect.* **96**, 171–174.

Smith, T., Charlwood, J. D., Kihonda, J., Mwankusye, S., Billingsley, P., Meuwissen, J., Lyimo, E., Takken, W., Teuscher, T. and Tanner, M. (1993). Absence of seasonal variation in malaria parasitaemia in an area of intense seasonal transmission. *Acta Trop.* **14**, 55–72.

Sutherst, R. W. (1993). Arthropods as disease vectors in a changing environment. *In* "Environmental Change and Human Health" (J. V. Lake, G. R. Bock and K. Ackrill, eds), pp. 124–145. *Ciba Foundation Symposium* **175**. Wiley, Chichester.

Walsh, J. F., Molyneux, D. H. and Birley, M. H. (1993). Deforestation: effects on vector-borne disease. *Parasitology* **106**, S55–S75.

Watts, D. M., Burke, D. S., Harrison, B. A., Whitmire, R. E. and Nisalak, A. (1987). Effect of temperature on the vector efficiency of *Aedes aegypti* for dengue 2 virus. *Am. J. Trop. Med. Hyg.* **36**, 143–152.

WHO (1993). "Dengue Prevention and Control". Resolution EB91/SR/13 of the Executive Board of the World Health Organisation.

8

Remote Sensing and the Changing Distribution of Tsetse Flies in Africa

D. J. ROGERS

I.	Introduction	178
II.	Analysing Species' Distributions	179
III.	Presence: Mapping Tsetse Flies in Africa	181
IV.	Precedence: The Example of History	182
V.	Paradigms: The Role of Biological and Statistical Models	182
VI.	Predictions: Preparing for the Future	190
VII.	Conclusions	191
	References	192

Abstract

This chapter examines the four categories of evidence used by researchers, and the activities they undertake, in order to understand the distribution of animals in space and time: presence, precedence, paradigms and predictions. Examples are given of each activity, drawn from experience with tsetse flies (*Glossina* spp.) in Africa.

Present-day distribution maps are often compendia of historical surveys, but provide the best-guesses available to us as the present time. Historical maps, if made at precise moments in time and especially if they precede or follow important ecological events such as disease panzootics, provide an insight into a species' range under either adverse or particularly favourable conditions. The ebb and flow of a species' distribution may be analysed to reveal the abiotic constraints operating on the species, and provide a stimulus for field work to test the predictions arising from such "hind-casting" exercises.

Models of species' distributions are of two main sorts, biological and statistical. The advantages of each are briefly mentioned and it is concluded that a synthesis of the two approaches will lead to the most rapid progress in understanding the major determinants of each species' range.

Arising from the statistical analysis of the distribution of *G. morsitans* and *G. pallidipes* in Kenya and Tanzania, several predictions are made of the impact of global warming of 1 and 3°C on these species' distributions. The predictions depend to a large extent upon the variables within the data set; the omission of only one of these can result in diametrically opposite conclusions. It is concluded that statistical analyses need to be reinforced with biological observations for progress to be made.

I. INTRODUCTION

Table 1 lists four "P"s associated with mapping insect distributions: presence, precedence, paradigms and predictions. Distribution maps tend to be amalgamations of surveys carried out by different people at different times and thus, although based on historical data, record the most likely present-day distribution of the species, a "best guess". They usually record the presence or absence of a species, rarely its abundance. Records of the past distributions of organisms ("precedence" in Table 1), when such distributions are different from the present-day ones, may provide useful insight into current distributions. These examples reveal the subset of the present-day distribution that the species occupied during a previous period of adversity or, alternatively, the area into which the species spread when conditions were particularly favourable. Statistical analysis of distribution maps, or more biologically based predictions of a species' distribution (the "paradigms" of Table 1), tend to be based on extensive or intensive studies respectively, each with strengths and weaknesses (Rogers and Randolph, 1993). The lack of useful biological data for many species forces us to rely heavily on statistical analyses of past and present distributions, which is not a particularly ideal situation when global environments are changing rapidly. The best analytical tools for statistical analysis are still debated, and some of them will be discussed in this chapter. The objective is to identify those environmental variables to which the species' distribution is particularly sensitive. Conclusions from the analyses must always be tested in field situations through careful monitoring of key variables over time, and progressive refinements of the models developed. At some stage the models become robust enough to forecast how a distribution pattern might change, given certain scenarios of future environmental change (the "predictions" of Table 1). The current interest in global warming has heightened our

8. Remote Sensing and Distribution of Tsetse Flies in Africa

Table 1. The four "P"s in understanding insect distributions

P	Insect	Activity
Presence	"is"	Observation
Precedence	"was"	Hind-casting
Paradigms	"could be"	Analysis/modelling
Predictions	"will be" ("might be") there	Guessing

awareness of our ignorance of the major determinants of the distributional limits of most species and stimulated a number of predictions which will almost certainly turn out to be wrong, though thought-provoking at the present time. Table 1 lists this exercise as "guessing" or, more charitably, of learning from our mistakes.

Examples will be given in this chapter of the four activities listed in Table 1, taking tsetse flies in Africa as an example, and showing how greater degrees of uncertainty appear at each stage.

II. ANALYSING SPECIES' DISTRIBUTIONS

Maps of a species' distribution can be analysed, and predictions made, in a number of ways. Early attempts to understand the distribution of both vertebrates (Bodenheimer, 1938) and invertebrates (Gaschen, 1945) involved the use of climate graphs, relating temperature and relative humidity (or rainfall) for each month of the year in areas where each species was known to occur. Gaschen, working on tsetse flies in west Africa, then carried out what is now known as climate matching, trying to predict areas of suitability for each species on the basis of the similarity of the local climate graphs to that of the areas in which each tsetse species occurred naturally. Gaschen also noted the correspondence between tsetse distributions, vegetation and the occurrence of different vertebrate species, hosts of tsetse, along a wet/dry south/north transect in west Africa. Only rarely have the biotic as well as the abiotic factors limiting species distribution been incorporated into such analyses. The natural successors to Gaschen's approach at the present time are models such as BIOCLIM which uses a more statistically based climate matching approach (Nix, 1986; Kohlmann et al., 1988) and CLIMEX (Sutherst and Maywald, 1985; Sutherst et al., Chapter 4, this volume) which seeks to define the abiotic limitations to species' distributions in terms of hot, cold, wet and dry stress indices in combination with a growth index that may incorporate the biotic constraints. Neither approach is particularly well documented in the literature at the present time, and it is not clear how CLIMEX's stress indices may be quantified in field situations since they lack any units of measurement.

A rather different approach to mapping species' distributions arose from quantitative community ecological studies where suites of predictor variables (generally measures of the abiotic environment) were associated with suites of criterion variables (such as species' assemblages) in an attempt to define the interactions between species within communities and their natural environments. Various forms of principal components analysis, logistic and multiple regressions and canonical correlation techniques have been applied to such results (Ter Braak and Prentice, 1988), in some cases especially modified for ecological situations (e.g. canonical correspondence analysis, Ter Braak, 1986; Hill, 1991). For some of these techniques the results are, at least in theory, fairly easy to interpret biologically. For example, partial regression coefficients in a multiple regression analysis may be taken to indicate the importance of each of the original variables, although covariation of the predictor variables often confuses the interpretation of multiple regression results. However, when the analyses involve the rotation of the original co-ordinate axes (as in principal components and canonical correlation analyses) an additional difficulty arises over the biological interpretation of the rotations involved. This is sometimes resolved by correlating the original variables with each of the rotated axes in turn. Whatever are the mechanics involved in these forms of data analysis, the objective is essentially the same in each case, that of data reduction. The biologist would like to produce the smallest subset of either the recorded predictor set or the transformed predictor set that adequately describes the distributions under study. The reasons for this are two-fold. First it is quite unreasonable to imagine that each of the predictor variables originally selected will be of equal importance in determining distributions; in throwing out the less important variables the biologist is drawn naturally to the more important ones, thus gaining biological insight. Secondly, in planning for future monitoring of a species' distribution, it is necessary to know which variables should be measured with great care, since it will be uneconomic to measure them all with equal precision.

It is argued elsewhere that biologists should urgently seek simpler rather than complex forms of data reduction since, with the advent of systems such as automatic weather stations and satellites capable of amassing raw data at an unprecedented rate, there is a danger that we shall be overwhelmed with data in the near future. If unprocessed as they are gathered, these data will be worthless (Rogers and Williams, 1994).

Any form of data reduction must be easy to interpret biologically if the potential users are to be the architects of the analytical techniques rather than the victims of them.

III. PRESENCE: MAPPING TSETSE FLIES IN AFRICA

Tsetse flies (*Glossina* species) are of sufficient economic importance in Africa to have been studied and mapped for at least 100 years. Our current maps (Ford and Katondo, 1977) are based on historical records going back to the beginning of the colonial era. A recent attempt to model the distribution of the tsetse *Glossina morsitans* in Zimbabwe (Rogers and Williams, 1993) revealed that the fly distribution map being used for the analysis (from Ford, 1971) was based on patchy records by explorers and colonial officers and a good deal of sensible interpolation by the map makers. Ford had relied heavily on the distribution maps published by Jack (Jack, 1914, 1933) where the recorded areas of fly presence occupy less than 20% of the total area thought to contain flies. It was necessary to use historical maps because two important events (a rinderpest panzootic at the end of the last century and a vigorous anti-tsetse control campaign in this century) have reduced the present-day tsetse distribution in Zimbabwe to a fraction of its former extent. Since the aim of the analysis was to discover those environmental conditions in which tsetse could survive it was important to use a distribution map predating these two events. The original map makers had appreciated enough about tsetse biology to suspect that the flies in Zimbabwe could not survive at higher altitudes associated with lower mean temperatures and had produced a distribution map which, in many places, followed the elevation contours associated with an annual mean temperature of 19.4°C. It was perhaps hardly surprising, therefore, that our statistical analyses identified temperature as a major limiting factor in Zimbabwe and produced predictions of fly distribution that were remarkably similar to the historical maps. Whether either the maps or the predictions were at all accurate will remain a mystery.

The distribution of the same species of tsetse, *G. morsitans*, in Kenya and Tanzania has been mapped much more recently and is more likely to be related to the contemporary climate data available for this region. The comparison of this species' distribution in Zimbabwe on the one hand and Kenya and Tanzania on the other therefore involves not only different places, but also different times (Rogers and Randolph, 1993).

In contrast to the dramatic changes that have taken place in the distribution of *G. morsitans* in Zimbabwe, the continental distribution of the riverine and forest-dwelling *palpalis* group has changed remarkably little over almost a century. Roubaud's 1909 map (Roubaud, 1909) is similar to that for the same species group in 1973 (Ford and Katondo, 1977; Rogers and Randolph, 1986). The greater change in the distribution of the savannah, rather than the riverine and forest, species of flies in Africa may be attributed to the greater human impact on the savannah regions, both their vegetation and their wild animals.

IV. PRECEDENCE: THE EXAMPLE OF HISTORY

The rinderpest panzootic that affected tsetse in Zimbabwe swept from Eritrea on the Red Sea coast to South Africa in the space of about 7 years, between 1889 and 1896 (Ford, 1971). As an introduced disease it eliminated more than 90% of domestic stock and, in all probability, a similar proportion of wild animals, especially bovids and suids. This in turn caused a massive reduction in the geographical extent of the savannah tsetse, especially the *morsitans* group, that relies heavily on these hosts. It now appears that the recovery of the wild game populations, and therefore the tsetse, took many decades. The colonial entomological literature on tsetse, especially that of East Africa, contains many records of fly advances at rates of a few kilometres a year, sustained over many years (e.g. Buxton, 1955). Figure 1 shows the distribution of *G. morsitans* in Tanzania in *c.* 1912/13 and 1937–40 (from Kjekshus, 1977). Figure 1(b) shows that between these two dates tsetse distributions expanded in many areas and the flies spread into the regions west of Lake Victoria and south-east of the Mbeya/Iringa line.

One interpretation of these maps is that the sudden disappearance of their hosts throughout the region caused the disappearance of flies from abiotically marginal habitats. If this is the case an analysis of the present-day distributions may be used to distinguish the optimal from the suboptimal parts of the flies' range, and therefore to identify the key abiotic limitations on tsetse in Tanzania. These may in turn be used to predict how tsetse distribution will change as environments change. Analysis and prediction of tsetse distributions are dealt with in the next two sections.

V. PARADIGMS: THE ROLE OF BIOLOGICAL AND STATISTICAL MODELS

Biological analysis of a species' distribution generally attempts to quantify the demographic parameters of birth, death, immigration and emigration within a conceptual framework of the balance between input and output population processes. In the case of tsetse, the conceptual framework

$$\sum_{n=1}^{n=12} f_n \geq \sum_{n=1}^{n=12} d.i.m._n \qquad (1)$$

(where f is log. monthly fertility, $d.i.m.$ is density independent mortality per month and $n = 1,12$ for the 12 months of the year) defines the distributional limits of a species as those areas where fertility exceeds or just balances density independent mortality on an annual basis (Rogers, 1979). It is therefore imagined within this framework that density dependence, the prime determinant of abundance within the distributional limits, has little or no effect on the limits themselves, because density is too low in such marginal areas for density dependence to be of any great importance (although the continued existence of a species in marginal areas requires there to be some form of negative feed-back on population size). Quantifying each side of Eq. (1) often takes many years of study. In the case of the tsetse the left hand side of the inequality, fertility, is sensitive to temperature, but appears to be unaffected by other environmental variables within the ranges normally experienced in Africa. The mortality on the right hand side, however, is related to both temperature and atmospheric moisture and requires careful field work over very long periods of time before the precise relationships between mortality and enironmental variables can be established. Because such studies are labour intensive they are carried out in relatively restricted areas, often well within the geographical boundaries of a species' range. In using the results of such studies to predict geographical ranges, therefore, two assumptions are made. The first is that a study of a species well within its geographical limits may be used to reveal the abiotic constraints on survival at the edges of the species' range; the second is that the constraints identified in one region operate throughout the continental range of the species. Despite these untested assumptions, it has been shown that analysis of data for *G. morsitans* from only two sites in Africa (Nigeria and Zambia) allows predictions to be made of the pan-African bioclimatic limits of this species (in fact three subspecies), and these predictions appear to coincide remarkably well with the distribution map of this species. A significantly greater proportion (>95%) of sites within the mapped distribution fall within the predicted bioclimatic limits than outside them, while a sample of sites outside the present limits is more or less equally divided, some sites falling within and the rest without the predicted limits (Rogers, 1979). One lesson to be learnt from this analysis is that whilst the biological approach may identify with reasonable precision those areas where flies cannot occur, because conditions are too extreme, flies will not always occur in areas

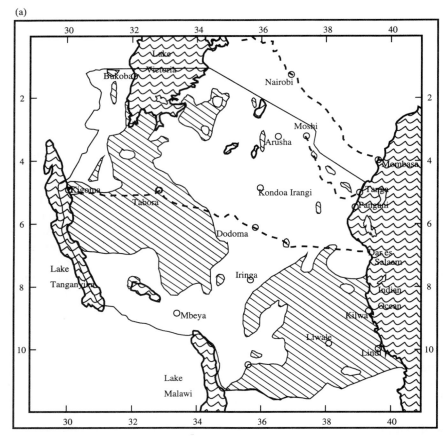

Fig. 1. The distribution of tsetse flies (probably *Glossina morsitans*) in Kenya and Tanzania in (a) 1913 and (b) 1937 is shown by the crosshatched areas on these maps (redrawn from Kjekshus, 1977). Notice the absence of flies west of Lake Victoria and in the region south-east of the Mbeya/Iringa line in (a) the possible result of the rinderpest panzootic that swept through this area between 1890 and 1896 and killed more than 90% of tsetse host animals (dashed lines are railways).

that are apparently suitable for them. Species do not always fully occupy their "ecological space". This conclusion has implications for the statistical analyses of a species' distribution, to which we now turn.

Species can be imagined to occur within a multivariate subset of environmental space defined by temperature, humidity, rainfall, etc. in their natural environments. Sites within the distributional range can be imagined to form a "cloud" of points in this multivariate space, while sites outside the distributional range will form a different cloud. If the

8. Remote Sensing and Distribution of Tsetse Flies in Africa

(b)

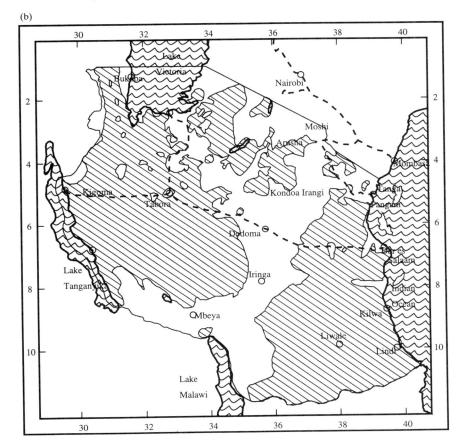

Fig. 1b.

multivariate means, the centroids, of the two clouds are separate, but the clouds are otherwise identical (i.e. in terms of the covariances of the variables) the technique of linear discriminant analysis may be used to define each of the clouds and a plane, the discriminant axis, which separates them. This technique can therefore be used to assign to a point on a map the probability with which the conditions at that point are suitable or unsuitable for the species in question (Green, 1978). If each point on the map is examined in turn, the end result is a map of suitability for the occurrence of the species throughout the region or continent.

In order to carry out discriminant analysis, therefore, information is required to define the clouds for presence and absence. This is usually gathered together within a geographical information system (GIS) where the observed distribution map and each of several environmental variables form a geographically coregistered data set which is sampled to give a "training set" defining the clouds. When the area considered is small, all points may contribute to the training set, but when it is large a subsample may be taken. By comparing the map of predicted probability of occurrence with the actual distribution map, four types of areas may be recognized. The first two are areas of correct predictions of presence or absence, the third is areas of incorrect predictions of presence ("false positives") and the fourth is areas of incorrect predictions of absence ("false negatives"). False positives may be examples of unoccupied environmental space referred to above and hence, in themselves, do not invalidate this approach. False negatives, however, indicate a potentially more serious situation; that the technique itself may be at fault. Fortunately, even with the relatively unsophisticated technique of linear discriminant analysis, the final percentage of false negative results is usually less than about 5%, and generally much less than the false positive figure. Both false positive and false negative predictions have important biological implications. False positive areas may well be at risk of invasion by tsetse from the nearest areas of fly presence. They may thus provide ecological corridors along which flies will move. Areas may be protected from re-invasion by placing barriers across such ecological corridors, if possible. False negative areas indicate places where the flies occur despite apparently adverse environmental conditions. It is possible that such areas represent an heterogeneous mixture of abiotically suitable and unsuitable habitats, in places good but on average bad, or biotically favourable places where, for example, an abundance of hosts more than balances the abiotically adverse conditions. In either case, false positive and false negative areas are places where biologists might learn more about the environmental relationships of the species under study, to refine future predictive modelling.

If false positive areas are indeed suitable for the species, they are inappropriately included in the areas of absence in the training set, and will therefore increase the size of the absence cloud of data points, resulting in a larger proportion of predicted areas of unsuitability. Discriminant analysis will therefore be somewhat conservative in its predictions of a species' presence.

There is often a considerable degree of overlap of the climatic conditions for sites where the flies are present or absent. This suggests that any technique that bases its predictions only on the areas of species' presence (i.e. defines only the environmental space occupied by the

8. Remote Sensing and Distribution of Tsetse Flies in Africa

species, not the areas not occupied by it) is likely to produce predictions of more widespread species' distribution than is in fact the case. It may therefore be preferable to adopt the techniques proposed here, in which one (or, eventually, a series) of non-tsetse areas serves to encompass and so restrict the areas of tsetse suitability in multivariate (and hence geographical) space (fig. 1 in Rogers and Randolph, 1993).

An example of the application of linear discriminant analysis is given in Plate 3 which records the distribution of the tsetse *G. morsitans* in Kenya and Tanzania as a series of horizontal lines. Subsets of approximately 150 data points, each from areas of fly presence and absence, formed a training set for the analysis. The relationships between the variables in the training set were then used to define the probability of occurrence of this species in each of the grid squares shown on the map (a total of c. 25 000 squares). The data used in the analysis are interpolated elevation and meteorological variables (temperature, rainfall, etc.) and remotely sensed normalized difference vegetation indices (NDVIs) from the NOAA series of satellites (Prince *et al.*, 1990; Rogers and Randolph, 1991). NDVIs have been shown to be related to the photosynthetic activity and biomass of vegetation, and are available as 10-day or monthly composite images that record the seasonal cycles of photosynthetic activity associated with rainfall. The overall accuracy of the map in Plate 3 is good, with 84% correct predictions, 13% false positives and 3% false negatives. The average values for each of the predictor variables used to produce this map are given in Table 2, from which it can be seen that the mean difference in temperature between areas of tsetse presence and absence in this part of Africa is less than 1°C, and mean vegetation indices differ by c. 0.2 (the theoretical range of NDVI is between -1 and $+1$, but for all practical purposes it varies between about -0.4 and $+0.7$). If these figures genuinely reflect the tsetse's sensitivity to its natural environment it can be predicted that the distribution of this species is likely to wax and wane with even slight changes in environmental conditions.

One of the problems encountered in exercises of this sort is that many of the temperature variables are interrelated, as are the NDVIs. As a result, different runs of the model, using different training sets, give slightly different answers in terms of the overall importance of each of the predictor variables, and their mean values. Nevertheless, the probability maps are remarkably constant.

It has been shown elsewhere that whereas a single predictor variable appears to be sufficient to describe the distribution of *G. morsitans* at the edge of its range, in Zimbabwe, a number of predictors are required to describe satisfactorily the distribution of this species in Kenya and Tanzania, as shown in Plate 3 (Rogers and Randolph, 1993). This is not surprising since, at the southern edge of the range of tsetse in Africa, conditions are clearly seasonally too cold for year-round survival of the flies. Within areas where this overriding restriction does not apply, the patchiness of the fly's distribution is likely to be determined by one limiting factor in one area and another elsewhere. Finally, the northern limits of the distribution of *G. morsitans* in Africa lie along the southern edge of the Sahel in west Africa. Here it is predicted that once again a single variable, this time high rather than low temperature, restricts the fly's range. Determining the geographically varying constraints on the distribution of widespread species such as *G. morsitans* is a major challenge for the future.

Plate 4 shows the result of analysing the distribution of *G. pallidipes* in Kenya and Tanzania (details in Table 2). This species has a more patchy distribution than does *G. morsitans*, concentrated in the more coastal regions of the two countries. Nevertheless, the overall fit of the model is almost as good as that for *G. morsitans*, with a rather higher percentage of false positives, but the same percentage of false negatives.

It is possible to use the results shown in Plate 3 to illuminate the historical changes recorded in Fig. 1 by setting different threshold probabilities within the analysis to delineate areas of fly presence from areas of absence. In these examples it is convenient to use just two colours, one for presence and one for absence. If the probability threshold is set very high the analysis will pick out only those areas which are very close to the centroid of the cloud of points associated with fly presence. It may be argued that these areas are the climatic "home" of the fly, to which it will retreat in adversity. As the threshold is lowered the area predicted to be suitable expands until, when it is 0.5, the area of suitability coincides with the areas shown in Plate 3 to have a probability of fly presence of 0.5 or greater. Plate 5 shows the result of setting the threshold to 0.95. In this plate the regions immediately south and west of Lake Victoria are no longer predicted to be suitable for the flies whilst the southern/central part of the eastern block of flies is also unsuitable, although it is almost surrounded by areas of suitability. This block shows a striking similarity in shape to the areas of the map in Fig. 1 inhabited by flies during the immediate post-rinderpest period, in 1912/13. It may therefore be possible to use "hind-casting" techniques to predict how the flies should have spread from these residual infestations, and to test these predictions against historical data on the routes and rates of spread of fly advances in the region (e.g. in Buxton, 1955).

Table 2. Average values of the predictor variables used in the discriminant analysis of the distributions of *Glossina morsitans*, and *G. pallidipes* in Kenya and Tanzania

	ND_{mean}[a]	T_{mmm}	ND_{min}	ND_{ran}	Elev.	T_{xmm}	T_{xmx}	T_{nmn}	ND_{max}	T_{nmm}
G. morsitans										
Abs[b]	0.24	22.2	0.02	0.50	1058	24.08	30.41	14.90	0.49	20.38
Pres	0.44	22.8	0.10	0.63	921	24.87	30.83	14.63	0.70	20.67
Rank	1	2	3	4	5	6	7	8	9	10
G. pallidipes										
Abs	0.28	22.37	0.03	0.51	1025	24.30	30.46	14.85	0.52	20.51
Pres	0.42	23.52	0.14	0.54	674	25.52	31.16	16.24	0.65	21.43
Rank	1	8	9	6	2	3	4	5	10	7

[a]Variables are elevation (Elev., in metres), normalized difference vegetation index (NDVI), monthly mean (ND_{mean}), minimum (ND_{min}), maximum (ND_{max}) and range (ND_{ran}), maximum of the monthly maximum temperature °C (T_{xmx}), maximum of the monthly mean temperature (T_{xmm}), mean of the monthly mean temperature (T_{mmm}), minimum of the monthly mean temperature (T_{nmm}) and minimum of the monthly minimum temperature (T_{nmn}).
[b]Abbreviations: Abs, sites where the vectors are absent; Pres, sites where the vectors are present; Rank refers to the order of importance of each variable (as determined by the analysis) for the production of the predicted distribution maps.

VI. PREDICTIONS: PREPARING FOR THE FUTURE

The most satisfactory way to predict how vector distributions will change in the future, under given scenarios of climate change, is to use the results from the biological approach outlined in the previous section to predict how the shifting balance between the two sides of the inequality in Eq. (1) translates to a changed spatial distribution over time. The statistical alternatives to this approach are easier to undertake, but more difficult to justify. Examples are given in Plates 6 and 7. Plate 6 shows the impact of an average increase in temperature of 1°C and 3°C (the average levels of increase predicted by global climate models for the years 2020 and 2100) on the distribution of *G. morsitans* in Kenya and Tanzania and Plate 7

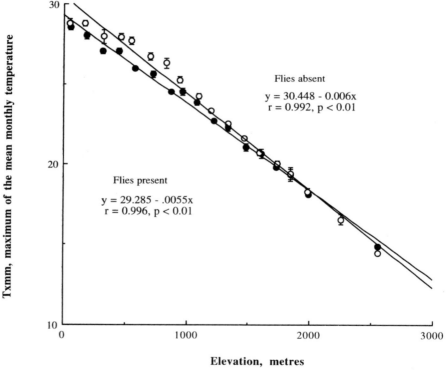

Fig. 2. The relationship between the maximum of the mean monthly temperature (T_{xmm}) and elevation for sites in Kenya and Tanzania where *G. pallidipes* is absent (open circles) or present (filled circles) (means of data assigned to temperature categories, $+/-1$ SE). Notice that at any particular elevation flies are present in cooler rather than warmer areas, explaining why increasing temperatures are predicted to cause the disappearance of this species from Kenya and Tanzania (Plate 7).

shows a similar result for *G. pallidipes*. Whereas the range of the first species is predicted to expand, that of the second disappears completely at the higher temperature. The reason for this is revealed in Fig. 2 which shows the relationship in the training set (drawn from the present-day distribution) between temperature and elevation for areas of presence and absence of *G. pallidipes*. It appears from Fig. 2 that at all elevations *G. pallidipes* occurs in cooler rather than warmer places. The difference between the two temperatures decreases with increasing elevation. If average temperatures increase, the presently cooler areas at each elevation will become as warm as the warmer areas are now and flies will disappear, as the map predicts. However, this prediction may be entirely erroneous. Plate 8 shows the prediction of the change in the distribution of *G. pallidipes* with a 1°C and 3°C increase in temperature when elevation is removed as a variable from the analysis (the fit to the present-day distribution when elevation is removed is only marginally worse than that shown in Plate 4). Here the range is predicted to expand, as does that of *G. morsitans*. Which, if either, of Plates 7 and 8 correctly predicts the impact of global warming on the distribution of *G. pallidipes* in east Africa is a matter for speculation, reinforcing the Scandinavian proverb "It is difficult to make predictions, especially about the future".

With so many problems involved in predicting the future distribution of vectors of tropical diseases, predicting the changing distributions of the diseases themselves adds several more layers of complexity to the task (Rogers and Packer, 1993). The scientific justification for these predictions is rarely as strong as the political pressures to which they are a response. A more sensible response to the same pressures is to try to understand the present-day patterns of the distribution of vectors and disease, a much neglected topic.

VII. CONCLUSIONS

The remarkable capacity of insects to survive in generally hostile terrestrial environments makes them good candidates for both intensive and extensive studies on the determinants of the distribution and abundance of animals in space and time. Their generally high fertility is balanced by a high generation mortality, much of which is density independent (Rogers, 1983). Even allowing for sampling errors, therefore, mortality rates are likely to be variable enough for their correlation with environmental factors to be demonstrable statistically. Whether or not the correlations arise from causation, they may be used with care to make predictions about the mortality of the same species in other places, or at other times. The more different these other places and times are from the present, however, the less reliable the predictions are likely to be. Carefully chosen networks of sites, as envisaged by the International Geosphere–Biosphere Programme (IGBP), may be used to monitor populations locally, building up a regional and continental picture of the biology of key species. Statistical methods may then be used to fill in the gaps between the monitoring sites.

Tsetse are often exceptions to general rules about insects. Larviparous reproduction is coupled with a very low average mortality rate (of no more than 3% per day) which probably varies less than that of some vertebrates. Nevertheless, progress has been made on both biological and statistical fronts in understanding the distribution of several key species of tsetse flies in Africa. It is suspected that the fly's response to local climate varies geographically (Rogers, 1990), as do the range and impact of natural enemies of tsetse (Swynnerton, 1936). Only by confronting statistical predictions with the results of carefully executed field work will we be able to refine our interpretation of the past, our understanding of the present and our predictions of the future.

Acknowledgements

The climatic data for East Africa were kindly provided by Dr Tim Robinson. Satellite sensor data were provided by the ARTEMIS program at the FAO Remote Sensing Centre, Rome, courtesy of Jelle Hielkema and Fred Snijders and by USAID/FEWS NASA GDFC, courtesy of Barry Henricksen. Dr Sarah Randolph kindly read and commented on the manuscript. Image and data processing were carried out using equipment provided by the NERC, project GR3/7524, and ODA project X0239. Finally, I thank the Symposium organizers, Nigel Stork and Richard Harrington, for their invitation to participate in this meeting.

REFERENCES

Bodenheimer, F. S. (1938). "Problems of Animal Ecology". Clarendon Press, Oxford.
Buxton, P. A. (1955). "The Natural History of Tsetse Flies". H. K. Lewis, London.
Ford, J. (1971). "The Role of the Trypanosomiases in African Ecology: a Study of the Tsetse Fly Problem". Clarendon Press, Oxford.
Ford, J. and Katondo, K. M. (1977). "The Distribution of Tsetse Flies in Africa". OAU, Cook, Hammond & Kell, Nairobi.
Gaschen, H. (1945). Les glossines de l'Afrique Occidentale Française. *Acta Tropica* Suppl. **2**, 1–131.
Green, P. E. (1978). "Analyzing Multivariate Data". Dryden Press, Hinsdale, IL.
Hill, M. O. (1991). Patterns of species distribution in Britain elucidated by canonical correspondence analysis. *J. Biogeogr.* **18**, 247–255.

Jack, R. W. (1914). Tsetse fly and big game in Southern Rhodesia. *Bull. Entomol. Res.* **5**, 97–110.

Jack, R. W. (1933). The Tsetse Fly Problem in Southern Rhodesia. *Rhodesia Agric. J. Bull.* **892**, 1–20.

Kjekshus, H. (1977). "Ecology Control and Economic Development in East African History". Heinemann, London.

Kohlmann, B., Nix, H. and Shaw, D. D. (1988). Environmental predictions and distributional limits of chromosomal taxa in the Australian grasshopper *Caledia captiva* (F.). *Oecologia* **75**, 483–493.

Nix, H. (1986). A biogeographic analysis of Australian elapid snakes. *In* "Atlas of Elapid Snakes of Australia" (R. Longmore, ed.), pp. 4–15. Australian Government Publishing Service, Canberra.

Prince, S. D., Justice, C. O. and Los, S. O. (1990). "Remote Sensing of the Sahelian Environment: a Review of the Current Status and Future Prospects". Directorate General VIII: Joint Research Centre Ispra Establishment: Technical Centre for Agricultural and Rural Development. Commission of the European Community, Brussels.

Rogers, D. J. (1979). Tsetse population dynamics and distribution: a new analytical approach. *J. Anim. Ecol.* **48**, 825–849.

Rogers, D. J. (1983). Pattern and process in large-scale animal movement. *In* "The Ecology of Animal Movement" (I. R. Swingland and P. J. Greenwood, eds), pp. 160–180. Clarendon Press, Oxford.

Rogers, D. J. (1990). A general model for tsetse populations. *Insect Sci. Appl.* **11**, 331–346.

Rogers, D. J. and Packer, M. J. (1993). Vector-borne diseases, models, and global change. *Lancet* **342**, 1282–1284.

Rogers, D. J. and Randolph, S. E. (1986). Distribution and abundance of tsetse flies. *J. Anim. Ecol.* **55**, 1007–1025.

Rogers, D. J. and Randolph, S. E. (1991). Mortality rates and population density of tsetse flies correlated with satellite imagery. *Nature* **351**, 739–741.

Rogers, D. J. and Randolph, S. E. (1993). Distribution of tsetse and ticks in Africa: past, present and future. *Parasitol. Today* **9**, 266–271.

Rogers, D. J. and Williams, B. G. (1993). Monitoring trypanosomiasis in space and time. *Parasitology* **106** (Suppl.), S77–S92.

Rogers, D. J. and Williams, B. G. (1994). Tsetse distribution in Africa: seeing the wood *and* the trees. *In* "Large-Scale Ecology and Conservation Biology" (P. J. Edwards, R. M. May and N. Webb, eds). Blackwell Scientific Publications, Oxford, in press.

Roubaud, M. E. (1909). "La *Glossina palpalis* R. Desv.; sa Biologie, son Rôle dans l'Etiologie des Trypanosomiases". Paris.

Sutherst, R. W. and Maywald, G. F. (1985). A computerised system for matching climates in ecology. *Agric. Ecol. Env.* **13**, 281–299.

Swynnerton, C. F. M. (1936). The tsetse flies of East Africa. *Trans. R. Entomol. Soc. London* **84**, 1–579.

Ter Braak, C. J. F. (1986). Canonical correspondence analysis: a new eigenvector technique for multivariate direct gradient analysis. *Ecology* **67**, 1167–1179.

Ter Braak, C. J. F. and Prentice, I. C. (1988). A theory of gradient analysis. *Adv. Ecol. Res.* **18**, 271–317.

Part III.
Changes in Gas and Pollutant Levels

9. The Impact of Elevated Atmospheric CO_2 on Insect Herbivores
   A. D. WATT, J. B. WHITTAKER, M. DOCHERTY, G. BROOKS, E. LINDSAY AND D. T. SALT
10. Insect Herbivores and Gaseous Air Pollutants – Current Knowledge and Predictions
    V. C. BROWN
11. Deficiency and Excess of Essential and Non-essential Metals in Terrestrial Insects
    S. P. HOPKIN
12. Chironomidae as Indicators of Water Quality – With a Comparison of the Chironomid Faunas of a Series of Contrasting Cumbrian Tarns
    L. C. V. PINDER AND D. J. MORLEY

9

The Impact of Elevated Atmospheric CO_2 on Insect Herbivores

A. D. WATT, J. B. WHITTAKER, M. DOCHERTY,
G. BROOKS, E. LINDSAY AND D. T. SALT

I. Introduction 198
II. Approaches to Studying the Impact of Elevated CO_2 on
   Insect Herbivores 198
III. Impact of Elevated CO_2 on Insect Performance and
   Abundance 202
IV. Response of Insects to Elevated CO_2 and Plant Chemistry 207
V. CO_2, Temperature and Other Factors 213
   References 215

Abstract

Atmospheric CO_2 is predicted to be double its current level in about 100 years. The impact of this rise on insect herbivores has been explored in studies on 15 insect species on 10 plant species, a total of 19 separate insect–plant interactions. These studies have tended to show that insect herbivores in elevated CO_2 grow more slowly, consume more plant material, take longer to develop, suffer heavier mortality, and are, therefore, likely to be less abundant in an elevated CO_2 environment. Chemical analysis of plant foliage has shown that elevated CO_2 causes a decrease in foliar nitrogen concentration, and the degree to which insect development is depressed in elevated CO_2 is correlated with the decrease in plant nitrogen. However, the research published to date has been biased towards leaf-chewing insects, and the impact of elevated CO_2 on the levels of antiherbivore chemicals is poorly understood. In addition, the combined impact of elevated CO_2 and other climatic and environmental impacts requires study.

I. INTRODUCTION

The concentration of CO_2 in the atmosphere is rising from its current level of about 355 μmol mol^{-1} at around 1.8 μmol mol^{-1} per annum (Houghton et al., 1992). It is estimated that atmospheric CO_2 will reach 600 μmol mol^{-1} before 2075, and be double the current level in about 100 years. This is due to fossil fuel consumption and land-use change, in particular tropical deforestation (Houghton et al., 1992). Research on the possible impact of elevated CO_2 has increased as more reliable scenarios for climate change have been formulated in recent years. Most of this research has concentrated on plant growth and physiology (Eamus and Jarvis, 1989; Bazzaz, 1990; Dahlman, 1993) but several studies on the impact of elevated CO_2 on insect herbivores have been and are now being carried out. The aim of this chapter is to review the published work on the impact of elevated CO_2 on insects. Reference is also made to some of the work on plant chemistry and phenology, conducted separately or in conjunction with studies on insect herbivores, because of their relevance to the impact of elevated CO_2 on insect herbivores.

It is not the aim of this chapter to review research on impacts of change in climate other than CO_2 although in the same period that atmospheric CO_2 is predicted to rise to 600 μmol mol^{-1}, average temperatures are predicted to rise by between 1 and 2.5°C (Houghton et al., 1992). The general topic of insects and climate change has already been dealt with in this book, and we limit ourselves to a brief discussion of the combined impacts of different aspects of environmental change.

II. APPROACHES TO STUDYING THE IMPACT OF ELEVATED CO_2 ON INSECT HERBIVORES

Since 1984, there have been 15 papers published containing original data on the effects of elevated CO_2 on plant-feeding insects (Table 1). Of the 15 insect species covered by these studies, the majority (eight) have been Lepidoptera, but two orthopteran, one coleopteran, three hemipteran, and one thysanopteran species have also been studied. Thus, although most emphasis has been placed on the impact of elevated CO_2 on plant-chewing insects, there has been some work on plant-sucking insects, but to date no published work on leaf-miners, gallers or borers.

Regarding the host-plants of these insects, the emphasis has been on agricultural crops (soybean *Glycine max*, lima bean *Phaseolus lunata*, tomato *Lycopersicon esculentum*, peppermint *Mentha piperita*, and cotton, *Gossypium hirsutum*) (Table 1). In addition, several papers have been published on *Junonia coenia*, which feeds on the perennial herbaceous weed *Plantago lanceolata*, two papers have been published on the grasshoppers which feed on the desert shrub sagebrush, *Artemisia tridentata*, and one paper on insects on the deciduous trees *Populus tremuloides*, *Quercus rubra*, and *Acer saccharum* (Table 1).

Experiments on the effects of elevated CO_2 on insects have been carried out using plants grown in either greenhouses with no temperature or light control (e.g. Tripp *et al.*, 1992), growth chambers with varying degrees of temperature control and full or partial artificial lighting (e.g. Fajer, 1989a; Johnson and Lincoln, 1990, 1991; Lindroth *et al.*, 1993), or in open-topped chambers, where plants were subjected to fluctuating temperatures, usually slightly higher than ambient (e.g. Butler *et al.*, 1986). Akey *et al.* (1988) carried out experiments in both open-topped chambers and completely open CO_2 enrichment field plots. In most studies, plants were exposed to two CO_2 concentrations, one at, or approximating to, ambient (usually around 350 μmol mol^{-1}), and another in the range 650–700 μmol mol^{-1}. In two studies a higher CO_2 concentration of around 1000 μmol mol^{-1} was used (Osbrink *et al.*, 1987; Tripp *et al.*, 1992), and in some cases an intermediate level (500 μmol mol^{-1}) or a pre-industrial level (270 μmol mol^{-1}) was used in addition to the 650–700 μmol mol^{-1} and ambient concentrations. For the purpose of summarizing results of these studies in this chapter, the results at "intermediate" and pre-industrial CO_2 concentrations were ignored, but the studies where the only elevated CO_2 treatment was 1000 μmol mol^{-1} were included.

In most studies on the effects of elevated CO_2, insect performance has been evaluated on excised plant material inside petri dishes or similar containers. However, some work has been done on whole plants (Osbrink *et al.*, 1987; Akey and Kimball, 1989; Fajer *et al.*, 1991a). Several aspects of insect performance have been considered. Most studies have included some measure of larval growth, either weight gain, weight at a certain stage of the life cycle, or relative growth rate (RGR). Several studies have also considered insect survival, larval development rates, consumption rates, approximate digestibility and related measures of food utilization, and one study also examined the effect of elevated CO_2 on insect fecundity (Fajer *et al.*, 1991a). In addition, direct measurements of insect abundance and rates of increase were made in two studies on *Bemisia tabaci* and other insects on cotton (Butler, 1985; Butler *et al.*, 1986), and on *Trialeurodes vaporariorum* on tomato (Tripp *et al.*, 1992).

In summary, the effects of elevated CO_2 have been studied on 15 insect herbivores on 10 plant species, a total of 19 separate insect–plant interactions (Table 1).

Table 1. Published studies on the impact of elevated CO_2 on insect herbivores. ([1]*Melanopus differentalis* only. [2]*Bemisia tabaci* only)

Authors	Insect species	Plant species	CO_2 concentration (μmol mol^{-1})
Lincoln *et al.*, 1984; Lincoln *et al.*, 1986	*Pseudoplusia includens* (Lepidoptera)	*Glycine max* (soybean)	350, 500, 650
Butler, 1985; Butler *et al.*, 1986[2]	*Bemisia tabaci*, *Empoasca* spp. (Hemiptera) *Frankliniella* sp. (Thysanoptera) *Chaetocnema ectypa* (Coleoptera) *Pectinophora gossypiella* (Lepidoptera)	*Gossypium hirsutum* (cotton)	ambient, 500, 650; 355, 500, 650
Osbrink *et al.*, 1987	*Trichoplusia ni* (Lepidoptera)	*Phaseolus lunata* (lima bean)	340, 1000
Akey *et al.*, 1988	*Pectinophora gossypiella* (Lepidoptera)	*Gossypium hirsutum* (cotton)	field- 360, 371 chambers- 522, 650
Akey and Kimball, 1989	*Spodoptera exigua* (Lepidoptera)	*Gossypium hirsutum*	320, 640

Reference	Insect	Host plant	CO$_2$ (ppm)
Fajer, 1989a; Fajer et al., 1989, 1991a	Junonia coenia (Lepidoptera)	Plantago lanceolata (plantain)	350, 700
Lincoln and Couvet, 1989	Spodoptera eridania (Lepidoptera)	Mentha piperita (peppermint)	350, 500, 650
Johnson and Lincoln, 1990, 1991[1]	Melanopus differentialis Melanopus sanguinipes (Orthoptera)	Artemisia tridentata	270, 350, 650; 350, 650
Tripp et al., 1992	Trialeurodes vaporariorum (Hemiptera)	Lycopersicon esculentum (tomato)	350, 1000
Lindroth et al., 1993	Lymantria dispar	Populus tremuloides (quaking aspen) Quercus rubra (red oak) Acer saccharum (sugar maple)	385, 642
	Lymantria dispar Lymantria dispar		
	Malacosoma disstria Malacosoma disstria Malacosoma disstria (Lepidoptera)	Populus tremuloides Quercus rubra Acer saccharum	

III. IMPACT OF ELEVATED CO_2 ON INSECT PERFORMANCE AND ABUNDANCE

Of the 13 cases where larval growth was measured as relative growth rate, only two studies, *Pseudoplusia includens* on *Glycine max* and *Lymantria dispar* on *Populus tremuloides*, both showing lower relative growth rates at elevated CO_2, produced statistically significant results (Lincoln et al., 1986; Lindroth et al., 1993). These and the results of the other studies are summarized in Fig. 1.

In the seven cases where larval growth was assessed as larval weights at the end of a given period (usually the end of an instar), only three produced significant results (Fig. 2). *Lymantria dispar* larvae were larger on *Quercus rubra* grown in elevated CO_2 and both *Juonia coenia* larvae on *Plantago lanceolata* and *L. dispar* larvae on *P. tremuloides* were smaller in elevated CO_2 (Fajer, 1989a; Lindroth et al., 1993). Pupal weights have been assessed in nine cases. In two of them pupae were significantly smaller on plants grown in elevated CO_2 (Fig. 3). Thus, elevated CO_2 appears generally to have a negative effect on larval growth and subsequent larval and pupal weight, although the majority of studies failed to produce statistically significant results (Figs 2–3). Nevertheless, of the seven cases where significant results were obtained, only the larval weight of *L. dispar* on *Quercus rubra* was greater on plants grown at elevated CO_2.

Including the four separate measurements of development presented by Fajer et al. (1989) for *Juonia coenia*, the effect of elevated CO_2 on the duration of development or some other measure of insect developmental rate has been assessed in 17 cases (Fig. 4). In 14 of these cases, insects took longer to develop on plants grown in elevated CO_2, and in three cases development was faster in elevated CO_2. The effect was statistically significant in only one of these latter cases: the duration of the fourth and fifth instars of *J. coenia* was approximately 1 day shorter in elevated CO_2 (Fajer et al., 1989). However, the duration of the first three instars was significantly longer in elevated CO_2, and the total duration of the larval instars was significantly longer for male (but not female) insects. Thus, although older *J. coenia* larvae appeared to compensate for the negative impact of elevated CO_2 experienced by younger larvae, this compensation was not enough to completely negate the effect on younger larvae. The other three cases where elevated CO_2 significantly reduced the rate of larval development was in a further study of *J. coenia* (Fajer et al., 1991a), *L. dispar* on *P. tremuloides* (Lindroth et al., 1993), and *Spodoptera exigua* on cotton (Akey and Kimball, 1989).

In contrast to observations on insect development, there have been

9. Impact of Elevated CO_2 on Insect Herbivores

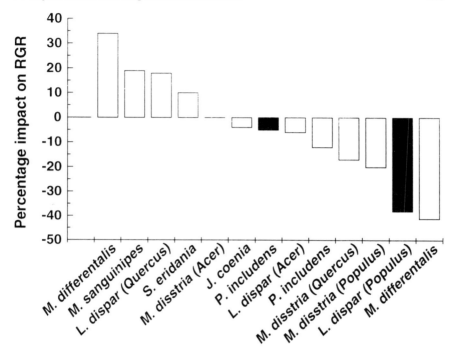

Fig. 1. Impact of elevated CO_2 on the relative growth rate of insect herbivores. Each column represents the percentage increase or decrease in RGR of insects reared under elevated CO_2 (significant differences due to the impact of elevated CO_2 ($P<0.05$) shown by solid histograms). Different host plants for *Lymantria dispar* and *Malacosoma disstria* (Lindroth *et al.*, 1993) are given in parentheses. See Table 1 for data sources.

only four cases where the impact of elevated CO_2 on insect survival has been assessed, and these include the three cases where the effect of CO_2 on *Trichoplusia ni* survival was studied on *Phaseolus lunata* receiving different amounts of a nutrient solution (Osbrink *et al.*, 1987). Statistically significant results were obtained in only one case: elevated CO_2 significantly reduced the survival of *J. coenia* larvae (Fajer *et al.*, 1989).

The impact of elevated CO_2 directly on the abundance of plant-feeding insects has been examined in three studies, involving six insect species (including *Bemisia tabaci* in two of these studies) (Fig. 5). In each of these cases, insect abundance was lower at elevated CO_2, and in four of these cases the difference was statistically significant, *Chaetocnema ectypa*, *Pectinophora gossypiella* and *Empoasca* spp. on cotton (Butler, 1985), and *Trialeurodes vaporariorum* on eight tomato cultivars (Tripp *et al.*, 1992).

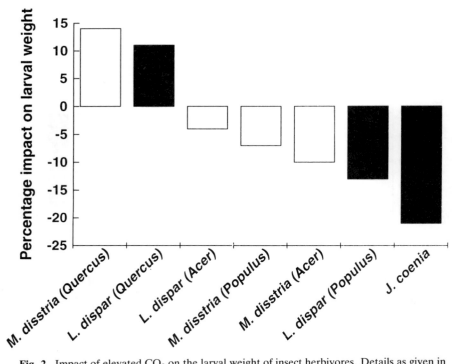

Fig. 2. Impact of elevated CO_2 on the larval weight of insect herbivores. Details as given in the legend to Fig. 1.

There have been several studies which have included an examination of the possibility that insect herbivores respond to the reduction in the quality of the foliage of host plants grown in elevated CO_2 by increasing their rate of food consumption. These studies showed that, on average, the consumption rate of larvae reared on plants grown in elevated CO_2 was 44% greater than on plants grown in ambient CO_2. There have been only two cases where food consumption decreased, both statistically non-significant, and 14 cases where consumption increased, all but three of them statistically significant (Fig. 6).

Thus, the bulk of the evidence from studies on the impact of elevated CO_2 indicates that herbivorous insects will grow more slowly, take longer to develop, suffer heavier mortality, and are, therefore, likely to be less abundant in the concentrations of atmospheric CO_2 predicted to occur within the next 100 years. However, the evidence from work on the effect of elevated CO_2 on the food consumption of insect larvae indicates that although insect pests are unlikely to become more abundant, they may

9. Impact of Elevated CO_2 on Insect Herbivores

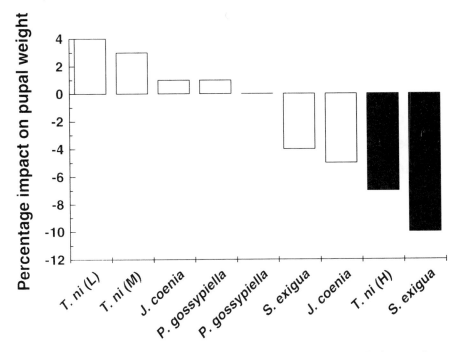

Fig. 3. Impact of elevated CO_2 on the pupal weight of insect herbivores. The impact of elevated CO_2 on *Trichoplusia ni* is given separately for larvae reared on *Phaseolus lunata* grown under low (L), medium (M), and high (H) nutrient regimes (Osbrink *et al.*, 1987). Other details as given in the legend to Fig. 1.

become more damaging, and negate part of the increase in crop productivity which is predicted to occur as a result of enhanced photosynthesis, reduced photorespiration and increased water-use efficiency of plants in elevated atmospheric CO_2 (Bazzaz, 1990).

Although these conclusions are based on the study of 18 insect–plant examples, their general applicability is reduced by the fact that most attention has been paid to leaf-chewing insects. Very little research has been done on the effects of elevated CO_2 on aphids and other plant-sucking insects, and leaf-mining, shoot-boring and plant-galling insects have, so far, been completely ignored. The research published to date on elevated CO_2 does not enable us to assess whether there are differences in the response of different insect groups to elevated CO_2 such as occurs in the response of different insect groups to drought stress (see reviews by Larsson, 1989; Price, 1991; Waring and Cobb, 1992; Watt, 1994). However, different insect groups are likely to be affected by

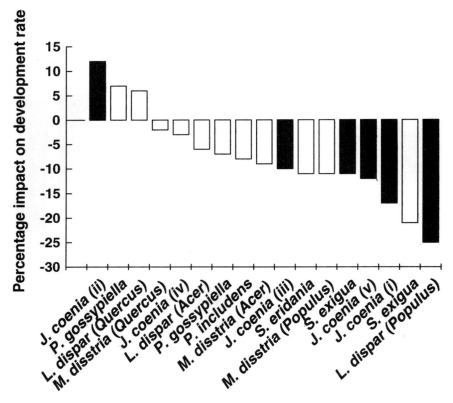

Fig. 4. Impact of elevated CO_2 on the development rate of insect herbivores. The impact of elevated CO_2 on *Juonia coenia* is given separately for instars 1–3 (i), instars 3–4 (ii), total male (iii), and total female (iv) development (Fajer *et al.*, 1989), and total development (v) (Fajer *et al.*, 1991a). Other details are given in the legend to Fig. 1.

elevated CO_2 to different degrees. In particular, sap-sucking insects will only be exposed to changes in plant chemistry that affect phloem sap. In addition, research has been biased towards insects on annual crops, but the impact of elevated CO_2 on other insects, particularly tree-dwelling species, is poorly understood. More attention also needs to be paid to the acclimation of plants to elevated CO_2 (e.g. Norby *et al.*, 1986) and the possibility that the impact of elevated CO_2 on herbivorous insects will change over time. Clearly, there is a need to study the impact of elevated CO_2 on a wider range of insects, following the example of Butler (1985) and Lindroth *et al.* (1993) by comparing different insect species on the same host plant, or the same insect species on different host plants.

9. Impact of Elevated CO_2 on Insect Herbivores

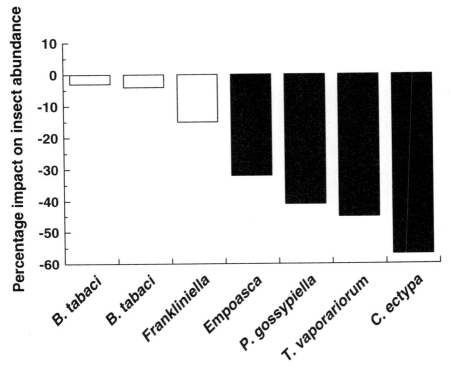

Fig. 5. Impact of elevated CO_2 on the abundance of insect herbivores. Details as given in the legend to Fig. 1.

Furthermore, more emphasis should be placed on insects feeding on acclimated plants, and to examine the interaction between elevated CO_2 and other factors such as plant nutrition (Osbrink *et al.*, 1987; Johnson and Lincoln, 1991), and drought stress (and see below).

IV. RESPONSE OF INSECTS TO ELEVATED CO_2 AND PLANT CHEMISTRY

The reason most frequently cited for the negative response of insect herbivores to elevated CO_2 is that plants, when grown in elevated CO_2 accumulate more carbon, the carbon to nitrogen ratio increases and herbivores are faced with nutritionally poorer host plants (e.g. Bazzaz, 1990; Lincoln *et al.*, 1993), nitrogen being the single most important nutritional chemical for insect herbivores (e.g. Mattson, 1980).

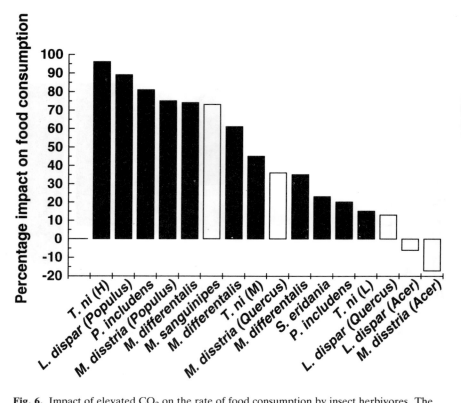

Fig. 6. Impact of elevated CO_2 on the rate of food consumption by insect herbivores. The impact of elevated CO_2 on *Trichoplusia ni* is given separately for larvae reared on *Phaseolus lunata* grown under low (L), medium (M), and high (H) nutrient regimes (Osbrink et al., 1987). Different host plants for *Lymantria dispar* and *Malacosoma disstria* (Lindroth et al., 1993) are given in parentheses. Other details as given in the legend to Fig. 1.

Interestingly, this "nitrogen dilution effect" can occur even in experiments when high levels of nitrogenous fertilizer are added, leading to the conclusion that, for some plants at least, the requirement for nitrogen is lower at elevated CO_2 (Conroy, 1992).

Most published studies on the impact of elevated CO_2 on insect herbivores have, therefore, included the measurement of nitrogen in plant foliage, and some have considered other aspects of the changes in plant chemistry. These studies have shown that plants do generally respond to elevated CO_2 with a reduction in the concentration of nitrogen in their foliage of up to 20–40% (Fig. 7). The plant species studied include cotton, tomato, soybean, lima bean (*Phaseolus lunata*), *Populus*

9. Impact of Elevated CO_2 on Insect Herbivores

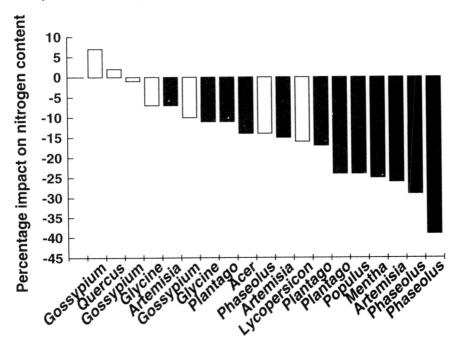

Fig. 7. Impact of elevated CO_2 on the nitrogen concentration of plant foliage or, in the case of *Gossypium*, seed. Details as given in the legend to Fig. 1.

tremuloides and *Acer saccharum*. The only exceptions, showing a small and statistically non-significant rise in nitrogen concentration, were the foliage of *Quercus rubra* (Lindroth *et al.*, 1993), and cotton seed (Akey *et al.*, 1988), suggesting that herbivores feeding on parts of plants other than foliage may not be affected in the same way as foliage-feeding insects. Other plant species which have been shown to respond to elevated CO_2 by a reduction in the concentration of nitrogen in their foliage include *Eucalyptus grandis* (Conroy *et al.*, 1992), and *Liriodendron tulipifera* (yellow poplar) (Norby *et al.*, 1992). In addition, analysis of herbarium specimens of species such as *Betula pendula* and *Pinus* spp. has shown a decline in foliar nitrogen concentrations since about 1750 (Peñuelas and Matamala, 1990). However, C_4 plants, such as maize, do not appear to respond in the same way (Conroy, 1992).

Apart from the nitrogen-dilution effect, an increase in carbon-based compounds in plants growing in elevated CO_2 may have an impact on insect herbivores, but this impact depends upon which type of carbon-based compound increases in concentration. An increase in starch may be beneficial because these compounds can enhance digestion, but structural carbohydrates, such as cellulose, can have a negative effect on digestion (Lincoln et al., 1993). The starch concentration of *Artemisia tridentata* leaves increased dramatically in elevated CO_2: by 78% in a low nutrient treatment and six-fold in a high nutrient treatment (Johnson and Lincoln, 1991). The same study showed that elevated CO_2 brought about increases in sugar concentration of 16 and 29% respectively. Further examples of marked increases in leaf starch concentration in response to elevated CO_2 are *Quercus rubra* (a two-fold increase), and *Populus tremuloides* (a three-fold increase) (Lindroth et al., 1993), but the same study showed no increase in the starch concentration of *Acer saccharum* leaves. On the evidence of that study, changes in leaf starch concentration play no significant role in the impact of elevated CO_2 on insect herbivores because there were no notable similarities in the response of *Lymantria dispar* or *Malacosoma disstria* on the two host plants with large increases in leaf starch concentration. The development and growth rates of *L. dispar* were significantly lower on *P. tremuloides* grown in elevated CO_2, but higher on *Q. rubra* in elevated CO_2 (Lindroth et al., 1993). Sugar concentration was not significantly affected by elevated CO_2 in the same study, but both hexose and sucrose concentrations were lower in *P. tremuloides*, and higher in *Q. rubra*, grown in elevated CO_2. It is also notable that the nitrogen concentration of these host plants was similarly affected by elevated CO_2 (significantly in the case of *P. tremuloides*) (Lindroth et al., 1993), further supporting the conclusion that it is the nitrogen dilution effect that is the main reason for the generally negative impact of elevated CO_2 on insect herbivores. Nevertheless, several other studies have shown large increases in the concentration of non-structural carbohydrates in the foliage of plants grown in elevated CO_2 (Lincoln et al., 1993), and their role in affecting insect herbivores on plants grown in CO_2 enriched conditions is still unclear. The same is also true for structural carbohydrates, although the limited amount of work on these compounds in relation to herbivory suggests that their concentration does not increase in elevated CO_2 (Johnson and Lincoln, 1990; Fajer et al., 1991a).

An increase in the carbon to nitrogen ratio of plant foliage is most likely to have a detrimental effect on insect herbivores if this leads to an increase in carbon-based allelochemicals. However, there is no convincing evidence that this occurs. The concentrations of iridoid glycosides and other carbon-based allelochemicals in *Plantago lanceolata* tend to decrease in elevated CO_2 (Fajer, 1989a; Fajer et al., 1989, 1991b, 1992), as do the allelochemicals in *Artemisia tridentata* (Johnson and Lincoln, 1990, 1991), and the concentration of volatile monoterpenes and sesquiterpenes in *Mentha piperita* is unaffected by CO_2 treatment (Lincoln and Couvet, 1989). These results have important implications for theories of plant defence mechanisms against herbivores, in particular the carbon/nutrient balance hypothesis (Fajer et al., 1992; Lincoln, 1993). This hypothesis proposes that the amounts of carbon-based allelochemicals in a plant are largely determined by the relative availability of carbon and nitrogen to the plant (Bryant et al., 1983). It predicts, for example, that in low nutrient conditions photosynthesis will not be reduced to the same extent as plant growth, resulting in the accumulation of "excess" carbohydrates (i.e. amounts greater than can be incorporated in increased plant growth) which will be incorporated instead into carbon-based allelochemicals (Bryant et al., 1983). The majority of the studies that have tested this prediction (and the related prediction that in low light conditions the production of carbon-based allelochemicals is reduced) have leant support to the carbon/nutrient balance hypothesis (Fajer et al., 1992). Another prediction of this hypothesis is that with the increase in carbon availability in a CO_2-enriched environment, the production of carbon-based allelochemicals will increase. However, as discussed above, the evidence does not support this prediction, implying that the carbon/nutrient balance hypothesis is not a good general model for plant defence against insects.

Studies on the impact of elevated CO_2 on the water concentration of plant foliage have shown generally a decline in foliar water concentration. Although the water concentration of soybean was significantly greater at elevated CO_2 (Lincoln et al., 1984, 1986), water concentration was significantly lower at elevated CO_2 in studies on *Plantago lanceolata* (Fajer, 1989a; Fajer et al., 1989), *Artemisia tridentata* (Johnson and Lincoln, 1991), *Phaseolus lunata* (Osbrink et al., 1987), *Populus tremuloides* and *Quercus rubra* (Lindroth et al., 1993). A decline in foliar water concentrations at elevated CO_2 probably results from an increase in the concentration of carbon compounds, rather than an absolute decrease in plant water content. However, since plants grown in elevated CO_2 have a greater water use efficiency (Bazzaz, 1990), plants grown under drought stress at elevated CO_2 may have a higher foliar water concentration than plants under drought stress at ambient CO_2.

The effect of elevated CO_2 on leaf toughness has been measured on only three host plants, *Populus tremuloides*, *Quercus rubra* and *Acer saccharum* (Lindroth et al., 1993). The leaf toughness of the first two tree species was significantly greater at elevated CO_2. In addition, specific leaf weight and leaf thickness, both indicators of leaf toughness, have generally been found to be greater at elevated CO_2 (Lincoln et al., 1993).

No studies have yet considered whether possible changes in plant

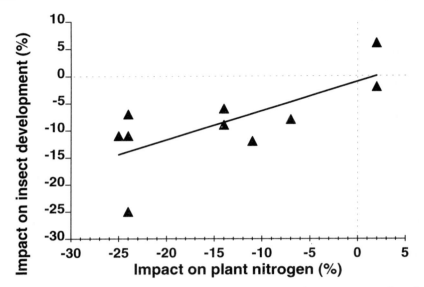

Fig. 8. Relationship between the impact of elevated CO_2 on the nitrogen concentration of plant foliage and the impact on insect development rate ($R = 0.71$, $P < 0.01$). Data sources as given in Table 1.

phenological development at elevated CO_2 affect the performance and abundance of insect herbivores, although direct and indirect observations of increased leaf toughness at elevated CO_2 may have resulted from more rapid leaf expansion and maturation. Experiments on *Picea sitchensis* have shown that at elevated CO_2, budburst is delayed and budset occurs earlier (Murray et al., 1994) (and see also below).

Overall, therefore, there is good evidence that the generally poorer performance of insects feeding on plants grown in elevated CO_2 is due to a decline in the nutritional quality of their host plants. The impact of elevated CO_2 on plant nitrogen alone appears to explain much of the variation in the response of insect development and food consumption to elevated CO_2 (Figs 8–9). The fate of the extra carbon acquired by plants in elevated CO_2 is poorly understood, and more work is needed on the effect of elevated CO_2 on both carbon-based and nitrogen-based allelochemicals. The role of elevated CO_2 on plant phenology, and the consequences this has for insects which are particularly prone to phenological asynchrony, also requires research.

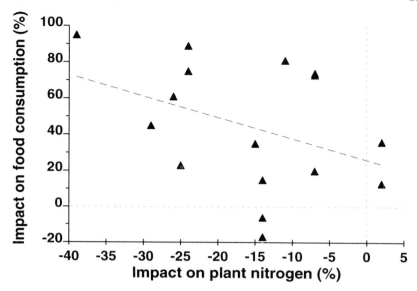

Fig. 9. Relationship between the impact of elevated CO_2 on the nitrogen concentration of plant foliage and the impact on insect consumption rate ($R = -0.38$, NS). Data sources as given in Table 1.

V. CO_2, TEMPERATURE AND OTHER FACTORS

The increase in atmospheric CO_2 concentration is only one aspect of environmental change. Global temperatures are predicted to rise (Bennetts, Chapter 3, this volume) and affect the abundance and distribution of insect herbivores and other species (e.g. Porter, Chapter 5; Rogers, Chapter 8; Sutherst *et al.*, Chapter 4, this volume). Pollutants, particularly ozone, SO_2 and NO_x are also likely to have a significant impact on insect herbivores (e.g. Brown, Chapter 10, this volume) and the disappearance and fragmentation of habitats are likely to have the most profound effects of all (e.g. Mawdsley and Stork, Chapter 14; Usher, Chapter 15, this volume).

Different authors have come to very different conclusions about the impact of elevated CO_2 on the abundance of insects, and, indeed, other organisms. The direct evidence of studies on insect abundance, and much of the indirect evidence from studies on insect performance have led to several writers concluding that insect herbivores will become less abundant as atmospheric CO_2 levels rise, and even suggesting that species extinctions are likely (Fajer, 1989b). This conclusion is further reinforced by the prediction that plant species extinctions are also forecast to occur (Possingham, 1993).

However, other writers have been sceptical about the role of elevated CO_2. For example, Landsberg and Smith (1992) argue that elevated CO_2 will have a minor impact on insect herbivores because the impact of elevated CO_2 on plants is only realized when plants are growing with a high level of water and nutrient supply. The evidence from the three studies which have examined the combined impact of elevated CO_2 and fertilizer regime support this conclusion. The impact of elevated CO_2 on survival, weight and food consumption of *Trichoplusia ni* larvae was most marked on lima bean grown under a high nutrient treatment (Osbrink et al., 1987), and similar results were obtained for *Spodoptera exigua* on cotton (Akey and Kimball, 1989), and *Melanopus differentalis* on sagebrush (Johnson and Lincoln, 1991).

One way in which temperature and elevated CO_2 may interact to affect insect herbivores is through their differential effects on plant and insect phenology and the consequent impact this will have on phenological asynchrony between insects and their host plants. Phenological asynchrony is particularly important for insects whose larvae start feeding around budburst, for example winter moth *Operophtera brumata* (Watt and McFarlane, 1991). Dewar and Watt (1992) showed that climatic warming of 1–2°C would increase the likelihood of asynchrony between winter moth larval emergence and Sitka spruce (*Picea sitchensis*) budburst. However, an increase in atmospheric CO_2 concentrations is also likely to result in a delay in the timing of budburst in Sitka spruce (Murray et al., 1994). This might negate the predicted impact of climatic warming on phenological asynchrony (Dewar and Watt, 1992). A further complication is that plant nutrient status is also likely to affect phenological asynchrony: the impact of elevated CO_2 on Sitka spruce phenology is greatest on plants receiving low nutrient treatments (Murray et al., 1994).

Thus there is an urgent need to study the combined effects of factors such as CO_2, drought, temperature, plant nutrient status, and pollutants such as ozone. Multi-impact studies are more appropriate than single-impact studies in predicting how herbivorous insects will respond to environmental change (Lawton, Chapter 1, this volume). However, attempts at multi-impact studies (e.g. the impacts of a pollutant (CO_2) and drought on *Elatobium abietinum* (Warrington and Whittaker, 1990)) have proved extremely difficult to manage and replicate adequately. Thus, the methodology for studying even individual factors has to be

9. Impact of Elevated CO_2 on Insect Herbivores

improved (e.g. Lincoln et al., 1993), and this can best be done by first studying different factors singly. To date, only a relatively small number of studies on the impact of elevated CO_2 have been carried out, and these studies have used a variety of methods. Thus, before deciding which insect species should be used for intensive study (Lawton, Chapter 1, this volume), more research, using standardized experimental regimes, has to be done to establish how different groups of insect herbivores respond to elevated CO_2.

Acknowledgements

We thank M. G. R. Cannell, A. Friend and M. Murray for constructive comments on this chapter.

REFERENCES

Akey, D. H. and Kimball, B. A. (1989). Growth and development of the beet armyworm on cotton grown in an enriched carbon dioxide atmosphere. *Southwestern Entomologist* **14**, 255–260.

Akey, D. H., Kimball, B. A. and Mauney, J. R. (1988). Growth and development of the pink bollworm, *Pectinophora gossypiella* (Lepidoptera: Gelechiidae) on bolls of cotton grown in enriched carbon dioxide atmospheres. *Environ. Entomol.* **17**, 452–455.

Bazzaz, F. A. (1990). The response of natural ecosystems to the rising global CO_2 levels. *Ann. Rev. Ecol. Syst.* **21**, 167–196.

Bryant, J. P., Chapin, F. S. and Klein, D. R. (1983). Carbon/nutrient balance of boreal plants in relation to vertebrate herbivory. *Oikos* **40**, 357–368.

Butler, G. D. (1985). Populations of several insects on cotton in open-top carbon dioxide enrichment chambers. *Southwestern Entomologist* **10**, 264–267.

Butler, G. D., Kimball, B. A. and Mauney, J. R. (1986). Populations of *Bemisia tabaci* (Homoptera: Aleyrodidae) on cotton grown in open-top field chambers enriched with CO_2. *Environ. Entomol.* **15**, 61–63.

Conroy, J. P. (1992). Influence of elevated CO_2 concentrations on plant nutrition. *Aust. J. Bot.* **40**, 445–456.

Conroy, J. P., Milham, P. J. and Barlow, E. W. R. (1992). Effect of nitrogen and phosphorus availability on the growth response of *Eucalyptus grandis* to high CO_2. *Plant, Cell Env.* **15**, 843–847.

Dahlman, R. C. (1993) CO_2 and plants: revisited. *Vegetatio* **104/105**, 339–355.

Dewar, R. C. and Watt, A. D. (1992). Predicted changes in the synchrony of larval emergence and budburst under climatic warming. *Oecologia* **89**, 557–559.

Eamus, D. and Jarvis, P. G. (1989). The direct effect of increase in the global atmospheric CO_2 concentration on natural and commercial temperate trees and forests. *Adv. Ecol. Res.* **19**, 1–55.

Fajer, E. D. (1989a) The effects of enriched CO_2 atmospheres on plant-insect herbivore interactions: Growth responses of larvae of the specialist butterfly, *Junonia coenia* (Lepidoptera: Nymphalidae). *Oecologia* **81**, 514–520.

Fajer, E. D. (1989b) How enriched carbon dioxide environment may alter biotic systems even in the absence of climatic changes. *Cons. Biol.* **3**, 318–319.

Fajer, E. D., Bowers, M. D. and Bazzaz, F. A. (1989). The effects of enriched carbon dioxide atmospheres on plant-insect herbivore interactions. *Science* **243**, 1198–1200.

Fajer, E. D., Bowers, M. D. and Bazzaz, F. A. (1991a). The effects of enriched CO_2 atmospheres on the buckeye butterfly *Junonia coenia*. *Ecology* **72**, 751–754.

Fajer, E. D., Bowers, M. D. and Bazzaz, F. A. (1991b). Performance and allocation patterns of the perennial herb, *Plantago lanceolata*, in response to simulated herbivory and elevated CO_2 environments. *Oecologia* **87**, 37–42.

Fajer, E. D., Bowers, M. D. and Bazzaz, F. A. (1992). The effect of nutrients and enriched CO_2 environments on production of carbon-based allelochemicals in *Plantago*: A test of the carbon/nutrient hypothesis. *Am. Nat.* **140**, 707–723.

Houghton, J. T., Callander, B. A. and Varney, S. K. (1992). "Climate Change 1992. The Supplementary Report to the IPCC Scientific Assessment". Cambridge University Press, Cambridge.

Johnson, R. H. and Lincoln, D. E. (1990). Sagebrush and grasshopper responses to atmospheric carbon dioxide concentration. *Oecologia* **84**, 103–110.

Johnson, R. H. and Lincoln, D. E. (1991). Sagebrush carbon allocation patterns and grasshopper nutrition: The influence of CO_2 enrichment and soil mineral limitation. *Oecologia* **87**, 127–134.

Landsberg, J. and Smith, M. S. (1992). A functional scheme for predicting the outbreak potential of herbivorous insects under global atmospheric change. *Aust. J. Bot.* **40**, 565–577.

Larsson, S. (1989). Stressful times for the plant stress-insect performance hypothesis. *Oikos* **56**, 277–283.

Lincoln, D. E. (1993). The influence of plant carbon dioxide and nutrient supply on susceptibility to insect herbivores. *Vegetatio* **104/105**, 273–280.

Lincoln, D. E. and Couvet, D. (1989). The effect of carbon supply on allocation to allelochemicals and caterpillar consumption of peppermint. *Oecologia* **78**, 112–114.

Lincoln, D. E., Sionit, N. and Strain, B. R. (1984). Growth and feeding of *Pseudoplusia includens* (Lepidoptera: Noctuidae) to host plants grown in controlled carbon dioxide atmospheres. *Environ. Entomol.* **13**, 1527–1530.

Lincoln, D. E., Couvet, D. and Sionit, N. (1986). Responses of an insect herbivore to host plants grown in carbon dioxide rich atmospheres. *Oecologia* **69**, 556–560.

Lincoln, D. E., Fajer, E. D. and Johnson, R. H. (1993). Plant-insect herbivore interactions in elevated CO_2 environments. *Trends Ecol. Evol.* **8**, 64–68.

Lindroth, R. L., Kinney, K. K. and Platz, C. L. (1993). Responses of deciduous trees to elevated atmospheric CO_2: productivity, phytochemistry, and insect performance. *Ecology* **74**, 763–777.

Mattson, W. J. (1980). Herbivory in relation to plant nitrogen content. *Ann. Rev. Ecol. Syst.* **11**, 119–161.

Murray, M. B., Smith, R. I., Leith, I. D., Fowler, D., Lee, H. S. J., Friend, A. D. and Jarvis, P. (1994). The effect of elevated CO_2, nutrition and climatic warming on bud phenology in Sitka spruce (*Picea sitchensis* (Bong.) Carr.) and its impact on frost tolerance. *New Phytol.*, in press.

Norby, R. J., O'Neill, E. G. and Luxmoore, R. J. (1986). Effects of atmospheric CO_2 enrichment on the growth and mineral nutrition of *Quercus alba* seedlings in nutrient-poor soils. *Plant Physiol.* **82**, 83–89.

Norby, R. J., Gunderson, C. A., Wullschleger, S. D., O'Neill, E. G. and McCracken, M. K. (1992). Productivity and compensatory responses of yellow-poplar trees in elevated CO_2. *Nature* **357**, 322–324.

Osbrink, W. L. A., Trimble, J. T. and Wagner, R. E. (1987). Host suitability of *Phaseolus lunata* for *Trichoplusia ni* (Lepidoptera: Noctuidae) in controlled carbon dioxide atmospheres. *Environ. Entomol.* **16**, 639–644.

Peñuelas, J. and Matamala, R. (1990). Changes in N and S leaf content, stomatal density and specific leaf area of 14 plant species during the last three centuries of CO_2 increase. *J. Exp. Bot.* **41**, 1119–1124.

Possingham, H. P. (1993). Impact of elevated atmospheric CO_2 on biodiversity: Mechanistic population-dynamic perspective. *Aust. J. Bot.* **41**, 11–21.

Price, P. W. (1991). The plant vigor hypothesis and herbivore attack. *Oikos* **62**, 244–251.

Tripp, K. E., Kroen, W. K., Peet, M. M. and Willits, D. H. (1992). Fewer whiteflies found on CO_2-enriched greenhouse tomatoes with high C:N ratios. *Hortscience* **27**, 1079–1080.

Waring, G. L. and Cobb, N. S. (1992). The impact of plant stress on herbivore population dynamics. *Insect–Plant Interactions* **4**, 167–226.

Warrington, S. and Whittaker, J. B. (1990). Interactions between Sitka spruce, the green spruce aphid, sulphur dioxide pollution and drought. *Environ. Poll.* **65**, 363–370.

Watt, A. D. (1994). The relevance of the stress hypothesis to insects feeding on tree foliage. *In* "Individuals, Populations and Patterns in Ecology" (S. R. Leather, A. D. Watt, N. J. Mills and K. F. A. Walters, eds), pp. 73–85. Intercept, Andover.

Watt, A. D. and McFarlane, A. M. (1991). Winter moth on Sitka spruce: synchrony of egg hatch and budburst, and its effect on larval survival. *Ecol. Entomol.* **16**, 387–390.

10

Insect Herbivores and Gaseous Air Pollutants – Current Knowledge and Predictions

V. C. BROWN

I. Introduction 220
II. The Gaseous Pollutants and Their Direct Effects on Plants 222
   A. Sulphur Dioxide 222
   B. Nitrogen Dioxide 223
   C. Ozone (O_3) 223
III. Correlative Evidence of Insect/Plant/Air Pollution Interactions 225
   A. Field Observations 225
   B. Insect Suction Trap Data 227
   C. Experimental Studies Along Gradients 227
IV. Experimental Manipulations of Air Pollutants 228
   A. Closed Chamber Studies 228
   B. Open-top Chamber Studies 239
   C. Field-based Experiments 240
V. Summary of Current Knowledge of the Effects of Air Pollutants on Insect Herbivores 242
VI. Predictions and Future Research 243
References 245

Abstract

Since the mid-nineteenth century, foresters and entomologists have suspected that there is a link between outbreaks of phytophagous insects and atmospheric pollution. There have been a great many field observations from polluted areas throughout Europe, and in general, these observations have reported increased insect populations in moderately polluted environments and decreases at highly polluted sites. This correlative evidence has prompted a growing interest in the experimental investigation of these interactions over the last decade. Much of this work has been comprehensively reviewed recently and rather than repeating these reviews this chapter aims to summarize our current knowledge about the effects of three major gaseous pollutants: sulphur dioxide (SO_2); nitrogen dioxide (NO_2) and ozone (O_3) on plant/insect interactions and to speculate on the likely effects of predicted changes in air pollution levels.

After describing the origins and occurrences of the three gases, the evidence for insect/plant/air pollution interactions is presented. Correlative evidence from field observations, insect suction trap data and experimental studies is discussed. This chapter concentrates on recent experimental work and results from a wide range of experimental techniques, from laboratory studies to large-scale field experiments, are presented. Results from studies with SO_2 and NO_2 suggest that these gases consistently increase the performance of herbivorous insects whilst the situation with O_3 is more complex with a range of responses possible. Finally, on the basis of predictions of future changes in the levels of these gaseous pollutants, due to factors such as changes in methods of electricity generation, transportation, legislation on emissions and industrialization in developing countries, and our current knowledge of the effects of these pollutants on plant/insect herbivore interactions, the likely consequences and areas in need of further research are outlined.

I. INTRODUCTION

Probably the earliest description of the effects of air pollution on insects and plants is by the diarist John Evelyn who in 1661 commented on the effects of smoke from coal burning in seventeenth century London: "It kills our bees and flowers abroad, suffering nothing in our gardens to bud, display themselves or ripen" (Evelyn, 1661). However, it is not until the nineteenth century that observational evidence on the effects of air pollutants on herbivorous insects begins to appear in the literature. Beling in 1831 (in Cramer, 1951) reported a seven-fold higher density in the populations of the larvae of the Lepidoptera, *Epinotia tedella*, on Norway spruce, around an iron foundry and in 1898, Gerlach coined the term "Rauch-rüsseler" (smoke-weevils) to describe increased populations of weevils on smoke-damaged Norway spruce (*Picea abies*) in Saxony (in Flückiger *et al.*, 1987). Since then, there have been a great many anecdotal field observations from polluted areas throughout Europe. It is not until the last decade or so that an experimental approach to studying the interactions between atmospheric pollutants, plants and their insect herbivores has been taken.

10. Insect Herbivores and Gaseous Air Pollutants

Most of the experimental studies have taken place at research institutions using facilities developed to study the direct effects of air pollutants on plants, and has therefore concentrated on gaseous air pollutants known to have important direct effects on vegetation. Much of the recent research with insect herbivores has been performed with aphids, primarily because of their economic importance as plant pests. Other advantages of using aphids in these studies include the ease of culturing large numbers of even-aged, clonal experimental animals throughout the year and the sensitivity of these phloem-feeding insects to small changes in the nutritional quality of their host plants. The parameter of aphid performance measured in many of the studies presented in this chapter is mean relative growth rate (MRGR) (van Emden, 1969):

$$\text{MRGR} = \frac{\ln \text{ final weight} - \ln \text{ initial weight}}{\text{growth period (days)}}$$

It is quick and simple to measure and for most aphid species investigated it is strongly, positively correlated with reproductive rate (Leather and Dixon, 1984).

Much of the observational and experimental evidence for these interactions has been reviewed recently. Reviews of the effects of air pollutants on herbivorous insects include: Alstad et al. (1982); Laurence et al. (1983); Hughes (1988); Riemer and Whittaker (1989); Whittaker and Warrington (1990); Heliövaara and Väisänen (1993). Other reviews have concentrated on insects feeding on trees, for example Baltensweiler (1985); Führer (1985); Hain (1987) and McNeill and Whittaker (1990), whilst other authors have combined a review of the effects of air pollutants on insect herbivores with those on plant pathogens, for example Hughes and Laurence (1984); Manning and Keane (1988) and Bell et al. (1993).

The aim of this chapter is not to attempt to repeat these comprehensive reviews, but to highlight the range of different techniques that have been used to investigate these interactions, and to summarize the main findings to date. Wherever possible, examples have been chosen from recent research that has not appeared in previous reviews. Finally the effects of a changing air pollution climate on insect herbivores is speculated upon, based on current knowledge and on predicted scenarios of future air pollution levels.

II. THE GASEOUS POLLUTANTS AND THEIR DIRECT EFFECTS ON PLANTS

The term "air pollution" covers a wide range of anthropogenic, atmospheric contaminants such as particulate matter, fluorides, ammonia, volatile organic compounds, carbon monoxide, acid rains and mists, hydrogen chloride, chlorine, radioactive substances, carbon dioxide and many more. However, this chapter concentrates on the effects of three major air pollutants (sulphur dioxide, nitrogen dioxide and ozone) which are of concern both in relation to their direct effects on plants and their effects on plant/insect herbivore interactions.

A. Sulphur Dioxide

Sulphur dioxide (SO_2) occurs in the atmosphere both naturally, mainly as a result of the volcanic activity, and anthropogenically as a result of burning fossil fuels, particularly coal and oil. Coal and fuel oil powered electricity generation and smeltering processes account for most of the SO_2 released in Britain. Petrol-engined motor vehicles do not produce significant amounts of SO_2 as the sulphur content of petrol is low (0.04%); however, diesel has a higher sulphur content (0.2%) and contributes more to emissions (QUARG, 1993).

Urban smogs in the 1950s, especially the great London smog in December 1952 (thought to be responsible for about 4000 human deaths), led to the Clean Air Acts (1956 and 1968), which resulted in a dramatic reduction in SO_2 concentrations in Britain. The introduction of tall stacks on power stations, the general decline in heavy industry in recent years and the switch from coal to natural gas as the main means of domestic heating have also contributed to lower SO_2 levels. The introduction of flue gas desulphurization, the use of natural gas (which has a low sulphur content) for electricity generation, and the increase in the importation of coal with a low sulphur content will further reduce SO_2 levels in the future. Nevertheless, SO_2 levels remain high over much of the East Midlands and South Yorkshire where the bulk of coal-fired electricity generation takes place.

SO_2 is a major air pollutant which can directly damage plants (e.g. Malhotra and Khan, 1984) and is of importance in areas where high concentrations occur (Ashmore, 1991). In gaseous form, SO_2 enters plants mainly through the stomata and can affect a wide range of plant physiological processes such as photosynthesis, respiration, enzyme activity and assimilate distribution (Darrell, 1989).

B. Nitrogen Dioxide

Unlike SO_2, most emissions of nitrogen dioxide (NO_2) are not the result of impurities in fuel, but are formed from the high temperature combination of natural atmospheric oxygen and nitrogen forming nitric oxide (NO).

$$N_2 + O_2 \Rightarrow 2NO$$

This becomes rapidly oxidized in the atmosphere to produce NO_2 either by reaction with oxygen:

$$2NO + O_2 \Rightarrow 2NO_2$$

or by reaction with ozone:

$$NO + O_3 \Leftrightarrow NO_2 + O_2$$

All combustion processes produce NO, and consequently NO_2, therefore unlike SO_2 the motor vehicle is a major source of the pollutant accounting for about 50% of UK emissions (UK PORG, 1990). Since this release is just above ground level it has a larger impact on terrestrial ecosystems than most other sources of this pollutant. Concentrations of NO_2 are highest in urban areas (two to four times those in rural areas) and very high concentrations can occur near major roads.

Like SO_2, a large number of deleterious effects on plants have been reported (e.g. Taylor and Eaton, 1966) and a wide range of physiological processes can be disrupted by this gas (Darrell, 1989). However, NO_2 differs from most other atmospheric pollutants in that nitrogen is often a limiting environmental factor so that beneficial as well as deleterious effects on plants have often been reported at low concentrations (Rowland et al., 1985).

C. Ozone (O_3)

The triatomic form of oxygen, ozone (O_3), is present in both the stratosphere (10–50 km above the Earth's surface) and the troposphere (0–10 km altitude) and its presence in each zone has important effects on the Earth's environment.

In the stratosphere, O_3 is formed by photochemical reactions with oxygen (O_2) and exists at a steady-state concentration. O_3 strongly absorbs ultraviolet radiation (wavelength 240–320 nm) and the stratospheric layer plays an important role, serving as a shield against the harmful effects of solar ultraviolet radiation on the biosphere. The deleterious effects of various O_3-destroying chemicals, in particular chlorofluorocarbons (CFCs), methane (CH_4) and nitrous oxide (N_2O) have been the subject of much recent scientific investigation and public debate. Cicerone (1987) provides a useful review of the chemistry of the formation of stratospheric O_3 and changes in its concentration due to human activities.

In the troposphere O_3 is not emitted into the atmosphere in significant amounts, but is derived from vertical mixing of stratospheric O_3 or by photochemical reactions. Photochemical production involves reactions between oxides of nitrogen and volatile hydrocarbons, originating both from natural sources and to a much larger extent from anthropogenic sources such as motor vehicles and other human activities involving high temperature combustion processes. For this reason O_3 is categorized as a secondary air pollutant. There is evidence that global tropospheric O_3 concentrations have increased by about 100% over the last 100 years and that in the absence of effective emission control measures they are likely to continue to do so, particularly at high- to mid-latitudes in the northern hemisphere (Hough and Derwent, 1990; Ashmore and Bell, 1991).

In the formation of tropospheric ozone, free oxygen atoms (O) are required in the reaction

$$O + O_2 + M \Rightarrow O_3 + M$$

where M is a molecule such as nitrogen or oxygen which can dissipate the energy released in the formation of O_3.

The free O atoms necessary for O_3 formation are produced mainly by the photodissociation of nitrogen dioxide (NO_2).

$$NO_2 + \text{u.v. light } (\lambda = 280\text{--}430 \text{ nm}) \Rightarrow NO + O$$

Although O_3 is formed by the combination of atomic O with diatomic O_2 in the troposphere, NO will react rapidly with O_3, forming NO_2 and O_2.

$$NO + O_3 \Rightarrow NO_2 + O_2$$

However, peroxy radicals $RO_2\cdot$ (where R is a hydrogen atom or an organic molecule), which are also produced by photochemical reactions, can reverse this reaction by oxidizing NO to NO_2 allowing the build up of O_3.

$$RO_2\cdot + NO \Rightarrow NO_2 + RO\cdot$$

Only the basic mechanisms of O_3 production in polluted air are outlined in the above reactions, fuller discussions of the complex atmospheric chemistry of O_3 formation are given by UK PORG (1987), Krupa and Manning (1988) and Wellburn (1988).

The formation of elevated tropospheric O_3 concentrations depends on sunlight and the presence of the pollutant precursors, nitrogen oxides and volatile hydrocarbons. In addition, high air temperatures promote certain of the reactions and low wind speeds prevent the dispersal of the pollutants. All these climatic conditions occur in temperate regions during periods of anticyclonic weather and it is during these conditions that elevated O_3 concentrations occur in the surface atmosphere in the UK (UK PORG, 1987).

Tropospheric O_3 is known to be highly phytotoxic when present in sufficient concentrations, with both acute and chronic effects on vegetation. In recent years the direct effects of O_3 on plants have been the subject of considerable research effort and effects on forests, crops and natural vegetation reported. Krupa and Manning (1988) provide a thorough and recent review of the subject. Interactions between O_3 and abiotic factors, such as drought and temperature, have also been well investigated and much of this work has been summarized by Guderian *et al.* (1984).

III. CORRELATIVE EVIDENCE OF INSECT/PLANT/AIR POLLUTION INTERACTIONS

A. Field Observations

As already noted, incidents of numerical changes in insect herbivore populations, apparently due to elevated levels of air pollutants, have been reported since the mid-nineteenth century. There are now many examples of field observations of this type, both from surveys comparing polluted and clean sites and from surveys along pollutant gradients, and the subject has been well covered in reviews (e.g. Alstad *et al.* (1982); Riemer and Whittaker, (1989)). Some recent examples of observations where industrial sources of pollution, motor vehicle emissions and O_3 all

Table 1. Examples of field observations of changes in incidence of insect herbivores at polluted locations

Insect species	Host plant	Effect	Reference
Leaf hoppers	Various spp.	Increased abundance near main road	Port (1981)
Aphids	Pinus sylvestris	Some spp. increased, others decreased in relation to SO_2 and hydrogen fluoride	Villemant (1981)
Aphids	Pinus sylvestris	Increased density near industrial plants	Heliövaara and Väisänen (1989a)
Aphids	Sorghum vulgare	Increased population density and insect weight in relation to SO_2	Rao et al. (1990)
Aphis pomi	Crataegus monogyna	Increased infestation near motorway	Flückiger et al. (1978)
Bufftip moth Goldtail moth	Fagus sylvatica Crataegus monogyna	Increased infestation near motorway	Port and Thompson (1980)
Bark beetles	Pinus ponderosa	Increased attack on O_3-damaged trees	Stark and Cobb (1969)
Sawflies	Picea abies	Increased population density	Wentzel (1965)
Scale insects	Fagus sylvatica	Complex modifications of development of infestation in relation to SO_2	Decourt et al. (1980)

appear to change the incidence of both sucking and chewing insects are shown in Table 1. In general, observational evidence of this type suggests that populations of herbivorous insects are usually increased in moderately polluted environments but at high pollution concentrations, usually close to the pollution source, they are often decreased.

Observational evidence, whilst of value in highlighting the potential importance of interactions between air pollutants and herbivorous insects, is circumstantial and of limited value in terms of proving causal relationships. Furthermore, changes in insect populations are poorly quantified in relation to pollution dose and are not appropriate for air pollutants such as O_3 which occur over large regions. Consequently, an experimental approach to studying these interactions has been developed in recent years.

B. Insect Suction Trap Data

Evidence for changes in natural populations of aphids in relation to ambient pollution levels has been sought by Houlden, McNeill and Bell (pers. comm.). They examined the numbers of aphid individuals for each species caught in insect suction traps operated by Rothamsted Experimental Station between 1976 and 1982 at 18 locations throughout Britain. Each site was ranked on a 1–5 scale in relation to local mean SO_2 concentrations. There was a remarkably skewed frequency distribution for the correlation coefficients of the 83 species analysed. Thirty-nine species showed significant ($P < 0.05$) positive correlations with SO_2 concentrations. Out of the remaining 43 species, 34 had non-significant positive correlations and in only one case was there a significant negative relationship. This adds weight to the view that ambient levels of air pollution may be causing substantial increases in the intensity of outbreaks of aphids, both on native vegetation and on agricultural crops.

C. Experimental Studies Along Gradients

In addition to surveying naturally occurring insect populations along pollution gradients, several workers have used such gradients in a more experimental manner.

In order to reduce variability in initial infestation rate, in studies on sawflies in Finland, Heliövaara and Väisänen (1989b) collected branches from Scots pine, *Pinus sylvestris*, along two 9-km transects away from an industrial complex emitting SO_2 and heavy metals. Then, in the laboratory, the branches were infested with four species of laboratory-reared sawfly larvae which were allowed to complete their development and pupate. They found that the cocoon size for two of the species studied was significantly reduced when kept on branches from near to the pollution source. In other studies, investigating sawfly larval mortality, Heliövaara et al. (1991) transferred sawfly colonies to plants along the transects and found that mortality due to nuclear polyhedrosis virus was increased near the industrial complex, whereas that due to parasitism was decreased. These results suggest that sawflies grow less well and are weaker at high pollution concentrations and also that parasitoids may be affected by high levels of pollutants.

Another experimental approach has been to place potted plants along gradients and to infest them with insects, both in the field or subsequently in the laboratory. An example of the first approach are the studies by Holopainen et al. (1993) who placed Scots pine and Norway spruce, *Picea abies*, saplings at five locations along a gradient (0.2–4.5 km) away from a pulp mill in Finland, which was emitting mostly SO_2. They infested the pines with two apterous females of the grey pine aphid, *Schizolachnus pineti*, and the spruces with two apterous females of the spruce shoot aphid, *Cinara pilicornis*, and allowed the aphid populations to develop. The populations of *S. pineti* did not show any significant change along the pollutant gradient except that those nearest to the mill declined faster late in the season. However, the populations of *C. pilicornis* did show a significant positive response with air pollution, in that reproduction was faster closer (0.2 and 0.5 km) to the mill. This study demonstrates that not all aphids necessarily respond in the same way to pollutants and, as will be shown elsewhere in this chapter, care should be taken when extrapolating results from one insect species to another.

Houlden, Bell and McNeill (pers. comm.) grew potted wheat, *Triticum aestivum*, and barley, *Hordeum vulgare*, plants at seven locations along a gradient of air pollution, declining from central London along a 37-km transect westward into the countryside. At the end of a 6-week exposure period, the plants were removed to a controlled temperature room and the MRGRs for *Metopolophium dirhodum* feeding on the barley, and *Rhopalosiphum padi* feeding on wheat, were determined. Both aphid species showed a marked increase in MRGR when feeding on plants from locations closer to London with significant positive correlations with both SO_2 and NO_2 concentrations. The data for *M. dirhodum* are shown in Fig. 1. and they strongly suggest that urban air pollution can increase the growth of this aphid.

IV. EXPERIMENTAL MANIPULATIONS OF AIR POLLUTANTS

A. Closed Chamber Studies

The types of closed chamber which have been used to study the effects of pollutants on herbivorous insects vary from perspex boxes housed either in glasshouses or more controlled environments, to larger enclosed systems such as whole glasshouses. They are often supplied with air which is filtered through activated charcoal which removes SO_2, NO_2 and O_3 but not NO. Other pollutant-removing materials such as "Purafil", which

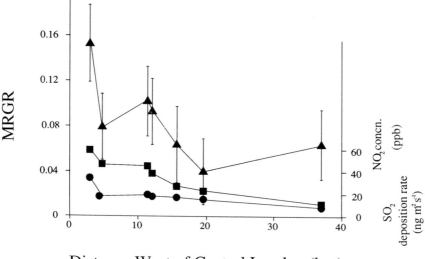

Fig. 1. MRGR (with 95% confidence limits) of *Metopolophium dirhodum* on barley grown for 6 weeks along an air pollution gradient westwards from Central London (Houlden, McNeill and Bell (pers. comm.)). ▲, MRGR; ■, SO_2 deposition rate; ●, NO_2 concentration.

converts NO to NO_2, facilitating its subsequent removal by charcoal, have also been used in closed chambers. Their main advantage for studying air pollution effects is that pollution concentrations can be very carefully controlled and depending on the location of the system, so can other environmental variables such as light, air temperature, humidity and air circulation. Chambers supplied with supplementary heating and lighting also enable experiments to be performed throughout the year or on plants from different climates.

1. Controlled fumigations

One of the earliest closed chamber fumigation experiments with insect herbivores was by Hughes *et al.* (1981) in which they demonstrated that Mexican bean beetles, *Epilachna varivestris*, preferred to feed on bean leaves, *Phaseolus vulgaris*, which had been previously fumigated for 7 days with SO_2 at a concentration of 150 ppb. They later showed (Hughes *et al.*, 1982) that beetles fed on soybean, *Glycine max*, leaves exposed to the pollutant had reduced developmental time, and increased adult weight and fecundity compared with beetles which fed on control plants.

The first aphid studies of this type were performed by Dohmen et al. (1984) and demonstrated that the MRGR of the black bean aphid, *Aphis fabae*, was significantly increased on *Vicia faba* plants which had been previously fumigated with either 150 ppb SO_2 or 200 ppb NO_2 for 7 days.

Since these experiments, many others have been performed with SO_2 and NO_2. As part of a major study of the effects of these two pollutants on aphid pests of agricultural crops, Houlden et al. (1990) prefumigated a range of crop plants for 7 h with either SO_2 or NO_2 at a concentration of 100 ppb. The results of 21 fumigation experiments using combinations of eight aphid species and six host plants are summarized in Table 2. The response to the pollutants is a increase in MRGR (7–75%) in all but two cases. Other studies with SO_2 have shown similar increases in insect performance, for example Warrington et al. (1987) showed that the fecundity of pea aphids, *Acyrthosiphon pisum*, was significantly increased on pea plants receiving either 50 or 80 ppb SO_2 for several weeks compared to controls which had not received the pollutant and this result is consistent with that from the shorter fumigation for *A. pisum* on peas shown in Table 2.

On trees, the green spruce aphid, *Elatobium abietinum*, has shown remarkable sensitivity to very short exposures to SO_2 when grown on potted saplings of its host plant, Sitka spruce, *Picea sitchensis*, which were previously fumigated with 100 ppb SO_2 for periods ranging from 30 min to 48 h (McNeill et al., 1987). The aphids showed a domed response with a significant increase in MRGR on plants receiving 30-min exposure and a peak in response on those given 3 h. A slight decline in response was seen at longer fumigation times. Longer term, population studies with this aphid, in which trees were exposed to a concentration of just 25 ppb for several weeks in glasshouses, show results consistent with the earlier study (Warrington and Whittaker, 1990). On plants that were infested with *E. abietinum*, the aphids became three times as abundant on fumigated trees compared to those on unfumigated controls.

Recently, the effects of SO_2 have been investigated with *Aphis fabae* feeding on a range of its alternative, wild, secondary host plants (Brown et al., 1993c). Increased MRGR was found on all seven plant species investigated and was significantly enhanced ($P<0.05$) on four of them.

Similarly, other closed chamber experiments with NO_2 have shown consistent increases in insect performance. For example, Fig. 2 shows the results from a series of closed chamber fumigation experiments with three species of conifer feeding aphid, *E. abietinum* and *C. pilicornis* feeding on *Picea sitchensis*, and *S. pineti* feeding on *Pinus sylvestris* (Brown et al., 1993a). All three species showed significantly increased ($P<0.05$) growth rates on saplings prefumigated with 100 ppb NO_2 for periods in excess of 24 h.

Table 2. Changes in MRGR of aphid pests of agricultural crops placed on host plants previously fumigated with 100 ppb SO_2 or NO_2 for 7 h (Houlden et al., 1990)

Aphid	Crop	Pollutant	% change in MRGR cf. control
Sitobion avenae	Wheat	SO_2	+28.2
		NO_2	+25.2
	Barley	SO_2	+32.2
		NO_2	+36.1
Metopolophium dirhodum	Wheat	SO_2	+15.2
		NO_2	+16.5
	Barley	SO_2	+19.5
		NO_2	+14.1
Rhopalosiphum padi	Wheat	SO_2	+35.1
		NO_2	+38.3
	Barley	SO_2	+40.4
		NO_2	+29.8
Aphis fabae	Broad beans	SO_2	+9.1
		NO_2	+8.5
Acyrthosiphon pisum	Peas	SO_2	+11.3
	Broad beans	SO_2	−12.4
		NO_2	−10.2
Macrosiphum albifrons	Lupins	SO_2	+34.7
		NO_2	+75.1
Myzus persicae	Brussels sprouts	SO_2	+6.8
Brevicoryne brassicae	Brussels sprouts	SO_2	+9.5

A study by Masters (pers. comm.) has shown that the MRGR of *Aphis fabae* was significantly increased when feeding on five different cultivars of *Vicia faba* which had been prefumigated with either 100 or 150 ppb NO_2. There were, however, significant differences due to plant variety in the degree of response shown by the aphid.

Recently, most closed chamber fumigation experiments have concentrated on the effects of O_3 on insect herbivores, mainly because O_3 is currently considered to be the most important pollutant directly affecting vegetation in N. America and W. Europe. Whilst a generally clear cut pattern of responses has emerged, with SO_2 and NO_2 appearing to increase consistently the performance of most insects so far investigated, the situation with O_3 is far more complex.

A number of experiments have recently been performed with chewing

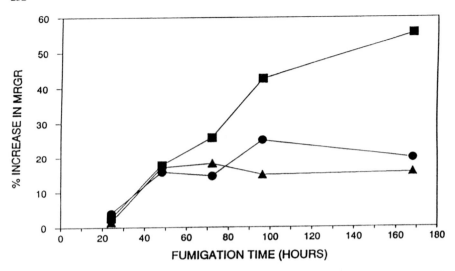

Fig. 2. Effect of prefumigation of 2-year-old saplings with 100 ppb NO_2 on the MRGR of *Elatobium abietinum* (■) and *Cinara pilicornis* (●) on Sitka spruce, and *Schizolachnus pineti* (▲) on Scots pine (Brown et al., 1993a).

insects and the results from most of these are summarized in Table 3. In general, these studies indicate that O_3 exposure of host plants leads to increased insect performance, particularly increased food consumption.

The situation with experiments on aphids, however, has not produced such a clear-cut pattern of results. Table 4 summarizes the results from many of these experiments and it can be seen that both positive and negative effects are reported, and most often the aphids showed no response at all.

Work with the spruce shoot aphid, *C. pilicornis*, (Brown et al., 1993a) has highlighted some of the complexities of interactions between O_3 fumigations and abiotic factors. In a series of experiments in which two year old *Picea sitchensis* trees were prefumigated with 100 ppb O_3 for a range of fumigation times a bewildering range of responses was shown by the aphid. There was a mixture of increases and decreases in MRGR and sometimes no response was observed; furthermore, both positive and negative responses occurred in separate experiments using the same fumigation period. It was only when the ambient temperatures at which the fumigations were performed were examined in relation to the response that a clear pattern emerged. The fumigations were conducted in closed chambers in glasshouse in spring and early summer, and little

Table 3. Examples of closed chamber O_3 fumigation experiments with chewing insects

Insect species	Host plant	Reported effects	Reference
Mexican bean beetle *Epilachna varivestris*	Soybean *Glycine max*	Increased feeding preference	Endress and Post (1985)
Dock leaf beetle *Gastrophysa viridula*	Broad-leaved dock *Rumex obtusifolius*	Increased growth, survival and reproduction. Decreased consumption	Whittaker *et al.* (1989)
Willow leaf beetle *Plagiodera versicolora*	Eastern cottonwood *Populus deltoides*	Increased consumption. Reduced fecundity	Coleman and Jones (1988a)
Tomato pinworm *Keiferia lycopersicilla*	Tomato *Lycopersicon esculentum*	Increased larval survival and developmental rate	Trumble *et al.* (1987)
Monarch butterfly *Danaus plexippus*	Bloodflower *Asclepias curassavica*	Increased growth rates and food preference	Bolsinger *et al.* (1992)
Gypsy moth *Lymantria dispar*	White oak *Quercus alba*	Increased feeding preference	Jeffords and Endress (1984)

Table 4. Examples of closed chamber O_3 fumigation experiments with aphids

Species	Host plant	Reported effects	Reference
Chaitophorus populicola	Eastern cottonwood	No effects	Coleman and Jones (1988b)
Acyrthosiphon pisum	Alfalfa Peas	No effects No effects	Elden *et al.* (1978) Whittaker *et al.* (1989)
Aphis rumicis	Broad-leaved dock	Increased and decreased growth rates – no clear dose response	Whittaker *et al.* (1989)
Elatobium abietinum	Sitka spruce	No effects	Brown *et al.* (1993a)
Cinara pilicornis	Sitka spruce	Increased and decreased growth rates	Brown *et al.* (1993a)
Schizolachnus pineti	Scots pine	No effects No effects	Holopainen *et al.* (1992) Brown *et al.* (1993a)
Aphis fabae	Broad beans	Decreased growth rates Increased and decreased growth rates	Dohmen (1988) Brown *et al.* (1992, 1993b)
Rhopalosiphum padi	Barley	Increased growth rates	Warrington (1989)
Metopolophium dirhodum	Barley Wheat	No effect Increased growth rates	Jackson (1991)

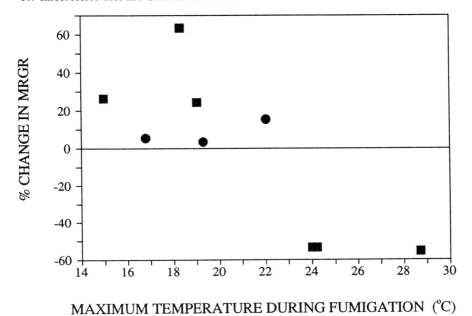

Fig. 3. Percentage change, compared with controls, in MRGR of *Cinara pilicornis* feeding on *Picea sitchensis* previously fumigated with 100 ppb O_3 at different temperatures (Brown et al., 1993a). ■, Significant change from controls ($P < 0.05$); ●, not significant.

control of temperature was possible, and maximum ambient temperatures ranged between 15 and 29°C. Figure 3 shows that for fumigations performed at maximum temperatures below about 20°C the MRGR was significantly increased whereas at above 22°C this pattern was reversed. This has strong implications for understanding the relevance of O_3 on this aphid in the field, as high O_3 is most likely to occur during hot weather, and thus may produce a negative impact on the pest. The results from these experiments also illustrate the need for rigorous environmental control in experiments with this pollutant.

The responses to O_3 fumigations have been most extensively studied with the black bean aphid, *Aphis fabae*. Dohmen (1988) showed decreases in the MRGR of this aphid on broad bean, *Vicia faba*, plants which had been prefumigated with 85 ppb O_3 for 48 h. However, more recent research has shown that O_3 does not always decrease the growth rate of this aphid. Brown et al. (1992) have carried out a wide range of closed chamber fumigation experiments with *A. fabae* on *V. faba*; all experiments were conducted in winter when full control of the temperatures in the chambers was possible. In one series of experiments

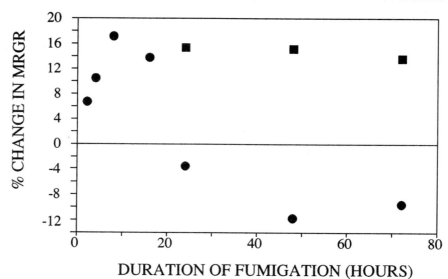

Fig. 4. Effect of continuous prefumigation (100 ppb O_3) and episodic fumigation (8 h 100 ppb O_3 + 16 h filtered air per day) of *Vicia faba* on the subsequent percentage change in MRGR compared with controls (Brown *et al.*, 1992). ●, continuous fumigation; ■, episodic fumigation.

the host plants were exposed to 10 ppb O_3 over periods from 4 to 72 h, administered either continuously or for 8 h per day. The effects on the MRGR of *A. fabae* nymphs subsequently grown on these plants are shown in Fig. 4. With the continuous fumigations, increases in MRGR were observed with fumigations up to 16 h, but longer fumigations resulted in a significant reduction in growth rate, in line with Dohmen's earlier results. In contrast the episodic fumigations produced an increase in MRGR in all cases. As O_3 occurs episodically, except at high elevations, these results suggest that the most likely effect of O_3 in the field would be to increase the performance of this aphid.

A further experiment (Brown *et al.*, 1992) has shown that the observed response by the aphid to a single O_3 episode is rather short-lived. Aphids placed on plants immediately post-fumigation showed increased MRGR, as in the previous series of experiments, but those placed on the plants 4 and 8 days later showed no significant response in MRGR.

In another experiment (Brown *et al.*, 1993b), plants were exposed to a low, "chronic" dose of 30 ppb O_3 for 8 h per day or received charcoal-filtered air for 3 weeks. Half of the plants in each treatment then received a single 8 h acute episode of 100 ppb O_3 before aphids were placed on the plants. The effect of the low O_3 treatment alone was to reduce significantly the MRGR of the aphids, whilst aphids on all plants which received the acute episode showed increased MRGR regardless of their previous treatment.

These experiments show that a very complex range of responses in aphid performance can occur with one insect species on one species of host plant, depending on the dose, exposure pattern and timing of the O_3 fumigation. In view of this, great caution should be exercised when considering the results from other closed-chamber experiments with aphids and O_3. In some studies, just one O_3 exposure regime has been used and sometimes the insects have not been placed on the plants until several days after the fumigation. In other studies, long-term performance parameters such as population growth have been measured. Given the possibility that responses to O_3 may be short lived, it is perhaps not surprising that many of these experiments have shown no response.

Most of the experiments described above were prefumigation experiments in which the plants were fumigated prior to being infested with insects. This strongly suggested that all of the observed responses are plant mediated rather than due to the direct effects of the pollutant on the insect. In order to establish whether there are direct effects on the insect and to separate these from plant-mediated effects, two methods have been used.

Firstly, the results of prefumigation experiments can be compared with those fumigation experiments in which the host plant is infested with insects before fumigation. There are few such comparisons: Brown *et al.* (1992) compared the effects of O_3 fumigation on *A. fabae* feeding on *V. faba* in this way and found no significant difference in the observed response by the aphid suggesting that O_3 did not directly affect *A. fabae* at a concentration of 100 ppb.

A second approach is to fumigate insects directly when they are feeding on artificial diets rather than on their host plants. Most studies on the direct effects of gaseous pollutants on insects have been performed with non-herbivorous species and most have used far higher pollutant concentrations than are likely to occur in the field. Much of this work has been reviewed (Riemer and Whittaker, 1989) and usually negative effects of pollutants are reported. Table 5 shows the results of experiments in which nymphs of *A. fabae* were fumigated with SO_2, NO_2 or O_3 whilst feeding on artificial diet and in all cases there was no significant effect of these pollutants directly on the aphid MRGR suggesting that high, yet realistic, concentrations of these pollutants do not affect the aphid. Similar studies with SO_2 and the grain aphid, *Sitobion avenae*, have also shown no direct

Table 5. Direct effects of gaseous pollutants on the MRGR of *Aphis fabae* feeding on artificial diet (NS; $P > 0.05$)

Gas	MRGR Treatment	MRGR Control	Significance	Reference
SO_2 (150 ppb)	0.395	0.394	NS	Dohmen et al. (1984)
NO_2 (200 ppb)	0.369	0.370	NS	Dohmen et al. (1984)
O_3 (100 ppb)	0.333	0.328	NS	Brown et al. (1992)

Table 6. MRGR of aphids reared on plants previously exposed to ambient or filtered air in Central London (Houlden et al., 1992)

Aphid species	Host plant	Change in MRGR cf. filtered air (%)
Aphis fabae	*Vicia faba*	+21
Acyrthosiphon pisum	*Vicia faba*	−13
Metopolophium dirhodum	*Triticum aestivum*	+41
Metopolophium dirhodum	*Hordeum vulgare*	+42
Rhopalosiphum padi	*Triticum aestivum*	+11

effect of the pollutant. However, studies on the direct effects of O_3 on the rose grain aphid. *M. dirhodum*, suggest that O_3 might on occasions have some direct deleterious effects on the aphid (Jackson, 1991).

2. Filtration experiments

Closed chambers have also been used to investigate the effects of ambient air pollutants on insect/plant systems. In the first investigation of this type, Dohmen et al. (1984), in a series of experiments, grew *V. faba* plants which were infested with *A. fabae* nymphs in closed chambers on a roof in Central London for 9–12 days. The plants received either ambient London air or charcoal-filtered air and the rate of population increase of the aphid was significantly higher in the ambient air.

In other experiments in Central London, Houlden et al. (1992) grew wheat, barley and broad beans in chambers receiving either ambient air or filtered air for 42 days. The plants were then removed and transferred to a controlled temperature room and infested with appropriate aphid species and their MRGR determined. Table 6 shows that the results from these experiments were similar to those obtained in earlier fumigation experiments with SO_2 and NO_2 (Table 2), with four out of the five aphid/host plant combinations showing large increases in MRGR. The exception was *Acyrthosiphon pisum* feeding on the beans which also showed decreased MRGR on this host in the fumigation experiments.

The results of both of these studies clearly show the potential for urban ambient air pollution to affect strongly the performance of aphids on crops.

B. Open-top Chamber Studies

Whilst experiments in closed chambers have been used to demonstrate the effects of air pollutants on a wide range of insect/host plant combinations they are highly artificial environments with major modifications of environmental conditions being imposed by the chambers. Various types of field chambers have been used to control concentrations of gaseous pollutants in studies on plants and in order to minimize the problems of overheating during daylight hours, cylindrical open-top field chambers have been developed (Heagle, 1989). They have been widely used to investigate the effects of air pollution on crops, trees and native vegetation. There are numerous different designs, all consisting of a framed cylinder covered in transparent material into which air can be blown and the concentrations of air pollutants controlled by filtration, fumigation or a combination of the two. They usually have no roof so that ambient rainfall and some light can enter the chamber, although some systems (semi open-top chambers) do exclude rainfall.

Their main advantages are that they allow research to be performed at near ambient conditions of temperature, light and humidity whilst retaining a good degree of control over the chemical composition of the air entering them, and they are much better suited to the long-term growth of plants than most closed chamber systems.

Workers in Switzerland have used semi open-top chambers to investigate the effects of ambient air pollution on several insect herbivores. For example, Bolsinger and Flückiger (1987), found that *Aphis fabae* increased in abundance on both *Viburnum opulus* and *Phaseolus vulgaris* in unfiltered air compared to charcoal-filtered air in chambers situated on a motorway verge. Studies in semi open-top chambers have also shown that the beech weevil, *Rhynchaenus fagi*, preferred beech foliage grown in non-filtered air rather than in filtered air (Hiltbrunner and Flückiger, 1992).

In the USA, Chappelka *et al.* (1988) grew two cultivars of soybean in open-top chambers exposed to filtered air, non-filtered air and non-filtered air with the addition of 30 or 60 ppb O_3. After 3 weeks they were infested with Mexican bean beetles, *Epilachna varivestris*, and although there was no effect of pollution treatment on feeding preference by the beetles, the level of defoliation on both cultivars was significantly, positively correlated with increasing O_3 concentration.

As part of the European Open-Top Chamber Network, investigating the effects of air pollution on crops throughout Europe, open-top chambers have been used to study the effects of rural air pollution on aphid pests of crops (Walsted-Kristiansen *et al.*, 1993). The effects of ambient air pollution and O_3 were investigated in the UK on *A. fabae* feeding on both *Vicia faba* and *Phaseolus vulgaris*, and in Denmark on *R. padi* on wheat.

In the UK studies, larger aphid colonies developed on individual plants in unfiltered air compared to filtered air, despite very low ambient pollutant levels. The MRGR of aphids was significantly higher and mortality significantly lower on plants in the unfiltered chambers. In later experiments, in which an additional O_3 fumigation treatment was included, increased growth rates were found on both non-filtered air- and non-filtered air plus ozone-treated *V. faba* and *P. vulgaris* plants, although decreases were found later in the season on *P. vulgaris*.

In the Danish studies, a higher number of aphids developed per tiller on wheat exposed to non-filtered air plus 30 ppb O_3 than on plants exposed to filtered or non-filtered air. Also, the aphids preferred to settle and had increased MRGRs (on occasions) on plants which had received non-filtered air or non-filtered air + O_3 compared with aphids on plants grown in filtered air.

The data from these two studies clearly demonstrate the potential for current levels of ambient pollution in rural areas to increase the performance of these important agricultural aphid pests.

C. Field-based Experiments

Although open-top chambers provide a much better physicochemical environment in which to study the effects of air pollutants on insect herbivores than closed chambers, they are still less than ideal in that they provide a very sheltered environment and the chamber sides can act as a barrier to the movement of insect herbivores and their natural enemies. Open-air, field systems eliminate the constraints imposed by chambers on insect movement and they have been used, like chamber systems, for both fumigation and filtration experiments.

Aminu-Kano *et al.* (1991) investigated the effects of SO_2 on the

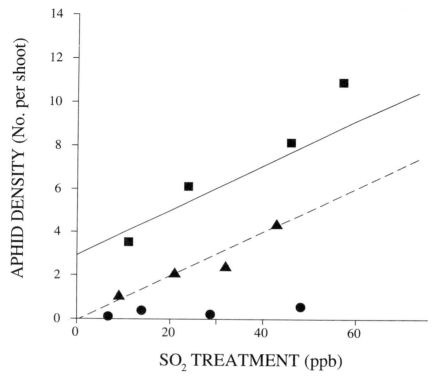

Fig. 5. Effect of field fumigation of winter wheat in 1984 (■——) and 1985 (▲---), and winter barley in 1986 (●) on the density of *Sitobion avenae* (Aminu-Kano *et al.*, 1991).

performance of the grain aphid, *Sitobion avenae*, on cereal crops in an open-air field fumigation system. The crops were exposed to elevated levels of SO_2. Figure 5 shows the effects of this treatment on the populations of the aphid which naturally infested the crop. The aphid populations on wheat in 1984 and 1985 were significantly, positively correlated with SO_2 concentration, whereas those on barley in 1986 were not, probably due to the very low infestation rate by *S. avenae* in this year. No effects of pollution treatment on natural enemies were found and it was suggested that changes in plant food quality were responsible for the observed aphid responses.

Holopainen *et al.* (1991) used an SO_2 field release system to investigate the effects of this pollutant on *Cinara pilicornis* feeding on Norway spruce. They found significantly increased populations in response to the pollution treatment, but there was no evidence of changes in natural enemies.

Recently, Croxford et al. (1993), have used an open-air field filtration system to investigate the effects of ambient air pollution on *Aphis fabae* on *Vicia faba*. With this system either charcoal-filtered air or non-filtered air is blown over the crop. No effects of filtration were found on the number of plants infested in this experiment, suggesting that the aphids did not actively select polluted plants. However, filtration significantly reduced the density of infestation per plant. Again there was no evidence for effects of filtration on rates of parasitism or predator populations and the observed change in infestation was attributed to altered plant food quality.

V. SUMMARY OF CURRENT KNOWLEDGE OF THE EFFECTS OF AIR POLLUTANTS ON INSECT HERBIVORES

A wide range of experimental techniques has been used to investigate the effects of air pollutants on insect herbivores in many studies. The results from most experiments with SO_2 or NO_2 show a high degree of consistency, with increased insect performance being found in almost all cases. This is in strong agreement with field observations and other correlative evidence of the effects of these two pollutants on insect herbivore populations. The situation with O_3 appears more complex, particularly the results obtained from closed chamber fumigation studies with aphids. Evidence from filtration studies on crops in rural areas, where O_3 is the main pollutant present, suggests that in the field this pollutant can also significantly increase insect performance.

There is little evidence for direct effects of realistic concentrations of the three major air pollutants on insect herbivores. There is also no experimental evidence for altered levels of natural enemies, although only a few studies in open-air systems have enabled this to be investigated. However, evidence from some field observations and experiments along pollution gradients suggests that direct effects on natural enemies may be important and clearly more experimental studies are required. The observed responses appear, in the main, to be due to an alteration in the quality of the host plant by the pollutants.

The possible mechanisms whereby pollutants change the host plant quality have been extensively reviewed (e.g. Lechowicz, 1987; Riemer and Whittaker, 1989; Whittaker and Warrington, 1990). Where the underlying mechanism has been investigated most studies with SO_2 or NO_2 have shown changes in the nitrogen balance of the host plant, particularly amino acid composition. These pollutants are usually detoxified and transported from plant leaves as amino acids (Heath, 1984) so this is a likely explanation. The understanding of the mechanisms by which O_3 alters host plants is less clear. Closed chamber (Brown et al., 1993b) and open-top chamber (Walsted-Kristiansen et al., 1993) studies with *Aphis fabae* have shown no consistent changes in the free amino acid composition of *Vicia faba* foliage. Similarly, no significant effects of O_3 on the free amino acid content were found on wheat in open-top chamber studies (Walsted-Kristiansen et al., 1993). Jackson (1991), however, did find that O_3 could affect the free amino acid content of wheat and barley in closed chamber studies with cereal aphids, and there are many examples of O_3-induced changes of plant amino acids from studies on the direct effects of O_3 on plants (e.g. Heath, 1984; Rowland et al., 1988).

Bolsinger and Flückiger (1989) have used a phloem exudation technique, rather than total leaf free amino acid analysis, to investigate the effects of ambient air pollution (where nitrogen oxides were the major pollutants present) on *Aphis fabae* on *Phaseolus* beans. They found a good correlation between the free amino acid composition of phloem exudates and the performance of this aphid on polluted plants and this may be a more sensitive method for investigating these mechanisms in studies with phloem-feeding insects.

However, there is a wide range of other factors which could affect either the nutritional status of host plants or their attractiveness to insects. For example, a recent study of the effects of O_3 on the monarch butterfly, *Danaus plexippus*, (Bolsinger et al., 1992) has suggested that O_3 may act by changing the feeding stimuli of the insect at the plant surface, rather than by changing the nutritional quality of the host plant.

Regardless of the underlying mechanisms, it is clear from research to date that these three major air pollutants can have large effects on herbivorous insect populations, both in the laboratory and the field, and that in most cases the effects on the insects are positive.

VI. PREDICTIONS AND FUTURE RESEARCH

For the reasons outlined in Section IIA, emissions of SO_2 in the UK are declining rapidly and will continue to do so in the foreseeable future (QUARG, 1993). The situation is much the same over most of Western Europe and North America. Consequently the impacts of this pollutant on herbivorous insects are also likely to decline. However, it should be noted that several of the studies reported in this review have shown that insect herbivores can respond positively to relatively low concentrations of this pollutant or to very short exposures, therefore SO_2 will probably continue to affect herbivorous insects in areas where local concentrations of the pollutant are high.

UK levels of NO_2 are also predicted to decline over the next two decades or so, mainly due to regulations restricting emissions from light motor vehicles, particularly regarding the fitting of closed-loop three way catalytic converters (QUARG, 1993). However, if the levels of motor traffic continue to rise at the current rate, despite catalysts, levels of NO_2 will also rise. It is probable that unless there is a major change in transportation use in Britain and other developed countries, or major technical advances in the control of motor vehicle emissions, NO_2 will remain an important influence, increasing insect populations in urban areas and near roads.

The scientific understanding of the complex atmospheric chemistry of tropospheric O_3 formation is not yet complete (UK PORG, 1987), but most predictions suggest that levels of this pollutant in Western Europe and North America will continue to increase (e.g. Ashmore and Bell, 1991) and consequently so will the effects on herbivorous insects.

It is not only the three pollutants discussed in this chapter which are predicted to change; as other chapters in this volume clearly point out, concentrations of carbon dioxide (CO_2) and other greenhouse gases are predicted to increase. The interaction of increased CO_2 concentrations with air pollutants on herbivorous insects and the effects of climate change on these systems has so far received little attention. Given the interaction with temperature seen in O_3 fumigation experiments with *Cinara pilicornis*, this is an area that merits further research.

Probably of most future concern in a global context are the effects of gaseous air pollutants on herbivorous insects outside of Western Europe and North America. Urbanization, industrialization and energy consumption are increasing rapidly in many developing countries. However, for various reasons pollution control is often inadequate (Ashmore, 1991). For example, SO_2 emissions in Thailand have been predicted to increase more than five-fold between 1985 and 2000 if no effective emission control occurs. Levels of the major pollutants, known to increase insect pest populations, are already high enough to cause the effects reported in this chapter in urban areas in many developing countries and virtually nothing is known about rural concentrations. However, there are virtually no data on the direct effects of air pollution on vegetation, let alone the effects on insect herbivores, in these countries. In one of the few published studies from developing countries, Rao et al. (1990), have observed increased aphid populations on sorghum apparently in response to quite low levels of SO_2 and NO_2. It is in developing countries that the impact of air-pollution induced increases in insect pest populations is likely to be the greatest. Whilst in most of the developed world there is surplus production of agricultural crops, food production is of prime national importance in most developing countries. Furthermore, because of climatic conditions, insect pests are already of more importance in these countries than in cooler climates. Also, urban populations in such countries often rely to a substantial extent on food grown in and around the city, so that the impact of air pollutants on insect pests will be even greater. It is clear that one of the greatest needs for future research on the effects of air pollutants on insect herbivores is in these areas.

Acknowledgements

I thank Dr Stuart McNeill, Dr Mike Ashmore and Professor Nigel Bell for their helpful criticism and encouragement whilst preparing this manuscript and for their permission to reproduce figures and data from their publications and from papers in preparation. I am also grateful for the efforts of all who have investigated the effects of gaseous pollutants on insect herbivores without whom this chapter would not have been possible.

REFERENCES

Alstad, D. N., Edmunds, G. G. and Weinstein, L. H. (1982). Effects of air pollutants on insect populations. *Ann. Rev. Entomol.* **27**, 369–384.

Aminu-Kano, M., McNeill, S. and Hails, R. S. (1991). Pollutant, plant and pest interactions: the grain aphid *Sitobion avenae* (F.). *Agric. Ecosystems Environ.* **33**, 233–243.

Ashmore, M. R. (1991). Air pollution and agriculture. *Outlook on Agriculture* **20**, 139–144.

Ashmore, M. R. and Bell, J. N. B. (1991). The role of ozone in global change. *Ann. Bot.* **67** (Suppl. 1), 39–48.

Baltensweiler, W. (1985). Waldsterben: forest pests and air pollution. *Z. Ang. Ent.* **99**, 77–85.

Bell, J. N. B., McNeill, S., Houlden, G., Brown, V. C. and Mansfield, P. J. (1993). Atmospheric change: effect on plant pests and diseases. *Parasitology* **106**, 11–24.

Bolsinger, M. and Flückiger, W. (1987). Enhanced aphid infestation at motorways: the role of ambient air pollution. *Entomol. Exp. Appl.* **45**, 237–243.

Bolsinger, M. and Flückiger, W. (1989). Ambient air pollution induced changes in amino acid pattern of phloem sap in host plants – relevance to aphid infestation. *Environ. Poll.* **56**, 209–235.

Bolsinger, M., Lier, M. E. and Hughes, P. R. (1992). Influence of ozone air pollution on plant–herbivore interaction. Part 2: Effects of ozone on feeding preference, growth and consumption rates of monarch butterflies (*Danaus plexippus*). *Environ. Poll.* **77**, 31–37.

Brown, V. C., McNeill, S. and Ashmore, M. R. (1992). The effects of ozone fumigation on the performance of the black bean aphid, *Aphis fabae* Scop., feeding on broad beans, *Vicia faba*. *Agric. Ecosystems Environ.* **38**, 71–78.

Brown, V. C., Ashmore, M. R. and McNeill, S. (1993a). Experimental investigations of the effects of air pollutants on aphids on coniferous trees. *Forstw. Cbl.* **112**, 128–132.

Brown, V. C., Ashmore, M. R. and McNeill, S. (1993b). Effects of chronic and acute ozone fumigation on the performance of the black bean aphid, *Aphis fabae*, feeding on *Vicia faba*. In "Effects of Air Pollution on Agricultural Crops in Europe." (H. J. Jäger, M. Unsworth, L. de Temmerman and P. Mathy, eds). *Air Pollution Research Report 46*, pp. 491–494. CEC, Brussels.

Brown, V. C., Gate, I. and Ashmore, M. R. (1993c). Effects of air pollution on the life cycle of *Aphis fabae*. In "Effects of Air Pollution on Agricultural Crops in Europe" (H. J. Jäger, M. Unsworth, L. de Temmerman and P. Mathy, eds). *Air Pollution Research Report 46*, pp. 487–490. CEC, Brussels.

Chappelka, A. H., Kraemer, M. E., Mebrahtu, T., Ragappa, M. and Benepal, P. S. (1988). The effects of ozone on soybean resistance to the Mexican bean beetle (*Epilachna varivestis* Mulsant). *Environ. Exp. Bot.* **28**, 53–60.

Cicerone, R. J. (1987). Changes in stratospheric ozone. *Science* **237**, 35–42.

Coleman, J. S. and Jones, G. C. (1988a). Plant stress and insect performance: Eastern cottonwood, ozone and a leaf beetle. *Oecologia* **76**, 57–61.

Coleman, J. S. and Jones, G. C. (1988b). Acute ozone stress on eastern cottonwood (*Populus deltoides* Bartr.) and the pest potential of the aphid, *Chaitophorus populicola* Thomas (Homoptera: Aphididae). *Environ. Entomol.* **17**, 207–212.

Cramer, H. H. (1951). De geographischen Grundlagen des Massenwechsels von *Epiblema tedella*. *Forstw. Cbl.* **70**, 42–53.

Croxford, A. C., Ashmore, M. R. and McNeill, S. (1993). Effects of ambient air pollution and aphid pests on *Vicia faba*. In "Effects of Air Pollution on Agricultural Crops in Europe" (H. J. Jäger, M. Unsworth, L. de Temmerman and P. Mathy, eds) *Air Pollution Research Report 46*, pp. 501–506, CEC, Brussels.

Darrell, N. M. (1989). The effect of air pollutants on physiological processes in plants. *Plant, Cell and Environ.* **12**, 1–30.

Decourt, N., Malphettes, C. B., Perrin, R. and Caron, D. (1980). La pollution soufrée limite-t-elle le développement de la malidie de l'écorce du hetre? (*Cryptococcus fagi*, *Nectria coccinea*). *Ann. Sci. For.* **37**, 135–145.

Dohmen, G. P. (1988). Indirect effects of air pollutants: changes in plant/parasite interactions. *Environ. Poll.* **53**, 197–207.

Dohmen, G. P., McNeill, S. and Bell, J. N. B. (1984). Air pollution increases *Aphis fabae* pest potential. *Nature* **307**, 52–53.

Elden, T. C., Howell, R. K. and Webb, R. E. (1978). Influence of ozone on pea aphid resistance in selected alfalfa strains. *J. Econ. Entomol.* **71**, 283–286.

van Emden, H. F. (1969). Plant resistance to *Myzus persicae* induced by a plant regulator and measured by aphid relative growth rate. *Entomol. Exp. Appl.* **12**, 125–131.

Endress, A. G. and Post, S. L. (1985). Altered feeding preference of Mexican bean beetle *Epilachna varivestis* for ozonated soybean foliage. *Environ. Poll. Series A.* **39**, 9–16.

Evelyn, J. (1661). "Fumifugium – or the Inconvenience of the Aere and Smoake of London Dissipated". Second reprint (1972), National Society for Clean Air.

Flückiger, W., Oertli, J. J. and Baltensweiler, W. (1978). Observations on aphid infestation on Hawthorn in the vicinity of a motorway. *Naturwissenschaften* **65**, 654–655.

Flückiger, W., Braun, S. and Bolsinger, M. (1987). Air pollution: Effect on hostplant – insect relationships. *In* "Air Pollution and Plant Metabolism" (S. Schulte-Hostende, N. M. Darrall, L. W. Blank and A. R. Wellburn, eds), pp. 366–380. Elsevier, London.

Führer, E. (1985). Air pollution and the incidence of forest insect problems. *Z. Ang. Ent.* **99**, 371–377.

Guderian, R., Tingey, D. T. and Rabe, R. (1984). Effects of photochemical oxidants on plants. *In* "Air Pollution by Photochemical Oxidants" (R. Guderian, ed.), Part 2, pp. 129–333. Springer-Verlag, Berlin.

Hain, F. P. (1987). Interactions of insects, trees and air pollutants. *Tree Physiol.* **3**, 93–102.

Heagle, A. S. (1989). Ozone and crop yield. *Ann. Rev. Phytopathol.* **27**, 397–423. Butterworths, London.

Heath, R. L. (1984). Air pollutant effects on biochemicals derived from metabolism: organic, fatty and amino-acids. *In* "Gaseous Air Pollutants and Plant Metabolism" (M. J. Koziol and F. R. Whatley, eds), pp. 275–290. Butterworths, London.

Heliövaara, K. and Väisänen, R. (1989a). Invertebrates of young Scots pine stands near the industrialized town of Harjavalta, Finland. *Silva Fennica* **23**, 13–19.

Heliövaara, K. and Väisänen, R. (1989b). Reduced cocoon size of diprionids (Hymenoptera) reared on pollutant affected pines. *J. Appl. Entomol.* **107**, 34–40.

Heliövaara, K. and Väisänen, R. (1993). "Insects and Pollution". CRC Press, Boca Raton, FL.

Heliövaara, K., Väisänen, R. and Varama, M. (1991). Larval mortality of pine sawflies (Hymenoptera: Diprionidae) in relation to pollution: a field experiment. *Entomophaga* **36**, 315–321.

Hiltbrunner, E. and Flückiger, W. (1992). Altered feeding preference of beech weevil, *Rhynchaenus fagi* L., for beech foliage under ambient air pollution. *Environ. Poll.* **75**, 333–336.

Holopainen, J. K., Kainulainen, E., Oksanen, J., Wulff, A. and Kärenlampi, L. (1991). Effect of exposure to fluoride, nitrogen compounds and SO_2 on the numbers of spruce shoot aphids on Norway spruce seedlings. *Oecologia* **86**, 51–56.

Holopainen, J. K., Mustaniemi, A., Kainulainen, E. and Oksanen, J. (1992). Reproduction of aphids and spider mites and levels of amino acids in conifer seedlings exposed to ozone. *In* "Responses of Forest Ecosystems to Environmental Change" (A. Teller, P. Mathy and J. N. R. Jeffers, eds), pp. 965–966. Elsevier, London.

Holopainen, J. K., Mustaniemi, A., Kainulainen, E., Satka, H. and Oksanen, J. (1993). Conifer aphids in an air-polluted environment I. Aphid density, growth and accumulation of sulphur and nitrogen by Scots pine and Norway spruce seedlings. *Environ. Poll.* **80**, 185–191.

Hough, A. M. and Derwent, R. G. (1990). Changes in the global concentration of tropospheric ozone due to human activities. *Nature* **344**, 645–648.

Houlden, G., McNeill, S., Aminu-Kano, M. and Bell, J. N. B. (1990). Air pollution and agricultural aphid pests. I: Fumigation experiments with SO_2 and NO_2. *Environ. Poll.* **67**, 305–314.

Houlden, G., McNeill, S., Craske, and Bell, J. N. B. (1992). Air pollution and agricultural aphid pests. II: Chamber filtration experiments. *Environ. Poll.* **72**, 45–55.

Hughes, P. R. (1988). Insect populations on host plants subjected to air pollutioin. *In* "Plant Stress–Insect Interactions" (E. A. Heinrichs, ed.), pp. 249–319. John Wiley, Chichester.

Hughes, P. R. and Laurence, J. A. (1984). Relationship of biochemical effects of air pollutants on plants to environmental problems: insect and microbial interactions. *In* "Gaseous Air Pollutants and Plant Metabolism." (M. J. Koziol and F. R. Whatley, eds), pp. 361–377. Butterworth, London.

Hughes, P. R., Potter, J. E. and Weinstein, L. H. (1981). Effects of air pollutants on plant-insect interactions: Reactions of the Mexican bean beetle to SO_2 fumigated pinto beans. *Environ. Entomol.* **10**, 741–744.

Hughes, P. R., Potter, J. E. and Weinstein, L. H. (1982). Effects of air pollutants on plant-insect interactions: Increased susceptibility of greenhouse grown soybeans to the Mexican bean beetle after exposure to SO_2. *Environ. Entomol.* **11**, 173–176.

Jackson, G. (1991). "The Effects of Ozone and Nitrogen Dioxide on Cereal/Aphid Interactions. PhD thesis, University of London.

Jeffords, M. R. and Endress, A. G. (1984). A possible role of ozone in tree defoliation by the gypsy moth (Lepidoptera: Lymantridae). *Environ. Entomol.* **13**, 1249–1252.

Krupa, S. V. and Manning, W. J. (1988). Atmospheric ozone: Formation and effects on vegetation. *Environ. Poll.* **50**, 101–137.

Laurence, J. A., Hughes, P. R., Weinstein, L. H., Geballe, G. T. and Smith, W. H. (1983). "Impact of Air Pollution on Plant-Pest Interactions: Implications of Current Research and Strategies for Future Studies". *Ecosystems Research Centre Report*, **20**, Cornell University.

Leather, S. R. and Dixon, A. F. G. (1984). Aphid growth and reproductive rates. *Ent. Exp. Appl.* **35**, 137–140.

Lechowicz, M. J. (1987). Resource allocation by plants under air pollution stress: implications for plant–pest–pathogen interactions. *Bot. Rev.* **53**, 281–300.

McNeill, S. and Whittaker, J. B. (1990). Air pollution and tree-dwelling aphids. *In* "Population Dynamics of Forest Insects" (A. D. Watt, S. R. Leather, M. D. Hunter and N. A. C. Kidd, eds), pp. 195–208. Intercept, Andover.

McNeill, S., Bell, J. N. B., Aminu-Kano, M. Houlden, G., Bullock, J. and Citrone, S. (1987). The interaction between air pollution and sucking insects. *In* "Acid Rain – Scientific and Technical Advances" (R. Perry, R. M. Harrison, J. N. B. Bell and J. N. Lester, eds), pp. 602–607. Selper, London.

Malhotra, S. S. and Khan, A. A. (1984). Biochemical and physiological impact of major pollutants. *In* "Air Pollution and Plant Life" (M. Treshow, ed.), pp. 113–157. John Wiley, Chichester.

Manning, W. J. and Keane, K. D. (1988). Effects of air pollutants on interactions between plants, insects and pathogens. *In* "Assessment of Crop Loss from Air Pollutants" (W. W. Heck, O. C. Taylor and D. T. Tingey, eds), pp. 365–386. Elsevier, London.

Port, G. R. and Thompson, J. R. (1980). Outbreaks of insect herbivores on plants along motorways in the United Kingdom. *J. Appl. Ecol.* **17**, 649–656.

QUARG (1993). "Urban Air Quality in the United Kingdom. The First Report of the Quality of Urban Air Review Group." Department of the Environment, London.

Rao, M. V., Gupta, G. K. and Dubey, P. S. (1990). Effects of relatively low ambient air pollution on total sucking insect populations on *Sorghum vulgare*. *Trop. Ecol.* **31**, 66–72.

Riemer, J. and Whittaker, J. B. (1989). Air pollution and insect herbivores: observed interactions and possible mechanisms. *In* "Insect–Plant Interactions. Vol. 1" (E. A. Bernays, ed.), pp. 73–105. CRC Press, Boca Raton, FL.

Rowland, A. J., Murray, A. J. S. and Wellburn, A. R. (1985). Oxides of nitrogen and their impact on vegetation. *Reviews on Environmental Health* **5**, 295–342.

Rowland, A. J., Borland, A. M. and Lea, P. J. (1988). Changes in amino acids, amides and proteins in response to air pollutants. *In* "Air Pollution and Plant Metabolism" (S. Schulte-Hostende, N. M. Darrall, L. W. Blank and A. R. Wellburn, eds), pp. 189–221. Elsevier, London.

Stark, R. W. and Cobb, F. W. (1969). Smog injury, root diseases and bark beetle damage in ponderosa pine. *Calif. Agric.* **23**, 13–15.

Taylor, G. E. and Eaton, F. M. (1966). Suppression of plant growth by nitrogen dioxide. *Plant Physiol.* **41**, 132–135.

Trumble, J. T., Hare, J. D. and McCool, P. M. (1987). Ozone-induced changes in host plant suitability: Interactions of *Keiferia lycopersicella* and *Lycopersicon esculentum*. *J. Chem. Ecol.* **13**, 203–218.

UK PORG. (1987). "Ozone in the United Kingdom. United Kingdom Photochemical Oxidants Review Group Interim Report." Department of the Environment, London.

UK PORG. (1990). "Oxides of Nitrogen in the United Kingdom. The Second Report of the United Kingdom Photochemical Oxidants Review Group." Department of the Environment, London.

Villemant, C. (1981). Influence de la pollution atmosphérique sur les populations d'aphides du pin sylvestre en Forêt de Roumare (Seine-Maritime). *Environ. Poll. (Ser. A).* **24**, 245–262.

Walsted-Kristiansen, L., Brown, V. C. and Ashmore, M. R. (1993). CEC assessment of biotic interactions. *In* "Effects of Air Pollution on Agricultural Crops in Europe" (H. J. Jäger, M. Unsworth, L. de Temmerman and P. Mathy, eds) *Air Pollution Research Report 46*, pp. 487–490. CEC, Brussels.

Warrington, S. (1989). Ozone enhances the growth rate of cereal aphids. *Agric. Ecosystems and Environ.* **26**, 65–68.

Warrington, S., Mansfield, T. A. and Whittaker, J. B. (1987). Effect of SO_2 on the reproduction of pea aphids, *Acyrthosiphon pisum*, and the impact of SO_2 and aphids on the growth and yield of peas. *Environ. Poll.* **48**, 285–296.

Warrington, S. and Whittaker, J. B. (1990). Interactions between Sitka spruce, the green spruce aphid, sulphur dioxide pollution and drought. *Environ. Poll.* **65**, 363–370.

Wellburn, A. R. (1988). "Air Pollution and Acid Rain: the Biological Impact". Longman, London.

Wentzel, K. F. (1965). Insekten als Immissionsfolgeschadingle. *Naturwissenschaften* **52**, 113.

Whittaker, J. B., Kristiansen, L. W., Mikkelsen, T. N. and Moore, R. (1989). Responses to ozone of insects feeding on a crop and a weed species. *Environ. Poll.* **62**, 89–101.

Whittaker, J. B. and Warrington, S. (1990). *In* "Pests, Pathogens and Plant Communities" (J. J. Burdon and S. R. Leather, eds), pp. 97–110. Blackwell Scientific Publications, Oxford.

11

Deficiency and Excess of Essential and Non-essential Metals in Terrestrial Insects

S. P. HOPKIN

I.	Introduction	252
II.	Natural Selection of the Elements	254
	A. Introduction	254
	B. Deficiency of Essential Metals	254
	C. Regulation of Metals	255
	D. Critical Concentrations of Metals	258
III.	Metal Pollution	259
	A. Introduction	259
	B. Community Effects	259
	C. Bioaccumulation	261
	D. Effects on Individuals and Species	262
	E. Genetic Effects	263
IV.	Conclusions	264
	References	266

Abstract

Essential elements such as iron, copper and zinc form integral parts of cellular processes and are required in trace amounts by insects if they are to grow and reproduce normally. The ability to regulate assimilation and excretion of these elements via the digestive tract was an essential prerequisite for their successful colonization of the land. However, when present at high concentrations due to pollution, these essential metals (and those that are not required such as cadmium, lead and mercury) may disrupt normal biochemistry. Sensitive insects may be killed by acute or chronic poisoning, or die from deficiency of an essential element through antagonism from a non-essential metal in the diet. Species that are tolerant to pollution may respond to the subsequent lack of competition and reach much higher population densities than in uncontaminated areas. These changes at the community level may lead to disruption of ecological processes such as plant litter decomposition.

Examples of these phenomena are described in terrestrial insects. Special attention is paid to Collembola and Diptera in which rapid evolution of resistance occurs in response to metal contamination.

I. INTRODUCTION

Metals are natural substances. With the exception of radioisotopes produced in nuclear reactors, all metals have been an integral part of the biosphere throughout evolution. Elements that are considered to be metals for the purpose of this review are shown in Fig. 1.

Metals are also non-biodegradable. Unlike many organic pesticides and pollutants, they cannot be broken down into less-harmful components. Detoxification consists of "hiding" active metal ions within a protein, or depositing them in an insoluble form in intracellular granules for long-term storage or excretion in the faeces.

A further important feature of some metals is their essentiality (Fig. 2). Metals such as copper and zinc have a window of essentiality within which

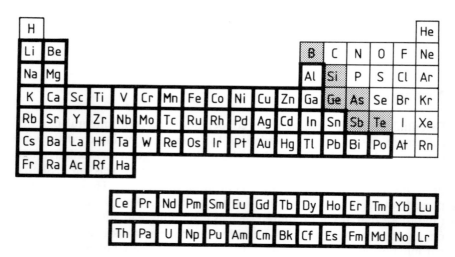

Fig. 1. Periodic table of the elements. Those considered to be metals are surrounded by bold lines. Metalloids (with properties of metals and non-metals) are shaded. From Hopkin (1989), by permission of Elsevier Applied Science.

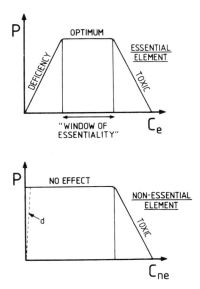

Fig. 2. Relationships between performance (P) (growth, fecundity, survival) and concentrations of an essential (C_e) or non-essential element (C_{ne}) in the diet of animals. Possible deficiency effects at ultra-trace levels (d) of an apparently non-essential element may be discovered as the sensitivities of analytical techniques are improved. From Hopkin (1989), by permission of Elsevier Applied Science.

dietary concentrations of the elements must be maintained. However, in terms of availability of these essential metals (and non-essential elements such as cadmium, mercury and lead), insects living in industrialized regions are in an environment that is much-changed from that in which regulatory processes for metals evolved.

In this review, the roles that metals play in the biology of terrestrial insects will be discussed. Most citations are to work published since the review of Hopkin (1989) which contains a comprehensive list of earlier literature. After considering essentiality (Section II), the responses of insects to metal pollution are examined (Section III). It is clear that adaptation to increased availability of metals involves the enhancement of existing regulatory and detoxification mechanisms, rather than the development of new systems. This has interesting implications for insect ecology and pest control which are discussed in the final part (Section IV).

II. NATURAL SELECTION OF THE ELEMENTS

A. Introduction

Life evolved in the sea some 3.5×10^9 years ago. The solubility of elements under the anaerobic conditions that existed at the time determined which metals could be incorporated into biochemical reactions. A natural selection of the elements took place where the choice of one element rather than a similar one was dictated by its availability, the ability of the organism to retain it, and its functional advantages relative to those of other metals (Williams, 1981). When the atmosphere changed from being chemically reducing, to one containing 20% oxygen, organisms had to develop aerobic biochemistry. One of the greatest changes was in the availability of iron which became almost completely insoluble. Thus proteins for storage (ferritin) and transport (transferrin) of this metal had to be developed (Locke and Leung, 1984; Locke and Nichol, 1992).

The earliest known hexapod is a collembolan *Rhyniella praecursor* Hirst & Maulik from the Rhynie chert formed in the Lower Devonian (Greenslade and Whalley, 1986). The earliest insect communities probably developed under algal mats and later, on low emergent vegetation (Little, 1990; Shear and Kukalová-Peck, 1990; Kukalová-Peck, 1991, 1992; Edwards and Selden, 1992). It is likely that the food of these early insects consisted of fungal hyphae, which were grazed directly, and bacteria that were stripped from fragments of decaying vegetation as they passed through the gut (Price, 1988).

Concentrations of metals in fungi are usually much higher than the substrate on which they are growing (Starling and Ross, 1991; Hopkin, 1993a). Thus from the time of the emergence on to land of the first insects, systems for metal regulation would have to have been well developed. The option of excreting unwanted elements directly to the surrounding water was no longer available. The ability to regulate assimilation and excretion of essential and non-essential metals via the digestive tract was an essential prerequisite for the successful colonization of the land by insects.

B. Deficiency of Essential Metals

Concentrations of some essential elements in the diets of terrestrial insects are quite low in comparison to minimum dietary requirements (Roth, 1992; Studier and Sevick, 1992). For example, the level of copper in the leaves of deciduous trees is often less than 10 μg g^{-1} dry weight (Guha and Mitchell, 1966; Flemming and Trevors, 1989; Markert, 1992). In some parts of the world, levels of copper (and other essential elements) in soils are so low that growth of plants is restricted (Boardman and McGuire, 1990a,b; Hunter et al., 1990). Deficiency diseases in livestock are well documented in such areas, but little attention has been paid to potential effects on invertebrates. Is there any evidence for metal deficiency in insects?

Creighton (1938) was the first author to examine this question. He reported very high mortality of lepidopteran larvae fed on foliage from copper- and zinc-deficient plants. There is clear experimental evidence for a minimum dietary requirement for copper, iron, manganese and zinc in aphids (Fig. 3; Dadd, 1967; Akey and Beck, 1972), copper and zinc in crickets (McFarlane, 1976), and zinc in beetles (Fig. 4; Fraenkel, 1959). Some insects go to great lengths to assimilate essential elements and to retain them once they are in the body. Bruchid beetles, for example, assimilate up to 90% of the copper and zinc from the seeds on which they feed (Ernst, 1992, 1993). The male of the noctuid moth *Heliothis virescens* (Fabricius) transfers 36% of its whole body zinc content to the female at the time of mating (Engebretson and Mason, 1980).

Thus, metal deficiency has been demonstrated in several insect groups in the laboratory and some species have evolved to maximize the assimilation and retention of essential elements. However, direct evidence is lacking of the possible effects of these phenomena on insect populations. Dietary switching in grasshoppers (Bernays and Bright, 1991) and other insects (Waldbauer and Friedman, 1991) may occur in response to nutrient deficiency, but there is as yet no evidence that such behaviour could be stimulated by inadequate levels of an essential metal in the food. This is clearly an area in need of more research.

C. Regulation of Metals

The pathways of metal detoxification in invertebrates are relatively well understood (for reviews see Hopkin, 1989, 1990; Hopkin et al., 1989; Beeby, 1991; Hare, 1992; Dallinger, 1993; Heliövaara and Väisänen, 1993). The epithelium of the midgut acts as a barrier separating the contents of the lumen from the haemolymph. Thus the midgut usually contains the highest concentrations of metals in insects (Fig. 5; Table 1; Lauverjat et al., 1989; Raes et al., 1992). A pool of available essential metals is maintained in the cells to supply the needs of normal biochemistry (e.g. 40% of total copper in the midgut of *Lucilia cuprina*

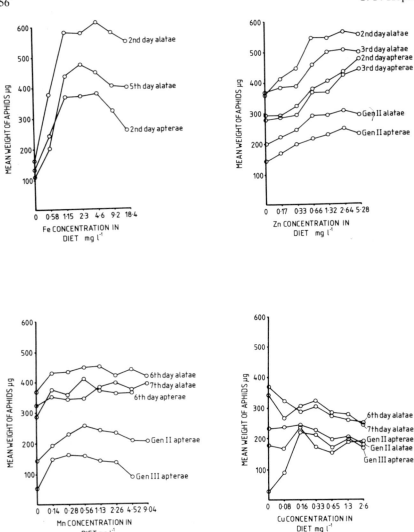

Fig. 3. Mean individual weights of the aphid *Myzus persicae* (Sulzer) after 8 days' growth on a synthetic diet containing a range of concentrations of iron, zinc, manganese and copper. 2nd, 3rd, 6th and 7th day apterae and alatae are larvae of the first generation on a synthetic diet weighed on the 8th day of growth. Gen II and Gen III are larvae of the second and third generations weighed when apterae had become adults on the best diets. Note the classic "bell-shaped" dose–response curve for iron. Redrawn from Dadd (1967) by permission of Pergamon Journals.

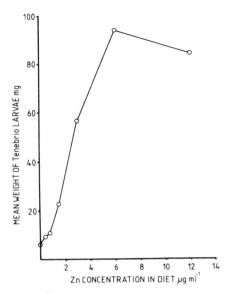

Fig. 4. Graph showing the mean individual weight of larvae of *Tenebrio molitor* L. reared for 11 weeks on a synthetic diet containing different levels of zinc ($n = 20$ for each treatment). The beetles require a concentration of zinc in their food of at least 6 μg g^{-1} (fresh weight) for normal growth. Based on data in Fraenkel (1959).

(Wiedemann) is water soluble – Waterhouse, 1945). Essential metals that are surplus to requirements, and non-essential metals, are prevented from reaching the other body tissues by storage in proteins and insoluble granules.

There are three main pathways of intracellular metal detoxification in insects (Hopkin, 1989). Metals following the *Type A pathway*, such as calcium, zinc and manganese, are stored as phosphates in concentrically structured granules (Ballan-Dufrançais *et al.*, 1980; Jeantet *et al.*, 1980). Metals following the *Type B pathway*, such as cadmium, mercury and copper, are bound initially in metallothionein proteins (Aoki *et al.*, 1984; Suzuki *et al.*, 1989; Cosson *et al.*, 1991; Theodore *et al.*, 1991), and are deposited eventually in specialized lysosomes as dense, sulphur-rich, insoluble bodies (Ballan-Dufrançais *et al.*, 1980; Jeantet *et al.*, 1980). The *Type C pathway* is followed by iron (Locke and Leung, 1984; Locke and Nichol, 1992). Iron is stored in ferritin. However, surplus iron may be deposited in modified lysosomes in a similar way to metals that follow the type B pathway. All metals stored in granules may be excreted in the faeces following lysis of the midgut cells.

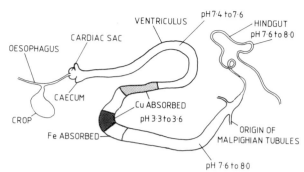

Fig. 5. Diagram of the gut of *Lucilia cuprina* larva (Diptera) showing pH of gut contents and regions of iron and copper absorption. Redrawn from R. F. Chapman, English Universities Press, London, "The Insects: Structure and Function", 1971, by permission of Hodder & Stoughton.

D. Critical Concentrations of Metals

Attempts have been made to set critical concentrations for metals in terrestrial ecosystems that will protect most species from the effects of pollution (Bengtsson and Tranvik, 1989; Van Straalen, 1993). However, this is extremely difficult to apply to insects owing to their diversity of feeding behaviours and habitats. Indeed, Hopkin (1993b) has speculated that under some circumstances, a critical concentration in soil that protects one species from poisoning may be below the minimum dietary requirement for another.

Table 1. Distribution of cadmium in larvae of the fleshfly *Sarcophaga peregrina* after feeding for 3 days on homogenized porcine liver containing 100 μg Cd g^{-1} as CdCl$_2$. From Aoki *et al.* (1984) by permission of Pergamon Press.

	Content (μg/larva)	Distribution (%)	Concentration (μg g^{-1} wet tissue)
Whole body	1.53 ± 0.19*	100	14.6 ± 2.2*
Digestive tract	1.37 ± 0.49*	90	301 ± 81*
Fat body	0.009†	0.6	0.1†
Mapighian tube	0.036†	2.3	9.0†
Trachea	ND		
Brain	ND		
Cuticle	0.062†	4.1	0.34†

*Means ± SD ($N = 5$).
†Mean value of pooled samples.
ND: Not detectable.

In the future, it is important that the windows of essentiality for metals such as copper and zinc are considered in the context of the biology of the animals one is attempting to protect. In the same way as the choice of a management regime on a nature reserve dictates which insect species will thrive and which will perish, we may have to choose a critical concentration that will benefit those species we wish to protect at the expense of others. At present we are nowhere near being able to make such a decision owing to the lack of information on the potential effects of metal deficiency on insects.

III. METAL POLLUTION

A. Introduction

Present-day insects in industrialized countries are living in a changed environment in comparison to that in which they evolved. Several metals are present in much greater concentrations in soils and vegetation than before the industrial revolution. For many metals such as cadmium, lead and mercury, the activities of Man contribute a much greater proportion of total global release to the atmosphere than emissions from volcanoes and other natural phenomena (Nriagu, 1988, 1990; Nriagu and Pacyna, 1988).

One of the largely unexplored side-effects of the use of metal-containing insecticides is their possible role in rectifying essential element deficiencies. Creighton (1938) remarked on the fact that populations of some insects show a temporary increase following application of Bordeaux mixture, a copper-containing pesticide that was used extensively before the development of organic insecticides. However, the greatest impact of metals on insect populations has undoubtedly been in response to excess from pollution rather than deficiency. The effects of metal pollution can be categorized into four main areas, community effects (Section IIIB), bioaccumulation (Section IIIC), effects on individuals and species (Section IIID), and genetics (Section IIIE).

B. Community Effects

When studying effects of metal pollution on communities, polluted sites are usually compared with an uncontaminated reference site which has similar characteristics (soil type, climate, vegetation etc.). However, unless the polluted site is heavily contaminated, it is difficult to attribute differences in insect communities to metal pollution rather than natural fluctuations.

One approach that has been developed extensively for freshwaters (Wright et al., 1989), but has received relatively little attention in insects (Stork and Eggleton, 1992), is to predict the probabilities of species occurring in a site and then to compare the predictions with the actual species present. In contaminated sites, it may be possible to attribute absences to metal pollution. However, such an approach relies on a very detailed knowledge of the fauna of many habitat types which is not presently available for most groups.

Despite these limitations, clear evidence of effects of metal pollution on insect communities has been obtained. Collembola are particularly susceptible and there are numerous examples of substantial changes in relative abundances of species near to sources of metal pollution (Fig. 6; Bengtsson et al., 1985; Bengtsson and Rundgren, 1988; Tranvik and Eijsackers, 1989; Tranvik et al., 1993).

C. Bioaccumulation

Insects which have evolved to deal with a wide range of metal concentrations in their diet (e.g. non-specialized carnivores with a large detoxification safety margin) are more likely to survive metal contamination of their food than those in which the levels are highly predictable and within narrow limits (e.g. sap-suckers with a narrow detoxification safety margin). For example, the pine bark bug *Aradus cinnamomeus* Panzer has a discontinuous gut. Under normal circumstances, levels of metals in the sap on which it feeds are so low that there is sufficient detoxification capacity for it to store all the metals ingested. However, this renders the species vulnerable to poisoning if its diet is contaminated. In Finland, *A. cinnamomeus* is absent from the close vicinity of a factory that emits substantial amounts of metal pollution (Heliövaara et al., 1987). Specimens collected from trees further away contain elevated concentrations of a range of metals including copper (Fig. 7).

Fig. 6. Abundance of (A) total Collembola, (B) *Onychiurus armatus* (Tullberg) (Collembola) and (C) *Isotoma olivacea* Tullberg (Collembola) in the 0–3 cm layer in lead-contaminated soils in the vicinity of a natural metalliferous outcrop in a Norwegian spruce forest. The concentrations of lead represent metal extracted from soil over 18 h in 0.1 M buffered acetic acid. Note that *O. armatus* is sensitive to lead pollution whereas *I. olivacea* reaches higher population densities in contaminated soils. Redrawn from Hagvar and Abrahamsen (1990) by permission of the Entomological Society of America.

11. Deficiency and Excess of Metals in Terrestrial Insects

(A)
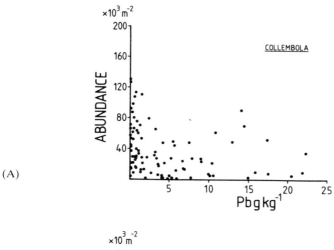

(B)
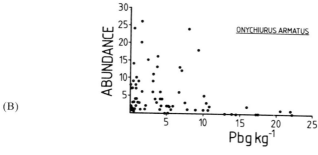

(C)
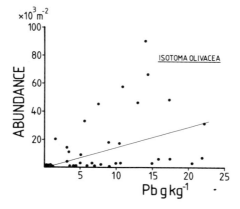

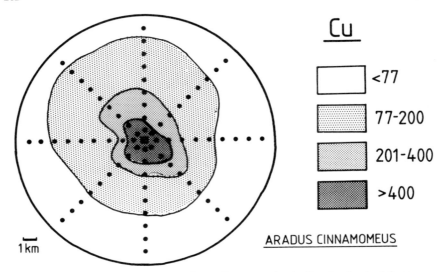

Fig. 7. Mean concentrations of copper ($\mu g\ g^{-1}$) expressed as isolines in pine bark bugs *Aradus cinnamomeus* around a factory (■) in Finland. Lines of dots indicate eight 9-km radial transects. Redrawn from Heliövaara and Väisänen (1987) by permission of Elsevier Applied Science.

Any metals that are in the body of an insect may be transferred to the next level in the food chain if that insect is eaten by a predator. However, it is important to recognize that there are large differences in the concentrations of metals between species in the same habitat (Heliövaara and Väisänen, 1989, 1990, 1991; Lindqvist, 1992; Janssen and Hogervorst, 1993). Indeed, it is impossible to generalize about the role of insects in food chain transfer of metals. A more useful concept is that of "critical pathways" where knowledge of the feeding biology and detoxification mechanisms enables one to predict which species will be most at risk from metal pollution (Beyer, 1986; Fagerström, 1991; Laskowski, 1991; Van Straalen and Ernst, 1991; Streit, 1992). Thus around the factory site in Finland described above, the concentrations of cadmium, copper, nickel and lead in larvae of the pine resin gall moth *Petrova resinella* (L.) were only about one tenth of those in *A. cinnamomeus*. Predators of *P. resinella* are clearly less at risk from metal poisoning than those that eat *A. cinnamomeus*.

D. Effects on Individuals and Species

Assimilation of non-essential metals above a critical concentration, and essential metals above the window of essentiality, may lead to decreased

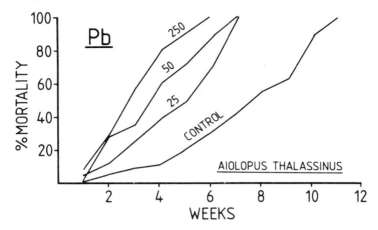

Fig. 8. Post-eclosion mortality of F1 adults of the grasshopper *Aiolopus thalassinus* (Fabricius) reared from eggs laid in soil treated with lead. The figures by each line represent the concentration of lead (μg g^{-1} dry weight) in the experimental soils. Redrawn from Schmidt *et al.* (1991) by permission of Elsevier Science Publishers.

growth and reproduction, and increased mortality (Fig. 8). If a sufficient proportion of a population is affected in the field, then the species will become extinct unless it can evolve resistance (see Section IIIE).

Some effects may be more subtle than simple poisoning. For example, detoxification of metals by insects in polluted sites involves expenditure of energy that could otherwise be used for purposes such as growth and reproduction. This may lead to longer development times (Creighton, 1938; Cohn *et al.*, 1992). A general weakening of individual insects within a species may make them more susceptible to parasites (Ortel *et al.*, 1993).

Effects may be more subtle than simple presence or absence. For example, Read *et al.* (1987) showed that the carabid beetle *Nebria brevicollis* (Fabricius) did not exhibit a summer diapause in a metal-contaminated woodland near to a primary cadmium, lead and zinc smelting works. This may have been due to the scarcity of prey that had been killed by metal pollution. A species of beetle discovered recently in Finland prospers in pollution-damaged conifers (Heliövaara *et al.*, 1990).

E. Genetic Effects

Some species of insects are able to *tolerate* higher environmental concentrations of metal pollutants than others because they are pre-adapted to cope with a wide range of metal concentrations in their food

(Section IIIC). However, for a population to be considered to be *resistant*, it must be genetically different from a non-resistant population.

At present, there is clear experimental evidence for the evolution of metal-resistant races of *Drosophila* (Maroni et al., 1987; Theodore et al., 1991) and Collembola (Posthuma, 1990; Frati et al., 1992; Posthuma et al., 1992, 1993a,b), although the phenomenon is certain to be detected in other groups of insects in the future. The topic has recently been reviewed comprehensively by Posthuma and Van Straalen (1993).

In Collembola, the basis of resistance to metal pollution is an increased ability to excrete assimilated metal (Posthuma et al., 1992). Metals are detoxified in insoluble intracellular granules in the midgut (Van Straalen et al., 1987). During moulting in Collembola, the epithelium of the midgut is shed along with the exoskeleton. Hence, an increased capacity for metal storage in granules leads to a greater excretion efficiency. In resistant populations, this adaptation is passed on to the next generation (Posthuma, 1990).

One way of increasing tolerance to pollutants is to increase the copy number or the transcription rate of the gene coding for detoxifying proteins. In the case of organophosphate insecticides, up to a 256-fold amplification of the genes coding for non-specific esterases that break down the insecticide have been found in mosquitoes (Devonshire and Field, 1991). For metals such as copper and cadmium in *Drosophila*, the genes that code for the metal-binding protein metallothionein may be duplicated up to four times. The metals are bound more rapidly in resistant animals following ingestion (Maroni et al., 1987; Theodore et al., 1991). Such amplification has been found in wild *Drosophila* populations and has probably evolved in response to the spraying of fruit trees with copper-containing fungicides. This provides a good example of how one species of insect has responded at the biochemical level to a changed environment.

IV. CONCLUSIONS

Metals have been present throughout the evolution of terrestrial insects. Different species have evolved a range of strategies to regulate internal concentrations of essential and non-essential metals that they ingest with their food. Since the industrial revolution, the availability of many of these elements has increased owing to pollution. Some species are able to survive because they are pre-adapted to cope with a wide range of metal concentrations in their diet (e.g. polyphages). Others are vulnerable to poisoning since the levels of metals in their food are normally within narrow limits (e.g. sap suckers). Some (e.g. *Drosophila*) have been able to evolve resistance to their changed environment by enhancing existing detoxification mechanisms.

Insects provide a route for transport of metals to higher levels in food chains. Indeed, some species may be suitable as biological monitors of pollution. (Samiullah, 1990; Hopkin, 1993c). However, it is important to recognize that the extent of metal accumulation is species-specific. Critical pathways of metal transfer should be identified based on basic biological knowledge of feeding behaviour, life histories and other ecological factors.

One largely unexplored area is the possible co-evolution of phytophagous insects and their food plants in terms of metal availability. One response of a plant to insect attack might be to restrict the availability of an essential metal such as copper, thus reducing the insects' performance. This possibility has been discussed in the literature with regard to major nutrients (Haukioja *et al.*, 1991; Lindroth *et al.*, 1991). However, I am not aware of essential metals being considered in this context. As Lawton (1986) remarked, "All three groups working on birch are agreed that other, as yet unknown, chemical changes probably take place in damaged leaves, and could be much more important for herbivores than crude observations on increases in total phenolics". Restriction of essential trace element availability by plants could be one such chemical change.

The opposite response of plants to insect attack would be to hyperaccumulate metals until their concentrations in the leaves reached toxic levels. Hyperaccumulation is well known in metallophytes (Baker and Proctor, 1990; Ernst *et al.*, 1992). In Central Queensland for example, leaves of *Stackhousia tryonii* Bailey may contain up to 2% nickel on a dry weight basis (Batianoff *et al.*, 1990). Do these plants accumulate such high levels of metals as a feeding deterrent?

The whole question of whether insects ever suffer from essential metal deficiency in the wild is almost completely unexplored. Many species show a slight increase in performance in response to low levels of contaminants in their diets. This is usually attributed to hormesis, the stimulation of growth by low levels of inhibitors (Stebbing, 1982; Van Ewick and Hoekstra, 1993). However, it is feasible that a slight supplementation of the diet of an insect, with say copper, could rectify a deficiency and improve performance.

If this conjecture is true, then new approaches to insect control may be possible. For example, varieties of crops could be selected to have a concentration of an essential metal that was below the window of essentiality for an insect pest, thus restricting attack. Pesticides could be developed that interfered with the biochemical regulation of essential metals in insects. In theory, the insect could not evolve resistance to a deficiency as there are no substitutes for the biochemical roles of essential elements.

It would seem reasonable to press ahead with attempts to reduce the emissions of metals to the environment. Metals are persistent pollutants and have extremely long residence times in soils. The continuous build-up of concentrations may have long-term consequences for insect populations which we are only just beginning to understand.

REFERENCES

Akey, D. H. and Beck, S. D. (1972). Nutrition of the pea aphid, *Acrythosiphon pisum*: requirements for trace metals, sulphur and cholesterol. *J. Insect Physiol.* **18**, 1901–1914.

Aoki, Y., Suzuki, K. T. and Kubota, K. (1984). Accumulation of cadmium and induction of its binding protein in the digestive tract of flesh fly (*Sarcophaga peregrina*) larvae. *Comp. Biochem. Physiol.* **77C**, 279–282.

Baker, A. J. M. and Proctor, J. (1990). The influence of cadmium, copper, lead and zinc on the distribution and evolution of metallophytes in the British Isles. *Pl. Syst. Evol.* **173**, 91–108.

Ballan-Dufrançais, C., Ruste, J. and Jeantet, A. Y. (1980). Quantitative electron probe microanalysis on insect exposed to mercury. I. Methods. An approach on the molecular form of the stored mercury. *Biol. Cell.* **39**, 317–324.

Batianoff, G. N., Reeves, R. D. and Specht, R. L. (1990). *Stackhousia tryonii* Bailey: a nickel-accumulating serpentinite-endemic species of Central Queensland. *Aust. J. Bot.* **38**, 121–130.

Beeby, A. (1991). Toxic metal uptake and essential metal regulation in terrestrial invertebrates: a review. *In* "Metal Ecotoxicology: Concepts and Applications" (M. C. Newman and A. W. McIntosh, eds), pp. 65–89. Lewis, Chelsea, MI.

Bengtsson, G. and Rundgren, S. (1988). The Gusum case: a brass mill and the distribution of soil Collembola. *Can. J. Zool.* **66**, 1518–1526.

Bengtsson, G. and Tranvik, L. (1989). Critical metal concentrations for forest soil invertebrates. *Water Air Soil Poll.* **47**, 381–417.

Bengtsson, G., Ohlsson, L. and Rundgren, S. (1985). Influence of fungi on growth and survival of *Onychiurus armatus* (Collembola) in a metal polluted soil. *Oecologia* **68**, 63–68.

Bernays, E. A. and Bright, K. L. (1991). Dietary mixing in grasshoppers: switching induced by nutritional imbalances in foods. *Entomol. Exp. Appl.* **61**, 247–253.

Beyer, W. N. (1986). A re-examination of biomagnification of metals in terrestrial food chains. *Environ. Toxicol. Chem.* **5**, 863–864.

Boardman, R. and McGuire, D. O. (1990a). The role of zinc in forestry. I. Zinc in forest environments, ecosystems and tree nutrition. *Forest. Ecol. Manag.* **37**, 167–205.

Boardman, R. and McGuire, D. O. (1990b). The role of zinc in forestry. II. Zinc deficiency and forest management: effect on yield and silviculture of *Pinus radiata* plantations in South Australia. *Forest. Ecol. Manag.* **37**, 207–218.

Cohn, J., Widzowski, D. V. and Cory-Slechta, D. A. (1992). Lead retards development of *Drosophila melanogaster*. *Comp. Biochem. Physiol.* **102C**, 45–49.

Cosson, R. P., Amiard-Triquet, C. and Amiard, J. C. (1991). Metallothioneins and detoxification. Is the use of detoxification protein for MTs a language abuse? *Water Air Soil Poll.* **57–58**, 555–567.

Creighton, J. T. (1938). Factors influencing insect abundance. *J. Econ. Entomol.* **31**, 735–739.

Dadd, R. H. (1967). Improvement of synthetic diet for the aphid *Myzus persicae* using plant juices, nucleic acid or trace metals. *J. Insect. Physiol.* **13**, 763–778.

Dallinger, R. (1993). Strategies of metal detoxification in terrestrial invertebrates. In "Ecotoxicology of Metals in Invertebrates" (R. Dallinger and P. S. Rainbow, eds), pp. 245–289. Lewis, Chelsea, MI.

Devonshire, A. L. and Field, L. M. (1991). Gene amplification and insecticide resistance. *Ann. Rev. Entomol.* **36**, 1–23.

Edwards, D. and Selden, P. A. (1992). The development of early terrestrial ecosystems. *Bot. J. Scot.* **46**, 337–366.

Engebretson, J. A. and Mason, W. H. (1980). Transfer of ^{65}Zn at mating in *Heliothis virescens*. *Environ. Entomol.* **9**, 119–121.

Ernst, W. H. O. (1992). Nutritional aspects in the development of *Bruchidius sahlbergi* (Coleoptera: Bruchidae) in seeds of *Acacia erioloba*. *J. Insect Physiol.* **38**, 831–838.

Ernst, W. H. O. (1993). Food consumption, life history and determinants of host range in the bruchid beetle *Specularius impressithorax* (Coleoptera: Bruchidae). *J. Stored Products Res.* **29**, 53–62.

Ernst, W. H. O., Verkleij, J. A. C. and Schat, H. (1992). Metal tolerance in plants. *Acta Botanica Neerlandica* **41**, 229–248.

Fagerström, T. (1991). Biomagnification in food chains and related concepts. *Oikos* **62**, 257–260.

Flemming, C. A. and Trevors, J. T. (1989). Copper toxicity and chemistry in the environment: a review. *Water Air Soil Poll.* **44**, 143–158.

Fraenkel, G. (1959). A historical and comparative survey of the dietary requirements of insects. *Ann. N. Y. Acad. Sci.* **77**, 267–274.

Frati, F., Fanciulli, P. P. and Posthuma, L. (1992). Allozyme variation in reference and metal-exposed natural populations of *Orchesella cincta* (Insecta: Collembola). *Biochem. Syst. Ecol.* **20**, 297–310.

Greenslade, P. and Whalley, P. E. S. (1986). The systematic position of *Rhyniella praecursor* Hirst and Maulik (Collembola), the earliest known hexapod. In "2nd International Symposium on Apterygota" (R. Dallai, ed.), pp. 319–323. University of Siena, Siena, Italy.

Guha, M. M. and Mitchell, R. L. (1966). Trace and major element composition of the leaves of some deciduous trees. *Pl. Soil* **24**, 90–112.

Hagvar, S. and Abrahamsen, G. (1990). Microarthropoda and Enchytraeidae (Oligochaeta) in naturally lead-contaminated soil: a gradient study. *Environ. Entomol.* **19**, 1263–1277.

Hare, L. (1992). Aquatic insects and trace metals: bioavailability, bioaccumulation, and toxicity. *Crit. Rev. Toxicol.* **22**, 327–369.

Haukioja, E., Ruohomaki, K., Suomela, J. and Vuorisalo, T. (1991). Nutritional quality as a defence against insect herbivores. *Forest. Ecol. Manag.* **39**, 237–245.

Heliövaara, K. and Väisänen, R. (1989). Between-species differences in heavy metal levels in four pine diprionids (Hymenoptera) along an air pollution gradient. *Environ. Poll.* **62**, 253–261.

Heliövaara, K. and Väisänen, R. (1990). Heavy metal contents in pupae of *Bupalus piniarius* (Lepidoptera: Geometridae) and *Panolis flammea* (Lepidoptera: Noctuidae) near an industrial source. *Environ. Entomol.* **19**, 481–485.

Heliövaara, K. and Väisänen, R. (1991). Bark beetles and associated species with high heavy metal tolerance. *J. Appl. Entomol.* **111**, 397–405.

Heliövaara, K. and Väisänen, R. (1993). "Insects and Pollution". CRC Press, Boca Raton, FL.
Heliövaara, K., Väisänen, R., Braunschweiler, H. and Lodenius, M. (1987). Heavy metal levels in two biennial pine insects with sap-sucking and gall-forming life styles. *Environ. Poll.* **48**, 13–23.
Heliövaara, K., Väisänen, R. and Mannerkoski, I. (1990). *Melanophila formaneki* (Jakobson) (Coleoptera, Buprestidae) new to Finland. *Entomologica Fenn.* **1**, 221–225.
Hopkin, S. P. (1989). "Ecophysiology of Metals in Terrestrial Invertebrates". Elsevier Applied Science, Barking, Essex.
Hopkin, S. P. (1990). Critical concentrations, pathways of detoxification and cellular ecotoxicology of metals in terrestrial arthropods. *Funct. Ecol.* **4**, 321–327.
Hopkin, S. P. (1993a). Deficiency and excess of copper in terrestrial isopods. In "Ecotoxicology of Metals in Invertebrates" (R. Dallinger and P. S. Rainbow, eds), pp. 359–382. Lewis, Chelsea, MI.
Hopkin, S. P. (1993b). Ecological implications of '95% protection levels' for metals in soils. *Oikos* **66**, 137–141.
Hopkin, S. P. (1993c). In situ biological monitoring of pollution in terrestrial and aquatic ecosystems. In "Handbook of Ecotoxicology, Vol. 1" (P. Calow, ed.), pp. 397–427. Blackwell, Oxford.
Hopkin, S. P., Hames, C. A. C. and Dray, A. (1989). X-ray microanalytical mapping of the intracellular distribution of pollutant metals. *Microsc. Anal.* **14**, 23–27.
Hunter, I. R., Hunter, J. A. C. and Nicholson, G. (1990). Current problems in the copper nutrition of radiata pine in New Zealand: a review. *Forest. Ecol. Manag.* **37**, 143–149.
Janssen, M. P. M. and Hogervorst, R. F. (1993). Metal accumulation in soil arthropods in relation to micro-nutrients. *Environ. Poll.* **79**, 181–189.
Jeantet, A. Y., Ballan-Dufrançais, C. and Ruste, J. (1980). Quantitative electron probe analysis on insect exposed to mercury. II. Involvement of the lysosomal system in detoxification process. *Biol. Cell.* **39**, 325–334.
Kukalová-Peck, J. (1991). Fossil history and the evolution of hexapod structures. In "The Insects of Australia" 2nd edn (I. D. Naumann and CSIRO, eds), pp. 141–179. Melbourne University Press, Melbourne.
Kukalová-Peck, J. (1992). The "Uniramia" do not exist: The ground plan of the Pterygota as revealed by Permian Diaphanopterodea from Russia (Insecta: Paleodictyopteroidea). *Can. J. Zool.* **70**, 236–255.
Laskowski, R. (1991). Are the top carnivores endangered by heavy metal biomagnification? *Oikos* **60**, 387–390.
Lauverjat, S., Ballan-Dufrançais, C. and Wegnez, M. (1989). Detoxification of cadmium. Ultrastructural study and electron-probe microanalysis of the midgut in a cadmium-resistant strain of *Drosophila melanogaster*. *Biol. Metals* **2**, 97–107.
Lawton, J. H. (1986). Food shortage in the midst of apparent plenty?: the case for birch-feeding insects. In "Proceedings of the 3rd European Congress of Entomology, Amsterdam, 24–29 August 1986" (H. H. W. Velthuis, ed.), pp. 219–228. Nederlandse Entomologische Vereniging.
Lindqvist, L. (1992). Accumulation of cadmium, copper and zinc in five species of phytophagous insects. *Environ. Entomol.* **21**, 160–163.
Lindroth, R. L., Barman, M. A. and Weisbrod, A. V. (1991). Nutrient deficiencies and the gypsy moth *Lymantria dispar*: effects on larval performance and detoxification enzyme activities. *J. Insect Physiol.* **37**, 45–52.
Little, C. (1990). "The Terrestrial Invasion: An Ecophysiological Approach to the Origins of Land Animals". Cambridge University Press, Cambridge.

Locke, M. and Leung, H. (1984). The induction and distribution of an insect ferritin – a new function for the endoplasmic reticulum. *Tiss. Cell* **16**, 739–766.
Locke, M. and Nichol, H. (1992). Iron economy in insects: transport, metabolism and storage. *Ann. Rev. Entomol.* **37**, 195–215.
Markert, B. (1992). Multi-element analysis in plant materials – analytical tools and biological questions. *In* "Biogeochemistry of Trace Metals" (D. C. Adriano, ed.), pp. 402–428. Lewis, Boca Raton, FL.
Maroni, G., Wise, J., Young, J. E. and Otto, E. (1987). Metallothionein gene duplications and metal tolerance in natural populations of *Drosophila melanogaster*. *Genetics* **117**, 739–744.
McFarlane, J. E. (1976). Influence of dietary copper and zinc on growth and reproduction of the house cricket (Orthoptera: Gryllidae). *Can. Entomol.* **108**, 387–390.
Nriagu, J. O. (1988). A silent epidemic of environmental metal poisoning? *Environ. Poll.* **50**, 139–161.
Nriagu, J. O. (1990). The rise and fall of leaded gasoline. *Sci. Tot. Environ.* **92**, 13–28.
Nriago, J. O. and Pacyna, J. M. (1988). Quantitative assessment of worldwide contamination of air, water and soils by trace metals. *Nature* **333**, 134–139.
Ortel, J., Gintenreiter, S. and Nopp, H. (1993). Metal bioaccumulation in a host insect (*Lymantria dispar* L., Lepidoptera) during development – ecotoxicological implications. *In* "Ecotoxicology of Metals in Invertebrates" (R. Dallinger and P. S. Rainbow, eds), pp. 401–425. Lewis, Chelsea, MI.
Posthuma, L. (1990). Genetic differentiation between populations of *Orchesella cincta* (Collembola) from heavy metal contaminated sites. *J. Appl. Ecol.* **27**, 609–622.
Posthuma, L. and Van Straalen, N. M. (1993). Heavy-metal adaptation in terrestrial invertebrates: a review of occurrence, genetics, physiology and ecological consequences. *Biochem. Physiol.* **106C**, 11–38.
Posthuma, L., Hogervorst, R. F. and Van Straalen, N. M. (1992). Adaptation to soil pollution by cadmium excretion in natural populations of *Orchesella cincta* (L.) (Collembola). *Arch. Environ. Contam. Toxicol.* **22**, 146–156.
Posthuma, L., Hogervorst, R. F., Joosse, E. N. G. and Van Straalen, N. M. (1993a). Genetic variation and covariation for characteristics associated with cadmium tolerance in natural populations of the springtail *Orchesella cincta* (L.). *Evolution* **47**, 619–631.
Posthuma, L., Verweij, R. A., Widianarko, B. and Zonneveld, C. (1993b). Life-history patterns in metal-adapted Collembola. *Oikos* **67**, 235–249.
Price, P. W. (1988). An overview of organismal interactions in ecosystems in evolutionary and ecological time. *Agric. Ecosyst. Environ.* **24**, 369–377.
Raes, H., Cornelis, R. and Rzeznik, U. (1992). Distribution, accumulation and depuration of administered lead in adult honeybees. *Sci. Tot. Environ.* **113**, 269–279.
Read, H. J., Wheater, C. P. and Martin, M. H. (1987). Aspects of the ecology of Carabidae (Coleoptera) from woodlands polluted by heavy metals. *Environ. Poll.* **48**, 61–76.
Roth, M. (1992). Metals in invertebrate animals of a forest ecosystem. *In* "Biogeochemistry of Trace Metals" (D. C. Adriano, ed.), pp. 299–328. Lewis, Boca Raton, FL.
Samiullah, Y. (1990). "Biological Monitoring of Environmental Contaminants: Animals". Monitoring and Assessment Research Centre, Kings College, University of London.
Schmidt, G. H., Ibrahim, N. M. H. and Abdallah, M. D. (1991). Toxicological studies on the long-term effects of heavy metals (Hg, Cd, Pb) in soil on the development of *Aiolopus thalassinus* (Fabr.) (Saltatoria, Acrididae). *Sci. Tot. Environ.* **107**, 109–133.
Shear, W. A. and Kukalová-Peck, J. (1990). The ecology of Paleozoic terrestrial arthropods: the fossil evidence. *Can. J. Zool.* **68**, 1807–1834.

Starling, A. P. and Ross, I. S. (1991). Uptake of zinc by *Penicillium notatum*. *Mycol. Res.* **95**, 712–714.

Stebbing, A. R. D. (1982). Hormesis – the stimulation of growth by low levels of inhibitors. *Sci. Tot. Environ.* **22**, 213–234.

Stork, N. E. and Eggleton, P. (1992). Invertebrates as determinants and indicators of soil quality. *Am. J. Alt. Agric.* **7**, 23–32.

Streit, B. (1992). Bioaccumulation processes in ecosystems. *Experentia* **48**, 955–970.

Studier, E. H. and Sevick, S. H. (1992). Live mass, water content, nitrogen and mineral levels in some insects from south-central lower Michigan. *Comp. Biochem. Physiol.* **103A**, 579–595.

Suzuki, K. T., Sunaga, H., Hatakeyama, S., Sumi, Y. and Suzuki, T. (1989). Differential binding of cadmium and copper to the same protein in a heavy metal tolerant species of mayfly (*Baetis thermicus*) larvae. *Comp. Biochem. Physiol.* **94C**, 99–103.

Theodore, L., Ho, A. S. and Maroni, G. (1991). Recent evolutionary history of the metallothionein gene *Mtn* in *Drosophila*. *Genet. Res.* **58**, 203–210.

Tranvik, L. and Eijsackers, H. (1989). On the advantage of *Folsomia fimetarioides* over *Isotomiella minor* (Collembola) in a metal polluted soil. *Oecologia* **80**, 195–200.

Tranvik, L., Bengtsson, G. and Rundgren, S. (1993). Relative abundance and resistance traits of two Collembola species under metal stress. *J. Appl. Ecol.* **30**, 43–52.

Van Ewijk, P. H. and Hoekstra, J. A. (1993). Calculation of the EC_{50} and its confidence interval when subtoxic stimulus is present. *Ecotox. Environ. Saf.* **25**, 25–32.

Van Straalen, N. M. (1993). Soil and sediment quality criteria derived from invertebrate toxicity data. *In* "Ecotoxicology of Metals in Invertebrates" (R. Dallinger and P. S. Rainbow, eds), pp. 427–441. Lewis, Chelsea, MI.

Van Straalen, N. M. and Ernst, W. H. O. (1991). Metal biomagnification may endanger species in critical pathways. *Oikos* **62**, 255–256.

Van Straalen, N. M., Burghouts, T. B. A., Doornhof, M. J., Groot, G. M., Janssen, M. P. M., Joosse, E. N. G., Van Meerendonk, J. H., Theeuwen, J. P. J. J., Verhoef, H. A. and Zoomer, H. R. (1987). Efficiency of lead and cadmium excretion in populations of *Orchesella cincta* (Collembola) from various contaminated forest soils. *J. Appl. Ecol.* **24**, 953–968.

Waldbauer, G. P. and Friedman, S. (1991). Self-selection of optimal diets by insects. *Ann. Rev. Entomol.* **36**, 43–63.

Waterhouse, D. F. (1945). Studies of the physiology and toxicology of blowflies. II. A quantitative investigation of the copper content of *Lucilia cuprina*. *Bull. Counc. Sci. Ind. Res., Melb.* **191**, 21–39.

Williams, R. J. P. (1981). Natural selection of the chemical elements. *Proc. R. Soc. Lond.* **213B**, 361–397.

Wright, J. F., Armitage, P. D., Furse, M. T. and Moss, D. (1989). Prediction of invertebrate communities using stream measurements. *Regul. Riv.: Res. Manag.* **4**, 147–155.

(a) Equilibrium response
10-year annual mean
Surface temperature change

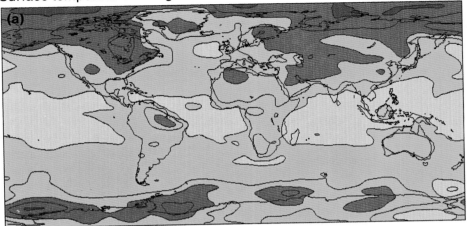

Plate 1. (a) The "equilibrium" response of surface temperature to a doubling of CO_2 predicted by an earlier version of the Hadley Centre GCM in which the ocean was represented by a 50-m thick layer of water. (See Chapter 3, Bennetts.)

(b) Coupled model
10 year annual mean (years 66 to 75)
Surface temperature change

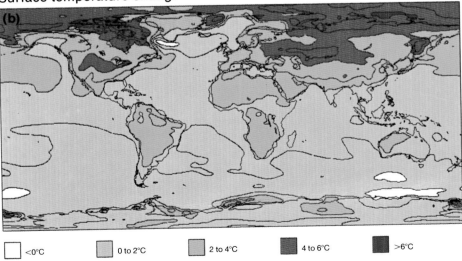

<0°C   0 to 2°C   2 to 4°C   4 to 6°C   >6°C

Plate 1. (b) The "transient" response from the Hadley Centre AOGCM in which the full depth of the ocean is included. Shown is the 10-year mean for years 66–75, i.e. centred on year 70 when the CO_2 atmospheric concentration is (in the experiment) double the present-day value. (See Chapter 3, Bennetts.)

(a) Coupled model
10-year DJF mean (years 66 to 75)
Surface temperature change

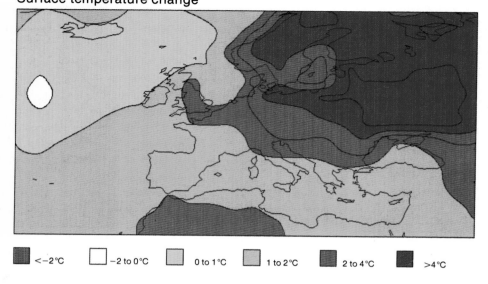

Coupled model
10-year JJA mean (years 66 to 75)
Surface temperature change

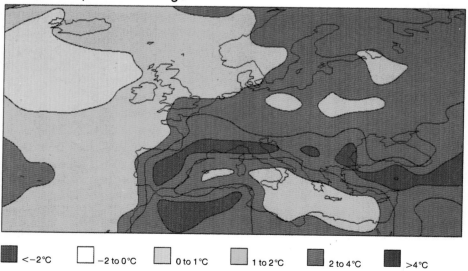

Plate 2. (a) Surface air temperature change in winter (DJF) and summer (JJA) from the Hadley Centre GCM, averaged over years 66–75; the time at which atmospheric concentrations of CO_2 are double present-day values. The data for western Europe have been taken from the global predictions and enlarged. (See Chapter 3, Bennetts.)

(b) Coupled model
10-year DJF mean (years 66 to 75)
Total precipitation change

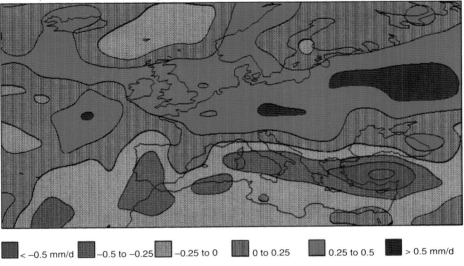

■ < –0.5 mm/d ■ –0.5 to –0.25 ▫ –0.25 to 0 ▫ 0 to 0.25 ▫ 0.25 to 0.5 ■ > 0.5 mm/d

Coupled model
10-year JJA mean (years 66 to 75)
Total precipitation change

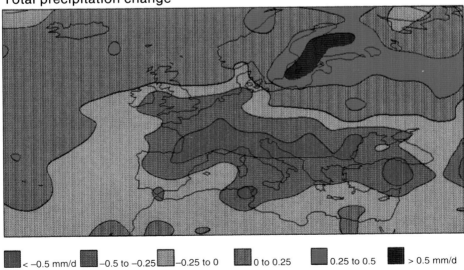

■ < –0.5 mm/d ■ –0.5 to –0.25 ▫ –0.25 to 0 ▫ 0 to 0.25 ▫ 0.25 to 0.5 ■ > 0.5 mm/d

Plate 2. (b) Precipitation change in winter (DJF) and summer (JJA) from the Hadley Centre GCM, averaged over years 66–75; the time at which atmospheric concentrations of CO_2 are double present-day values. The data for western Europe have been taken from the global predictions and enlarged. (See Chapter 3, Bennetts.)

(c) Coupled model
10-year DJF mean (years 66 to 75)
Soil moisture change

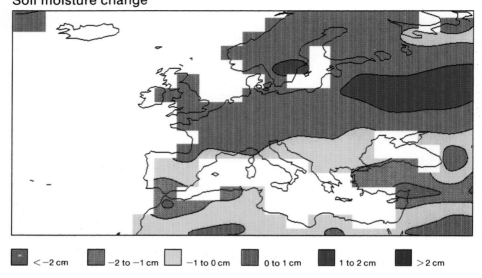

Coupled model
10-year JJA mean (years 66 to 75)
Soil moisture change

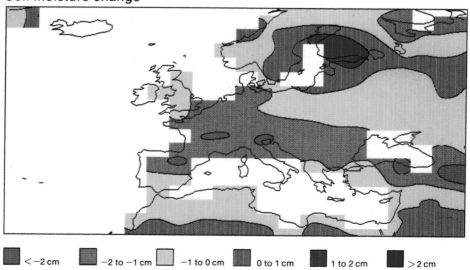

Plate 2. (c) Soil moisture changes in winter (DJF) and summer (JJA) from the Hadley Centre GCM, averaged over years 66–75; the time at which atmospheric concentrations of CO_2 are double present-day values. The data for western Europe have been taken from the global predictions and enlarged. (See Chapter 3, Bennetts.)

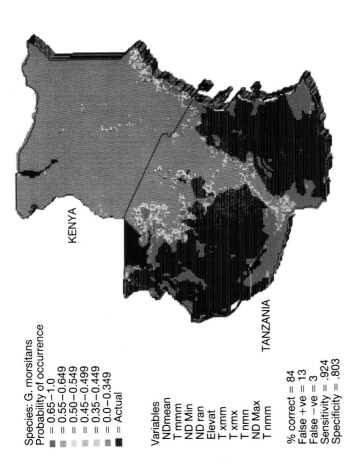

Plate 3. Application of linear discriminant analysis to describe the distribution of *Glossina morsitans* in Kenya and Tanzania. Training samples of >140 data points each for areas of fly presence and fly absence were selected from the total sample of >25000 data points and the relationships between the variables within the training set were used to assign to each point the probability of occurrence of this species. These probabilities are colour coded from red (low probability) to green (high probability) according to the scale on the plate. The variables used in making this prediction are listed in order of their importance (determined within the analysis by the size of the Mahalanobis distance separating areas of presence from areas of absence (Green, 1978): for full details see Table 2. The results in terms of percentage of correct predictions, percentage of false positives (false predictions of presence), percentage of false negatives (false predictions of absence), sensitivity (ability correctly to predict presence) and specificity (ability correctly to predict absence) are also given. (See Chapter 8, Rogers.)

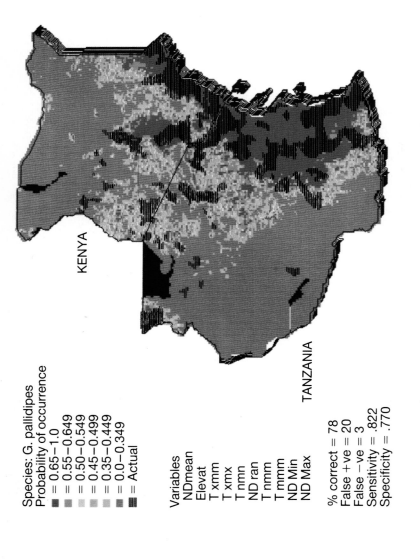

Plate 4. Application of linear discriminant analysis to describe the distribution of *Glossina pallidipes* in Kenya and Tanzania. This species has a more patchy distribution than does *G. morsitans* (Plate 3), possibly an artefact of the poorer sampling methods available for it. Other details as for Plate 3, and in Table 2. (See Chapter 8, Rogers.)

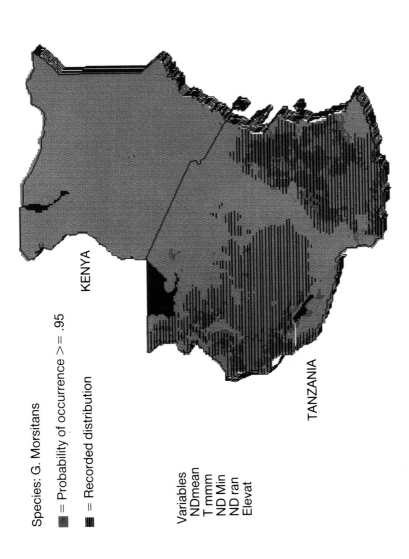

Plate 5. When the threshold probability for occurrence of *G. morsitans* in Kenya and Tanzania (from Plate 3) is raised from 0.5 to 0.95 the analysis picks out the areas shown here in green. These areas are closest to the centroid of the distribution of this tsetse's sites within the training set, and are therefore presumably ideal from the flies' point of view. Notice the predicted unsuitability of most areas west of Lake Victoria, and the similarity in shape of the eastern predictions to the area occupied by this species in 1912 (in Fig. 1(a)), after the rinderpest panzootic at the end of the last century. (See Chapter 8, Rogers.)

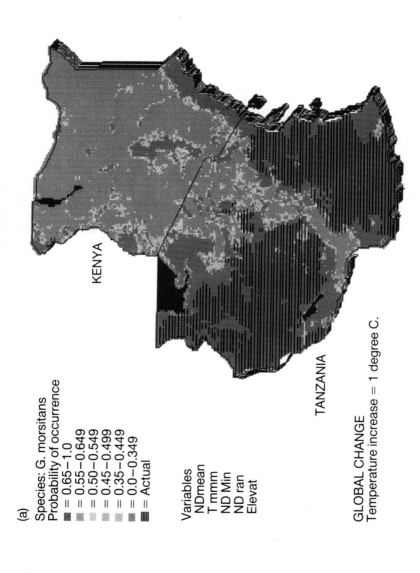

Plate 6. (a) The predicted impact of increasing average temperatures by 1°C on the areas of suitability for the tsetse *G. morsitans* in Kenya and Tanzania. The predictor variable set is listed on the left of the Plate. (See Chapter 8, Rogers.)

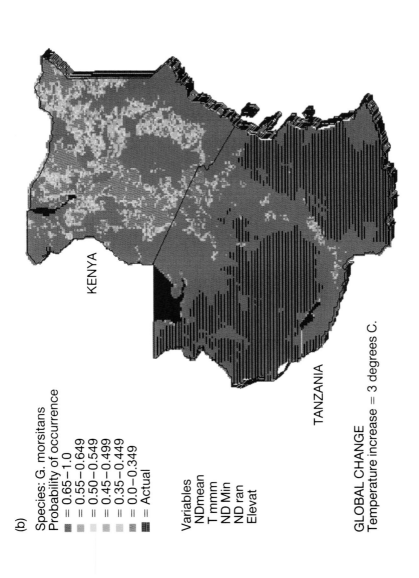

Plate 6. (b) The predicted impact of increasing average temperatures by 3°C on the areas of suitability for the tsetse *G. morsitans* in Kenya and Tanzania. The predictor variable set is listed on the left of the Plate. (See Chapter 8, Rogers.)

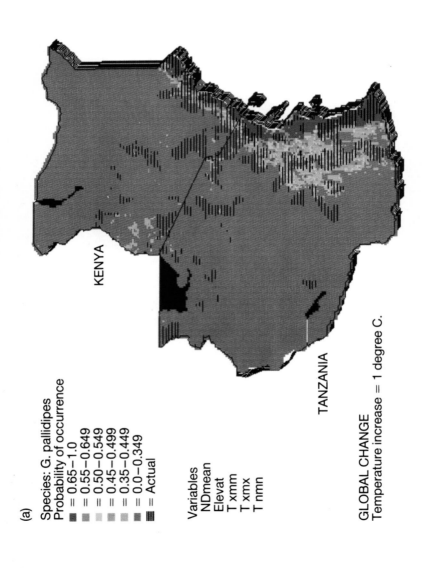

Plate 7. (a) The predicted impact of increasing average temperatures by 1°C on the areas of suitability for the tsetse *G. pallidipes* in Kenya and Tanzania. The predictor variable set is listed on the left of the Plate. (See Chapter 8, Rogers.)

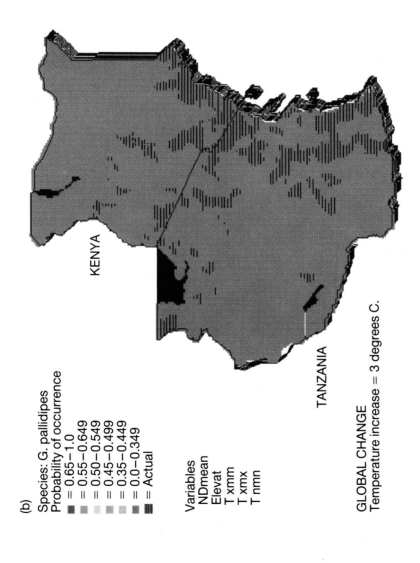

Plate 7. (b) The predicted impact of increasing average temperatures by 3°C on the areas of suitability for the tsetse *G. pallidipes* in Kenya and Tanzania. A 3°C temperature increase apparently causes the elimination of this species from the map. The predictor variable set is listed on the left of the Plate. (See Chapter 8, Rogers.)

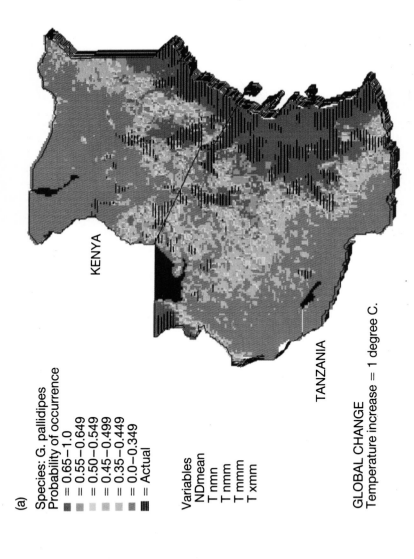

Plate 8. (a) When elevation is removed from the predictor variable set in Plate 7, and the analysis is rerun without it, the prediction of the impact of an increase in temperature of 1°C on *G. pallidipes*' distribution in Kenya and Tanzania is quite different from that shown in Plate 7. This highlights the need for careful biological studies to explore the relationships revealed by statistical analysis of distribution data. (See Chapter 8, Rogers.)

(b)

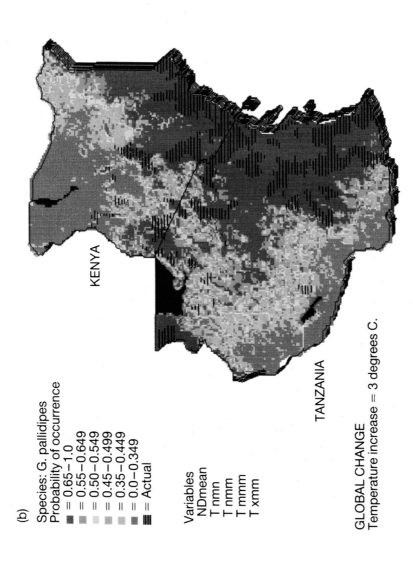

Species: G. pallidipes
Probability of occurrence
■ = 0.65–1.0
■ = 0.55–0.649
■ = 0.50–0.549
■ = 0.45–0.499
■ = 0.35–0.449
■ = 0.0–0.349
▮ = Actual

Variables
NDmean
T nmn
T nmm
T mmm
T xmm

GLOBAL CHANGE
Temperature increase = 3 degrees C.

Plate 8. (b) When elevation is removed from the predictor variable set in Plate 7, and the analysis is rerun without it, the prediction of the impact of an increase in temperature of 3°C on *G. pallidipes*' distribution in Kenya and Tanzania is quite different from that shown in Plate 7. This highlights the need for careful biological studies to explore the relationships revealed by statistical analysis of distribution data. (See Chapter 8, Rogers.)

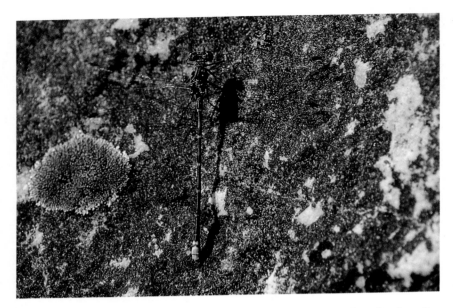

Plate 9. *Chlorolestes apricans* – an extremely rare Gondwana relict from the Cape Province. It was only discovered in the early 1970s. Just a few populations remain, and none of these is in a reserve. Will this species see the twenty-first century after millions of years of survival? (See Chapter 13, Samways.)

Plate 10. *Ecchlorolestes peringueyi* – this Gondwana relict species had not been seen since 1949, and was thought to be extinct, until rediscovered in 1993. It occurs in two small streams in a remote and protected area of the Cape Province. By chance, it has survived above waterfalls where the introduced invasive predator, the rainbow trout, cannot penetrate (see Fig. 4). (See Chapter 13, Samways.)

Plate 11. (a) Xerces blue butterfly; (b) Cinyras butterfly; (c) Pitt Island Longhorn Borer; (d) Banks Peninsula Speargrass weevil and (e) Volutine Stoneyran Tabanid fly. (See Chapter 14, Mawdsley and Stork.)

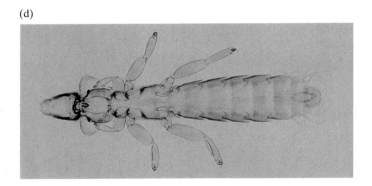

Plate 12. Examples of extinct insects. (a) Lord Howe Island stick insect; (b) St Helena earwig; (c) Passenger pigeon louse (glass slide) and (d) Passenger pigeon louse (photograph). (See Chapter 14, Mawdsley and Stork.)

12

Chironomidae as Indicators of Water Quality – With a Comparison of the Chironomid Faunas of a Series of Contrasting Cumbrian Tarns

L. C. V. PINDER AND D. J. MORLEY

I.	Introduction	272
II.	Biotic Indices and River Quality	273
III.	Chironomidae and Water Quality	274
	A. Chironomids and River Water Quality	275
	B. Chironomids and Lacustrine Environments	276
IV.	Chironomids as Palaeoecological Indicators	277
V.	Deformities Induced by Pollution	279
VI.	The Chironomidae of Lake District Tarns	280
	A. Geology of the Lake District	280
	B. Sampling Sites	281
	C. Methods	282
	D. Results	284
VII.	Discussion	287
	References	290

Abstract

Indices of water quality, derived from communities of aquatic invertebrates, are reviewed briefly and a more detailed appraisal is made of the value of Chironomidae as indicators of water quality in rivers and in systems of lake classification. Remains of larval Chironomidae are useful as indicators of historical change in lake environments and some examples from the literature are provided. Studies of characteristic deformities in chironomid larvae, that result from exposure to pollutants, are reviewed. The chironomid communities of 25 tarns in the English Lake District (Cumbria) are described using collections of pupal exuviae. Tarns were chosen to represent a wide range of pH and alkalinity, as well as covering a range of surface area, depth and altitude. The tarns are classified into five groups on the basis of their pH and alkalinity.

Generally, the number of species recorded was greatest in tarns with pH between 6.3 and 7.5. Somewhat fewer species were found in very alkaline tarns but species number was lowest in the most acid tarns, where there was no measurable alkalinity. The chironomid faunas of acid and very soft-water tarns were characterized by dominance of *Heterotrissocladius grimshawi* and *Macropelopia adaucta*. The remaining tarns, classified as soft-, medium-hard or hard-water, were dominated by *Chironomus* spp. Species that consistently occurred in the most alkaline (hard-water) tarns, but not in the other groups, included *Chironomus anthracinus*, *Endochironomus impar* or *E. dispar* and *Phaenopsectra flavipes*. *Procladius sagittalis* and *Zavrelimyia melanura* also were found only in the hard-water tarns, but were not present in all of them.

I. INTRODUCTION

Many groups of organisms have been used as indicators of water quality or environmental change in freshwater bodies, including algae, macrophytes, protozoa and fish, but aquatic invertebrates have been used most extensively, especially in flowing waters. The value of freshwater insects and other invertebrates as indicators of water quality in rivers and streams was recognized very early in this century and a variety of biotic indices has been devised, based on the different tolerance levels of aquatic invertebrates to pollution, mainly organic pollution. The earliest were developed in Europe, arising from the Saprobien system of Kolkwitz and Marsson (1909). They recognized four stages in the oxidation of organic matter – polysaprobic, α-mesosaprobic, β-mesosaprobic and oligosaprobic – and observed that groups of indicator species, including insects, could be associated with each stage. One of the characteristic species in this scheme is the chironomid *Chironomus riparius* (Meigen) (= *C. thummi* Kieffer). High densities of this species in rivers are generally indicative of a high content of organic matter. For example, Gower and Buckland (1978) showed that eutrophic conditions below a sewage outfall resulted in the replacement of *Polypedilum laetum* (Meigen) by *C. riparius* as the dominant invertebrate and there are many other recorded instances of *C. riparius* occurring in abundance in organically polluted rivers and streams (e.g. Thienemann, 1954; Hynes, 1960; Learner and Edwards, 1966; Hawkes and Davies, 1971; Pinder and Farr, 1987).

Insects, in particular Chironomidae, have also played a prominent role in biological systems of classifying lakes (e.g. Thienemann, 1915; Brundin, 1949) and are proving to be very useful in palaeoecological studies of lake sediments (e.g. Brodin and Gransberg, 1993; Warwick, 1980; Walker *et al.*, 1991). In this chapter we review briefly biotic systems of assessing water quality in rivers, using invertebrates. We go on to describe a variety of ways in which the family Chironomidae has been useful, or appears to have the potential to be useful, in detecting pollution, classifying lakes and providing a record of historical changes in lacustrine environments. Finally, using new data, we compare the chironomid faunas of 25 tarns in the English Lake District (Cumbria), covering a wide range of acidity and alkalinity.

II. BIOTIC INDICES AND RIVER QUALITY

The saprobien system has been used widely in continental Europe over many years but has never gained general acceptance in Britain or North America (Mason, 1981). Alternative indices have been developed in Britain of which two of the best known are the Trent Biotic Index (TBI) (Woodiwiss, 1964) and the Chandler Score (Chandler, 1970). Both were developed for use in specific rivers or river types but have been applied, with various modifications, over a much wider range of conditions than their originators had envisaged. More recently an index, known as the BMWP (Biological Monitoring Working Party) Score (National Water Council, 1981) was devised for general use throughout British rivers and has gained wide acceptance in Britain.

Essentially, all of these scores or indices weight insect and other macroinvertebrate taxa of rivers and streams according to their known (or in some cases, supposed) tolerance to organic pollution. Among aquatic insects the majority of mayfly (Ephemeroptera) and stonefly (Plecoptera) taxa are considered to be intolerant of organic pollution and, when present, contribute a high value to the overall score. Taxa that are regarded as tolerant, on the other hand, such as *C. riparius*, or sometimes the entire family, Chironomidae, make only a small contribution to the total score.

The use of biotic scores as measures of water quality, begs the question "what score should be expected in the absence of pollution?". Physical and chemical differences between rivers, or reaches of the same river, can have a profound influence on the community that develops, even in the absence of pollution. One way around this problem is to express indices, not as a total score but as an average score per taxon (ASPT). This is estimated by dividing the total score by the number of taxa that contribute to it. Armitage et al. (1983) and Pinder et al. (1987), for example, have shown that the ASPT version of the BMWP score is substantially less variable, both spatially and temporally, than the total BMWP score.

An extremely thorough study of communities of unpolluted rivers throughout Britain has also led to the development of a model known as RIVPACS (River InVertebrate Prediction and Classification System) which allows the macroinvertebrate fauna that would occur at a given site, in the absence of pollution, to be predicted from a small number of environmental variables (e.g. Wright et al., 1989, 1993). One of a number of uses for this system is to provide a baseline against which observed values can be assessed. It has also been used, for example, to assess macroinvertebrate responses to river regulation (Armitage et al., 1987).

The level of identification that is demanded varies substantially between the various indices. The BMWP score necessitates identification to family level only, and thus implicitly assumes that all species within a family have a similar tolerance to pollution. Although this is not true, certainly for the Chironomidae with around 600 British species, in practice the system appears to serve its intended purpose, of providing a broad assessment of river water quality.

III. CHIRONOMIDAE AND WATER QUALITY

By virtue of its high level of species diversity and ubiquitous distribution, the family Chironomidae has the potential to be extremely useful in the biological assessment of water quality and classification of water bodies. Tables of the ecological tolerances of freshwater Insecta of the Nearctic were provided by Beck (1977) and Roback (1974). As Roback pointed out, the Diptera "represented mainly by the Chironomidae, has species tolerant of many chemical extremes". In his tabulations, relating to the Nearctic fauna, 13 species of Diptera, all of them Chironomidae, were tolerant of pH in excess of 8.5. Below pH 4.5, with the exception of one tipulid, all tolerant Diptera were Chironomidae. At sites with alkalinity in excess of 210 ppm only Chironomidae, with the exception of *Palpomyia* (Ceratopogonidae) are represented and the same is true where chloride concentration exceeds 1000 ppm. Roback also found that six of seven species of Diptera recorded from brackish situations were Chironomidae and where iron exceeded 5 ppm the only macro-invertebrates were three species of Chironomidae. In situations with dissolved oxygen concentrations of less than 4 ppm, 13 of 15 species of Diptera that were present were Chironomidae. With sulphate concentration exceeding 400 ppm, eight species of Chironomidae were present, along with *Palpomyia* sp. (Ceratopogonidae) and a horsefly (*Tabanus* sp.). In situations with a high BOD (biochemical oxygen demand) only chironomids (notably *C. riparius*) and oligochaetes were able to survive.

Worldwide, species of *Chironomus*, in particular, have been shown to be tolerant of a wide range of pollutants. They are, for example, unusually tolerant of highly acid conditions. *C. riparius* has been found in ponds with a pH of 2.8 (Havas and Hutchinson, 1982) while *C. plumosus* (L.) has been recorded from acid strip-mine lakes with pH 2.3 (Harp and Campbell, 1967) and Yamamoto (1986) found a species of *Chironomus* in a volcanic lake in Japan with a recorded pH of only 1.4.

It has been suggested that it is the buffering capacity of haemoglobin, possessed by all *Chironomus* species (and other members of the tribe Chironomini), that allows these species to colonize such a variety of habitats that are unsuitable for most other groups. However, this cannot provide the whole explanation, since some Orthocladiinae, without haemoglobin, have also been shown to tolerate pollution by heavy metals. For example, *Cricotopus bicinctus* (Meigen) was found by Surber (1959) to be exceptionally resistant to electroplating wastes containing chromium and copper as well as cyanides, while Armitage and Blackburn (1985) found another orthocladiine to be one of the few invertebrates living in a stream contaminated by zinc from old mine workings.

A. Chironomids and River Water Quality

In spite of all the indications that the Chironomidae have the potential to be extremely useful as indicators of water quality, they have not been used extensively in systems of assessing water quality in rivers. That this potential has not been fully exploited is largely due to the difficulties associated with specific identification of larvae and the paucity of information on the ecological tolerances of the vast majority of species. The first of these problems has been eased by the publication of keys and diagnoses to Holarctic Chironomidae (Wiederholm, 1983, 1986, 1989), which allow the identification of all stages to genus. In many cases, specific identification still requires knowledge of the adult or pupal stage and often requires familiarity with a diffuse and extensive literature. Adult males of many of the British species are identifiable using Pinder's (1976) keys while Langton's (1984, 1991) keys allow identification of the pupae of the majority of western Palaearctic species.

One noteworthy attempt to exploit the potential of the Chironomidae in assessing the quality of rivers is the work of Wilson and colleagues (e.g. Wilson and McGill, 1979). They utilized collections of pupal exuviae as a "quick and easy means of studying the spatial distribution of Chironomidae in rivers". Thienemann (1910) first suggested using collections of exuviae to determine the chironomid community of a water body, but Wilson and associates first attempted to correlate the data with water quality (e.g. Wilson and Bright, 1973; Wilson and McGill, 1977). Using this technique, Wilson and Wilson (1985) carried out a survey of the River Rhine and selected tributaries, from the Swiss Alps to the North Sea, thereby demonstrating the relative ease with which samples can be gathered. Exuviae are also more likely to be identifiable to species than larvae, thanks largely to Langton's studies (1984, 1991), but the method still suffers from the disadvantage that little is known of the ecological limits and tolerances of the majority of species. Nevertheless, Wilson and Wilson (1985) were able to draw general conclusions about the fauna of the Rhine. For example, the fauna generally becomes less diverse and "of poorer quality throughout the Oberrhein sections" and "faunal changes are demonstrated that are related to the inflows of the River Neckar and River Main".

B. Chironomids and Lacustrine Environments

While there has been limited success in using chironomids as indicators of river water quality they have played a major role in systems of lake classification. Brundin (1958) was able to show that such systems, largely developed in Europe and North America, are also valid for much of the southern hemisphere, although this was disputed by Forsyth (1978) in respect of a series of New Zealand lakes which he investigated.

The early history of lake classification on the basis of chironomid fauna has been reviewed by Brundin (1949) and by Brinkhurst (1974) and summarized by Sæther (1979). It is therefore only necessary to provide brief details here. As with so much of modern limnology the origins go back to the work of Thienemann. In his early studies, Thienemann (1913) noted that lake faunas were dominated by *Chironomus* spp. in northern Germany and Denmark, while in subalpine lakes the dominant taxa belonged to the Tanytarsini, with *Chironomus* playing only a minor role. Later, Thienemann (1915) discovered that the fauna of lakes in nutrient-poor catchments of the Eifel region of Germany was also dominated by Tanytarsini (*Lauterbornia coracina* Kieffer) (= *Micropsectra coracina*), whereas nearby, shallower lakes in nutrient-rich catchments were dominated by *Chironomus anthracinus* Zetterstedt. Thienemann believed that the differences between faunal characteristics of lakes could be explained in relation to different oxygen concentrations. In the deeper, nutrient-poor lakes of the Eifel region, oxygen concentrations were between 7.77 and 8.16 ml l^{-1}, while in the nutrient-rich lakes they lay between 0.94 and 3.49 ml l^{-1}.

This explanation remained largely unchallenged for many years. However, Warwick (1975) and Wiederholm (1976) both believed that food availability, not oxygen status, was the major factor determining chironomid dominance. Their view was strongly reinforced by Sæther's (1979) studies. In Lake Winnipeg oxygen is plentiful everywhere but the chironomid fauna differs markedly in different parts of the lake, reflecting differing degrees of eutrophication. From his extensive knowledge of lake faunas, in both Europe and North America, Sæther (1979) identified 15 characteristic chironomid communities associated with rather narrow trophic bands, six in each of the oligotrophic and eutrophic ranges and three in the mesotrophic range. Sæther (1979) also provided a dichotomous key to the 15 trophic levels, based on their chironomid communities and showed there to be a close correlation between chironomid associations and concentrations of both total phosphorus and chlorophyll *a*. According to Sæther (1979), oxygen concentration only becomes important in relation to chironomid communities when levels are very low, such as in lakes of advanced eutrophication.

As a result of studies of the impacts of lake acidification on chironomid faunas, Brodin (1990) also considered that trophic level seems to be a more important factor influencing species composition, than pH and concluded that the fauna of acidified lakes was very similar to that of non-acidified lakes with similarly low trophic levels.

IV. CHIRONOMIDS AS PALAEOECOLOGICAL INDICATORS

The responsiveness of chironomid faunas, directly or indirectly, to changes in water quality and the long-term persistence of useful taxonomic characters associated with the head capsule and mouthparts, has led to the widespread use of their remains in palaeoecological studies of lakes. A good deal of such work has been carried out in Scandinavia (e.g. Alhonen and Haavisto, 1969; Wiederholm and Eriksson, 1979; Brodin, 1986), Germany (Hofmann, 1971), and North America (e.g. Warwick, 1980) but relatively few studies have been done in Britain (Bryce, 1962; Brodin, 1990; Brodin and Gransberg, 1993). Such studies permit comparison of faunal changes, over long periods of time, to be compared with the impacts of recent, anthropogenic changes, allowing the significance of the latter to be more fully appreciated (Brodin, 1986). For example, the accumulation of sediment in Lake Flarken in southern Sweden has proceeded almost without disturbance throughout all postglacial periods and Brodin (1986) was able to divide the fauna into eight phases, corresponding with changes in environmental conditions, primarily brought about by major changes in the humidity of the climate. However, the most dramatic changes, by far, occurred during recent decades, associated with anthropogenic acidification. This has resulted in a great impoverishment of the fauna and total dominance by two species of *Chironomus*. As Brodin (1986) points out, dominance by *Chironomus* is as often the result of strong acidification as of eutrophication, although different species of *Chironomus* are involved.

Another detailed study was carried out by Warwick (1980) on sediments from the Bay of Quinte, Lake Ontario. The initial objectives of this study were to determine whether significant changes had occurred in the sequence of chironomid remains, reflecting lake-type succession, and to determine whether such changes were associated with changes in trophic status. As the study progressed, Warwick found that the sedimentary record of chironomid remains was capable of providing the answers to many more questions. Cultures that predated colonization by British and French settlers, the indigenous Iroquoian and Hopewellian cultures, had also left their mark on the sediment record by influencing conditions in the bay and its catchment. However, massive changes in the fauna were clearly associated with the arrival of European colonists and the development of their communities up to the present day. Despite showing influences of human activities, the fauna had continued to be typical of oligotrophic conditions, throughout the periods preceding European influences on the catchment. The initial activities of European settlers included clearance of forests and poor agricultural practices. These resulted in increased levels of erosion and input of mineral sediments to the lake. In spite of a contemporary increase in nutrient input, oligotrophic conditions persisted because of the rapid accumulation of fine mineral sediments. As the rate of addition of mineral sediments declined and the population surrounding the bay increased further, there followed a period of rapid eutrophication.

Warwick (1989) and others (e.g. Hann *et al.*, 1992) argue that chironomid remains are not useful indicators of climatic conditions and, for example, that changes in the chironomid fauna associated with the late glacial sequence are explained by indirect effects, such as changes in rates of sediment accumulation. In contrast, Walker and associates (Walker and Matthewes, 1989; Walker, 1991; Walker *et al.*, 1991) believe that late-glacial changes in chironomid faunas were, partly at least, the direct and predictable consequence of changes in temperature (see also

Chapter 17, this volume). Walker et al. (1991) go so far as stating that chironomids are excellent indicators of past climate and developed a model to infer historic temperatures from the remains of Chironomidae in unstratified lakes. The controversy surrounding this aspect of interpreting the fossil record continues to attract vigorous discussion (e.g. Hann et al., 1992) and an excellent summary of this whole field of research, and the debate surrounding it, is provided by Walker (in Armitage et al., 1994).

V. DEFORMITIES INDUCED BY POLLUTION

Hamilton and Sæther (1971) reported the occurrence of severely deformed chironomid larvae in Lake Erie and two lakes in the Okanagan valley of British Columbia. The occurrence of deformed mouthparts is not particularly unusual in chironomids but in this case the authors remarked on the fact that the deformities were often asymmetric and that the teeth of both the mentum and mandibles were often twisted. The head capsule itself was also often heavily pigmented and thickened and, in more severe cases, the body walls of the thorax and abdomen were also thickened. The deformed larvae were usually taken from restricted areas of the lakes. Those from Lake Erie were all from close to the mouth of the Maumee River which carries effluents from the area near Toledo. All deformed larvae were *Chironomus* spp. and no normal *Chironomus* were found at any of the stations involved, whereas samples from elsewhere in the lake were normal. Three genera were involved in the Okanagan Lakes, *Chironomus*, *Stictochironomus* and *Procladius*, although in each case more than half the specimens were normal. The largest collection of deformed larvae was obtained close to the inflow of the Okanagan river. Other deformed individuals were found near known sources of pollution, but all samples taken from the neighbourhood of sewage outfalls, with the exception of a single specimen, were normal. Hamilton and Sæther (1971) speculated that the deformities were the consequence of severe physiological disturbance caused by exposure to industrial, or possibly agricultural pollutants.

Warwick (1980) also found similarly deformed larval remains in the Bay of Quinte, but only in the most recently deposited sediments. Since then Warwick, in particular, has made an extensive study of such deformities (e.g. Warwick, 1989, 1991). They appear to have the potential to be extremely useful as indicators of sediment contamination but substantially more research is required.

It is perhaps worth noting that deformities may also sometimes occur in other stages but these have received little attention. Some years ago the first author found a number of *Cryptochironomus* adults, at a single location, with two, or even three spurs on the tibial comb instead of the usual single spur. As in the case of larval deformities these were usually asymmetrical. The deformity was never found on adults from other locations and circumstantial evidence suggested that it may have been induced through exposure of larvae or pupae to chemical contamination of the ditch in which they developed, which received small quantities of effluent from a nearby analytical laboratory.

VI. THE CHIRONOMIDAE OF LAKE DISTRICT TARNS

While there have been numerous studies of the chironomid faunas of natural lakes in continental Europe and North America there have been relatively few published studies of chironomids from lakes in the British Isles. Notable exceptions are the research on Loch Leven, in Scotland (Charles *et al.*, 1974; Maitland and Hudspith, 1974), Lough Neagh in Ireland (e.g. Carter, 1977) and Malham Tarn in England (Bryce, 1962). Surprisingly, in spite of the very long history of detailed limnological research in the English Lake District, there is a dearth of information on the Chironomidae of this region – no published list of species exists for even one of the region's large number of lakes and tarns. Clearly this is a significant omission, especially in view of the increasing interest in the sedimentary record provided by the remains of Chironomidae. Therefore, during 1986, Chironomidae were sampled from a series of 25 tarns, exhibiting a range of chemical and physical characteristics (Table 1). Some of the results of this investigation are presented here, for the first time.

A. Geology of the Lake District

The English Lake District consists of a central dome of palaeozoic rocks, 400–500 million years old, surrounded by a variety of younger rocks. The oldest rocks, laid down during the Ordovician period, are in the northern part of the central dome and are of two types. The oldest and most northerly are the Skiddaw slates, which are soft and easily weathered, except where they have been metamorphosed by heat or pressure. The central part of the region is also Ordovician, but of volcanic origin. These rocks are hard and include slates and porphyrines, notably the green slates for which the Lake District is well known. The southern boundary of this series is formed by a narrow band of limestone, south of which are the Silurian slates, shales and grits. In addition there are igneous

12. Chironomidae as Indicators of Water Quality

Table 1. List of tarns that were sampled, with their main characteristics

Tarn name and group	Altitude (m)	Area (m^2)	Max depth (m)	pH	Alkalinity
Group 1					
Floutern	381	15 100	4	4.6	0
Muncaster	155	21 900	12	4.8	0
Parkgate	58	34 000	1	4.4	0
Red Tarn	503	7100	1	5.1	0
Group 2					
Blackbeck	472	9700	2	5.9	27
Cogra Moss	225	158 090	4	6.4	49
Harrop	290	15 400	6	5.4	4
Seathwaite	366	267 000	26	5.1	8
Three Dubs	206	15 810	3	6.3	82
Watendlath	258	40 000	17	6.3	89
Group 3					
Brotherswater	160	200 000	15	6.7	188
Eskdale Green	58	23 600	3.8	6.3	120
Wise Een	195	39 850	3	6.9	197
Yew Tree	107	38 000	2	6.5	200
Group 4					
Gurnal Dubs	290	32 000	14	6.9	254
Holehird	110	9300	10	7.4	722
Moss Eccles	180	33 050	3	7.0	359
Rather Heath	132	21 200	7	7.2	758
Silver Howe	370	3500	1.5	7.4	243
Tarn Hows	188	144 000	9	6.9	239
Group 5					
Cunswick	145	84 000	15	8.0	3180
Moss Side	100	7600	2	7.8	1083
Parkside	70	7600	2	8.1	2500
Standing	29	24 000	4	8.1	3100
Urswick	38	48 000	16	8.2	3240

intrusions in several regions. A more detailed account of the geology of the region is provided by Fryer (1991), in his account of the natural history of the Lake District.

B. Sampling sites

The varied geology of the Lake District gives rise to tarns and lakes exhibiting a range of water chemistry, from those with pH as low as 4.7 and no measurable alkalinity, to those with alkalinities exceeding 1000 and a pH of 7 or more (Sutcliffe, 1983). These have been divided into five broad types, as follows:

Group 1: ACID – No measurable alkalinity; pH < 5.4
Group 2: VERY SOFT – Alkalinity 1–100; pH 5.4–7.0
Group 3: SOFT – Alkalinity 100–200; pH 6.0–7.4
Group 4: MEDIUM HARD – Alkalinity 200–1000; pH 6.0–10
Group 5: HARD – Alkalinity > 1000; pH 7.1–10

Using chemical data provided by Carrick and Sutcliffe (1982), a series of 25 tarns, representing each of the above types, was selected for the chironomid sampling programme. A list of sites, with their chief characteristics is given in Table 1. Among the selected tarns, alkalinity ranged from 0 to 3240 $\mu E\ l^{-1}$ and pH from 4.6 to 8.2. Maximum depth was between 1 and 26 m and as far as possible a range of depths is included within each group of selected tarns. Similarly, each group of tarns covers a range of altitudes, the absolute range being from 29 to 525 m above sea level, and a range of areas from 3500 m^2 in the case of the smallest tarn, Silver Howe, to 267 000 m^2 in the case of Seathwaite. None of these physical characteristics is significantly correlated with pH (e.g. Fig. 1A and B), although there is a weak trend towards increased acidity with greater altitude. Alkalinity and pH are, however, very closely related, as would be expected.

C. Methods

Each tarn was sampled on five occasions, during March, April, June, August and October of 1986. On each occasion samples of pupal exuviae were obtained using the technique described by Wilson and McGill (1979). When the adult midge is ready to eclose it swims to the surface, still enclosed in its pupal skin. After emergence of the imago, the skins remain floating for up to 1 or 2 days, depending on conditions, and large numbers become stranded on lee shores. Samples of exuviae were collected by simply scraping them up or using a skim net. The samples were first preserved in 4% formaldehyde for transporting back to the laboratory, where they were washed through a 5-mm mesh sieve, to remove coarse debris. The exuviae, retained in another sieve of 150-μm mesh, were then preserved in 70% alcohol to await sorting and identification.

When samples were spread over a white tray, it was possible, with experience, to identify many species under a low power microscope, using

12. Chironomidae as Indicators of Water Quality

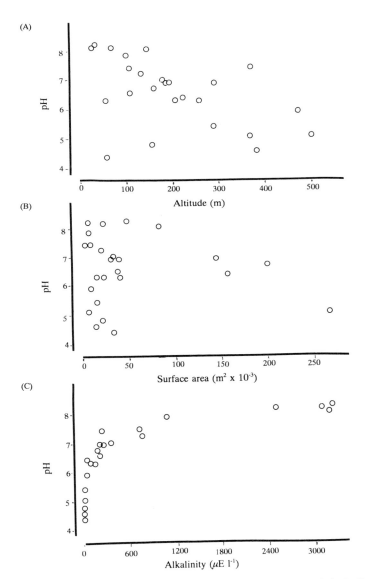

Fig. 1. Relationship between pH and (A) altitude, (B) surface area and (C) alkalinity of 25 tarns in the English Lake District.

the keys by Langton (1984, 1991). More difficult or unfamiliar species were mounted in euparal, after dehydration in 100% alcohol, for examination using a compound microscope. A subsample of 200 exuviae was identified from each sample.

D. Results

1. Species richness

A total of 106 species was recorded from the 25 tarns, of which 22 were found in the most acid group (Group 1), 40 in the very soft group (Group 2), 50 in the soft-water group (Group 3) and 64 in tarns with medium-hard water (Group 4) but only 33 in the hard-water tarns (Group 5). Species richness is low in tarns with very low alkalinity and increases until alkalinity reaches 200–300 μE l^{-1}. At higher alkalinities species richness declines (Fig. 2B).

Species richness shows little relationship with pH (Fig. 2C) below a pH of about 6.0. Between pH 6.0 and 7.5, species richness increases markedly but declines again at still higher pH values, corresponding with the depression in species number noted in respect of very high alkalinities.

The number of species that was recorded showed no significant relationship with either surface area or depth of tarn. This is probably a reflection of the fact that the majority of species inhabit littoral and sublittoral regions, where habitat diversity is greatest. There was, however, a significant, negative correlation between species richness and altitude ($r = -0.485$; $P < 0.05$) reflecting the fact that the more acid tarns tended to be on the higher ground and the alkaline tarns on the lower, carboniferous substrata, but this was not true in every case. A low number of species was recorded for all tarns that were more than 300 m above sea level and a high number of species was recorded from only one tarn over 200 m.

One of the Group 4 tarns, indicated by an arrow in Fig. 2C, had a much lower than expected number of species. This was Silver Howe Tarn, which has the smallest area of any of the 25 tarns and is also at a higher altitude (270 m) than any other tarn in Groups 3, 4 or 5 (Table 1).

2. Indicator species and communities

Species that occurred in every tarn within a specific category are listed in Table 2. Only four species occurred in all five acid tarns, three in all of

12. Chironomidae as Indicators of Water Quality

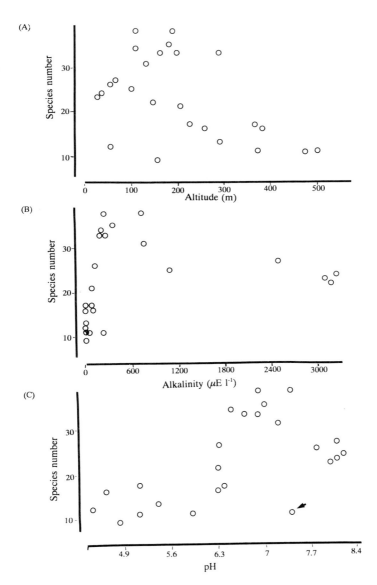

Fig. 2. Relationship between number of chironomid species (diversity) and (A) altitude, (B) alkalinity and (C) pH of 25 tarns in the English Lake District.

Table 2. Chironomid species that were present in all of the tarns within a specified category

Group 1 Acid
Ablabesmyia monilis
Macropelopia adaucta
Heterotrissocladius grimshawi
Pagastiella orophila

Group 2 Very Soft
Ablabesmyia monilis
Macropelopia adaucta
Heterotrissocladius grimshawi
Psectrocladius obvius and/or
Psectrocladius platypus

Group 3 Soft
Ablabesmyia monilis
Procladius flavifrons
Cricotopus intersectus
Chironomus plumosus
Chironomus riparius
Glyptotendipes pallens

Group 4 Medium Hard
Ablabesmyia monilis
Ablabesmyia longistyla
Cricotopus intersectus
Orthocladius consobrinus
Psectrocladius obvius
Chironomus plumosus
Chironomus riparius
Glyptotendipes pallens
Microtendipes sp.
Pseudochironomus prasinatus

Group 5 Very Hard
Ablabesmyia monilis
Procladius sagittalis
Zavrelimyia melanura
Cricotopus intersectus
Orthocladius consobrinus
Chironomus anthracinus
Chironomus plumosus
Chironomus riparius
Endochironomus impar/dispar
Endochironomus tendens
Glyptotendipes pallens
Phaenopsectra flavipes
Polypedilum sp.
Stictochironomus sticticus
Tanytarsus sp.

the very soft-water tarns, five in all of the soft waters, 12 in all medium-hard waters and 13 in all of the hard-water tarns. A few species, such as *Chironomus plumosus* and *C. riparius* were found in all categories of tarn. *Ablabesmyia monilis* (L.) was also collected from most of the 25 tarns.

The acid and very soft tarns (Groups 1 and 2) were characterized by the dominance of *Heterotrissocladius grimshawi* (Edwards), together with *Macropelopia adaucta* Kieffer (syn. *M. goetghebueri* (Kieffer). While *H.*

grimshawi was not found in any of the other three categories of tarn, *M. adaucta* did occur, but in low numbers, in two of the Group 4 tarns. *Pagastiella orophila* (Edwards) also occurred in all of the very acid tarns but only in low numbers. This species did not occur in tarns of any of the other four categories.

Psectrocladius obvius (Walker) is a useful indicator of very soft water. The occurrence of this species in large numbers, in association with *H. grimshawi*, and *M. adaucta* and possibly *Psectrocladius platypus* (Edwards) is indicative of very soft (Group 2) but not acid (Group 1) waters.

The faunas of Groups 3 and 4, the soft and medium-hard tarns were very similar. Although the Group 4 tarns held a generally much more diverse chironomid fauna, most of the recorded species were not consistently found in all tarns of these groups. Like the hard-water (Group 5) tarns, Groups 3 and 4 are dominated by *Chironomus* species, which occur only in very low abundance in the lower alkalinity groups. Other species which consistently appeared in these two types of tarn and in Group 5 included *Cricotopus intersectus* (Staeger) and *Glyptotendipes pallens* (Meigen). *Procladius flavifrons* (Edwards) was fairly abundant in all of the Group 3 tarns but occurred in low numbers in (or was absent from) more alkaline tarns. Additional species to those found in Groups 3 and 4, which consistently appeared in Group 5, included *Chironomus anthracinus* Zetterstedt, *Endochironomus impar* (Walker) or *E. dispar* (Meigen) (these species cannot be distinguished as pupae) and *Phaenopsectra flavipes* (Meigen). *Procladius sagittalis* (Kieffer) and *Zavrelimyia melanura* (Meigen) were present only in Group 5 tarns, but not in all of them. This information is summarized in Table 3.

VII. DISCUSSION

The concentrations of most ions in Lake District tarns varies in a pronounced seasonal manner (Sutcliffe, 1983), (alkalinity is generally lowest in winter) which makes comparisons between tarns difficult, unless they are based on a substantial number of measurements. Where possible, the values for pH and alkalinity quoted in Table 1, are the means of a range of values presented by Carrick and Sutcliffe (1982). However, in a number of cases, they represent spot samples taken at the same time as the chironomid samples.

Havas (1981) stated that at pH values below 5.5, many aquatic animals experience difficulties with calcium regulation while below pH 5 problems also occur with respect to regulation of sodium. In waters of very low alkalinity, the concentrations of certain ions become physiologically

Table 3. Indicator species for Lakeland tarns

Species	Group 1	Group 2	Group 3	Group 4	Group 5
Protanypus morio	*				
Pagastiella orophila	—				
Heterotrissocladius grimshawi	+++++	+++++			
Macropelopia adaucta	++++	++			
Psectrocladius obvius		+++++			
Chironomus riparius	—	—	++	—	
C. plumosus	—	—	++	++	
Cricotopus intersectus			+++++	+++++	+++++
Glyptotendipes pallens			++	++	++
Orthocladius consobrinus		—	++	++	+++++
Endochironomus tendens			—		++
Chironomus anthracinus				—	++
Endochironomus impar					++
Phaenopsectra flavipes					++
Procladius sagittalis					—
Zavrelimyia melanura					—

* Useful indicator but not always present.
— Usually present in low numbers.
++ Present at all sites in group, usually contributing between 1 and 10% of total catch.
+++++ Present at all sites in group, usually contributing >10% of total catch.

important for some invertebrates, because they must actively take in ions from very dilute water (Sutcliffe, 1983). Low pH may then impose a further, fatal stress because the small, mobile H^+ competes with larger cations at the ion transporting sites. Havas and Hutchinson (1982) studied the relative sensitivity to pH, of a range of invertebrates, including some chironomids, in ponds with pH ranging from 2.8 to 8.2. Sensitive species were incapable of maintaining high levels of sodium at low environmental pH. Many chironomids are tolerant of a wide range of pH, between 6 and 9 (Roback, 1974) but outside of this range species-richness is reduced (Raddum and Sæther, 1981; Simpson, 1983). In the present study diversity was reduced below pH 6.3 and above pH 7.5.

Among Chironomidae, tolerance of acidic conditions is most commonly encountered among the tribe Chironomini, especially among *Chironomus* spp. Out of seven species considered by Roback (1974) to be tolerant of acid conditions, five belonged to the Chironomini. Brodin (1986) also noted that *Chironomus* spp. are often associated with acidification. However, in the Lake District Tarns, *Chironomus* was virtually absent from the acid and soft waters, but abundant in the three more alkaline groups.

The more acid tarns were characterized by dominance of *Heterotrissocladius grimshawi* (Orthocladiinae) and *Macropelopia adaucta* (Tanypodinae). *M. adaucta* is a typically acidophilous species and was abundant in five acidified Norwegian lakes, studied by Raddum and Sæther (1981), with a pH range between 4.47 and 6.25. *H. grimshawi*, however, occurred in significant numbers only in clear water, (as opposed to humic) less acid sites. Elsewhere, *H. marcidus* (Walker) was generally more common. *Pagastiella orophila*, which occurred only in the most acid Lake District tarns was abundant in all of the Norwegian lakes.

Raddum and Sæther (1981) compared the trophic levels of the five Norwegian lakes, measured by chlorophyll a, with trophic levels inferred on the basis of their chironomid communities (using Sæther's (1979) method). Chironomid associations in the most acidified lakes, indicated a trophic level lower than that shown by the chlorophyll measurements, which suggests a direct effect of acidification on the community, independent of trophic level.

It is likely that, in tarns with very low alkalinity, pH has a direct, physiologically induced impact on the chironomid community. Unfortunately no direct measure of trophic level is available for the Lake District tarns. However, at least some of the tarns with high alkalinities were strongly eutrophic and it is possible that the reduced number of species in the very alkaline tarns results from periodically reduced oxygen concentrations.

Acknowledgements

The authors acknowledge the constructive comments of P. S. Cranston on an earlier version of this chapter and thank Sue Smith for help in preparing the figures. We also thank G. Fryer, D. G. George, E. Y. Haworth and A. G. Hildrew for helpful advice and valuable discussions in respect of the Lakeland tarns study, which was undertaken while D. J. M. was the beneficiary of a "Frost Studentship", awarded by the Freshwater Biological Association.

REFERENCES

Alhonen, P. and Haavisto, M. L. (1969). The biostratigraphical history of Lake Otalampi in southern Finland, with special reference to the remains of subfossil midge fauna. *Bull. Geol. Soc. Finland* **41**, 157–164.
Armitage, P. D. and Blackburn, J. H. (1985). Chironomidae in a Pennine stream system receiving mine drainage and organic enrichment. *Hydrobiologia* **121**, 165–172.
Armitage, P. D., Moss, D., Wright, J. F. and Furse, M. T. (1983). The performance of a new biological water quality score system based on macroinvertebrates over a wide range of unpolluted running-water sites. *Wat. Res.* **17**, 333–347.
Armitage, P. D., Gunn, R. J. M., Furse, M. T., Wright, J. F. and Moss, D. (1987). The use of prediction to assess macroinvertebrate response to river regulation. *Hydrobiologia* **144**, 25–32.
Armitage, P. D., Cranston, P. S. and Pinder, L. C. V. (1994). "Chironomidae – Biology and Ecology of Non-biting Midges". Chapman & Hall, London; in press.
Beck, W. M. (1977). "Environmental Requirements and Pollution Tolerance of Common Freshwater Chironomidae". *United States Environmental Protection Agency* EPA-600/4-77-024, pp. 260.
Brinkhurst, R. C. (1974). "The Benthos of Lakes". Macmillan Press, London.
Brodin, Y. W. (1986). The postglacial history of Lake Flarken, southern Sweden, interpreted from subfossil insect remains. *Int. Revue ges. Hydrobiol. Hydrogr.* **71**, 371–432.
Brodin, Y. W. (1990). Midge fauna development in acidified lakes in northern Europe. *Phil. Trans. R. Soc. Series B* **327**, 295–298.
Brodin, Y. W. and Gransberg, M. (1993). Responses of insects, especially Chironomidae (Diptera), and mites to 130 years of acidification in a Scottish lake. *Hydrobiologia* **250**, 201–212.
Brundin, L. (1949). Chironomiden und andere Bodentiere der Südschwedischen Urgebirgsseen. Ein Beitrag zur Kenntniss der bodenfaunistischen Charakterzüge schwedischer oligotropher Seen. *Rep. Inst. Freshwat. Res. Drottningholm* **30**, 1–914.
Brundin, L. (1958). The bottom faunistical lake type system and its application to the southern hemisphere. Moreover a theory of glacial erosion as a factor of productivity in lakes and oceans. *Verh. Int. Verein. Theor. Angew. Limnol.* **13**, 288–297.
Bryce, D. (1962). Chironomidae (Diptera) from freshwater sediments, with special reference to Malham Tarn (Yorks.). *Trans. Soc. Brit. Ent.* **15**, 41–54.
Carrick, T. R. and Sutcliffe, D. W. (1982). Concentrations of major ions in lakes and tarns of the English lake District (1953–1978). *Occ. Publs. Freshwat. Biol. Ass.* **16**, 1–168.
Carter, C. E. (1977). The recent history of the chironomid fauna of Lough Neagh, from the analysis of remains in sediment cores. *Freshwat. Biol.* **7**, 415–423.
Chandler, J. R. (1970). A biological approach to water quality management. *Wat. Poll. Contr.* **69**, 415–422.

Charles, N., East, K., Brown, D., Gray, M. C. and Murray, E. (1974). The production of larval Chironomidae in the mud at Loch Leven, Kinross. *Proc. R. Soc. Edinb. B.* **74**, 241–258.

Forsyth, D. J. (1978). Benthic macroinvertebrates in seven New Zealand lakes. *N. Z. Jl Mar. Freshwat. Res.* **12**, 41–49.

Fryer, G. (1991). "A Natural History of the Lakes, Tarns and Streams of the English Lake District". Freshwater Biological Association, Ambleside, England.

Gower, A. M. and Buckland, P. J. (1978). Water quality and occurrence of *Chironomus riparius* Meigen (Diptera: Chironomidae) in a stream receiving sewage effluent. *Freshwat. Biol.* **8**, 153–164.

Hamilton, A. L. and Sæther, O. A. (1971). The occurrence of characteristic deformities in the chironomid larvae of several Canadian lakes. *Can. Ent.* **103**, 363–368.

Hann, B. J., Warner, B. G. and Warwick, W. F. (1992). Aquatic invertebrates and climate change: a comment on Walker *et al.* (1991). *Can. J. Fish. Aquat. Sci.* **49**, 1274–1276.

Harp, G. L. and Campbell, R. S. (1967). The distribution of *Tendipes plumosus* (Linné) in mineral acid water. *Limnol. Oceanogr.* **12**, 260–263.

Havas, M. (1981). Physiological response of aquatic animals to low pH. *In* "Effects of Acidic Precipitation on Benthos" (R. Singer, ed.), pp. 49–65. North American Benthological Society, Springfield.

Havas, M. and Hutchinson, T. C. (1982). Aquatic invertebrates from the Smoking Hills, N.W.T.: Effect of pH and metals on mortality. *Can. J. Fish. Aquat. Sci.* **39**, 890–903.

Hawkes, H. A. and Davies, L. J. (1971). Some effects of organic enrichment on benthic invertebrate communities in stream riffles. *In* "The Scientific Management of Animal and Plant Communities for Conservation" (E. Duffey and A. S. Watts, eds), pp. 271–293. Blackwell Scientific Publications, Oxford.

Hofmann, W. (1971). Zur Taxonomie und Palökologie subfossiler Chironomide (Dipt.) in Seesedimenten. *Arch. Hydrobiol.* **6**, 1–50.

Hynes, H. B. N. (1960). "The Biology of Polluted Waters". Liverpool University Press, Liverpool.

Kolkwitz, R. and Marsson, M. (1909). Ökologie der tierischen Saprobien. *Int. Revue. ges. Hydrobiol. Hydrogr.* **2**, 125–152.

Langton, P. H. (1984). "A Key to Pupal Exuviae of British Chironomidae". Published privately, P. H. Langton.

Langton, P. H. (1991). "A Key to Pupal Exuviae of West Palaearctic Chironomidae". Published privately, P. H. Langton.

Learner, M. A. and Edwards, R. W. (1966). The distribution of the midge *Chironomus riparius* in a polluted river system and its environs. *Air Water Pollution* **10**, 757–768.

Maitland, P. S. and Hudspith, P. M. G. (1974). The zoobenthos of Loch Leven, Kinross, and estimates of its production in the sandy littoral area during 1970 and 1971. *Proc. R. Soc. Edinb. B* **74**, 219–239.

Mason, C. F. (1981). "Biology of Freshwater Pollution". Longman, London.

National Water Council (1981). "River Quality: the 1980 Survey and Future Outlook". National Water Council, London.

Pinder, L. C. V. (1976). "A Key to Adult Males of the British Chironomidae". *Scient. Publs Freshwat. Biol. Ass.* **37**.

Pinder, L. C. V. and Farr, I. S. (1987). Biological surveillance of water quality – 3. The influence of organic enrichment on the macroinvertebrate fauna of small chalk streams. *Arch. Hydrobiol.* **109**, 619–637.

Pinder, L. C. V., Ladle, M., Gledhill, T., Bass, J. A. B. and Matthews, A. M. (1987). Biological surveillance of water quality – 1. A comparison of macroinvertebrate surveillance methods in relation to assessment of water quality, in a chalk stream. *Arch. Hydrobiol.* **109**, 207–226.

Raddum, G. G and Sæther, O. A. (1981). Chironomid communities in Norwegian lakes with different degrees of acidification. *Verh. int. Verein. theor. angew. Limnol.* **21**, 399–405.

Roback, S. S. (1974). Insects (Arthropoda: Insecta). *In* Pollution ecology of freshwater invertebrates (C. W. Hart and S. L. H. Fuller, eds), pp. 313–376. Academic Press, New York.

Sæther, O. A. (1979). Chironomid communities as water quality indicators. *Holarct. Ecol.* **2**, 65–74.

Simpson, K. W. (1983). Communities of Chironomidae (Diptera) from an acid-stressed headwater stream in the Adirondack mountains, New York. *Mem. Am. Entomol. Soc.* **34**, 315–327.

Surber, E. W. (1959). *Cricotopus bicinctus*, a midgefly resistant to electroplating wastes. *Trans. Am. Fish. Soc.* **88**, 11–116.

Sutcliffe, D. W. (1983). Acid precipitation and its effects on aquatic systems in the English Lake District (Cumbria). *Rep. Freshwat. Biol. Ass.* **51**, 30–62.

Thienemann, A. (1910). Die Sammeln von Puppenhäuten der Chironomiden. *Arch. Hydrobiol.* **6**, 213–214.

Thienemann, A. (1913). Der Zusammenhang zwischen dem Sauerstoffgehalt des Tiefenwassers und der Zusammensetzung der Tiefenfauna unserer Seen. *Int. Revue. ges. Hydrobiol. Hydrogr.* **6**, 243–249.

Thienemann, A. (1915). Die Chironomiden der Eifelmaare. Mit Beschreibung neuer. *Verh. naturh. Ver. Preuss. Rheinl.* **72**, 1–58.

Thienemann, A. (1954). *Chironomus*. Leben, Verbreitung und wirtschaftliche Bedeutung der Chironomiden. *Binnengewässer* **20**, 1–834.

Walker, I. R. (1991). Modern assemblages of arctic and alpine Chironomidae as analogues for late-glacial communities. *Hydrobiologia* **214**, 223–227.

Walker, I. R. and Matthewes, R. W. (1989). Early post glacial chironomid succession in southwestern British Columbia, Canada, and its paleoenvironmental significance. *Paleolimnology.* **2**, 1–14.

Walker, I. R., Smol, J. P., Engstrom, D. R. and Birks, H. J. B. (1991). An assessment of Chironomidae as quantitative indicators of past climatic change. *Can. J. Fish. Aquat. Sci.* **48**, 975–987.

Warwick, W. F. (1975). The impact of man on the Bay of Quinte, Lake Ontario, as shown by the subfossil chironomid succession (Chironomidae, Diptera). *Verh. Int. Verein. Theor. Angew. Limnol.* **19**, 3134–3141.

Warwick, W. F. (1980). Palaeolimnology of the Bay of Quinte, Lake Ontario: 2800 years of cultural influence. *Can. Bull. Fish. Aquat. Sci.* **206**, 1–117.

Warwick, W. F. (1989). Morphological deformities in larvae of *Procladius* Skuse (Diptera: Chironomidae) and their biomonitoring potential. *Can. J. Fish. Aquat. Sci.* **46**, 1255–1270.

Warwick, W. F. (1991). Indexing deformities in ligulae and antennae of *Procladius* larvae (Diptera: Chironomidae): application to contaminant-stressed environments. *Can. J. Fish. Aquat. Sci.* **48**, 1151–1166.

Wiederholm, T. (1976). Long-term changes in the profundal benthos of Lake Mälaren. *Verh. Int. Ver. Theor. Angew. Limnol.* **20**, 818–824.

Wiederholm, T. (1983). Chironomidae of the Holarctic region. Keys and diagnoses Part 1 – Larvae. *Entomologica Scand. Suppl.* **19**, 1–457.

Wiederholm, T. (1986). Chironomidae of the Holarctic region. Keys and diagnoses Part 2 – Pupae. *Entomologica Scand. Suppl.* **28**, 1–482.

Wiederholm, T. (1989). Chironomidae of the Holarctic region. Keys and diagnoses Part 3 – Adult males. *Entomologica Scand. Suppl.* **34**, 1–532.

Wiederholm, T. and Eriksson, (1979). Subfossil chironomids as evidence of eutrophication in Ekoln bay, Central Sweden. *Hydrobiologia* **62**, 195–208.

Wilson, R. S. and Bright, P. L. (1973). The use of chironomid pupal exuviae for characterising streams. *Freshwat. Biol.* **3**, 283–302.

Wilson, R. S. and McGill, J. D. (1977). A new method of monitoring water quality in a stream receiving sewage effluent, using chironomid pupal exuviae. *Wat. Res.* **11**, 959–962.

Wilson, R. S. and McGill, J. D. (1979). The use of chironomid pupal exuviae for biological surveillance of water quality. *Water Data Unit Technical Memorandum* **18**, 1–20.

Wilson, R. S. and Wilson, S. E. (1985). A survey of the distribution of Chironomidae (Diptera, Insecta) of the River Rhine by sampling pupal exuviae. *Hydrobiol. Bull.* **18**, 119–132.

Woodiwiss, F. S. (1964). The biological system of stream classification used by the Trent River Board. *Chem. Ind.* **11**, 443–437.

Wright, J. F., Armitage, P. D., Furse, M. T. and Moss, D. (1989). Prediction of invertebrate communities using stream measurements. *Reg. Rivs. Res. Mgt.* **4**, 147–155.

Wright, J. F., Furse, M. T. and Armitage, P. D. (1993). RIVPACS – a technique for evaluating the biological quality of rivers in the U.K. *Eur. Water Poll. Control* **3**, 15–25.

Yamamoto, M. (1986). Study of the Japanese *Chironomus* inhabiting high acidic water (Diptera, Chironomidae) I. *Kontyû* **54**, 324–332.

Part IV.
Changes in Land Use

13. Southern Hemisphere Insects: Their Variety and the Environmental Pressures upon Them
    M. J. SAMWAYS
14. Species Extinctions in Insects: Ecological and Biogeographical Considerations
    N. A. MAWDSLEY AND N. E. STORK
15. A World of Change: Land-use Patterns and Arthropod Communities
    M. B. USHER
16. Insects as Indicators of Land-use Change: A European Perspective, Focusing on Moths and Ground Beetles
    M. L. LUFF AND I. P. WOIWOD

13

Southern Hemisphere Insects: Their Variety and the Environmental Pressures upon Them

MICHAEL J. SAMWAYS

I. Introduction 298
II. The World Setting 298
   A. Land Distribution and Climatic History 298
   B. Evolution of Southern Biodiversity 299
   C. Insect Species Richness and Endemism in the South 301
   D. Tracking of Plant Species Richness by Insect Species 302
III. The Impacts 303
   A. Fire 303
   B. Hunting, Pastoralism and Livestock Rearing 305
   C. Arable Agriculture 306
   D. Deforestation 307
   E. Introduced Forest Patches 309
   F. Water Impoundments 309
   G. Exotic Invasives 310
   H. Urbanization 311
   I. Global Changes 311
IV. Conclusions and Future Perspectives 313
V. Summary 315
   References 316

Abstract

Only about one third of the world's landmass lies in the southern hemisphere. Except for Antarctica, the diversity of insect life in the south is high. Freedom from extensive Pleistocene ice sheets and the highly varied topographies have probably contributed towards this high diversity. The southern tips of the continents lie closer to the equator than is generally thought, with virtually no land at 50° S. The mostly tropical and subtropical soils are generally shallow and fragile. Many of the plants, but rarely the insects, are resistant to regular natural and man-induced fires.

Extensive and occasionally intensive megaherbivore grazing and browsing have historically greatly influenced the landscape and the insects that live there. Threats to southern insect habitats include planting of exotic trees, overgrazing by domestic animals, river impoundment and the impact of exotic weeds and animals. In this rapidly changing world environment, it is important to recognize the quantity and quality of biodiversity in the far south. Major research thrusts should be to selectively inventorize this diversity, to identify good indicators of particular environmental changes, determine responses to changing land use, to determine the relationships between native mammals and insects, and to identify threatened species and landscapes.

I. INTRODUCTION

The anthropogenic environmental changes taking place in the southern hemisphere are as severe as those in the north. Yet, in terms of biodiversity per unit area, the south is comparatively much richer. The environmental impacts in the south are therefore of great conservation concern. Further, as insects are the most speciose component of biodiversity, it is particularly relevant to overview the threats upon them in the southern hemisphere.

The aim here is first to speculate on the ancient and recent physical conditions that have generated and maintained this insect variety. Then this variety is appraised in terms of world biodiversity conservation. Finally, the various types of major anthropogenic impacts are considered. Inevitably, space here precludes a detailed review of the entire topic, and so emphasis is given to the main trends affecting insects in the changing environment of the southern hemisphere. By way of personal experience, most examples come from southern Africa and then from Australia. Biogeographical trends aside, few relevant research findings are available for temperate South America. The information from Africa, Australia and New Zealand points to many of the general patterns of land use and impacts upon the insect fauna being similar across the southern continents, the special case of Antarctica excepted. Some of these major patterns are reviewed here.

II. THE WORLD SETTING

A. Land Distribution and Climatic History

The far South, which so often takes third place after the North and the Tropics in biodiversity discussions, is a biogeographer's paradise

(Darlington, 1965). Rich in many endemic species, the southern hemisphere is now coming of age in biodiversity overviews (Platnick, 1991; Samways, 1992a) and in biotic inventories (Groombridge, 1992).

Why should the world's qualitative biodiversity (genetic uniqueness) be pear-shaped, bulging in the tropics and the South? And what are the threats to this valuable biota in the changing environment?

Only about one-third of the Earth's land is in the southern hemisphere. In the North, 39% of the Earth is land, whereas in the South it is only 19%. Further, about 29% of the southern hemisphere land surface is Antarctica, and poor in terrestrial invertebrate species.

The southern extremities of the southern continents are mostly cold to warm temperate. At 50°S there is little land, in stark contrast with the land-bulge at 50°N (Duxbury and Duxbury, 1991). Apart from Antarctica, climates in the South are mostly moderate, and support a rich insect fauna. Relative to the North, this has been the case for millennia, with most of the South only experiencing, at coldest, Pleistocene periglacial conditions at high elevations (Tyson, 1986). Not that the climate has been constant. The southern Cape was 5–6°C cooler 17–21 kyr ago, while the mean Holocene maximum 5–8 kyr ago was higher than at present (Partridge et al., 1990). At that time, the Cape was drier but the Kalahari wetter. Only Antarctica and the most southern tip of South America experienced true glaciation in the Pleistocene.

A significant feature of the South is that the blocks of land are widely separated (Fig. 1) and have been so for a long time. Pangaea began to break up about 200 myr ago. By 180 myr ago the rift between northern Laurasia and southern Gondwanaland had taken place, with Gondwanaland itself beginning to break up. During the Tertiary and Quaternary, successive new groups of plants and animals had moved into the South from the tropics and subtropics. There is evidence of taxon cycles (see Wilson, 1992) taking place, with the tropical African savanna for example, have acted as a species generator for grasshoppers (Jago, 1973).

B. Evolution of Southern Biodiversity

The climatically moderate setting in the South before and during the Pleistocene appears to have been synergistic with a highly variable topography in providing conditions for long-term species survival and for generation of many new species. The climate has been predictably adverse, rather than stochastically adverse. These conditions and various elevations, aspects of slope and nutrient-poor soils together have been instrumental in generation of the high biodiversity.

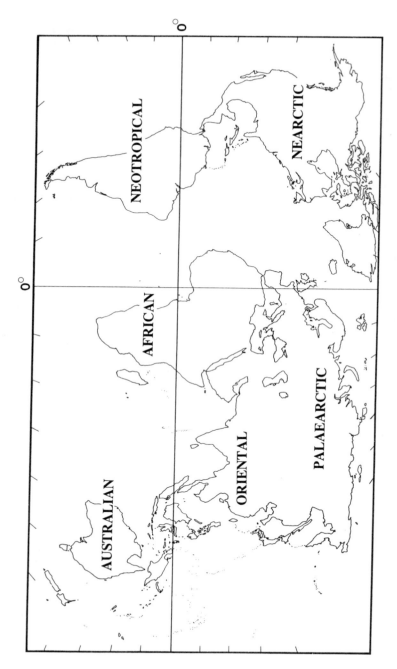

Fig. 1. One map of the world, with zoogeographical regions. A characteristic feature of the South is the wide spacing of the continents.

In Australia, the highest levels of invertebrate endemism are often in the most predictably severe biotopes (Greenslade, 1983). Collembola diversity is greatest in relatively stable, favourable biotopes such as lowland eucalypt forest (Greenslade and New, 1991). The Collembola are also extremely sensitive to human disturbance. Interestingly, a similar situation exists with the succulent plants of the Karoo in South Africa (Lovegrove, 1993).

The Cape Floristic region is particularly rich in plant species, with 8550 in 90 000 km^2. This is at least comparable with equal-sized areas in the equatorial forest regions. This richness derives from high species turnover across the land surface (beta diversity) (Cowling, 1992). Spatial scale is significant here, as comparative values vary, depending on whether, for example, 1-m^2 or 1000-m^2 quadrats are used. At the 1000 m^2 level, the South African succulent Karoo communities (up to 113 species 1000-m^{-2}) and the Brazilian cerrado savanna (230 species 1000 m^{-2}) are very rich, but not as rich as the highest figures for neotropical lowland rain forests (173 and 365 species 1000 m^{-2}) (Cowling et al., 1989).

Topography and flora can greatly influence the presence or absence of insect species. In the Cape, one side of the mountain may have a different biota from the other. Also, many of these plant and insect species have extremely restricted ranges. Many lycaenid butterflies occur only at single localities of less than a few hectares (Henning and Henning, 1989). Where more than one colony of a butterfly species has been found, these are often widely separated, sometimes comprising a distinct subspecies, for example *Argyrocupha malagrida cedrusmontana* Dickson and Stephen and *A. m. maryae* Dickson and Henning (Fig. 2).

C. Insect Species Richness and Endemism in the South

Estimates of insect species richness in the South are very rough, with an estimated two-thirds of the insect fauna still to be taxonomically described. Australia has at least 125 000 species (Greenslade and New, 1991) and probably twice this number, while southern Africa, south of the Limpopo, possibly has about 200 000–250 000 species (Samways, 1994). Many of these species are wide endemics to a region or narrow endemics to a locality. Some percentage endemism estimates for southern Africa are: Masarine Vespidae 100% (Gess, 1992), Apoidea 85% (Eardley, 1989), Tettigonioidea about 75% (Ragge, 1974), Acridoidea 47% (Johnsen, 1987) and Odonata 18% (Samways, 1992b). In Australia, percentage endemism may be even higher, and at the generic level is 44% in the Grylloidea, 100% in the Tettigoniidae, and 90% in the Acridoidea

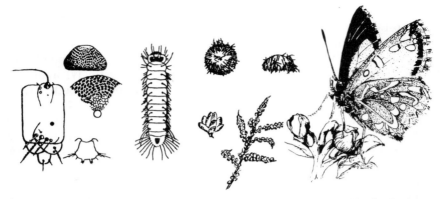

Fig. 2. The lycaenid butterfly *Argyrocupha malagrida malagrida* restricted in distribution to one small, tennis-court sized area in the South African Cape Peninsula. This is one of several subspecies, all with highly restricted ranges. Such narrow endemics are not uncommon in parts of the southern hemisphere. From Henning and Henning (1989).

(Rentz, 1991). In the South, even endemic families are not unusual, with Australia supporting three endemic beetle families (Rhinorphipidae, Acanthocnemidae and Lamingtoniidae).

The differences between the acridid percentage figure for Australia (90%) and South Africa (47%) are interesting. Australia has more of an island fauna, while South Africa is strongly influenced by a southward spread of species from the tropical heart of the continent (Jago, 1973).

The South, like the North, is today under increasingly severe pressure from human impact. But unlike the North, this impact has come about relatively recently, mostly over the last 200 years or so. The prehistorical, historical and geographical features of the South mean that it has a different but overlapping suite of threats from those in the North or in the tropics.

D. Tracking of Plant Species Richness by Insect Species

The plant species-rich biomes in the South do not necessarily automatically imply very high insect species richness through co-evolutionary relationships. Cottrell's (1985) studies on Cape butterflies have shown that none of the genera represents a faunal element that has co-evolved with the Capensis floral element. It appears that speciation has come about through allopatry, with much conservatism in the plant–insect relationship.

External feeders tend to be generalists, while internal feeders tend to specialization (Gaston et al., 1992) and, on this basis, the evolution of some endophagous insects appears to have been influenced by the endemic plant taxa (Wright, 1993). The highly endemic masarine wasps show a marked difference in major choices of flower taxa between geographic regions, with strong correlations between areas of species richness of these insects and their forage plants (Gess, 1992). These wasps are also more narrowly endemic than their forage plants.

These plant-rich southern biomes nevertheless have high insect species richnesses, but not proportionately high. This seems to be a world trend, with the insect-to-plant ratio decreasing with increasing plant species density (Gaston, 1992; Samways, 1994).

The reasons for insect species richness not tracking plant species richness in the Cape are probably many, complex and interrelated. They include poor nutrient returns from the plants, relatively simple plant community architecture (especially through paucity of trees), highly changeable and occasionally harsh weather (not climate), and low plant apparency. This low apparency is through restricted ranges and/or low density. It is also through investment by many of the plants in a large amount of below-ground tissue, such as bulbs and corms. Additionally, some plants are serotinous, retaining their seeds and dispersing them only after fire. A fire may kill most of the above-ground tissue, leaving only the seeds or the geophytic tissue which, for many insects, reduces the plant to zero apparency.

III. THE IMPACTS

A. Fire

Natural fires are common in the southern hemisphere savannas; scleromorphic forests and shrublands (Gill et al., 1981; Booysen and Tainton, 1984; Trabaud, 1987). Lightning strikes may reach over $10 \text{ km}^2 \text{ yr}^{-1}$ and are responsible for at least 12% of the fires in South Africa. Man-made fires are even more frequent, accounting for at least 64% of fires in South African ecosystems. Man-made fire is of great antiquity in North America, as well as in South Africa and Australia, meaning that today it is difficult to determine the extent to which present-day ecosystems have been shaped by natural or human fire.

Plants have many evolutionary adaptations to fire. Usually a particular part of the plant survives. This may be the trunk, root, corm, bulb or seed. Insects in contrast do not show specific adaptations to fire, although melanic crypsis on post-burn ground is not uncommon. Larger species

Fig. 3. Natural burns are part of the dynamics of the savannah biome. Today the burns, both natural and man-made, create a mosaic pattern across the fragmented landscape. This is the case even in wilderness areas in the Natal Drakensberg of South Africa shown here. In such areas, burns often do not spread extensively because of landscape features such as dirt tracks.

tend to fly ahead of the fire and small species go beneath rocks and into soil crevices. Grassland fires have almost no effect on soil temperature during the burn, but the temperature regimes on post-burn soil versus unburnt vary considerably. Burning removes most of the physical shelter which has implications for predation, disease, and developmental rates, and also removes the vegetation cover which provides insulation against the winter cold. Winters in the temperate South can be cold and windless, with topography responsible for distinct cold-air drainage patterns. Without this protective cover, mortality can be high (Samways, 1990). Inevitably this means that the fauna is much poorer in both species and individuals on burnt ground (Fig. 3). Conversely, when land is not burnt for very long periods, some species of grasshopper become more abundant while others become locally extinct (Chambers, 1993). The overall grasshopper diversity is maintained by a variety of vegetation seral stages which arises from burns occurring sporadically in different areas at different times.

Although man-made fires in ecosystems are a major disturbance feature, the problem facing insects in the changing environment, at least in southern Africa, is not increased fire pressure but reduced pressure. This reduction is mostly the result of landscape fragmentation, from roads and agricultural development, isolating patches from extensive burns. In these patches, there is a change in plant community structure with an increase in percentage of woody species. This fragmentation problem, particularly in Australia and Africa, means that instead of large expanses of fire-rejuvenated grasslands there is now a patchwork of distinct blocks of land in various stages of post-burn succession (Fig. 3).

B. Hunting, Pastoralism and Livestock Rearing

The hunting by man of large herbivores has been widespread throughout the southern hemisphere for many centuries. In the middle of the last century this sustainable utilization was superseded by intensive hunting using firearms. The inevitable reduction in megaherbivore populations changed the character of the landscape. This was accentuated by the erection of fences, introduction of more intensive pastoralism and the confinement of game to relatively small reserve patches. This is an important consideration, as savannas cover about one third of the world's land surface, including half the area of Africa, Australia and South America, and about 10% of tropical Asia (Skarpe, 1991). In southern Africa alone, it comprises 959 000 km^2, 46% of the subregion (Scholtz and Chown, 1993).

Grazing and browsing by wild and domesticated animals and the impact of these activities on insect populations is little researched. The effect of these vertebrates on grasses and shrubs has been hotly debated, and the apparent overgrazing may be due more to seasonal rainfall and long-term climate cycles than to the effect from domesticated animals (Hoffman and Cowling, 1990; O'Connor, 1993). There are also geographical differences. In the eastern Cape for example, there has been a clear shift from subtropical thicket to karoo shrubs, which means that palatable (to domestic animals) evergreen perennial trees, shrubs and succulents have been replaced by annual weedy plants and relatively unpalatable shrubs. Other changes are also evident. Many of the succulents and shrubs do not have soil-stored seeds, which means that even when grazing pressure is reduced, recovery is poor (Lovegrove, 1993).

There are also other problems, delicate stem and leaf structures are damaged by livestock trampling, and soil erosion is enhanced. The trampling and top soil loss is particularly harmful to the seed bank. Yet on the positive side, the hills on the landscape are reservoirs for plant propagules. These geographical features are relatively resistant to the grazing pressure which is highest in the surrounding lowland matrix. Chambers (1993) used mowing to simulate grazing, and showed that this treatment does not alter the grasshopper assemblage patterns as much as burning, but nevertheless is substantial.

Also of concern in the South is the impact of cattle on the delicate riparian vegetation and on the aquatic insects. The extremely rare Gondwana relict damselfly *Chlorolestes apricans* Wilmot (Plate 9) which is confined to a few localities on a few streams in the eastern Cape is threatened, among other things, by cattle breaking the bushy bank vegetation which is used for perching and oviposition.

With the replacement of wild megaherbivores by domestic grazers there has been a shift in some insect assemblages. Dung beetles, many of which are intimately associated with particular dung types in particular biotopes, is one group that has shown some major changes. The huge scarabaeid *Circellium bacchus* (Fabricius), once widespread in southern Africa, is now restricted to a small area of the eastern and southern Cape (Scholtz and Coles, 1991). One reason for its range decrease has been a reduction in buffalo (*Syncerus caffer* (Sparrman)) numbers. This demographic response has also been aggravated by direct habitat destruction such as removal of thick bush (Scholtz and Chown, 1993). Yet other dung beetle species, such as three scarabaeid *Euoniticellus* species, have increased their ranges as a result of bush removal for creation of pastures (Davis, 1993).

C. Arable Agriculture

Relatively little of the southern hemisphere has truly arable soil. In South Africa, only 11.4% of the land is cultivated or under permanent crops, compared with 69% under natural pasture (Anonymous, 1989). Although the impact on the overall insect fauna is probably little in terms of risks of widespread extinctions, local impacts can be severe. With the success of the wine industry, water extraction in streams of the south-western Cape is reducing the summer flow to such an extent that the Odonata assemblages is modified. The extremely localized Gondwana relict damselfly *Ecchlorolestes peringueyi* (Ris) (Plate 10) has, as a result, become extinct in the main streams, surviving only in a few montane feeder streams. The impact may be multifaceted, with introduced rainbow trout (*Oncorhyncus mykiss*, Baun) strongly suspected as having caused population extinctions, with the damselfly now surviving only above waterfalls out of reach of the fish (Fig. 4).

13. Southern Hemisphere Insects: Variety and Environmental Pressure

Fig. 4. Invasive organisms have caused problems throughout the southern hemisphere. Some of these organisms were deliberately introduced, such as the rainbow trout which has penetrated many of the colder streams of South Africa. The Gondwana relict damselfly, *Ecchlorolestes peringueyi*, which is an extremely rare Cape endemic, now only survives in the small streams above the waterfalls (shown here) beyond the reach of the trout.

There are many other impacts caused by agriculture, and not all these impacts reduce populations. The polyphagous citrus thrips *Scirtothrips aurantii* Faure occurs in high numbers in citrus orchards, which, in winter are green patches in a matrix of inhospitable brown and dormant vegetation (Samways, 1986).

D. Deforestation

Norton (1991) points out that prior to the arrival of people of Polynesian origin, New Zealand was largely forested below the climatic treeline. In contrast, southern Africa was probably never extensively forested and was always naturally fragmented (Lawes, 1990). These African remnant patches often support highly localized insect species. Deforestation therefore not only fragments populations but also risks extirpation of species. The delicate damselfly *Ecchlorolestes nylephtha* Barnard only occurs in the organic pools and streams in the darkness of full forest canopy in the southern Cape. Even moderate removal of the tree canopy eliminates local populations.

Tree cutting in the South has never been so widespread as in the North. Nevertheless, there has been damage. In South Africa, over 59% of the magnificent Knysna Forest has been felled, while in Australia the eucalypt forests have been largely disturbed from many causes, including "nutrient pollution" from human activities (Specht, 1990). In New Zealand, the forest area has now been reduced to 23% (Mark and McSweeny, 1990).

There is an urgent need to protect fully these southern forest fragments. They are too small for even minor utilization of their resources, and generally they are not extensive enough to sustain rotational logging as has been proposed for North American forests (Harris, 1984). Full protection and minimal utilization is the only viable option for maintenance of the biodiversity in these south-temperate woodlands.

From an invertebrate conservation viewpoint, these remnant forest patches have had a long geological history, and are important species reserves for some forest specialists such as Onychophora and pneumorid Orthoptera. Species equilibrium has probably been reached where allele preservation and rates of extinction have been stabilized. Potential problems such as inbreeding depression, demographic stochasticity, and adverse effect of environmental stochasticity and catastrophes, have been overcome through evolutionary time. Although there are no data on the insect faunal richness of these forest patches relative to their size, it is quite likely that each is at its optimum and maximum species richness. By this is meant that destruction of even a small area of reserve is likely to remove certain of these habitat specialists and reduce the viable area for some of the generalists.

From a conservation point of view, these indigenous forest patches must be left intact. In much of the far South this is now becoming the *status quo*. One problem, however, is that these patches are often being viewed in complete isolation. External impact, whether roads, planting of exotic trees or agricultural development, is taking place up to the edges of the original forests. As a result, the quality of the edge is influenced, with a likely adverse impact on dispersal and occasional gene flow between the scattered patches.

So often forest patches are considered as boxes with species either inside or outside. In reality, different species are scattered at various distances away from the edges and clearings. This is particularly so for bushy plants and also for many insects that feed on or live in them. This natural variegation is now being converted to sharper-edged fragmentation with concurrent inhibition to insect movement (e.g. Thomas, 1991) and more severe (i.e. more widespread and more definite) genetic bottlenecks (see Brakefield, 1991).

E. Introduced Forest Patches

The planting of exotic trees, particularly pines, and in South Africa, also eucalypts, is now a major feature of the landscape. Although the total area under timber is not large (about 1.3% in South Africa), the localized impacts can be severe. Impacts are indirect (e.g. changing water drainage patterns, altering stream water quality) and direct (e.g. changing biotope). As the South has so many localized endemic insect species, it means that the influence of exotic trees is disproportionately large, especially in mountainous areas where endemism is often highest and afforestation is significant. Donnelly and Giliomee (1985) showed that pine plantations (*Pinus radiata* D. Don) dramatically reduced the local ant species richness in the southwestern Cape. Afforestation is also one of the major threats to localized lycaenids.

Planting of exotics is also influential on insect movement. Different insect species react to landscape fragmentation in quite different ways. This was demonstrated with butterflies in South Africa (Wood and Samways, 1991). Particularly evident was that tall, introduced trees affected almost all species adversely, in causing them to change flight direction and follow the tree line. Viewed from above, the insects are being forced to now take zig-zag or even return paths rather than moving straight across the land surface.

The impact of tall, introduced trees has a strong influence on the settling behaviour of herbivorous insects such as the South African citrus psyllid *Trioza erytreae* (Del Guercio) (Samways and Manicom, 1983). For grasshoppers, the influence from pine trees can be over 30 m into the surrounding matrix (Samways and Moore, 1991). This means that to maintain the presence of the native insects and to permit gene flow, it is essential that gaps between plantation blocks are at least 80 m. This recognizes that these corridors are not necessarily movement corridors. They may only be elongated home bases or even inhibitory filters for some species.

F. Water Impoundments

Virtually all major river systems throughout the savanna regions of the South are impounded. Inevitably this has had a major impact on the local distribution patterns of riverine insects (Davies and Day, 1986), and changing the assemblage profile of bank-inhabiting species. Although damming of small streams can increase the local insect diversity and also provide conditions for more extensive and less patchy distribution of species, this really only holds for the ecologically eurytopic and widely dispersing species (Samways, 1989). Water impoundments generally do not benefit the localized stenotopic species, which are best conserved by maintaining ecosystems in as near as possible pristine conditions.

G. Exotic Invasives

The threat posed by accidentally and deliberately introduced organisms, is real and increasing, both in the North and the South. It is not a phenomenon that relates so much to latitude as to type of organism and the size of the land mass. Small islands have been particularly impoverished by vertebrates, as well as by other compounding factors such as habitat destruction. Rats (*Rattus rattus* Linnaeus) probably eliminated the Lord Howe Island stick insect (*Dryococelus australis* (Montrouzier)) when they went ashore from a wrecked yacht in 1918 (Howarth and Ramsay, 1991). The impact of rodents is well illustrated by the unplanned natural experiment when, before 1964, rats established on Big South Cape Island, New Zealand. They used the mooring line of a fishing boat and devastated the biota, exterminating the large flightless weevil *Hadramphus stilborcarpae* Kuschel. It exists today only on the small rat-free islands (Howarth and Ramsay, 1991).

Continental faunal extinction rates may be higher than is thought, because the area to cover in searches is greater than on islands. Also, on continents the fragmentation of populations into subpopulations renders the species subject to future adverse impacts. This is the realm of island biogeography (e.g. Spellerberg, 1991), the basic tenets of which are no different in the South than in the North.

Howarth (1991) sees biocontrol agents as having a major impact on indigenous faunas, with mostly only circumstantial evidence for biocontrol being the prime extinction factor. Samways (1988) points out that biocontrol may have been synergistic with habitat destruction in causing insect extinctions, and that accidentally established predators may have been more harmful than deliberately introduced ones. Unquestionably, habitat destruction has been, and will continue to be, the greatest threat to insect species survival.

Weeds are rarely either particularly fragmenting to populations or the cause of extinctions. Indeed, Stewart (1993) has shown that the aquatic weed *Eichhornia crassipes* (Solms.) increases the local odonatan diversity in an African savanna river. But he also points out that this, and other aquatic plants, including indigenous ones, change the flow in rivers which, in turn, alters the general species assemblage.

H. Urbanization

Towns and cities of the South are of fairly recent origin compared with many of those of the Old-World North. The impact of a city is not merely the built up area but also the surrounding resource supply area. Many southern hemisphere towns have areas of relatively low human impact nearby. Quantitative data on the relationship between insects and town development are almost non-existent in the South compared with in the North (see Frankie and Ehler, 1978; Sukopp and Werner, 1982; Gilbert, 1989). This is an important and urgent research area, particularly as some taxa, such as the Eltham copper (*Paralucia pyrodiscus lucida* Crosby) near Melbourne in Australia (New, 1991), are known to be threatened by housing development. Similarly, the lycaenid *Argyrocupha malagrida malagrida* (Wallengren) appears to occur only in one small area the size of a tennis court in the Cape Peninsula. It is indirectly threatened by urbanization through fires during the flight period and by invasion of weeds (Henning and Henning, 1989).

These rich southern faunas are being spatially repatterned by the recent urban developments. In an urban botanic garden in South Africa, 821 species of macro-arthropods were sampled in 1500 m^2 within 1 week (Clark and Samways, 1994). Interestingly, a horticulturally landscaped site did not differ significantly in numbers of arthropod species from a nearby site with indigenous plants. However, the fragmentation enhanced the beta diversity, that is, the artificial patchiness did not so much impoverish the fauna as change its species composition. This work also illustrated the importance of scale when monitoring these faunal pattern changes. At the scale of the urban garden, cicindelid beetles were particularly sensitive to physical modification of biotopes, whereas butterflies generally had home ranges that were too great for them to be sufficiently sensitive indicators at this level of measurement.

I. Global Changes

Global climate change is synergistic with landscape modification and habitat destruction. Most discussions have focused on the North (e.g.

Peters and Lovejoy, 1992). The effects of global climate change upon ecosystems is unlikely to be linear, and as there is considerable natural variability, distinguishing trends from non-human effects will be difficult. Also, the various factors from enhanced warming to hyperacidic precipitation are together likely to compound the adverse effects.

Although there are various predictions that by the year 2030 the mean average global temperature will increase by 1–3°C, there is considerable uncertainty about rainfall patterns and rhythms. Theoretically, southern boundaries will move south by about 200–300 km. This is an order of magnitude faster than natural changes in the past, and so neither vegetation nor insects are likely to track this rate of change. This is particularly so as source areas of biotic populations have been reduced in size and also as human landscape modification would filter movements.

Similar problems face animals and plants when compensating for enhanced warming by shifts in elevation. As high mountains are among the least disturbed of ecosystems, some species may be able to move up 500 m in response to a 3°C change. This again is purely theoretical because ultraviolet radiation levels, wind exposure and changing precipitation patterns will all interplay. Further, as mountains are roughly conical, land surface is less with increasing elevation. Also, only foothill species benefit. Those living on the tops of mountains, such as the flightless lucanid stag beetles, *Colophon* spp., which inhabit the mountain peaks in the Cape, have nowhere to go. Of course it is a similar problem for those species that live on flat land, especially as mountains are also a proportionately scarce landscape type and may not be near enough to benefit a flat-land species.

The interrelations between various aspects of climate change are beyond this review, but there is one point that is especially relevant. The tips of the southern continents are relatively close to the south polar vortex. This is a system of strong west to east winds that encircle the polar cap, and prevent ozone-rich air from lower latitudes reaching the polar cap. As a result there is a build-up of ozone concentration outside the vortex, particularly about 50–60°S. At the polar cap, however, there is a hole (Scourfield *et al.*, 1990). Although there are marked seasonal changes in ozone levels, being lowest in March–April and highest in September–October, there was nevertheless an overall depletion in South Africa of 45% in 1989 (Scourfield *et al.*, 1990). Clearly, the Southern continents may bear the brunt of depleted ozone. The effect on insects is likely to be principally via the plants becoming increasingly stressed, especially from pigment destruction. This is inevitable, especially as Antarctic phytoplankton are already UV-stressed by these recent changes (Voytek, 1990).

IV. CONCLUSIONS AND FUTURE PERSPECTIVES

In the South, the human impact has been recent, intense and localized. This is not unlike North America, but quite different from much of Europe and Asia. In the Old World, the longer history of gradual landscape change has meant that insect population response and conservation relates more to managed landscapes, whether grassland or forest (e.g. Peterken, 1991; Thomas, 1991). Only in the remotest areas are biotopes still relatively intact (e.g. China (Anonymous, 1992)). Indeed, the concepts of conservation in the South are generally much more akin to those in North America, particularly the USA, than in Britain (cf. Henderson, 1992). In the South, the idea of "wilderness" still very much applies, with management being principally restricted to maintaining the naturalness of areas. This landscape approach is important as it is an umbrella for "protecting" the large number of species, with twice as many undescribed as described.

Evidence from some of the better known taxa suggests that many species in the South have much smaller geographical ranges than at the same latitude in the North. This is tantamount to an extension of Rapaport's Rule from the tropics to the South (Stevens, 1989). Narrow endemism (occurrence at one locality) and wide endemism (confined to a region) are particularly evident in the South. This is especially so for species of the mountains, forests and rivers. Savanna species are, or were before man, generally more widespread, as this is the matrix and large-scale movement corridor for long-term diffusive dispersal.

These points have an important bearing on responses of insects to the changing environment. It is the insects of the natural remnant forest patches and those restricted to particular topographic and aquatic sites that are under most pressure. The impacts on them are both proximal (localized human disturbance) and distal (changing atmospheric conditions), with various and varying synergism.

Landscapes of great taxic and geological interest need to be identified and the threat levels determined. At the same time, inventories and distribution maps need to be refined. The landscape umbrella approach is not mutually exclusive of the autecological Red Data Book assessments using sensitive taxa. For example, the lowland fynbos of the Cape has threats ranging from invasion by the exotic Argentine ant (*Iridomyrmex humilis* Mayr) through weeds such as Port Jackson (*Acacia saligna* (Labill.)) to fires, urban development, agroforestry and agricultural development. This emphasizes the value of wildlife preserves, which are particularly important for many insects. The two southwestern Cape damselflies *Ecchlorolestes peringueyi* and *E. nylephtha*, formerly thought to be extinct from a suite of impacts but rediscovered in 1993, are now considered to be relatively safe in wildlife sanctuaries.

Concepts such as hedgerow conservation (see Fry and Lonsdale, 1991) and conservation headlands (Dover, 1991), which are important in the North (especially Europe), are not of prime concern for insect conservation in the South. Of much greater significance is the preservation of pristine, or at least near-pristine, landscapes with their local indigenous faunas. This is not to ignore the findings from the North, which in the South, also can be used to enhance population levels of the more eurytopic, widespread species, and occasionally the localized, rare ones.

Yet, in the South, 'typicalness' (Usher, 1986) also needs to be conserved. For the South, this is the savannah. One African savanna lycaenid species (*Lepidochrysops hypopolia* Trimen) has not been seen since 1879 and is thought to be extinct. Many others are highly localized, such as *Lepidochrysops lotana* Swanepoel, discovered in 1959, yet not seen in recent years (Henning and Henning, 1989). A similar situation exists in Australia. Species in the moth genera *Synemon* and *Pterolocera*, in the Hesperiidae and in the acridoid grasshoppers, and which are associated with native grasses such as *Danthonia*, *Themeda* and *Stipa*, are threatened (Hill and Michaelis, 1988).

In some respects, the montane and forest endemic insects are less threatened than the narrow-range savanna species. The mountains inherently give protection by being of little agricultural and urban value, yet of high aesthetic appeal. These reasons together stimulate the creation of reserves. Similarly, forests are also now fairly well protected as they too are pleasing to the eye, conspicuous and rich in architecture and life. It is the insects of the lowlands, where there is good soil and least aesthetic appeal, that are most threatened.

No matter in which biomes these endemics occur, they are particularly susceptible to future changes for three interrelated reasons. They are not usually mobile at the regional scale; they are restricted to one area with particular edaphic, atmospheric and vegetational characteristics; and, in the event of major global climatic change, have neither the capacity to shift geographical locations nor the physiological tolerance or resilience to survive at the same location.

Indeed, it may be the synergistic effect of climate change with habitat destruction that will be the greatest cause of extinctions. Yet to date, there is little evidence (cf. Wells *et al.*, 1983; Hill and Michaelis; 1988; Henning and Henning 1989; Greenslade and New, 1991) that many species extinctions have occurred in the South. In South Africa, the only listed extinctions are two lycaenids. One of those, the riparian *Deloneura immoculata* Trimen appears to have become extinct naturally sometime soon after 1864, and this may also have been the case with the other species, the savanna lycaenid *L. hypopolia*. Careful searches and observations are necessary, as some species are so scarce (e.g. the lycaenid *Aslauga australis* Cottrell, the zygopteran *Chlorolestes apricans* Wilmot) that landscape modification and subsequent loss of the species may go unnoticed. Updating inventories is an important and urgent task, particularly for environmentally sensitive and conspicuous indicator species.

In summary, whereas the tropics is suffering devastating local and, in time, global impacts, the South is primarily threatened now by the global situation. However, generalizations are dangerous. The South needs to conserve landscapes, both typical and unique, and to establish which insects are then protected by this approach and which also are good monitors of change. It is imperative that we look at the interface between wild land and agricultural and urban development, as this is where the greatest stresses are, and where a rational voice for insect conservation is most needed.

V. SUMMARY

The southern hemisphere is rich in many endemic species. The land masses are proportionately much smaller in the South than in the North, making genetically unique biodiversity per unit area extremely high. Relative to the glaciation in the North, historically the climate has been fairly moderate. The absence of extreme climatic adversity, combined with varied topography appear to have generated considerable plant and insect diversity. Even quite small differences in elevation or slope are reflected by substantial species turnover. However, insect species richness on a world comparative basis does not track that of plants, which for the Cape Floristic Kingdom is almost the highest in the world.

Fire has been an important natural impact across the southern hemisphere. Man-made fires have intensified the impact, but conversely man's prevention of fire in some areas has dramatically influenced ecosystems and insect communities. Extensive changes to landscapes have come about through pastoralism and livestock rearing. Not only has this caused distinctive changes in vegetation character but has also affected abundances and ranges of certain insects. Arable farming covers only about one tenth of the land surface in southern Africa. This impact is localized and intense, impacting on certain narrow endemic insect species. Deforestation has been variably intensive across the South. As the natural forests are relatively small in extent, and are rich endemic sinks, their full protection is essential. Introduced forest patches cover only a small area, but are severe in impact, and a major threat to some localized endemic insect species. Water impoundments are widespread in the comparatively dry South. Although causing range infilling for many eurytopic species, these improvements have fragmented populations of stenotopic species. Some exotic invasive animals have caused population extinctions of several insects. In contrast, exotic plants although detrimental to several native plants, have impacted variably upon indigenous insect population levels. Urbanization has impacted severely on some localized endemic taxa. All the above impacts will probably be adversely synergistic with global changes in their adverse impacts. Thinning of the ozone shield is of particular concern in the South. It is an urgent task to monitor changes in population levels of some of these southern insects as the environmental impacts become increasingly acute locally and regionally.

Acknowledgements

Thanks to Dr Mike Laws for commenting on the manuscript. Financial support from the Royal Entomological Society, the Foundation for Research Development and the University of Natal Research Fund is gratefully acknowledged. Ms Pamela Sweet kindly processed the manuscript.

REFERENCES

Anonymous (1989). "Agriculture in South Africa", 4th edn. Chris van Rensburg Publications, Johannesburg.
Anonymous (1992). "Biodiversity in China: Status and Conservation Needs". Science Press, Beijing.
Brakefield, P. M. (1991). Genetics and the conservation of invertebrates. In "The Scientific Management of Temperate Communities for Conservation" (I. F. Spellerberg, F. B. Goldsmith and M. G. Morris, eds), pp. 45–79. Blackwell Scientific Publications, Oxford.
Booysen, P. de V. and Tainton, N. M. (eds) (1984). "Ecological Effects of Fire in South African Ecosystems". Springer-Verlag, Berlin.
Chambers, B. Q. (1993). "Influences of Burning and Mowing on Grasshopper (Orthoptera) Assemblages in the the Tall Grassveld of Natal". MSc thesis, University of Natal.
Clark, T. E. and Samways, M. J. (1994). Ecological landscaping for conservation of macro-arthropod diversity in a southern hemisphere (South African) urban botanic garden. In "Habitat Creation and Wildlife Conservation in Post-industrial and Urban Habitats" (J. Rieley and S. Page, eds), Packard, Chichester, England.
Cottrell, C. B. (1985). The absence of coevolutionary associations with Capensis Floral Element plants in the larval/plant relationships of Southwestern Cape butteflies. In "Species and Speciation" (E. S. Vrba, ed.), pp. 115–124. Transvaal Museum, Pretoria.

Cowling, R. (ed.) (1992). "The Ecology of Fynbos: Nutrients, Fire and Diversity". Oxford University Press, Cape Town.
Cowling, R. M., Gibbs Russell, G. E., Hoffman, M. T. and Hilton-Taylor, C. (1989). Patterns of plant species diversity in southern Africa. In "Biotic Diversity in Southern Africa: Concepts and Conservation" (B. J. Huntley, ed.), pp. 19–50. Oxford University Press, Cape Town.
Darlington, P. J. (1965). "Biogeography of the Southern End of the World". Harvard University Press, Cambridge, MA.
Davis, A. L. V. (1993). Alpha-diversity patterns of dung beetle assemblages (Coleoptera: Scarabaeidae, Aphodiidae, Staphylinidae, Histeridae, Hydrophilidae) in the winter rainfall region of South Africa. *Afr. Ent.* **1**, 67–80.
Davies, B. R. and Day, J. A. (eds) (1986). "The Biology and Conservation of South Africa's Vanishing Waters." The Centre for Extra-Mural Studies, University of Cape Town.
Donnelly, D. and Giliomee, J. H. (1985). Community structure of epigaeic ants in a pine plantation and in a newly burnt fynbos. *J. ent. Soc. sth Afr* **48**, 259–265.
Dover, J. W. (1991). The conservation of insects on arable farmland. In "The Conservation of Insects and their Habitats" (N. M. Collins and J. A. Thomas, eds), pp. 293–318. *15th Symposium of the Royal Entomological Society of London*. Academic Press, London.
Duxbury, A. C. and Duxbury, A. B. (1991). "An Introduction to the World's Oceans", 3rd edn. Wm. C. Brown, Dubuque, USA.
Eardley, C. D. (1989). Diversity and endemism of southern African bees. *Plant Protection News* **18**, 1–2.
Frankie, G. W. and Ehler, L. E. (1978). Ecology of insects in urban environments. *Ann. Rev. Entomol.* **23**, 367–287.
Fry, R. and Lonsdale, E. (eds) (1991). "Habitat Conservation for Insects – a Neglected Green Issue". Amateur Entomologists' Society, Middlesex, England.
Gaston, K. J. (1992). Regional numbers of insect and plant species. *Funct. Ecol.* **6**, 243–247.
Gaston, K. J., Reavey, D. and Valladares, G. R. (1992). Intimacy and fidelity: internal and external feeding by the British microlepidoptera. *Ecol. Entomol.* **17**, 86–88.
Gess, S. K. (1992). Biogeography of the masarine wasps (Hymenoptera: Vespidae: Masarinae), with particular emphasis on the southern African taxa and on correlations between masarine and forage plant distributions. *J. Biogeogr.* **19**, 491–503.
Gilbert, O. L. (1989). "The Ecology of Urban Habitats". Chapman & Hall, London.
Gill, A. M., Groves, R. H. and Noble, I. R. (eds) (1981). "Fire and the Australian Biota". Australian Academy of Science, Canberra.
Greenslade, P. J. M. (1983). Adversity selection and the habitat templet. *Am. Nat.* **122**, 352–365.
Greenslade, P. and New, T. R. (1991). Australia: Conservation of a continental insect fauna. In "The Conservation of Insects and their Habitats" (N. M. Collins and J. A. Thomas, eds), pp. 33–70. *15th Symposium of the Royal Entomological Society of London.* Academic Press, London.
Groombridge, B. (ed.) (1992). "Global Biodiversity: Status of the Earth's Living Resources". Chapman & Hall, London.
Harris, L. D. (1984). "The Fragmented Forest". The University of Chicago Press, Chicago.
Henderson, N. (1992). Wilderness and the nature conservation ideal: Britain, Canada, and the United States contrasted. *Ambio* **21**, 394–399.
Henning, S. F. and Henning, G. A. (1989). "South African Red Data Book – Butterflies". *South African National Scientific Programmes Report* **158**. Foundation for Research Development, Pretoria.

Hill, L. and Michaelis, F. B. (1988). "Conservation of Insects and Related Wildlife". *Occasional Paper* **13**. Australian National Parks and Wildlife Service, Canberra.

Hoffman, M. T. and Cowling, R. M. (1990). Vegetation change in the semi-arid eastern Karoo over the last 200 years: an expanding Karoo – fact of fiction? *S. Afr. J. Sci.* **86**, 286–294.

Howarth, F. G. (1991). Environmental impacts of classical biological control. *Ann. Rev. Entomol.* **36**, 485–509.

Howarth, F. G. and Ramsay, G. W. (1991). The conservation of island insects and their habitats. *In* "The Conservation of Insects and their Habitats" (N. M. Collins and J. A. Thomas, eds), pp. 71–107. *15th Symposium of the Royal Entomological Society of London.* Academic Press, London.

Jago, N. D. (1973). The genesis and nature of tropical forest and savanna grasshopper faunas, with special reference to Africa. *In* "Tropical Forest Ecosystems in Africa and South America: A Comparative Review" (B. J. Meggers, E. S. Ayensu and D. Duckworth, eds), pp. 187–196. Smithsonian Institution Press, Washington, DC.

Johnsen, P. (1987). The status of the South African Acridoidea s.l. (Orthoptera: Caelifera). *In* "Evolutionary Biology of Orthopteroid Insects" (B. M. Baccetti, ed.), pp. 293–295. Ellis Horwood, Chichester.

Lawes, M. J. (1990). The distribution of the samango monkey (*Cercopithecus mitis erythrachus*: Peters, 1852 and *Cercopithecus mitis labiatus*: I. Geoffroy, 1843) and forest history in southern Africa. *J. Biogeogr.* **17**, 669–680.

Lovegrove, B. G. (1993). "The Living Deserts of Southern Africa". Struik, Cape Town.

Mark, A. F. and McSweeney, G. D. (1990). Patterns of impoverishment in natural communities: case history studies in forest ecosystems – New Zealand. *In* "The Earth in Transition: Patterns and Processes of Biotic Impoverishment" (G. M. Woodwell, ed.), pp. 151–176. Cambridge University Press, Cambridge.

New, T. R. (1991). "Butterfly Conservation". Oxford University Press, Melbourne.

Norton, D. A. (1991). Scientific basis for the conservation management of New Zealand plant communities. *In* "The Scientific Management of Temperate Communities for Conservation" (I. F. Spellerberg, F. B. Goldsmith and M. G. Morris, eds), pp. 369–381. Blackwell Scientific Publications, Oxford.

O'Connor, T. G. (1993). The influence of rainfall and grazing on the demography of some African savanna grasses: a matrix modelling approach. *J. Appl. Ecol.* **30**, 119–132.

Partridge, T. C., Avery, D. M., Botha, G. A., Brink, J. S., Deacon, J., Herbert, R. S., Maud, R. R., Scholtz, A., Scott, L., Talma, A. S. and Vagel, J. C. (1990). Late Pleistocene and Holocene climatic change in southern Africa. *S. Afr. J. Sci.* **86**, 302–305.

Platnick, N. I. (1991). Patterns of biodiversity: tropical vs temperate. *J. Nat. Hist.* **25**, 1083–1088.

Peterken, G. F. (1991). Ecological issues in the management of woodland nature reserves. *In* "Scientific Management of Temperate Communities for Conservation" (I. F. Spellerberg, F. B. Goldsmith and M. G. Morris, eds), pp. 245–291. Blackwell Scientific Publications, Oxford.

Peters, R. L. and Lovejoy, T. E. (eds) (1992). "Global Warming and Biological Diversity". Yale University Press, New Haven.

Ragge, D. R. (1974). Class Insecta, Order Orthoptera: Tettigonioidea. *In* "Taxonomy of the Hexapoda" (W. G. H. Coaton, ed.), pp. 37–38. *Entomology Memoir* **38**. Department of Agriculture of the Union of South Africa, Pretoria.

Rentz, D. C. F. (1991). Orthoptera. *In* "The Insects of Australia", Vol. 1, pp. 369–393. Melbourne University Press, Melbourne.

Samways, M. J. (1986). Spatial distribution of Scirtothrips aurantii Faure (Thysanoptera:

Thripidae) and threshold level for one per cent damage on citrus fruit based on trapping with fluorescent yellow sticky traps. *Bull. entomol. Res.* **76**, 649–659.

Samways, M. J. (1988). Classical biological control and insect conservation: are they compatible? *Envir. Cons.* **15**, 349–354.

Samways, M. J. (1989). Farm dams as nature reserves for dragonflies (Odonata) at various altitudes in the Natal Drakensberg mountains, South Africa. *Biol. Cons.* **48**, 181–187.

Samways, M. J. (1990). Land forms and winter habitat refugia in the conservation of montane grasshoppers. *Cons. Biol.* **4**, 375–382.

Samways, M. J. (1992a). Some comparative insect conservation issues of north temperate, tropical and south temperate landscapes. *Agric. Ecosyst. Envir.* **40**, 137–154.

Samways, M. J. (1992b). Dragonfly conservation in South Africa: a biogeographical perspective. *Odonatologica* **21**, 165–180.

Samways, M. J. (1994). "Insect Conservation Biology". Chapman & Hall, London.

Samways, M. J. and Manicom, B. Q. (1983). Immigration, frequency distributions and dispersion patterns of the psyllid *Trioza erytreae* (Del Guercio) in a citrus orchard. *J. Appl. Ecol.* **20**, 463–472.

Samways, M. J. and Moore, S. D. (1991). Influence of exotic conifer patches on grasshopper (Orthoptera) assemblages in a grassland matrix at a recreational resort, Natal, South Africa. *Biol. Cons.* **57**, 205–219.

Scholtz, C. H. and Chown, S. (1993). Insect conservation and extensive agriculture: the savanna of southern Africa. *In* "Perspectives on Insect Conservation" (K. J. Gaston, T. R. New and M. J. Samways, eds), pp. 75–95. Intercept Press, Andover.

Scholtz, C. H. and Coles, K. S. (1991). Description of the larva of *Circellium bacchus* (Fabricius) (Coleoptera: Scarabaeidae: Scarabaeinae). *J. entomol. Soc. sth Afr.* **54**, 261–264.

Scourfield, M. W. J., Bodeker, G., Barker, M. D., Diab, R. D. and Salter, L. F. (1990). Ozone: the South African connection. *S. Afr. J. Sci.* **86**, 279–281.

Skarpe, G. (1991). Impact of grazing in savanna ecosystems. *Ambio* **20**, 351–356.

Specht, R. L. (1990). Changes in the eucalypt forests of Australia as a result of human disturbance. *In* "The Earth in Transition: Patterns and Processes of Biotic Impoverishment". (G. M. Woodwell, ed.), pp. 177–197. Cambridge University Press, Cambridge.

Spellerberg, I. F. (1991). Biogeographical basis of conservation. *In* "The Scientific Management of Temperate Communities for Conservation" (I. F. Spellerberg, F. B. Goldsmith and M. G. Morris, eds), pp. 293–322. Blackwell Scientific Publications, Oxford.

Stevens, G. C. (1989). The latitudinal gradient in geographical range: how so many species coexist in the tropics. *Am. Nat.* **133**, 240–256.

Stewart, D. A. B. (1993). "Dragonfly Assemblage Composition Relative to Local Environmental Conditions of the Southern Rivers of the Kruger National Park". Unpublished MSc thesis, University of Natal.

Sukopp, H. and Werner, P. (1982). "Nature in Cities". Council of Europe, Strasbourg.

Thomas, J. A. (1991). Rare species conservation: case studies of European butterflies. *In* "The Scienfitic Management of Temperate Communities for Conservation" (I. F. Spellerberg, F. B. Goldsmith and M. G. Morris, eds), pp. 149–197. Blackwell Scientific Publications, Oxford.

Trabaud, L. (ed.) (1987). "The Role of Fire in Ecological Systems". SPB Academic Publishing, The Hague.

Tyson, P. D. (1986). "Climatic Change and Variability in Southern Africa". Oxford University Press, Cape Town.

Usher, M. B. (1986). Wildlife conservation evaluation: attributes, criteria and values. *In*

"Wildlife Conservation Evaluation" (M. B. Usher, ed.), pp. 3–44. Chapman & Hall, London.

Voytek, M. A. (1990). Addressing the biological effects of decreased ozone on the Antarctic environment. *Ambio* **19**, 52–61.

Wells, S. M., Pyle, R. M. and Collins, N. M. (1983). "The IUCN Invertebrate Red Data Book". IUCN, Gland, Switzerland.

Wilson, E. O. (1992). "The Diversity of Life". The Belknap Press of Harvard University, MA.

Wood, P. A. and Samways, M. J. (1991). Landscape element pattern and continuity of butterfly flight paths in an ecologically landscaped botanic garden, Natal, South Africa. *Biol. Cons.* **58**, 149–166.

Wright, M. G. (1993). Insect conservation in the African Cape Fynbos, with special reference to endophagous insects. *In* "Perspectives on Insect Conservation" (K. J. Gaston, T. R. New and M. J. Samways, eds), pp. 97–110. Intercept Press, Andover.

14

Species Extinctions in Insects: Ecological and Biogeographical Considerations

NICK A. MAWDSLEY AND NIGEL E. STORK

I. Introduction	322
II. Insect Extinctions: Present State of Knowledge	326
III. Extinction: Defining the Parameters	329
A. Extinctions and the Evolutionary Past	329
B. Extinctions in Ecological Time: Local and Global Extinction	330
IV. The External Threats Affecting Extinction in Insects	335
A. Theoretical Aspects and Ultimate Causes	335
B. The External Threats of Extinction for Insects	336
C. The Threats and Habitats of British Beetles and Butterflies	337
V. The Intrinsic Vulnerability of Insects to Extinction	341
A. Ecological Correlates of Extinction in Insects	341
B. Ecological Factors and the RDB Status of British Butterflies	346
C. Phylogenetic Aspects of RDB Status in British Beetles	348
VI. Extinction Rates, Community Vulnerability and Biogeography	351
A. Community Vulnerability	351
B. Predicting the Geography of Extinctions	353
VII. Relative Extinction Rates of Invertebrates and Vertebrates: Quantifying Global Species Extinctions in Insects	354
VIII. Extinction and Co-extinction	358
IX. Concluding Remarks	359
References	361

Abstract

Extinction, the irreversible loss of a species or population, is perhaps the most serious of all the effects of the human population upon the natural world. The threat of species extinction forms the prime motivation for conservation resulting in action to protect habitats and species and which ultimately requires a thorough understanding of the extinction process and patterns of extinction between taxa in different areas and habitats. But although study of these factors and patterns is a relatively recent endeavour, little is now understood about the important parameters that relate to the probability of extinction for various taxa. Here we review current understanding of the extinction process from both an evolutionary and an ecological viewpoint and analyse the threats and intrinsic characteristics that relate to the probability of extinction in insects, specifically British beetles and butterflies.

The threats and habitats of butterflies and beetles differ to the extent that effective conservation of both will require different but co-ordinated strategies. We note that despite these differences, the proportion of each of these groups classified under the various Red Data Book (RDB) categories is similar. Characterization of the intrinsic traits of British butterflies that are correlated with extinction probability (measured in terms of RDB status) shows no clear pattern although voltinism, vagility, feeding specificity, larval growth, over-wintering stage and habitat specificity, all show significant correlations with distribution. It is not clear how distribution and range size relate to the probability of extinction. Furthermore, the observation that species-poor beetle families tend to have a higher proportion of species listed in the Red Data Book suggests a phylogenetic component to extinction.

We use data on the threatened and endangered status of a number of vertebrate and invertebrate taxa to estimate a relative extinction rate between insect and vertebrate taxa. This suggests that broadly speaking the probability of extinction of British birds is of the order of seven times that of British insects. All things being equal, it is likely that tropical insect faunas will show higher levels of extinction following disturbance. Empirical data on population trends and distribution is required on a regional basis to account for the effects of past history and the present and future patterns of land use in different regions thereby allowing predictions of the insect faunas of the future to be made.

I. INTRODUCTION

That the world is in the process of unprecedented change is beyond doubt. Human activities, driven by an exponentially increasing human population, are destroying and degrading many of the world's natural wilderness areas and other wildlife habitats. The combined effects of

14. Species Extinctions in Insects

habitat loss and fragmentation, pollution, climate change and the introduction of exotic species represent a potentially devastating threat to the world's biological diversity. Indeed, some suggest that we are already witnessing a mass extinction of species comparable to any in the evolutionary past (Ehrlich and Wilson, 1991; Soulé, 1991; Myers, 1987, 1993). Biological degradation is a feature common to all nations, yet some consider that there is insufficient scientific evidence to support notions of mass extinction (Mann, 1991; Brown and Brown, 1992). Certainly there is great variation between current estimates for future rates of species extinction (Table 1, after Reid, 1992; Smith et al., 1993a, Mace, 1994), and an obvious lack of empirical data. The estimation of species extinction rates is inevitably open to the difficulties of predicting a global phenomenon on very limited data (Smith et al., 1993a,b). Nevertheless, these estimates are a potent way for biologists to convey to the rest of society the state of the natural world.

Current estimates of global species richness are in the order of 5–15 million species (Gaston, 1991; Hammond, 1992; May, 1992; Stork, 1993). With as many as 80% of these being insects, potentially the greatest number of extinctions *may* come from this group and from other megadiverse groups such as the terrestrial nematodes, marine invertebrates and fungi (Hawksworth, 1991, 1992). In the short term it is likely that large-bodied taxa such as birds and mammals are the most seriously threatened (Shaffer, 1981, 1987; Belovsky, 1987; Pimm, 1991), but despite their undoubted importance as conservation flagships, it would seem unwise to simply use vertebrates as umbrella species for invertebrates (Janzen, 1987; Hafernik 1992; Prendergast et al., 1993). The theory and practice of species conservation now includes objective ways of prioritizing taxa for conservation that emphasize the full phylogenetic diversity of life (Vane-Wright et al., 1991; Pressey et al., 1993; Forey et al., 1994). Simply stated, the conservation of invertebrate life should rank alongside vertebrate conservation.

There are several major problems associated with targeting invertebrates for conservation and extinction assessment. Much of the existing theoretical framework for investigating species extinctions is based on studies of large vertebrates and plants (e.g. Pimm et al., 1988; Laurance, 1991; Newmark, 1991; McDonald and Brown, 1992, Tracy and George, 1992; Gustafsson, 1994) but the basic distributions of invertebrates are very different from those of vertebrate and plant taxa (Prendergast et al., 1993, but see Abbott, 1974; Dinerstein and Wikramanayake, 1993), with many rare insects having highly specific microhabitat requirements (Hammond and Harding, 1991; Thomas, J. A., 1991). Changes in plant and insect diversity during succession, with peak diversity at mid-

Table 1. Estimated rates of extinction ([1]after Reid, 1992; Smith et al., 1993a; Mace, 1994). Note that the estimated rates include species "committed" to extinction (see text and Heywood et al., 1994)

Estimate	% global loss per decade	Method of estimation	Reference
One million species between 1975 and 2000	4	Extrapolation of past exponentially increasing trend	Myers (1979)[1]
15–20% of species between 1980 and 2000	8–11	Estimated species–area curve; forest loss based on Global 2000 projections	Lovejoy (1980)[1]
50% of species by 2000 or soon after – 100% by 2010–2025	20–30	Various assumptions	Ehrlich and Ehrlich (1981)
9% extinction by 2000	7–8	Estimates based on Lovejoy's calculations using Lanly's (1982) estimates of forest loss	Lugo (1988)
12% of plant species in neotropics, 15% of bird species in Amazon basin	–	Species-area curve ($z = 0.25$)	Simberloff (1986)[1]
2000 plant species per year in tropics and subtropics	8	Loss of half the species in area likely to be deforested by 2015	Raven (1987)[1]
25% of species between 1985 and 2015	9	As above	Raven (1988a,b)[1]
At least 7% of plant species	7	Half of species lost over next decade in 10 "hot spots" covering 3.5% of forest area	Myers (1988)[1]
0.2–0.3% per year	2–6	Half of rain forest species assumed lost in tropical rain forests to be local endemics and becoming extinct with forest loss	Wilson (1988, 1989, 1993)

Table 1. – contd.

Estimate	% global loss per decade	Method of estimation	Reference
2–13% loss between 1990 and 2015	1–5	Species-area curve ($0.15 < z < 0.35$); range includes current rate of forest loss and 50% increase	Reid (1992)
Red Data Books: selected taxa: 50% extinct in 50–100 years (palms), 300–400 years (birds and mammals)	1–10	Extrapolating current recorded extinction rates and by the dynamics of threatened status	Smith et al. (1993)
Red Data Books: selected vertebrate taxa	0.6–5	Fitting of exponential extinction functions based on IUCN categories of threat	Mace (1994)

successional and climax phases, respectively (Southwood et al., 1979), illustrate the marked differences between plants and insects in their interactions with their habitat. The responses of vertebrate and insect communities to anthropogenic perturbations will, no doubt, reflect the inherent differences in their community structure and regulation (Schoener, 1986), and the scale at which they interact with the environment.

Another obstacle to the study of insect extinctions is the general negative perception of insects as undesirable pests and "creepy-crawlies" (Kim, 1993; Kellert, 1993), so that the total scientific effort expended on insects compared to other groups, particularly megavertebrates, far belies their importance (Gaston and May, 1992). The result is that insects, arguably the most successful group of animals that have ever existed, are also one of the most poorly known terrestrial groups, with a great wealth of natural history and new species awaiting to be discovered and named (Wheeler, 1990; Soulé, 1990; Gaston and Mound, 1993). An understanding of the threat of extinction within these groups is greatly hindered by the paucity of empirical data relating to both their geographical distributions and general biology. Insects display many different life styles and perform a great variety of critical roles in ecosystems from pollination and seed predation to simply being links in the trophic web (Janzen, 1987; Wilson, 1987; Miller, 1993).

The extinctions of species is basically the antithesis of conservation, and attempts at insect conservation have shown the need for detailed autecological information (Ehrlich and Murphy, 1987; Brussard, 1991; Thomas, J. A., 1984, 1991; Warren, 1991). A corollary of this is that predicting the extinction of specific insects is immensely difficult (Thomas and Morris, 1994). Despite this there may be some general ecological principles that we can apply in identifying the taxa (and their geographical locations) which are most at risk (Lawton, 1993, 1994). To date, assessment of the overall risk to the global insect fauna has been achieved only by linking species extinction to somewhat general parameters such as reduction in the area of natural habitat (Table 1; Boecklen and Simberloff, 1986; Simberloff, 1992) driven ultimately by growth in the human population. While these provide us with a useful notion of the scale of the problem of global extinction, they do not provide much of an understanding of the patterns, causes and consequences of extinction in insects, especially at the scale of regions and landscapes (but see Kangas, 1991).

Here we examine how real the impending threat of extinction is for most insects. We are less concerned with the fundamentally important *absolute extinction rates*, for this is heavily dependent upon another uncertain quantity – the world's total number of species, but rather how *relative extinction rates* vary both between different taxa and geographically, and what factors govern the susceptibility of different taxa to extinction. We ask:

1. What are the threats causing the decline and extinction of insects?
2. How do these threats vary between insect taxa?
3. What are the biological and ecological features of insect species susceptible to extinction?
4. How might the extinction rate vary geographically over the globe?
5. How does the relative rate for insects compare with other taxa such as birds and mammals?

II. INSECT EXTINCTIONS: PRESENT STATE OF KNOWLEDGE

Global extinction of a species is immensely difficult to ascertain (Diamond, 1987; Ehrlich and Wilson, 1991). A species is officially accepted as having become extinct when no living individuals have been

14. Species Extinctions in Insects

Table 2. Comparison of the number of species recorded extinct since 1600 for mammals, birds and insects. (Data from WCMC (1992)

Taxa	Number of species recorded extinct since 1600	Global species total	Percentage (%) of fauna extinct
Mammals	59	4327	1.4
Birds	116	9672	1.2
Insects	61	10^6–10^7	0.006–0.0006

seen for 50 years (IUCN, 1990). This and other problems associated with documenting extinctions limit our absolute knowledge of species extinctions in all plant and animal groups. Even large and seemingly obvious birds and mammals apparently can be overlooked by biologists (Diamond, 1985). For example, in 1992 three species of primates were discovered in Brazil (Ferrari and Lopes, 1992; Mittermeier et al., 1992; Queiroz, 1992), and a new species of bovid was found in Vietnam (MacKinnon, 1993). If it is so difficult to discover these large species, then confirming when the last individual of a threatened insect species has disappeared is virtually an impossible task. For example, of the 1 million described species of insects, we really know next to nothing about their geographical distributions. About 400 000 of these are beetles, and one study has indicated that perhaps 40% of beetle species are known from just single localities (Stork and Hine, unpublished). Consequently, empirical studies of extinction in insects are generally only feasible at the local level for most taxa.

Not surprisingly only 61 species of insects (less than 0.01% of described species) are recorded as having become extinct since 1600 (Table 2) compared to 116 birds, (1.2% of 9672 described species) and 59 mammal species (1.4% of 4327). (WCMC, 1992; Smith et al., 1993a,b). This difference of two orders of magnitude is likely to have as a significant component the inequality in recording effort between insects and vertebrates at a global level. Undoubtedly predictions of absolute extinction rates in insects would be greatly improved by having better understanding of how the threatened status of insects and vertebrates compare (Section VII).

The geographical and taxonomic distribution of recorded extinctions reflects more the distribution of entomologists than extinct species (e.g. Gaston and May, 1992). Of the 61 species of insect recorded as extinct since 1600, about half are Lepidoptera, perhaps the most well known of the large insect orders (Table 3). In addition, most of the extinct species

Table 3. Summary of recorded species extinctions for insects: (a) the distribution of recorded extinctions amongst the insect orders, and between islands and continents, and (b) its geographical distribution (data from WCMC, 1992). Note: [a]the recorded extinction of two species on St Helena, *Labidura herculeana* (Dermaptera) and *Aplothorax burchelli* (Coleoptera) may be premature

(a)

Order	Number of species	Island	Continent
Ephemeroptera	2	–	2
Orthoptera	1	–	1
Phasmoptera	1	1	–
Dermaptera	1[a]	1[a]	–
Plecoptera	1	–	1
Homoptera	2	2	–
Coleoptera	10[a]	10[a]	–
Diptera	3	2	1
Trichoptera	4	–	4
Lepidoptera	33	32	1
Hymenoptera	3	3	–
Total	61	51	10

(b)

Islands/continents	Number of species
N. America (USA)	9
Europe (Germany)	1
New Zealand	1
Chatham Is. (NZ)	1
Stephen Is. (NZ)	1
Lord Howe Is. (Australia)	1
Fiji	1
Taiwan	1
Mauritius	1
St Helena	2[a]
Hawaii	42
Total	61

recorded are from islands, with only two species being recorded as extinct from continental regions. This probably reflects both the increased vulnerability of island populations to extinction (MacArthur and Wilson, 1967; Williamson, 1981) coupled with the greater ease in surveying islands. Comparison with the geographical distribution of extinct species of birds and mammals around the world (e.g. see figs 16.7, 16.9 and 16.10 in WCMC, 1992), shows that this is typical of other animal groups.

14. Species Extinctions in Insects   329

However, atypical is the relatively high proportion of extinct species (41 of a total of 61 species) from Hawaii compared to the proportion from other islands. Many other extinctions have no doubt gone unrecorded from other equally disturbed islands for which there is no knowledge of the fauna prior to human colonization (Smith et al., 1993b; Pimm et al., 1994).

Information on the cause of the demise of the 61 recorded insect extinctions is variable and thus forms a poor dataset upon which we can derive an analysis for insects in general. Much more data are available on the threatened status of insects, at least from some well-defined geographical areas such as the British Isles; these data are explored later in this chapter.

In summary, direct measurement of global extinction for most insect species is impossible. We see at least three approaches that may be useful (a) the study of anthropogenic rarity and local extinction and its relation to global extinction; (b) assessment of the threatened status of species as a surrogate for extinction (Smith et al., 1993a, Mace, 1994); (c) assessment of the relative extinction rates in vertebrate and invertebrate taxa.

III. EXTINCTION: DEFINING THE PARAMETERS

A. Extinctions and the Evolutionary Past

The extinction of species is a natural evolutionary process that can be traced throughout geological time for groups that are well preserved as fossils. Most taxa that have ever existed are now extinct, yet for many taxa including insects and plants, a higher level of speciation has resulted in diversity being greater now than at any other period in the geological past (Benton, 1986; Valentine, 1990).

Study of the fossil record has revealed that the extinction rate has varied greatly through time. At five points in the geological past major peaks have occurred in the extinction rate of a wide range of taxa, representing so called "mass extinction" event (see figs 1–2 in Raup and Sepkoski, 1982). For example, 245 million years ago at the end of the Permian, estimates point to between a 77 and 96% extinction of species (Jablonski, 1986), and similarly the end of the Cretaceous also saw a 60–80% reduction in global species richness (Jablonski, 1991). These events were usually followed by increased speciation rates (Allmon et al., 1993) with recovery taking several million years (Jackson, 1994). In spite of the scale of these catastrophic events, Jablonski (1994) suggests that about 90% of extinctions may have occurred outside these periods of mass extinctions. Indeed, other authors have questioned the existence and scale of mass extinction events (Patterson and Smith, 1989).

The characteristics of a mass extinction event are that it affects a wide range of taxa over a large area and a relatively short timescale. However, in each mass extinction the taxa involved appear to have been somewhat different and insects only seem to have been particularly affected at the end of the Permian (245 MYBP) (Labandeira and Sepkoski, 1993). The evidence for insects and plants is relatively poor, but although plants appear to have been affected by the end-Cretaceous event, the concentration of its effects at higher latitudes is thought to have limited the impact upon insects because of their inherent adaptations to a seasonal climate (Briggs *et al.*, 1988; Coope, Chapter 2, this volume). Interpretation of a patchy fossil record makes it extremely difficult to make absolute comparisons, especially for assessing geographical variation in extinction rates. For example, several authors have proposed that the Cretaceous extinction event had a latitudinal trend with extinction being more severe in the tropics (e.g. Jablonski, 1986). However, recent analysis of fossil bivalve faunas suggests that the end-Cretaceous event was globally uniform, and dependent upon the taxa analysed (Raup and Jablonski, 1993). Other studies indicate that latitudinal trends in species richness in evolutionary terms may be solely due to higher speciation rates in the tropics (Rohde, 1992; Jablonski, 1993), and, importantly, that extinction rates have no latitudinal pattern (Raup and Jablonski, 1993). These points need to be explored further with contemporary biotas because of their importance in predicting the future for the species-rich tropics.

A final conclusion reached from the study of the fossil record is that the species richness *per se* of a taxon does not directly affect its vulnerability to extinction. Instead the correlation of traits within species-rich clades, and of endemism with species richness is thought to disadvantage species-rich taxa, making the probability of extinction of all their species relatively higher (Jablonski, 1986; Vermeij, 1986). However, a recent study of insect families in the fossil record by Labandeira and Sepkoski (1993) concluded that, in contrast, the great diversity of insects is attributable to low extinction rates of families in evolutionary time relative to tetrapods and bivalves. Excluding the effect of differing taxonomic concepts between these three groups, this result suggests that insect taxa are more resistant to extinction than these other groups.

B. Extinctions in Ecological Time: Local and Global Extinction

In their review of the causes of extinction, Terborgh and Winter (1980) considered this field so new that they could not suggest any general reviews of the subject. Yet, paradoxically, the extinction of local populations forms part of one of the most influential theories in ecology. MacArthur and Wilson's (1967) equilibrium theory of island biogeography identified extinction rates on islands to be dependent upon population size and island area despite a lack of empirical evidence to support this (Williamson, 1981). Furthermore, the distance of an island from a source of immigrants has also been included as an important determinant of extinction rates through the "rescue effect" (Brown and Kodric-Brown, 1977). The variables, "area" and "distance" describe the fragmentation of habitats, and these have been central to how ecologists have approached the question of extinction, both in term of species–area relationships (Diamond and May, 1976; Boecklen and Simberloff, 1986; Simberloff, 1992) and metapopulation dynamics (Levins, 1970; Hanski, 1989; Gilpin and Hanski, 1991; Gotelli and Kelley, 1993).

At the population level the effect of fragmentation is initially to convert large, essentially closed populations, into metapopulations composed of smaller populations linked by dispersal. In time, further habitat loss and isolation of these populations will result in the breakdown of metapopulation structure, and local populations will become closed, greatly increasing the risk of local and global extinction (Harrison, 1991). Central to predicting extinction rates is understanding whether natural populations exist as metapopulations whose local populations have an equal probability of extinction (the Levins type model) or as small closed populations. Metapopulations, and in time, isolated closed populations, will become much more prevalent in the fragmented landscapes of the world with high human population densities.

One criticism of MacArthur and Wilson's theory which, in turn, relates to the predictive power of area and distance concerning extinction rates, is that it does not take into account the nature and quality of habitats that are found in an area (Williamson, 1981). Undoubtedly these factors will affect extinction rates, giving high rates for habitat-specialists whose habitats are relatively rare (Pimm, 1991). This is emphasized by Thomas, J. A. (1991), who highlights the importance of the quality of the habitat in preserving populations of rare butterflies in Britain. Here, as in much of Europe, the majority of species have shown dramatic declines; in Suffolk, for example, 42% of species have been lost since about 1850 (Thomas, J. A. 1991). Many of the declines are of species whose larvae are specialized on early short-lived grassland seral stages that often correspond to a warm microclimate. Extinction and decline in these butterflies seems to be a result of being highly specialized on a dynamic microhabitat whose continued presence at a site is dependent upon the disturbance associated with traditional management practices. The cessation of these management practices has resulted in a lowered carrying capacity of the habitat and a lack of spatial and temporal continuity of sites; coupled with the limited dispersal ability of many butterfly species, decline then appears inevitable (Thomas and Morris, 1994).

1. Local population extinction

Metapopulations may typify the spatial population structure of many species of insect, especially those dependent on patchy resources such as the butterflies (Ehrlich and Murphy, 1987; Harrison et al., 1988; Hafernik, 1992; Thomas and Jones, 1993). But perhaps more importantly, local population extinctions appear to be a relatively common occurrence especially in small habitat patches (Williamson, 1981; McCauley, 1989; Harrison, 1991; Hanski and Gilpin, 1991; Thomas, J. A., 1991; Pollard and Yates, 1992; Warren, 1993). For example, of 2248 colonies of the 29 key nationally and regionally rare or declining butterfly species in southern England, 384 (17%) are now recorded as extinct, and mostly in the last 40 years (Warren, 1993). Similar levels of local extinction have been recorded over even shorter timescales for rare and threatened species such as the silver-spotted skipper, *Hesperia comma* L. For this species, local extinction occurred on 17% of habitat patches on the North and South Downs in just a 9-year period. However, this was more than compensated for by the new colonization of vacant habitat patches (Thomas and Jones, 1993).

Thomas and Jones' (1993) study highlights an important feature of metapopulations in that the probability of local extinction is related not only to the area of the local habitat patch and distance to the nearest source of colonists, but also to the distance of unoccupied suitable habitat patches (see also Thomas et al., 1992). Extinction of *H. comma* is made more probable if there is a loss of suitable habitat independent of whether it supports a population or not. These unoccupied habitat patches may be colonized only in "good years" but nevertheless these "sink populations" may be crucial to the size and persistence of metapopulations (Howe et al., 1991). Local extinctions will be much more common in sink populations than core populations, especially in populations with a core-satellite metapopulation structure (Harrison, 1991). Extinction of local satellite (i.e. sink) populations may represent a natural equilibrium, and the probability of global extinction will crucially be determined by the persistence of the core (i.e. source) populations.

Most species of butterfly have low dispersal rates and, indeed, 80% are classified as living in closed populations (Thomas, J. A., 1991). The isolation of local populations, common to species in decline as habitat fragmentation intensifies, is a non-equilibrium situation, and local extinction of these populations is unlikely to be reversed by recolonization. By acting as stepping stones for dispersal, the spatial configuration of suitable habitat (occupied *and* unoccupied) and the ecology of dispersal of species act as key factors in governing the extinction or persistence of species; for *H. comma* a distance of about 10 km between habitat patches is enough to limit its spread (Thomas and Jones, 1993). For other species such as *P. argus*, *Mellicta athalia* (heath fritillary) and *Thymelicus acteon* (Lulworth skipper) smaller distances down to 1 km or so appear enough to limit their colonization (Thomas *et al.*, 1992). Large-scale patterns of habitat alteration, landscape dynamics and species vagility in fragmented landscapes (Welch, 1990) are therefore critical factors in assessing extinction probabilities.

2. Global population extinctions

Global extinction of a species occurs when the last population disappears, and therefore requires either: (a) extinction processes to act synchronously throughout the core range of a species; or (b) in a non-equilibrium metapopulation in decline, a lack of recolonization may result in global extinction following sequential local population extinction over a long period of time. The former process reflects regional environmental effects, such as weather patterns, food availability, large-scale habitat loss, etc., whereas the latter emphasizes the phenomenon of "latent extinction" (Sutton and Collins, 1991). Described by Janzen (1986) as the "living dead", these are species in decline which persist as long-lived individuals or vulnerable populations but which have an extremely low probability of long-term persistence. We have no idea of what proportion of the world's species may be in such a position. A worrying indicator is that the extinction rate of colonies of Scarce/Red Data Book (RDB) British butterflies appears to be as great on protected as on unprotected sites (Warren, 1993), perhaps reflecting the process of latent extinction. Current estimates of extinction rates correctly include species that are "committed" to extinction as extinct because, although the final extinction event has not yet taken place, the probability of their survival is extremely low (Heywood *et al.*, 1994).

Recent work on rates of extinction of local populations and the spatial correlation of population dynamics between local populations gives some indication of the spatial variation in the probability of extinction. Of the 29 species of butterfly assessed by Warren (1993), over two-fifths had local extinction rates of less than 10% of colonies over a period of 40 years or so, and the frequency distribution of the prevalence of local

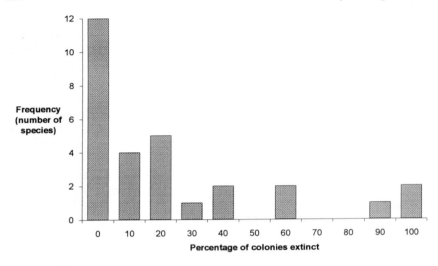

Fig. 1. Frequency distribution of local extinction for 29 species of rare and declining butterfly species in southern England (data from Warren, 1993). Despite high overall levels of local extinction in butterfly colonies (local populations), the majority of local extinctions are concentrated in a few species such as *Nymphalis polychloros* (large tortoiseshell – RDB1): 100% of colonies; *Carterocephalus palaemon* (chequered skipper – RDB4): 100% of colonies; *Argynnis adippe* (high brown fritillary – RDB2): 98% of colonies; *Hesperia comma* (silver-spotted skipper – RDB3): 63% of colonies; and *Boloria euphrosyne* (pearl-bordered fritillary – Scarce): 63% of colonies.

extinction within species is skewed to the right (Fig. 1). Therefore, although local extinction was relatively common, occurring in nearly one-fifth of all colonies, it tends to be concentrated in five species (17%) which have suffered losses of over 50% of their local populations. Clearly, in most species the probability of extinction varies greatly between their populations, but a considerable proportion of species have lost many local populations. This may well be because general habitat changes (e.g. cessation of grazing) are equally likely to occur in different parts of a species' distribution and are therefore equally likely to affect populations. Furthermore, Hanski and Woiwod (1993) found a high degree of spatial synchrony in the local population dynamics of a large number of aphid and moth species between localities in Britain separated by up to 800 km. Similarly abundances and ranges of butterflies are spatially correlated between Britain and The Netherlands (Pollard *et al.*, 1993). Such spatial correlation of population dynamics and local extinctions will increase the probability of species extinction by reducing the persistence times of metapopulations (Quinn and Hastings, 1987; Harrison and Quinn, 1989; Hanski, 1989), and will also affect closed populations in a similar way by reducing the benefits of "risk-spreading" against environmental stochasticity (den Boer, 1981, 1990). Appreciation of these considerations highlights the importance of local action in globally planned efforts to preserve biological diversity.

IV. THE EXTERNAL THREATS AFFECTING EXTINCTION IN INSECTS

The extinction of a species is a multifaceted process – the product of a multitude of direct and indirect factors which act on a species' population dynamics to maintain the death rate higher than the birth rate. These factors may be grouped into "extrinsic" abiotic and biotic factors and "intrinsic" species-specific factors (WCMC, 1992).

A. Theoretical Aspects and Ultimate Causes

The extrinsic threats to a species' persistence come from the numerous abiotic and biotic forces acting upon its population dynamics, and may be deterministic or stochastic in nature (Gilpin and Soulé, 1986). Deterministic factors act in a quantifiable and predictable way. For example, predation from introduced rats on islands around New Zealand has brought about the probable global extinction of the Pitt Island longhorn beetle (*Xylotoles costatus* Pascoe) and Banks Peninsula speargrass weevil (*Hadramphus tuberculatus* Pascoe) (Howarth and Ramsay, 1991), and a lack of overlap in the occurrence of the ant *Myrmica sabuleti* Meinert and wild thyme has resulted in the extinction of the large blue butterfly *Maculinea arion* in Britain (Thomas, 1976). Ultimately though, it seems that the most pervasive deterministic threat to species is the loss and fragmentation of habitat (Soulé, 1986).

Stochastic factors, in contrast, are unpredictable random events and so are difficult to guard against. Conservation biologists have defined four stochastic factors that dominate the dynamics of small populations. Environmental stochasticity, or changes in the environment related to the weather or regional climate, and its extreme manifestation as catastrophe, act in concert with deterministic factors to reduce population sizes to the low levels where demographic and genetic stochasticity may act. These two factors represent the effects of random birth and death processes and inbreeding respectively, which may ultimately drive a species to extinction (Gilpin and Soulé, 1986; Soulé, 1987). Insect populations are susceptible to extinction from environmental stochasticity, but it is difficult to generalize about their response to extreme conditions (Ehrlich et al., 1980). Despite a rich theory concerning the effects of inbreeding upon small populations of insects, empirical studies are lacking. However, a loss of heterozygosity has been recorded in small isolated populations of spittlebugs *Philaenus spumarius* (Homoptera: Cercopidae) on the Isles of Scilly (Brakefield, 1991). Although genetic effects might be expected to be seen in populations of 10–30 individuals, these are only expected to be of significance if sustained over many generations, and in such cases would probably be secondary to specific ecological or management factors (Brakefield, 1991).

The importance of the deterministic and stochastic factors, representing a variety of external threats to the survival of species, has only been analysed through theoretical modelling in terms of "minimum viable populations" (e.g. Goodman, 1987). They have yet to be verified empirically (Brussard, 1991). Although, anthropogenic threats represent only one set of extinction factors, they are often the proximate causes of extinction (Terborgh and Winter, 1980) and as such should demand the most attention from biologists.

B. The External Threats of Extinction for Insects

Hafernik (1992) has considered the extinction of insects in the United States in terms of Diamond's (1989) "evil quartet" of mechanisms responsible for most extinctions: (a) habitat destruction and fragmentation; (b) introduced species; (c) chains of extinctions; and (d) overkill (over-collecting). In the United States, the rate of recent extinctions is roughly parallel to the pattern of human population growth (Hafernik, 1992). The conversion of natural habitat for urban development, agriculture and recreation has not only resulted in extinctions from the direct area and edge effects of habitat loss (e.g. the satyr butterfly *Cercyonis sthenele sthenele*, endemic to the San Fransisco Peninsula), but also from the less obvious collapse of metapopulations. In almost all cases the "evil four" do not act alone, and a species' population may be reduced by a combination of these with the potential for synergistic interactions (Myers, 1987). Anthropogenic effects on invertebrate assemblages are thus the creation of both a new abiotic and biotic environment for any given species, and, in particular, an environment and disturbance regime to which it may not be evolutionarily adapted (Connell, 1978; Jackson, 1994).

C. The Threats and Habitats of British Beetles and Butterflies

The British insect fauna is better recorded than that of any other nation. Data on the threatened status, habitats and threats of two contrasting groups of British insects – the beetles and butterflies – are included in Shirt (1987), who covered 13 746 of the estimated 22 000–23 000 British insect species, and Hyman and Parsons (1992), who assessed the status and habitats of all species in 71 of the 98 families of British beetles. The latter volume records those species of Notable and Red Data Book (RDB) status (hereafter known as "listed beetles"), providing a database of about 1043 species – roughly one-third of all species of the 71 families considered. These data are used here with RDB status being considered as a surrogate for extinction probability in order to investigate future threats and expected patterns of extinction (Smith et al., 1993a; Mace, 1994). We term RDB species as "threatened" and other species as "unthreatened". We recognize that there are problems with this approach because these data were not collected specifically for this purpose and that Red Data Book (RDB) listing of species is designed to promote action to avert extinction (Mace, 1994). However, in the absence of any better information, the data are clearly of value for such analysis.

We can compare beetles and butterflies in terms of the number of threatened and unthreatened species supported by different habitat types in the UK (Fig. 3). Beetles and butterflies are very different in many respects (e.g. morphology, species richness, life styles, feeding habits), and it is no surprise that their patterns of habitat distribution are very different. The listed beetles are primarily associated with woodland, non-chalk grassland and riparian sites, whereas the butterflies are well represented on chalk grassland, at woodland edges, hedgerows and on heathlands (Table 4). For the listed beetles, endangered and extinct species are not randomly distributed amongst habitats, but are concentrated in a few key habitats ($\chi^2 = 78.3$; $n = 17$; $P<0.01$). Particularly ancient woodlands, and to a lesser extent sand dunes, riparian sites and bogs, have more endangered and extinct species than expected by chance alone.

As a consequence of their divergent habitat preferences, beetles and butterflies are likely to have different detailed anthropogenic threats acting upon them. However, when we consider the broad threats acting on British butterflies and beetles recorded in Shirt (1987), Emmet and Heath (1989) and Hyman and Parsons (1992), the three most important threats to both taxa are loss of natural broadleaf woodland, agricultural expansion and changes in management practices (= undermanagement for butterflies) (Table 5). Moreover the proportions of species in the

Table 4. British habitats ranked in terms of their species richness of beetles and butterflies (a) listed beetles (Hyman and Parsons, 1992) and all butterflies, (b) endangered *plus* extinct beetles, and threatened butterflies

(a)

Habitat rank	Listed beetles	All butterflies
1	Deciduous woodland	Chalk grassland
2	Other grassland	Woodland edge
3	Ancient woodland	Hedgerow
4	Coastal shingle	Heathland
5	Riparian	Other grassland
6	Sand dunes	Woodland (closed)

(b)

Habitat rank	Endangered *plus* extinct beetles	Threatened butterflies
1	Ancient woodland	Chalk grassland
2	Deciduous woodland	Woodland edge
3	Other grassland	Other grassland
4	Sand dunes	Hedgerow
5	Riparian	Heathland
6	Bog	Fen

threatened categories of these two taxa are extremely similar (Fig. 2). There may be several reasons for this. The intrinsic biological qualities of the two taxa may be such that they both respond in a similar fashion to human disturbance. It is possible that this observed pattern is simply an artefact of the difficulties associated with the classification of rarity in insects. Additionally, the resolution of this threats analysis is broad – loss of natural broadleaf woodland, agricultural expansion and changes in management practices (= undermanagement for butterflies) covers a wide range of specific threats.

These similarities in Fig. 2 are testament to the widespread effects that woodland loss, agricultural expansion and "poor" management have had on the British fauna. But large-scale, non-specific threats such as acidification and climate change (Dobson *et al.*, 1989; Dennis and Shreeve, 1991) may also help explain the similarities between butterflies and beetles described above. For example, it is known that 26% of land

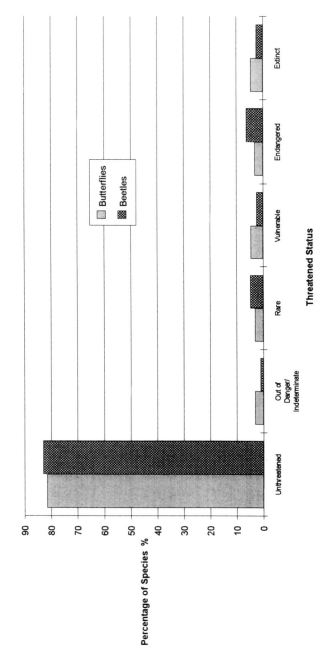

Fig. 2. A comparison of the threatened status of British butterflies and British beetles listed in Hyman and Parsons (1992) (data from Shirt, 1987; Hyman and Parsons, 1992).

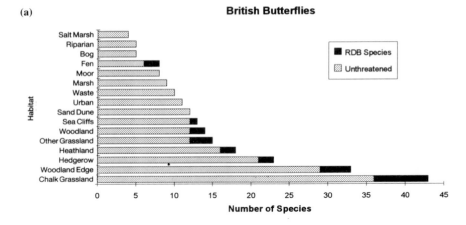

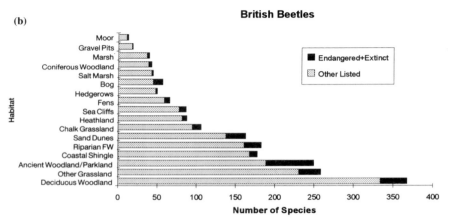

Fig. 3. The species richness of Red Data Book and unthreatened species in various habitats for (a) British butterflies and (b) listed British beetles (data from Shirt, 1987; Emmet and Heath, 1989; Hyman and Parsons, 1992). Note that the figure for butterflies is not directly comparable to that for the beetles.

area under SSSI (site of special scientific interest) status is showing signs of acidification from acid rainfall (Bisset and Farmer, 1993), and this is likely to have deleterious effects upon almost all habitats, especially forests (Pitelka and Raynal, 1989), yet it is not mentioned as a threat to beetles or butterflies.

Further breakdown of these results for three families of beetles indicate that the threats vary considerably for different taxonomic groups (Fig. 5). For longhorn beetles (Cerambycidae) (32 threatened or scarce species

Table 5. The external threats acting upon listed British beetles (Hyman and Parsons, 1992) and all butterflies ranked in terms of the number of species threatened

Threat rank	Listed beetles	All butterflies
1	Management practices	Agricultural expansion
2	Deforestation	Deforestation
3	Agricultural conversion	Undermanagement
4	Natural succession	Afforestation
5	Removal of dead wood	Urbanization

from a British total of 65) deforestation, removal of dead wood and management practices are the main threats. This reflects the fact that these insects live in forests feeding predominantly as larvae under the bark of dead and living timber. For weevils (Curculionidae) (243 of 440) deforestation and management practices are again the main threats but a wide range of other threats are also important. This reflects the fact that weevils feed on woody and non-woody plants in a wide range of habitats (see below). In contrast, deforestation is only a minor threat to ground beetles (Carabidae) (175 of 361) and, in addition to changes in management practices, and natural succession, other important threats are agricultural conversion, river drainage and engineering. The latter threats reflect the importance of open land and riparian habitats for many carabids. A detailed account of the factors involved in the decline and extinction of British beetles through historical time and in the problems with determining trends and causes is given by Hammond (1974). These differences within the beetles and between the butterflies and beetles demonstrate how different taxa are subject to different specific threats because of their differing ecologies and habitat preferences. This argues against the assumption that conservation of well-known flagship organisms serves poorly known groups as well (Yen, 1987; Prendergast et al., 1993).

V. THE INTRINSIC VULNERABILITY OF INSECTS TO EXTINCTION

A. Ecological Correlates of Extinction in Insects

Why is it that, given a common environment and habitat, some species appear to be more susceptible to extinction than others? Population size is intuitively important (Terborgh and Winter, 1980; Leigh, 1981;

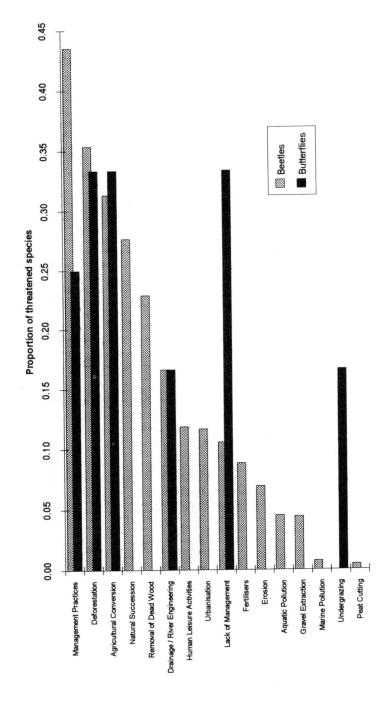

Fig. 4. The proportion of British butterfly and listed British beetle species under specific threats causing declines in populations. Data from Shirt (1987), Emmet and Heath (1989) and Hyman and Parsons (1992).

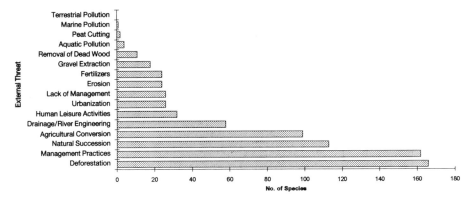
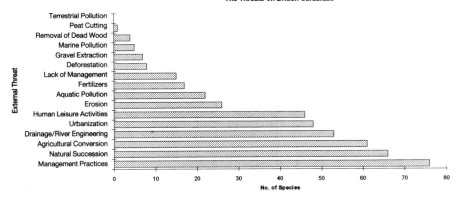
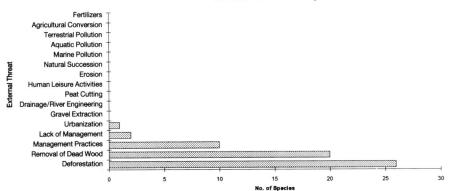

Fig. 5. An analysis of the threats acting upon species of (top) listed weevils (Curculionidae), (middle) listed ground beetles (Carabidae) and (lower) listed longhorn beetles (Cerambycidae).

Goodman, 1987; Pimm *et al.*, 1988), but equally it is clear that there are many species that are naturally rare and unthreatened (Rabinowitz *et al.*, 1986; Schoener, 1987; Gaston, 1994). In a similar fashion, endemic species with narrow distributions are thought to be most at risk (e.g. Thomas, C. D. and Mallorie, 1985; Thomas, C. D., 1991). But again we can identify species with small ranges that are relatively unthreatened, and more worryingly, we can identify once widespread species that have inexplicably and rapidly declined (Hafernik, 1992; Vermeij, 1993). A striking case is the American burying beetle *Nicrophorous americanus*. Once widespread in 32 states in America, and a generalist carrion-feeder, it is now endangered and recorded from just two widely separated localities (Amaral and Morse, 1990; cited in Hafernik, 1992).

There are two further patterns in nature that give cause for concern with respect to extinction. First, the frequency distributions of abundance and geographic range size tell us that most species have relatively small ranges (Fig. 6a); Gaston, 1990; Pagel *et al.*, 1991) and are found in relatively low numbers (May, 1975; Gray, 1987; Tokeshi, 1993). Second, for many species (except perhaps habitat specialists (Gaston and Lawton, 1990)) there is a direct positive relationship between regional distribution and local abundance (e.g. Brown, 1984; Gaston and Lawton, 1988, 1990; Hanski *et al.*, 1993). This means that if a species range is reduced through habitat loss it is also likely to undergo a reduction in its local population size (and vice versa) (Lawton, 1993).

Some have tried to determine the intrinsic features of organisms that make them more susceptible to extinction (e.g. Terborgh and Winter, 1980; Belovsky, 1987; Karr, 1982a,b; Laurance, 1991; Newmark, 1991; Pimm, 1991; Tracy and George, 1992; Gustafsson, 1994). For tropical forest birds, for example, the relative abundance in the forest and the habitat specificity and use of disturbed habitats are important determinants of local extinction (Newmark, 1991), while for non-flying mammals, Laurance (1991) found that natural rarity does not appear to be important, leaving abundance in disturbed habitats as the best predictor of the probability of extinction. To date no such analysis has been made for insects, and here we examine the ecological correlates of susceptibility to extinction of British butterflies. We use the threatened status of the species in Shirt (1987) as a measure of that susceptibility.

Fig. 6. (opposite) The frequency distribution of (top) range size, (middle) habitat specificity and (lower) voltinism for all British butterflies and those in the Red Data Book (Shirt, 1987). The butterfly species listed as threatened in the British Red Data Book (Shirt, 1987) tend to have small geographical ranges, narrow habitat specificity and low voltinism.

14. Species Extinctions in Insects

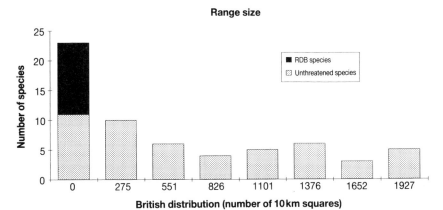

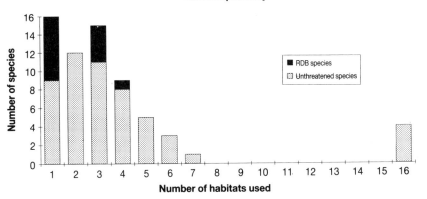

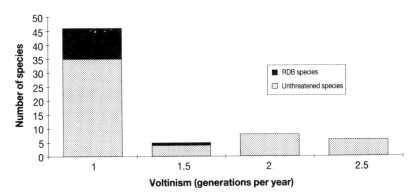

There are many intrinsic factors likely to determine extinction rates and these include: body size (Pimm et al., 1988; Tracy and George, 1992); temporal variability (Pimm et al., 1988; Schoener and Spiller, 1992; Pollard and Yates, 1992; Tracy and George, 1992; Pimm, 1993); dispersal ability (Pimm, 1991; generation time (Fowler and MacMahon, 1982; Pimm, 1991); intrinsic rate of increase, "r" (Richter-Dynn and Goel, 1972; Belovsky, 1987; Pollard and Yates, 1992); and host-specificity (Ehrlich and Murphy, 1987). These factors form a web of relationships that determines the distribution and abundance of species (Gaston and Lawton, 1988), which together are central to the risk of extinction. We can again use British data to explore the relationship between extinction and these variables.

B. Ecological Factors and the RDB Status of British Butterflies

Of the 65 species of British butterfly, 12 are listed in the British Red Data Book (RDB) for insects (Shirt, 1987). Using data on the distribution, body size, feeding habits and other intrinsic life-history characteristics of all the British butterfly species given in Emmett and Heath (1989), Bink (1992), and Kudrna (1986), we looked for intrinsic traits that correlated with RDB status measured from 1 (unthreatened) to 5 (extinct). The life-history traits discussed by Bink (1992) are ranked on a scale from 1 to 9 for each species in the Dutch fauna. However, all British species are found in The Netherlands, and only data for traits that were unlikely to vary between countries were used. For details on sources see Table 6. We predicted that the following characteristics would make a species more likely to become extinct: small distributional range, large body size, low local abundance, low dispersal ability, low adult tolerance to the physical environment, fastidious female oviposition behaviour, low rates of larval growth, low fecundity, few generations per year (low voltinism), high habitat specificity and high specialization with respect to host plants. For the following characteristics we were uncertain as to the effects: male courtship behaviour (solitary–territorial) and over-wintering stage. Many of these variables are correlated making it difficult to separate with certainty those factors involved in correlations with extinction vulnerability.

Red Data Book status is positively correlated with the two measures of distribution – the number of 10 km squares occupied and the number of different habitats occupied (Table 6 and see Fig. 6(a,b)). No other variable shows a significant result. The dependence upon distribution is expected because it is an important criterion used to assess RDB status.

14. Species Extinctions in Insects

Number of habitats occupied is a measure of habitat specificity and its significance here mirrors its importance in tropical birds and mammals (Laurance, 1991; Newmark, 1991).

Figure 7 illustrates the frequency distribution of RDB and unthreatened species with body size, dispersal ability and host-specificity. Although none of these is correlated with RDB status, British distribution of butterfly species is positively correlated with voltinism, dispersal ability, feeding specificity, larval growth and over-wintering stage (egg [rare]–larva–pupa–adult [common]), and negatively correlated with habitat specificity (Table 6). Thus a geographically rare butterfly species probably has just one generation a year, is a poor disperser, feeds on a taxonomically restricted group of plants, grows slowly as a larva, over-winters as an egg and is found in only one or two habitats. These conclusions closely mirror those of Hodgson (1993), who additionally classified the British butterflies into eight groups based on their ecological characteristics. Of these groups, five were butterflies of relatively unproductive habitats (45 species), two of which contained a high proportion (average 79%) of rare species with narrow distributions (15

Table 6. Spearman's rank correlation coefficients for the independent variables of Red Data Book status and British distribution with a range of dependent variables representing some potentially important intrinsic ecological and behavioural traits of British butterflies. Data from [1]Emmet and Heath (1989), [2]Kudrna (1986), [3]Bink (1992).

Dependent variable	Red Data Book status	British distribution
Body size[1]	0.21	0.08
British distribution[1]	−0.65**	—
European distribution (number of countries)[2]	−0.23	0.68***
Male mating behaviour[3]	0.15	−0.17
Female ovipositing behaviour[3]	0.12	0.07
Local abundance (Netherlands)[3]	0.1	0.1
Dispersal ability[3]	0.06	0.41***
Adult tolerance to physical stress[3]	0	0.12
Larval growth rate[3]	−0.11	0.35**
Fecundity[3]	0.22	0.24*
Voltinism[1]	−0.25*	0.44***
Number of habitats occupied[1]	−0.31**	0.66***
Food type (herb–tree)[1]	0.19	−0.24
Feeding specificity (high–low)	−0.1	0.43***
Over-wintering stage (egg–adult)[1]	−0.09	0.3**

*$P<0.10$; **$P<0.05$; ***$P<0.01$.

species). However, the Red Data Book species were distributed widely between seven of the eight groups, and not concentrated in the two groups with many rare butterflies. The important question that needs answering, and for which there is apparently no straightforward answer (see above; Vermeij, 1993), is: "How does geographical rarity and small range size relate to the probability of extinction?"

C. Phylogenetic Aspects of RDB Status in British Beetles

The results for butterflies suggest that it is difficult to generalize about basic biological traits that influence extinction. Intrinsic factors are only part of the extinction recipe and phylogenetic or habitat-specific factors may play a large role in determining extinction. There are clear relationships between phylogeny and ecological patterns (Harvey and Pagel, 1992; Eggleton and Vane-Wright, 1994) and it is likely that phylogeny also will have an important role in determining patterns of extinction (Nee et al., 1994). Here we ask how membership of a particular taxon (in this case a British beetle family; Hyman and Parsons, 1992) and the size of this taxon affects the likelihood of extinction (i.e. threatened status). The distribution of species richness amongst beetle families usually is skewed, with most families having an intermediate number of species and just a few being highly species-rich (Fig. 8; Anderson, 1974). We find that for the beetle families in Hyman and Parsons (1992) there is a significant relationship between the proportions of species listed as threatened within a family and the number of species in that family (Fig. 9). Species-poor families tend to have proportionally more species listed than species-rich families. Holloway et al. (1992) also found that the smaller taxonomic groups of Bornean moths were particularly vulnerable to the effects of logging because they contained a high proportion of endemics. This indicates that there is indeed a phylogenetic component to threatened status and probably extinction, with species-poor families being at greater risk. This is in contrast to some fossil assemblages which have shown higher extinction rates in species-rich families, primarily due to the phylogenetic correlation of traits amongst species within clades (e.g. Jablonski, 1986).

Fig. 7. The frequency distribution of (a) dispersal ability, (b) host-specificity and (c) body size for all British butterflies and those in the Red Data Book (Shirt, 1987). Species listed in the British Red Data Book (Shirt, 1987) show no apparent relationships with these parameters. Data from Bink (1992).

14. Species Extinctions in Insects

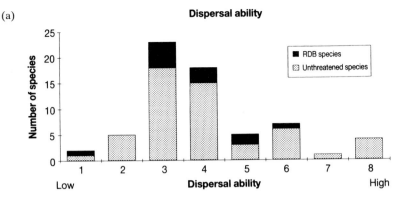

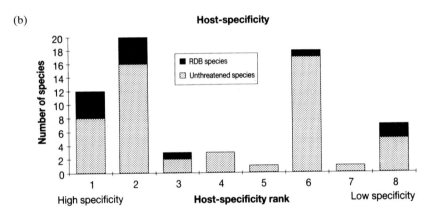

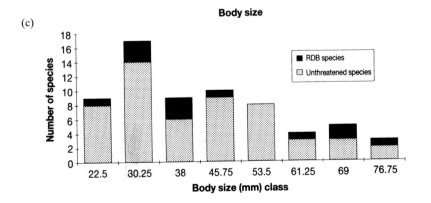

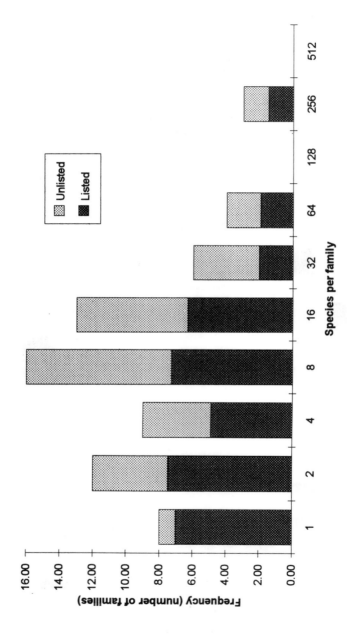

Fig. 8. Frequency distribution of the number of species per beetle family in the 71 British beetle families considered in Hyman and Parsons (1992). The species-poor families appear most at risk, while the intermediate and species-rich families (8–512 species per family) all have less than 50% of their species listed.

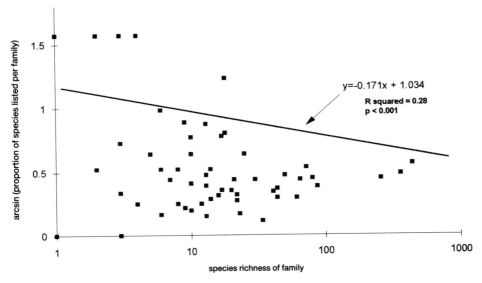

Fig. 9. The relationship between the size of a beetle family in numbers of species and the proportion that are listed in Hyman and Parsons (1992). There is a significant relationship suggesting that species of small families are more threatened than those of large families.

VI. EXTINCTION RATES, COMMUNITY VULNERABILITY AND BIOGEOGRAPHY

A. Community Vulnerability

For extinction of a species to occur the extrinsic threats need to occur throughout its core range (see Lawton, 1993). High levels of such threats in an area with a fauna intrinsically susceptible to extinction will result in high extinction rates (Witting and Loeschcke, 1993). However, two additional factors that will affect the absolute rate of species extinction are the geographical patterns of species richness and endemism, and their congruence with intrinsic and extrinsic factors (Fig. 10). Hence extinction rates will depend upon the degree of overlap in the geographical distribution of extrinsic threats, intrinsic traits, endemism and species richness. Integration of these four factors for a given area provides a notion of the vulnerability of that community to extinction.

The relationships between species richness, endemism and the intrinsic and extrinsic factors are undoubtedly complex. Generalizations on these

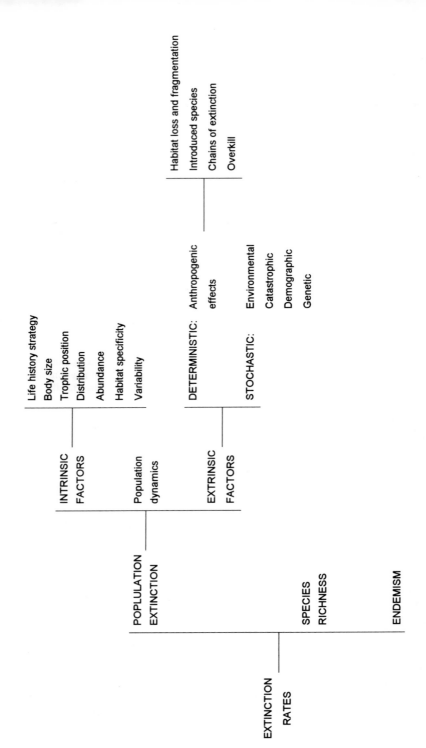

Fig. 10. Schematic figure of the factors affecting extinction and extinction rates.

relationships are by no means straightforward (e.g. Pimm, 1991), but there are some points that can be identified. For one, the degree of endemism of a region and its species richness often appear to be correlated (WCMC, 1992). For example, regions rich in plant species, such as those with extensive tropical forest and the Cape floristic region, tend to have a high proportion of endemics (Gentry, 1986; Cowling et al., 1992). However such broad-scale relationships for most insect groups are poorly known and are more likely to apply to larger areas only. At a smaller spatial scale, areas of highest subspecies, endemism for aposematic butterflies within the Brazilian Amazon region do not correlate directly with the areas of high local species diversity (Brown, 1991; Brown and Brown, 1992). A similar result was found for butterflies in Britain, where rare species do not show a high overlap with local diversity "hotspots" (Prendergast et al., 1993; but see Thomas and Mallorie, 1985). While this is beneficial in terms of risk-spreading, it makes the practicalities of identifying and conserving areas with rare species all the more difficult.

Second, it is likely that on continents many areas of endemism and high levels of extrinsic threats do not directly overlap. Continental areas of endemism tend to be isolated or marginal habitats such as mountains or extreme marginal habitats (Brown and Gibson, 1983; Gentry, 1986). Human population densities and anthropogenic disturbance tends to be less in these areas than in the lowland areas (Hunter and Yonzon, 1993). Indeed, upland areas are better protected than lowland areas in many parts of the world (Dinerstein and Wikramanayake, 1993; Hunter and Yonzon, 1993; but see Dodson and Gentry, 1991). While these areas may protect a relatively small number of localized endemic species from extinction, the majority of species in richer, lowland habitats often remain unprotected (Dinerstein and Wikramanayake, 1993; Hunter and Yonzon, 1993). This is the opposite situation for inhabited islands where high numbers of endemics overlap with human activities (Howarth and Ramsay, 1991; WCMC, 1992; Smith et al., 1993b; Pimm et al., 1994).

B. Predicting the Geography of Extinctions

Using the notion of community vulnerability, we can ask where are extinctions most likely to occur. Such a question may be answered on a country-by-country basis in terms of conservation planning and administration (e.g. National Biodiversity Action Plans). An example of this approach is the analysis of the Indo-Pacific by Dinerstein and Wikramanayake (1993). For each country they integrated the deforestation rate and the extent of remaining forest (extrinsic threat) with species richness and endemism in a variety of animal and plant taxa. They sorted countries into four classes based on the area of forest that are currently protected and the predicted area of unprotected forest that will remain in 10 years' time. These classes represent a conservation threat (or potential) index which sorts countries into those having a relatively low threat to those with a high threat to species survival and conservation. Their analyses identified China, Bangladesh, the Philippines, Vietnam and Tonga as the countries with highest threats to conservation. Of these China and the Philippines also rank highest in terms of species richness and endemism. These countries clearly stand out as the most likely to suffer high extinction rates. For the Philippines these general predictions are reflected in a detailed study of the skippers (Lepidoptera: Hesperiidae) (de Jong and Treadaway, 1993). Of the 151 known Philippine species, 23% are endemic. The high rate of deforestation leads these authors to conclude that 50% of the fauna is endangered and may become extinct in 10–15 years.

With the majority of the world's species concentrated at low latitudes, the tropical regions of the world are immensely important to biodiversity conservation and the analysis of extinction risks. Although there are major biological and biogeographical differences between the three tropical regions, there are some generalizations that we can make about tropical and temperate faunas with respect to the persistence of species. Tropical regions tend to have higher species richness for many taxa, species with smaller ranges (Rappoport's rule) (Rappoport, 1982; Stevens, 1989), species that are more habitat specific (Stevens, 1992), species with lower population densities (Elton, 1973, 1975; Currie and Fritz, 1993), and high human population growth rates above the global average (Tuckwell and Koziol, 1993). Conventional wisdom associates these features with relatively high extinction rates, hence predictions of higher extinction rates for the tropics.

VII. RELATIVE EXTINCTION RATES OF INVERTEBRATES AND VERTEBRATES: QUANTIFYING GLOBAL SPECIES EXTINCTIONS IN INSECTS

Despite current interest in present and future global extinction rates (Reid, 1992; Smith et al., 1993a,b), there have been few explicit estimates of the extinction rate for insects. We have attempted to estimate this by comparing insects with some of the better known vertebrate taxa. Table 7 shows the total number of species in the world and in Britain for five

Table 7. In this table "relative rates of extinction" are estimated using data on the threatened status of British animals. Here "threatened" includes all Red Data Book categories; "endangered" is RDB category 1. Columns 7 and 8 are derived from comparing the relative percentages between taxa in columns 4 and 6, respectively. Note that the percentage of British birds threatened represents an overestimate on account of the data used to calculate this figure – see text for further details. [a]There are more than 22 000 species of insects in Britain but Shirt (1987) discusses the RDB status of only some of these; [b]only terrestrial, freshwater and brackish Mollusca considered. Data in columns 1–6 from Shirt (1987), Bratton (1991), Batten (1990), NCC (1989), Kierney (1994), and Knox (1992)

	1 Number of British species	2 Number of threatened British species	3 Number of endangered British species	4 % of British species threatened	5 % of world species threatened	6 % of British species endangered	7 Relative extinction rate: threatened (from (4))	8 Relative extinction rate: endangered (from (6))	9 Mean relative extinction rate (from (8) and (9))
Mammals	55	16	No data	29.1	11	–	2.3	–	2.3
Birds	210	117	76	55.7	11	36.2	4.3	9.8	7.1
Insects	13 746[a]	1786	506	12.9	0.07	3.7	1	1	1
Araneae	622	86	22	13.8	–	3.5	1.1	0.9	1
Mollusca	210[b]	30	10	14.2	0.4	4.8	1.1	0.8	1

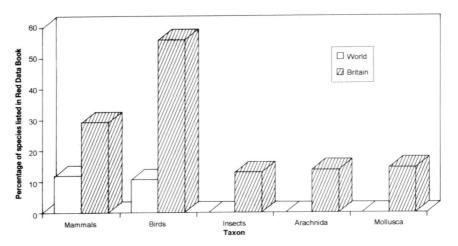

Fig. 11. A comparison of vertebrate and invertebrate taxa and the proportion of their species listed as Red Data species both in Britain (shaded) and in the world (unshaded).

groups: birds, mammals, insects, molluscs and spiders. It also shows the number of threatened and endangered species. There is a striking difference in the relative proportions of threatened and endangered species for Britain and the world for the invertebrate groups. This reflects how little we know of invertebrates worldwide by comparison to the British fauna and by comparison to vertebrate groups, worldwide (Fig. 11). If we assume that the accuracy of the Red Data Book status is more or less similar for different groups of the British fauna (see below), and that the threatened status of an organism is a suitable surrogate for extinction, then we can determine *relative rates of extinction* from the proportions of threatened and endangered species. The relative value for threatened birds to threatened insects is 4.3 (Table 7), inferring that a British breeding bird is, on average, 4.3 times as likely to become extinct as a British insect. Using the figures for endangered species gives a higher relative extinction rate for birds/insects of 9.8, because on average, of the threatened species in each group a greater proportion of birds are perceived to be endangered than insects. The values for molluscs and spiders are similar to those for insects. It must be stressed that the data for birds in Batten *et al.* (1990) have not been assessed according to IUCN guidelines for Britain, making direct comparisons somewhat precarious. Because these relative rates between birds and insects depend upon a figure of 55% of British birds being threatened, undoubtedly an overestimate of the threat of national extinction of British birds, these

14. Species Extinctions in Insects

data represent an upper limit to the relative extinction rates between birds and insects. A national Red Data List for birds would allow a more accurate calculation to be made.

Smith et al. (1993a) also used Red Data Book status in two alternative ways to estimate extinction rates for groups such as birds, mammals, palms and other plants, but they were unable to provide estimates for insects. First they examined the number of species added to the list of extinctions in a 4-year period (1986–1990) for animals on the Red Lists, and a 2-year period for plants (1990–1992) based on data in the World Conservation Monitoring Centre database. Using these data and the estimated number of species worldwide for a range of taxa, they estimated the extinction rates in terms of the time required for 50% of species to go extinct. For birds the answer was 1500 years, and for mammals 6500 years. Second, they suggested that net changes in the Red Data Book status of species (from "rare", "vulnerable", "endangered", to "probably extinct" – in increasing order of threatened status) could be used to estimate species extinctions. For example, 50% extinction values for birds, mammals and palms from this method are 350, 250 and 70 years respectively – much greater than using their first method (Smith et al., 1993a).

We have used the relative extinction rates (RER) outlined above to estimate species extinction rates for insects. Since about 1% of birds and mammals have become extinct since 1600 (WCMC, 1992), a mean RER of 7.1 for birds/insects would suggest that 0.14% of insects have become extinct in the same period of time. Given an average global estimate of 8 million species of insects this would indicate that 11 200 species of insects have become extinct since 1600. Our present database for insect extinctions then could be out by a factor of almost 200! Smith et al. (1993a) predict a 12–55-fold future increase in the extinction rate of birds in the next 300 years. Our RER value would predict that over the next 300 years a further 100 000–500 000 species of insect may become extinct: the equivalent of between 7 and 30 species every week. Although this estimate is about an order of magnitude less than other estimates it is still a huge number of species – roughly half the total already described by taxonomists. Given that less than 7500 new species of insects are described each year (Hammond, 1992), this suggests that the chances of description and extinction for an undescribed species of insect are not too different!

Such estimates of global extinction rates for insects based on RERs take on board all the problems and errors associated with the estimates for the original taxon used to scale the estimate for insects. Moreover, they make many assumptions about the patterns of the relative status of different organisms around the world. These patterns in themselves might be seriously affected by differences in many important factors discussed above such as species richness, endemism, the intrinsic biology of different taxa around the world (Section VI) and by historical differences between regions. Britain with its large number of species on the edge of their range is certainly not representative of many other regions. Nevertheless we feel that RERs may be of considerable use, particularly in the absence of any other direct predictor of extinction rates for insects. We also see a value in this method as it is possible to make backpredictions using RERs. For example, Shirt (1987) includes 99 of the 13 746 (0.7%) British species of insects considered in his Red Data Book as unseen since 1900 (i.e. extinct). Using an RER of 7.1, this would suggest that 5.0%, equivalent to 11 species, of British breeding birds should also not have been seen since 1900. Sharrock (1974) states that in the period 1800–1949, 11 bird species have become extinct in Britain and Ireland, with two later recolonizing. For mammals, our prediction of one species becoming extinct since 1900 is borne out by the loss of the greater horseshoe bat.

VIII. EXTINCTION AND CO-EXTINCTION

Stork and Lyal (1993) recently drew attention to the plight of ectoparasitic insects such as Phthiraptera (lice) and Siphonaptera (fleas) which may have become extinct with the loss of some of the 200 or more species of birds and mammals that act as their hosts. For example, although WCMC (1992) lists no species of these insect groups as extinct, at least two species of lice (*Columbicola extinctus* Malcolm and *Campanulotes defectus* Tendeiro) suffered the same demise as their host, the passenger pigeon, *Ectopistes migratorius* (Linnaeus). Since lice are generally very host specific, it is very likely that many more species of this group have similarly become extinct. Fleas are less host specific, but some species may have become extinct or are endangered.

These examples are merely the tip of an iceberg as many other invertebrates and vertebrates are more or less loosely tied to species that are threatened with extinction. Just how many species are threatened through this continuum of species associations is of course far from clear but the problem should not be ignored in estimates of extinction rates. Some species of insects, particularly pollinators and parasitoids (but also decomposers, such as termites), may be keystone species, and their loss may have critical effects upon the whole biotic community.

IX. CONCLUDING REMARKS

In this chapter we have attempted to review what is known about extinction processes and extinction rates with respect to insects. We have drawn attention to the extreme lack of data for insects on species richness, distribution and abundance, as well as numbers of species that have become extinct in recent years and those likely to become extinct in the near future. In spite of this paucity of data, we have examined ways in which the threat and likelihood of extinction of a species can be determined, both theoretically and empirically, through analyses of the extrinsic threats to species and the intrinsic qualities of species. Our examinations of the threatened UK beetle and butterfly faunas indicate that such species vulnerability analyses are profitable in determining the most threatened habitats (in terms of species conservation) and the most serious threats to species, but that uncovering the ecological attributes of species that make them prone to extinction is a much more difficult task. On a larger scale we suggest that by incorporating other data, such as land use and human population density, such species vulnerability analyses can be expanded to include other factors such as endemism and species richness.

Estimation of insect extinction rates has so far almost entirely relied on the species–area model coupled with estimates of loss of tropical forests. We have introduced a new model based on knowledge of the relative proportions of endangered species for different groups for well-known faunas and floras: relative extinction rates. This, coupled with methods of estimating extinction rates using changes in the Red Data Book Status of threatened species (Smith *et al.*, 1993a) has allowed us to make the first direct estimates of past species extinctions, and predictions for future extinctions of insects. The utility of the relative extinction rate model requires further critical examination but may prove useful for the future.

We have not directly addressed the issues of tropical forest destruction and fragmentation (an excellent review is given in Whitmore and Sayer (1992)), although we see this as perhaps the most pressing area for future research. As Lugo (1988) noted, most species–area based estimates of extinctions which use data on tropical forest loss assume total loss of accompanying species. Such estimates ignore the fact that tropical forests are affected to varying degrees by logging or burning; from complete clearance to little more than "normal disturbance". We do not know how seriously affected many species of insects are by such land-use change, nor what proportion of natural forests species can survive in both secondary forests of varying ages and agricultural land. Critically we do not have empirical data to model the relationship between habitat loss and species richness at a regional scale. Brown and Brown (1992) provide a graphical model of the effects of habitat alteration on different components of a regional fauna. The important points are that small-scale disturbance can increase measured local species richness (but probably not at the regional level) by creating a greater mix of habitats, and that with any level of disturbance there are forest specialists that will inevitably suffer. In spite of the huge loss of natural forests in, for example, the Atlantic forests of Brazil or the forests of the Indonesian Moluccas, no species of the relatively well-studied group, the butterflies, is known to have disappeared (Brown, 1991; Brown and Brown, 1992). These specific faunas appear to be adapted to cope with such habitat loss and fragmentation. Empirical data representative of other locations and taxa are desperately needed to assess the generality of these results.

We see studies of the role of secondary forests (the forests of the future) in maintaining biodiversity as an important step towards understanding the effects of land-use change on species extinctions and loss of genetic diversity (e.g. Lugo, 1988; Pimental et al., 1992). For some groups of insects, such as ants in leaf litter, differences in forest type, as typified by primary forest, secondary forest, and cocoa plantation, do not appear to affect their species richness or composition (Belshaw and Bolton, 1993), whereas other soil and non-soil based groups, such as termites and moths, show significant differences between primary and secondary habitats (Holloway et al., 1992; Eggleton et al., 1995). How typical ants, termites and moths are of other tropical forest insects is unknown.

Finally, our chapter has dealt solely with loss of insect diversity as represented at the species level. How much is being lost at the genetic level is virtually unknown, but the loss of local populations is perhaps the most prevalent effect of changing land use. The preservation of species in a few restricted localities while reducing species extinction rates does not reduce the erosion of biodiversity at the genetic, landscape and functional levels. The worst present-day scenarios predict the loss of 50% or more of the world's species in the space of 1000–2000 years, the majority of this happening in just a few centuries (Ehrlich and Wilson, 1991). Previous mass extinctions appear to have been greater and to have resulted in biological impoverishment lasting for millions of years, yet undoubtedly local and regional impoverishment will precede and far outweigh global species extinction.

Acknowledgements

Many thanks to Victoria Best and Tom Rungge who helped collate the British butterfly and beetle data. We are grateful to Paul Eggleton and John Lawton for helpful discussion of the manuscript, and to John Davis, Gabor Lovei, Mark Parsons, Andy Strathdee and Jeremy Thomas who read and commented on the manuscript.

References

Abbott, I. (1974). Numbers of plant, insect and land bird species on nineteen remote islands in the Southern Hemisphere. *Biol. J. Linn. Soc.* **6**, 143–152.
Allmon, W. D., Rosenberg, G., Portell, R. W. and Schindler, K. S. (1993). Diversity of Atlantic coastal plain molluses since the Pliocene. *Science* **260**, 1626–1629.
Amaral, M. and Morse, L. (1990). Reintroducing the American burying beetle. *Endangered Sp. Tech. Bull.* **15**, 3.
Anderson, S. (1974). Patterns in faunal evolution. *Quart. Rev. Biol.* **49**, 311–332.
Batten, L. A. *et al.* (1990) "Red Data Birds in Britain: Action for Rare, Threatened and Important Species". NCC and RSPB, T. and A. D. Poyser, London.
Belshaw, R. and Bolton, B. (1993). The effect of forest disturbance in the leaf litter ant fauna in Ghana. *Biod. Cons.* **2**, 656–666.
Belovsky, G. E. (1987). Extinction models and mammalian persistence. *In* "Viable Populations for Conservation" (M. E. Soulé, ed.), pp. 35–58. Cambridge University Press, Cambridge.
Benton, M. J. (1986) The evolutionary significance of mass extinction. *Trends in Ecology and Evolution* **1**, 127–130.
Bink, F. A. (1992). "Ecologische Atlas van de Dagvlinders van Noordwest–Europa". Haarlem, Schuyt.
Bissett, K. and Farmer, A. (1993). "SSSIs in England at Risk from Acid Rain". *English Nature Science Report* **15**. English Nature, Peterborough.
Boecklen, W. J. and Simberloff, D. (1986). Area-based extinction models in conservation. *In* "Dynamics of Extinction" (D. K. Elliot, ed.), pp. 247–276. Wiley, New York.
Brakefield, P. M. (1991). Genetics and the conservation of invertebrates. *In* "The Scientific Management of Temperate Communities for Conservation". (I. F. Spellerberg, F. B. Goldsmith and M. G. Morris, eds), pp. 45–79. Blackwell Scientific Publications, Oxford.
Bratton, J. H. (1991). "British Red Data Books. 3. Invertebrates Other Than Insects". Joint Nature Conservation Committee.
Briggs, D. E. G., Fortey, R. A. and Clarkson, E. N. K. (1988). Extinction and the fossil record of the arthropods. *In* "Extinction and Survival in the Fossil Record" (G. P. Larwood, ed.), pp. 171–209. *Systematics Association Special Volume* **34**. Clarendon Press, Oxford.
Brown, J. H. (1984). On the relationship between abundance and distribution of species. *Am. Nat.* **124**, 255–279.
Brown, J. H. and Gibson, A. (1983). "Biogeography". C. V. Mosby, St Louis.
Brown, J. H. and Kodric-Brown, A. (1977). Turnover rates in insular biogeography: effect of immigration on extinction. *Ecology* **58**, 445–449.
Brown, K. S. (1991). Conservation of neotropical environments: insects as indicators. *In* "The Conservation of Insects and Their Habitats" (N. M. Collins and J. A. Thomas, eds), pp. 350–404. *15th Symposium of the Royal Entomological Society of London*. Academic Press, London.

Brown, K. S. and Brown, G. G. (1992). Habitat alteration and species loss in Brazilian forests. In "Tropical Deforestation and Species Extinction" (T. C. Whitmore and J. A. Sayer, eds), pp. 119–142. Chapman & Hall, London.
Brussard, P. F. (1991). The role of ecology in biological conservation. *Ecol. Appl.* **1**, 6–12.
Connell, J. H. (1978). Diversity in tropical rain forests and coral reefs. *Science* **199**, 1302–1310.
Coope, G. R. (1995) The effects of Quaternary climate changes on insect populations: lessons from the past. In "Insects in a Changing Environment" (R. Harrington and N. E. Stork, eds), pp. 29–47. Academic Press, London.
Cowling, R. M., Holmes, P. M. and Rebelo, A. G. (1992). Plant diversity and endemism. In "The Ecology of Fynbos" (R. M. Cowling, ed.), pp. 62–112. Oxford University Press, Cape Town.
Currie, D. J. and Fritz, J. T. (1993). Global patterns of animal abundance and species energy use. *Oikos* **67**, 56–68.
de Jong, R. and Treadaway, C. G. (1993). The Hesperiidae (Lepidoptera) of the Phillipines. *Zool. Verh. Leiden.* **288**, 1–125.
Den Boer, P. J. (1981). On the survival of populations in a heterogeneous and variable environment. *Oecologia* **50**, 39–53.
Den Boer, P. J. (1990). The survival value of dispersal in terrestrial arthropods. *Biol. Cons.* **54**, 175–192.
Dennis, R. L. H. and Shreeve, T. G. (1991). Climatic change and the British butterfly fauna: opportunities and constraints. *Biol. Cons.* **55**, 1–16.
Diamond, J. M. (1975). Assembly of species communities. In "Ecology and Evolution of Communities" (M. J. Cody and J. M. Diamond, eds), pp. 342–444. Harvard University Press, Cambridge, MA.
Diamond, J. M. (1985). How many unknown species are yet to be discovered? *Nature* **315**, 538–539.
Diamond, J. M. (1987). Extant unless proven extinct? Or, extinct unless proven extant? *Cons. Biol.* **1**, 77–79.
Diamond, J. M. (1989). The present, past and future of human-caused extinctions. *Phil. Trans. R. Soc. Lond. B.* **325**, 469–477.
Diamond, J. M. and May, R. M. (1976). Island biogeography and the design of nature reserves. In "Theoretical Ecology: Principles and Applications" (R. M. May, ed.), pp. 163–186. Blackwell Scientific Publications, Oxford.
Dinerstein, E. and Wikramanayake, E. D. (1993). Beyond "hostspots": how to prioritize investments to conserve biodiversity in the Indo-Pacific region. *Cons. Biol.* **7**, 53–65.
Dobson, A., Jolly, A. and Rubenstein, D. (1989). The greenhouse effect and biological diversity. *TREE* **4**, 64–68.
Dodson, C. H. and Gentry, A. H. (1991). Biological extinction in western Ecuador. *Ann. Miss. Bot. Gard.* **78**, 273–305.
Eggleton, P. and Vane-Wright, R. I. (eds) (1994) "Phylogenetics and Ecology". Academic Press, London.
Eggleton, P., Bignell, D. E., Sands, W. A., Waite, B., Wood, T. G. and Lawton, J. H. (1994). The species richness of termites (Isoptera) under differing levels of forest disturbance in the Mbalmayo forest reserve, southern Cameroon. *J. Trop. Ecol.* **11**, 1–14.
Ehrlich, P. and Ehrlich, A. (1981). "Extinction. The Causes of the Disappearance of Species". Random House, New York.
Ehrlich, P. R. and Murphy, D. D. (1987). Conservation lessons from long-term studies of checkerspot butterflies. *Cons. Biol.* **1**, 122–131.
Ehrlich, P. R. and Wilson, E. O. (1991). Biodiversity studies: science and policy. *Science* **253**, 758–762.

Ehrlich, P. R., Murphy, D. D., Singer, M. C., Sherwood, C. B., White, R. T. and Broen, I. C. (1980). Extinction, reduction, stability and increase: the responses of checkerspot butterfly (*Euphydryas*) populations to California drought. *Oecologia* **46**, 101–105.
Elton, C. S. (1973). The structure of invertebrate populations inside neotropical rain forest. *J. Anim. Ecol.* **42**, 55–104.
Elton, C. S. (1975). Conservation and the low population density of invertebrates inside neotropical rain forest. *Biol. Cons.* **7**, 3–15.
Emmett, A. M. and Heath, J. (1989) "The Butterflies". Harley Books, Colchester.
Ferrari, S. F. and Lopes, M. A. (1992). A new species of marmoset, genus *Callithrix* Erxleben, 1777 (Callitrichidae, Primates), from western Brazilian Amazonia. *Goel. Zool.* **11**, 1–13.
Forey, P., Humphries, C. J. and Vane-Wright, R. I. (eds) (1994). "Systematics and Conservation Evaluation". *Systematics Association Special Volume*. Oxford University Press, Oxford.
Fowler, C. W. and MacMahon, J. A. (1982). Selective extinction and speciation: their influence on the structure and functioning of communities and ecosystems. *Am. Nat.* **199**, 480–498.
Gaston, K. J. (1990). Patterns in the geographical ranges of species. *Biol. Rev.* **65**, 105–129.
Gaston, K. J. (1991). The magnitude of global insect species richness. *Cons. Biol.* **5**, 283–296.
Gaston, K. J. (1994). "Rarity". Chapman & Hall, London.
Gaston, K. J. and Lawton, J. H. (1988). Patterns in the distribution and abundance of insect populations. *Nature* **331**, 709–712.
Gaston, K. J. and Lawton, J. H. (1990). Effects of scale and habitat on the relationship between regional distribution and local abundance. *Oikos* **58**, 329–335.
Gaston, K. J. and May, R. M. (1992). Taxonomy of taxonomists. *Nature* **356**, 281–282.
Gaston, K. J. and Mound, L. A. (1993). Taxonomy, hypothesis testing and the biodiversity crisis. *Proc. R. Soc. Lond. B.* **251**, 139–142.
Gentry, A. H. (1986). Endemism in tropical versus temperate plant communities. *In* "Conservation Biology. The Science of Scarcity and Diversity". (M. E. Soulé, ed.), pp. 153–181. Sinauer Associates, Sunderland, MA.
Gilpin, M. E. and Hanski, I. (1991). "Metapopulation Dynamics: Empirical and Theoretical Investigations". Academic Press, New York.
Gilpin, M. E. and Soulé, M. E. (1986). Minimum viable population: processes of species extinction. *In* "Conservation Biology. The Science of Scarcity and Diversity" (M. E. Soulé, ed.), pp. 19–34. Sinauer Associates, Sunderland, MA.
Goodman, D. (1987). The demography of chance extinction. *In* "Viable Populations for Conservation" (M. E. Soulé, ed.), pp. 11–34. Cambridge University Press, Cambridge.
Gotelli, N. J. and Kelley, W. G. (1993). A general model of metapopulation dynamics. *Oikos* **68**, 36–44.
Gray, J. S. (1987). Species-abundance patterns. *In* "Organization of Communities: Past and Present" (J. H. R. Gee and P. S. Giller, eds), pp. 53–67. Blackwell, Oxford.
Gustafsson, L. (1994). A comparison of biological characteristics and distribution between Swedish threatened and non-threatened forest vascular plants. *Ecography* **17**, 39–49.
Hafernik, J. E. J. (1992). Threats to invertebrate biodiversity: implications for conservation strategies. *In* "Conservation Biology. The Theory and Practice of Nature Conservation, Preservation and Management" (P. L Fielder and S. K. Jain, eds), pp. 172–195. Chapman & Hall, London.
Hammond, P. M. (1974). Changes in the British Coleopterous fauna. *In* "The Changing Flora and Fauna of Britain" (D. L. Hawksworth, ed.), pp. 323–369. Academic Press for the Systematics Association, London.

Hammond, P. M. (1992). Species inventory. *In* "Global Biodiversity, Status of the Earth's Living Resources" (B. Groombridge, ed.), pp. 17–39. Chapman & Hall, London.

Hammond, P. M. and Harding, P. T. (1991). Saproxylic invertebrate assemblages in British woodlands: their conservation significance and its evaluation. *In* "Pollard and Veteran Tree Management" (H. J. Read, ed.), pp. 30–37. Richmond Publications, Slough.

Hanski, I. (1989). Metapopulation dynamics: does it help to have more of the same? *TREE* **4**, 113–114.

Hanski, I. and Gilpin, M. E. (1991). Metapopulation dynamics: brief history and conceptual domain. *Biol. J. Linn. Soc.* **42**, 73–88.

Hanski, I. and Woiwod, I. P. (1993). Spatial synchrony in the dynamics of moth and aphid populations. *J. Anim. Ecol.* **62**, 656–668.

Hanski, I., Kouki, J. and Halkka, A. (1993). Three explanations of the positive relationship between distribution and abundance of species. *In* "Species Diversity in Ecological Communities: Historical and Geographical Perspectives" (R. E. Ricklefs and D. Schluter, eds), pp. 108–116. University of Chicago Press, Chicago.

Harrison, S. (1991). Local extinction in a metapopulation context: an empirical evaluation. *Biol. J. Linn. Soc.* **42**, 73–88.

Harrison, S. and Quinn, J. F. (1989). Correlated environments and the persistence of metapopulations. *Oikos* **56**, 293–298.

Harrison, S., Murphy, D. D. and Ehrlich, P. R. (1988). Distribution of the bay checkerspot butterfly, Euphydryas editha bayensis: evidence for a metapopulation model. *Am. Nat.* **132**, 360–382.

Harvey, P. H. and Pagel, M. D. (1992). "The Comparative Method". Oxford University Press, Oxford.

Hawksworth, D. L. (ed.) (1991). "The Biodiversity of Microorganisms and Invertebrates: its Role in Sustainable Agriculture". CAB International, Wallingford.

Hawksworth, D. L. (1992). Biodiversity in microorganisms and its role in ecosystem function. *In* "Biodiversity and Global Change" (O. T. Solbrig, H. M. van Emden and P. G. W. J. van Oordt, eds), pp. 83–93. IUBS, Paris.

Heywood, V. H., Mace, G. M., May, R. M. and Stuart, S. N. (1994). Uncertainties in extinction rates. *Nature* **368**, 105.

Hodgson, J. G. (1993). Commonness and rarity in British butterflies. *J. Appl. Ecol.* **30**, 407–427.

Holloway, J. D., Kirk-Spriggs, A. H. and Chey, V. K. (1992). The responses of some rain forest groups to logging and conversion to plantation. *Phil. Trans. R. Soc. B.*, **335**, 425–436.

Howarth, F. G. and Ramsay, G. W. (1991). *In* "Conservation of Insects and Their Habitats" (N. M. Collins and J. A. Thomas, eds), pp. 71–107. *15th Symposium of the Royal Entomological Society of London.* Academic Press, London.

Howe, R. W., Davis, G. J. and Mosca, V. (1991). The demographic significance of "sink" populations. *Biol. Cons.* **57**, 239–255.

Hunter, M. L. and Yonzon, P. (1993). Altitudinal distributions of birds, mammals, people, forests, and parks in Nepal. *Cons. Biol.* **7**, 420–423.

Hyman, P. S. and Parsons, M. S. (eds) (1992). "A Review of the Scarce and Threatened Coleoptera of Great Britain. Part 1". The UK Joint Nature Conservation Committee, Peterborough.

IUCN (1990). "1990 IUCN Red List of Threatened Animals" WCMC/IUCN Species Survival Commission/ICBP, Gland and Cambridge.

Jablonski, D. (1986). Causes and consequences of mass extinctions: a comparative approach. *In* "Dynamics of Extinction" (Elliot, D. K. ed.), pp. 183–229. Wiley, New York.

Jablonski, D. (1991). Extinctions: a palaeontological perspective. *Science* **253**, 754–757.
Jablonski, D. (1993). The tropics as a source of evolutionary novelty through geological time. *Nature* **364**, 142–144.
Jablonski, D. (1994). Extinctions in the fossil record. *Phil. Trans. R. Soc. Lond. B.*, in press.
Jackson, J. B. C. (1994). Constancy and change of life in the sea. *Phil. Trans. R. Soc. Lond. B.*, in press.
Janzen, D. H. (1986). The eternal external threat. *In* "Conservation Biology. The Science of Scarcity and Diversity" (M. E. Soulé, ed.), pp. 286–303. Sinauer Associates, Sunderland, MA.
Janzen, D. H. (1987). The insect diversity of a Costa Rican dry forest: why keep it and how. *Biol. J. Linn. Soc.* **30**, 343–356.
Kangas, P. (1991). Macroscopic minimodels of deforestation and diversity. *Ecol. Model.* **57**, 277–294.
Karr, J. R. (1982a). Population variability and extinction in the avifauna of a tropical land bridge island. *Ecology* **63**, 1975–1978.
Karr, J. R. (1982b). Avian extinctions on Barro Colorado Island, Panama: a reassessment. *Am. Nat.* **119**, 220–239.
Kellert, S. R. (1993). Values and perspectives of invertebrates. *Cons. Biol.* **7**, 845–855.
Kierney, M. (1994). "Atlas of the Land and Freshwater Molluscs of Britain and Ireland". Harley Books, in press.
Kim, K. C. (1993). Biodiversity, conservation and inventory: why insects matter. *Biod. Cons.* **2**, 191–214.
Knox, A. G. (1992). "Checklist of Birds of Britain and Ireland" British Ornithologists' Union, Tring.
Kudrna, O. (ed.) (1986). Aspects of the conservation of butterflies in Europe. *Butterflies of Europe*, Vol. 8. AULA-Verlag, Wiesbaden.
Labandeira, C. C. and Sepkoski, J. J. (1993). Insect diversity and the fossil record. *Science* **261**, 310–315.
Lanly, J. P. (1982). Tropical Forest Resources. Forestry Paper No. 30. Food and Agriculture Organisation of the United Nations, Rome.
Laurance, W. F. (1991). Ecological correlates of extinction proneness in Australian tropical rain forest mammals. *Cons. Biol.* **5**, 79–89.
Lawton, J. H. (1993). Range, population abundance and conservation. *TREE* **8**, 409–413.
Lawton, J. H. (1994). Population dynamic principles. *Phil. Trans. R. Soc. Lond. B.*, in press.
Leigh, E. G. J. (1981). The average lifetime of a population in a varying environment. *J. Theor. Biol.* **90**, 213–239.
Levins, R. (1970). Extinction. *Lect. Math. Life Sci.* **2**, 77–107.
Lovejoy, T. E. (1980). A projection of species extinctions. *In* "Council on Environmental Quality (CEQ)" The Global 2000 Report to the President, Vol. CEQ, Washington D.C., pp. 328–331.
Lugo, A. E. (1988). Estimating reductions in the diversity of tropical forest species. *In* "Biodiversity" (E. O. Wilson and F. M. Peter, eds), pp. 58–70, National Academic Press, Washington.
MacArthur, R. H. and Wilson, E. O. (1967). "The Theory of Island Biogeography". Princeton University Press, Princeton, NJ.
Mace, G. M. (1994). Classifying threatened species: means and ends. *Phil. Trans. R. Soc. Lond. B.* **344**, 91–97.
MacKinnon, J. (1993). A new species of living bovid from Vietnam. *Nature* **363**, 443–445.

Mann, C. C. (1991). Extinction: are ecologists crying wolf? *Science* **253**, 736–738.
May, R. M. (1975). Patterns of species abundance and diversity. In "Ecology and Evolution of Communities" (M. L. Cody and J. M. Diamond, eds), pp. 81–120. Belknap Press of Harvard University, Cambridge, MA.
May, R. M. (1992). How many species inhabit the earth? *Sci. Am.* **October**, 18–24.
McCauley, D. E. (1991). Genetic consequences of local population extinction and recolonization. *TREE* **6**, 5–8.
McDonald, K. A. and Brown, J. H. (1992). Using montane mammals to model extinctions due to global change. *Cons. Biol.* **6**, 409–415.
Miller, J. C. (1993). Insect natural history, multi-species interactions and biodiversity in ecosystems. *Biod. Cons.* **2**, 233–241.
Mittermeier, R. A., Schwartz, M. and Ayres, J. M. (1992). A new species of marmoset, genus *Callithrix* Erxleben, 1777 (Callitrichidae, Primates) from the Rio Maues region, State of Amazonas, central Brazilian Amazonia. *Goel. Zool.* **14**, 1–17.
Myers, N. (1979). "The Sinking Ark. A New Look at the Problem of Disappearing Species". Pergamon, New York.
Myers, N. (1987). The extinction spasm impending: synergisms at work. *Cons. Biol.* **1**, 15–21.
Myers, N. (1988) Threatened biotas: "hotspots" in tropical forests. *Environ.* **8**, 1–20.
Myers, N. (1993). Questions of mass extinction. *Biod. Cons.* **2**, 2–17.
NCC (1989). "Guidelines for the Selection of Biological Sites of Special Scientific Interest: Rationale, Operational Approach and Criteria. NCC, Peterborough.
Nee, S. Holmes, E. C., May, R. M. and Harvey, P. H. (1994). Extinction rates can be estimated from molecular phylogenies. *Phil. Trans. R. Soc. Lond. B.*, in press.
Newmark, W. D. (1991). Tropical forest fragmentation and the local extinction of understory birds in the eastern Usambara mountains, Tanzania. *Cons. Biol.* **5**, 67–78.
Pagel, M. D., May, R. M. and Collie, A. R. (1991). Ecological aspects of the geographical distribution and diversity of mammalian species. *Am. Nat.* **137**, 791–815.
Patterson, C. and Smith, A. B. (1989). Periodicity and extinction: the role of systematics. *Ecology* **70**, 802–811.
Pimental, D., Stachow, U., Takacs, D. A., Brubaker, H. W., Dumas, A. R., Meaney, J. J., O'Neil, J. A. S., Onsi, D. E. and Corzilius, D. B. (1992). Conserving biological diversity in agricultural/forestry systems. *Bioscience* **42**, 354–362.
Pimm, S. L. (1991). "The Balance of Nature? Ecological Issues in the Conservation of Species and Communities". Chicago University Press, Chicago.
Pimm, S. L. (1993). Life on an intermittent edge. *TREE* **8**, 45–46.
Pimm, S. L., Jones, H. L. and Diamond, J. (1988). On the risk of extinction. *Am. Nat.* **132**, 757–785.
Pimm, S. L., Moulton, M. P. and Justice, L. J. (1994). Bird extinction in the central Pacific. *Phil. Trans. R. Soc. Lond. B.*, in press.
Pitelka, L. F. and Raynal, D. J. (1989). Forest decline and acidic deposition. *Ecology* **70**, 2–10.
Pollard, E. and Yates, T. J. (1992). The extinction and foundation of local butterfly populations in relation to population variability and other factors. *Ecol. Entomol.* **17**, 249–254.
Pollard, E., Van Swaay, C. A. M. and Yates, T. J. (1993). Changes in butterfly numbers in Britain and The Netherlands, 1990–91. *Ecol. Entomol.* **18**, 93–94.
Prendergast, J. R., Quinn, R. M., Lawton, J. H. Eversham, B. C. and Gibbons, D. W. (1993). Rare species, the coincidence of diversity hotspots and conservation strategies. *Nature* **365**, 335–337.

Pressey, R. L., Humphries, C. J., Margules, C. R., Vane-Wright, R. I. and Williams, P. H. (1993). Beyond opportunism: key principles for systematic reserve selection. *TREE* **8**, 124–128.
Quinn, J. F. and Hastings, A. (1987). Extinction in subdivided habitats. *Cons. Biol.* **1**, 198–208.
Queiroz, H. L. (1992). A new species of capuchin monkey, genus *Cebus* Erxleben, 1777 (Cebidae: Primates) from eastern Brazilian Amazonia. *Goel. Zool.* **15**, 1–13.
Rabinowitz, D., Cairns, S. and Dillon, T. (1986). Seven forms of rarity and their frequency in the flora of the British Isles. "Conservation Biology – The Science of Scarcity and Diversity" (M. E. Soulé, ed.), pp. 182–204. Sinauer Associates, Sunderland, MA.
Rappoport, E. H. (1982). "Areography: Geographical Strategies of Species". Pergamon, Oxford.
Raup, D. M. and Jablonski, D. (1993). Geography of end-Cretaceous marine bivalve extinctions. *Science* **260**, 971–973.
Raup, D. M. and Sepkoski, J. J. (1982). Mass extinctions in the marine fossil record. *Science* **215**, 1501–1503.
Raven, P. H. (1987). The scope of the plant conservation problem world-wide. *In* "Botanic Gardens and the World Conservation Strategy" (D. Bramwell, O. Hamann, V. Heywood and H. Synge, eds), pp. 19–29. Academic Press, London.
Raven, P. H. (1988a). Biological resources and global stability. *In* "Evolution and Coadaptation in Biotic Communities" (S. Kawano, J. H. Connell and H. Hidaka, eds), pp. 3–27. University of Tokyo Press, Tokyo.
Raven, P. H. (1988b). Our diminishing tropical forests. *In* "Biodiversity" (E. O. Wilson and F. M. Peter, eds), pp. 119–122. National Academy Press, Washington, DC.
Reid, W. V. (1992). How many species will there be? *In* "Tropical Deforestation and Species Extinction" (T. C. Whitmore and J. A. Sayer, eds), pp. 55–73. Chapman & Hall, London.
Richter-Dyn, N. and Goel, R. S. (1972). On the extinction of a colonizing species. *Theor. Pop. Biol.* **3**, 406–433.
Rohde, K. (1992). Latitudinal gradients in species diversity: the search for the primary cause. *Oikos* **65**, 514–527.
Schoener, T. W. (1986). Patterns in terrestrial vertebrate versus invertebrate communities: do systematic differences in regularity exist? *In* "Community Ecology" (J. Diamond and T. J. Case, eds), pp. 556–586. Harper & Row, New York.
Schoener, T. W. (1987). The geographical distribution of rarity. *Oecologia* **74**, 161–173.
Schoener, T. W. and Spiller, D. A. (1992). Is extinction rate related to temporal variability in population size? An empirical answer for orb spiders. *Am. Nat.* **139**, 1176–1207.
Shaffer, M. L. (1981). Minimum population sizes for conservation. *Bioscience* **31**, 131–134.
Shaffer, M. (1987). Minimum viable populations: coping with uncertainty. *In* "Viable Populations for Conservation" (M. E. Soulé, ed.), pp. 69–86. Cambridge University Press, Cambridge.
Sharrock, J. T. R. (1974). The changing status of breeding birds in Britain and Ireland. *In* "The Changing Flora and Fauna of Britain" (D. L. Hawksworth, ed.), pp. 203–220. Academic Press for the Systematics Association, London.
Shirt, D. B. (ed.) (1987). "British Red Data Books. 2. Insects". Nature Conservancy Council, Peterborough.
Simberloff, D. (1986). Are we on the verge of a mass extinction in tropical rain forests? *In* "Dynamics of Extinction" (D. K. Elliot, ed.), pp. 165–180. Wiley, New York.
Simberloff, D. (1992). Do species-area curves predict extinction in fragmented forest? *In* "Tropical Deforestation and Species Extinction" (T. C. Whitmore and J. A. Sayer, eds), pp. 75–89. Chapman & Hall, London.

Smith, F. D. M., May, R. M., Pellew, R., Johnson, T. H. and Walter, K. S. (1993a). Estimating extinction rates. *Nature* **364**, 494–496.
Smith, F. D. M., May, R. M., Pellew, R., Johnson, T. H. and Walter, K. R. (1993b). How much do we know about the current extinction rate. *TREE* **8**, 375–378.
Soulé, M. E. (ed.) (1986). "Conservation Biology. The Science of Scarcity and Diversity". Sinauer Associates, Sunderland, MA.
Soulé, M. E. (1987). "Viable Populations for Conservation". Cambridge University Press, Cambridge.
Soulé, M. E. (1990). The real work of systematics. *Ann. Miss. Bot. Gard.* **77**, 4–12.
Soulé, M. E. (1991). Conservation: tactics for a constant crisis. *Science* **253**, 744–750.
Southwood, T. R. E., Brown, V. K. and Reader, P. M. (1979). The relationships of plant and insect diversities in succession. *Biol. J. Linn. Soc.* **12**, 327–348.
Stevens, G. C. (1989). The latitudinal gradient in geographical range: how so many species coexist in the tropics. *Am. Nat.* **133**, 240–256.
Stevens, G. C. (1992). The elevational gradient in altitudinal range: an extension of Rappoport's latitudinal rule to altitude. *Am. Nat.* **140**, 893–911.
Stork, N. E. (1993). How many species are there? *Biod. Cons.* **2**, 215–232.
Stork, N. E. and Lyal, C. J. C. (1993). Extinction or "co-extinction" rates? *Nature* **366**, 307.
Sutton, S. L. and Collins, N. M. (1991). Insects and tropical forest conservation. In "Conservation of Insects and their Habitats". (N. M. Collins and J. A. Thomas, eds), pp. 405–424. Academic Press, London.
Terborgh, J. and Winter, B. (1980). Some causes of extinction, "Conservation Biology: an Evolutionary–Ecological Perspective" (M. E. Soulé, ed.), pp. 119–133. Sinauer Associates, Sunderland, Mass.
Thomas, C. D. (1991). Habitat use and geographic range of butterflies from the wet lowlands of Costa Rica. *Biol. Cons.* **55**, 269–281.
Thomas, C. D. and Jones, T. M. (1993). Partial recovery of a skipper butterfly (*Hesperia comma*) from population refuges: lessons for conservation in a fragmented landscape. *J. Anim. Ecol.* **62**, 472–481.
Thomas, C. D. and Mallorie, H. C. (1985). Rarity, species richness and conservation: butterflies of the Atlas Mountains in Morocco. *Biol. Cons.* **33**, 95–117.
Thomas, C. D., Thomas, J. A. and Warren, M. S. (1992). Distribution of occupied and vacant butterfly habitats in fragmented landscapes. *Oecologia* **92**, 563–567.
Thomas, J. A. (1976). "The Ecology and Conservation of the Large Blue Butterfly". Institute of Terrestrial Ecology.
Thomas, J. A. (1984). The conservation of butterflies in temperate countries: past efforts and lessons for the future. In "The Biology of Butterflies" (R. I. Vane-Wright and P. R. Ackery, eds), pp. 333–353. Academic Press, London.
Thomas, J. A. (1991). Rare species conservation: case studies of European butterflies. In "The Scientific Management of Temperate Communities for Conservation" (I. F. Spellerberg, F. B. Goldsmith and M. G. Morris, eds), pp. 149–197. Blackwell Scientific Publications, Oxford.
Thomas, J. A. and Morris, M. G. (1994). Patterns, mechanisms and rates of extinction among invertebrates in the United Kingdom. *Phil. Trans. R. Soc. Lond. B.* **344**, 47–54.
Tokeshi, M. (1993). Species abundance patterns and community structure. *Adv. Ecol. Res.* **24**, 111–186.
Tracy, C. R. and George, T. L. (1992). On the determinants of extinction. *Am. Nat.* **139**, 102–122.
Tuckwell, H. C. and Koziol, J. A. (1993). World and regional populations. *BioSystems* **31**, 59–63.

Valentine, J. W. (1990). The fossil record: a sampler of life's diversity. *Phil. Trans. R. Soc. Lond. B.* **330**, 261–268.

Vane-Wright, R. I., Humphries, C. J. and Williams, P. H. (1991). What to protect? – Systematics and the Agony of Choice. *Biol. Cons.* **55**, 235–254.

Vermeij, G. J. (1986). Survival during biotic crises: the properties and evolutionary significance of refuges. *In* "Dynamics of Extinction" (D. Elliott, ed.), pp. 231–246. Wiley, New York.

Vermeij, G. J. (1993). Biogeography of recently extinct marine species: Implications for conservation. *Cons. Biol.* **7**, 391–397.

Warren, M. S. (1991). The successful conservation of an endangered species, the Heath fritillary butterfly *Mellicta athalia*, in Britain. *Biol. Cons.* **55**, 27–56.

Warren, M. S. (1993). A review of butterfly conservation in central southern Britain: I. Protection, evaluation and extinction on prime sites. *Biol. Cons.* **64**, 25–35.

WCMC (1992) "Global Biodiversity: Status of the Earth's Living Resources". Chapman & Hall, London.

Welch, R. C. (1990). Dispersal of invertebrates in the agricultural environment. *In* "Species Dispersal in Agricultural Habitats". (R. G. H. Bunce and D. C. Howard, eds), pp. 203–218. Belhaven Press.

Wheeler, Q. D. (1990). Insect diversity and cladistic constraints. *Ann. Entomol. Soc. Am.*, **83**, 1031–1047.

Whitmore, T. C. and Sayer, J. A. (1992). "Tropical Deforestation and Species Extinction". IUCN and Chapman & Hall, London.

Williamson, M. (1981). "Island Populations". Oxford University Press, Oxford.

Wilson, E. O. (1987). The little things that run the world (the importance and conservation of invertebrates). *Cons. Biol.* **1**, 344–346.

Wilson, E. O. (1988). The current state of biological diversity. *In* "Biodiversity" (E. O. Wilson and F. M. Peter, eds), pp. 3–18. National Academy Press, Washington DC.

Wilson, E. O. (1989). Threats to biodiversity. *Sci. Am.* **September**, 108–116.

Wilson, E. O. (1993). "The Diversity of Life". Belknap Press, Harvard, MA.

Witting, L. and Loeschke, V. (1993). Biodiversity conservation: reserve optimization or loss minimization? *TREE* **8**, 417.

Yen, A. L. (1987). A preliminary assessment of the correlation between plant, vertebrate and coleoptera communities in the Victorian mallee. *In* "The Role of Invertebrates in Conservation and Biological Survey" J. D. Majer (ed.), pp. 73–88. Perth, Western Australia.

15

A World of Change: Land-Use Patterns and Arthropod Communities

MICHAEL B. USHER

I. Introduction	372
II. The Extent of Land-use Changes	373
A. Patterns of Change	373
B. The Effects on Arthropod Assemblages	379
III. Subtle Changes within Land Uses	384
A. Grassland Invertebrates: the Effects of Grazing Mammals	386
B. Woodland Invertebrates: the Effects of Forest Management	387
IV. Discussion	391
References	394

Abstract

What is the extent of land-use change? Data will be drawn from the first phase of the National Countryside Monitoring Scheme (NCMS), a scheme devised to look at gross changes in the Scottish countryside between the 1940s and 1970s. Results indicate large changes over this 30-year period; the greatest increases have been in young forest plantations (2584 km^2) and coniferous plantations (2135 km^2), and the greatest reductions in heather moorland (2741 km^2), unimproved grassland (1194 km^2) and blanket mire (1064 km^2). Despite the size of these reductions, they may not represent the greatest proportional reductions, which are for coniferous woodland (51.4%), standing natural water (25.3%), lowland raised mire (23.1%), heather moorland (17.8%), broadleaved woodland (14.0%) and parkland (13.5%).

Gross changes in land use, as demonstrated by NCMS in Scotland, have an important effect on the structure and composition of invertebrate communities. These changes will be discussed in relation to the arthropods likely to be affected. The two most important aspects of change relate to agricultural and forestry land use, and changes between these. Changes in these land uses also have major impacts on the communities of arthropods in the freshwater that drains such areas of land.

The more subtle changes in the arthropod communities are more interesting. What happens when the grazing pressure is varied, either by increasing or decreasing the number of sheep, cattle, red deer, etc? What are the effects of set-aside, or of differing agricultural practices, such as retaining headlands between or within fields? What are the effects of changing the tree species of woodlands, or of using mixtures of species rather than monocultures? Does the forest management regime, especially the intensity of thinning or the method of regeneration, have any identifiable effects? What are the effects of the general acidification of the environment on the arthropods? There are many questions that can be posed, but generally there are few answers in the research literature. This chapter reviews a selection of such questions.

Change is a feature of natural ecosystems and communities. Change has been induced by the seasonal cycle of autumn, winter, spring and summer in the northern temperate zone. Change has been induced by the year-to-year variability in all of the climatic influences. Change is inherent with the long-term cycles in climate over periods of one or more centuries. Change results from the natural catastrophic processes of fire, flood and wind, each inducing local successions in the arthropod communities. Change has been induced by the way that the human community has used the land, varying the pattern on a year-to-year or decade-to-decade basis. Change is being induced by pollution, so often arising outwith the ecosystem that is being influenced. To understand such changes we need to explore how whole systems work rather than targeting research at individual species or components of the system. Conservation of invertebrate diversity will ultimately depend on the conservation of the whole, functional system rather than on selected management of larger or smaller fragments of that system.

I. INTRODUCTION

If land-use change is recognized as a dominant feature of environmental change in the northern hemisphere, then this has hardly been reflected in recent research publications which have generally targeted climate change. As a first step in understanding land-use change it is important to know just how much change is actually occurring to the land cover within defined geographical areas and over defined periods of time. The relationships between land cover and the arthropods that are supported can then be viewed from two perspectives. One perspective is to look at the processes of change, the socioeconomic factors that determine the patterns of human use of land, and infer from them what gross changes are likely to occur. A knowledge of the arthropod communities of broad land cover types – for example, coniferous woodland, rye grass ley, winter barley fields – can then be used to predict which arthropods might be present after a change in land cover has occurred. The other perspective is to view the process from the point of view of the arthropods. Will the food resources remain after the change, will the habitat structure still be satisfactory, how rapidly has the change occurred, will there be a process of colonization and succession leading from an old to a new community, etc.?

As well as the gross changes in land use, grazing to forestry, wetland to industrial complex, etc., it is also interesting to investigate slight rather than gross changes in land-use patterns. What are the effects of changes in the grazing pressure on grasslands or of changes from one arable crop to another? How does the management of a forest – thinning regime, choice of tree species or rotational length – affect the assemblage of arthropods? One immediate conclusion from thinking about such questions is that there is relatively little research that has focused on such subtle changes in land use and their effect on arthropod species and assemblages.

The aims of this chapter are, therefore, to characterize the gross and subtle changes in land use and to speculate on how these may affect the arthropod populations and communities. Such aspects of land-use change inevitably must be set into a context of environmental change, whether of a regular short-term nature (i.e. the annual cycle of seasons in the north temperate zone); of the year-to-year variability in climate (e.g. temperature and precipitation); of the variability over centuries or millennia as the world's climate has changed (especially since the last Ice Age); of the catastrophic events of fire, flood and wind; or of the suspected human-induced changes in climate and to consequent effects on species, habitats and land use, especially in agriculture and changing patterns of crops (Parry, 1990; Parry *et al.*, 1990).

II. THE EXTENT OF LAND-USE CHANGES

A. Patterns of Change

Land use has changed; haymeadows and hedgerows have been replaced by silage meadows and fields without the patchwork of boundaries, rough grazing land and peat bogs have been replaced by conifer plantations, etc. Romantic notions of the environment of our childhood are, however, hardly sufficient to indicate just how much change there has been. One study that has attempted to quantify land cover change is the National Countryside Monitoring Scheme (NCMS), which is based on photo-interpretation, comparing aerial photographs taken in the 1940s, 1970s and late 1980s. The results for a number of land cover types and linear features are shown for the whole of Scotland (Table 1). The data demonstrate clearly the change in land cover from rough grazing (unimproved grassland and heather moorland in particular) to forestry (coniferous and young plantations especially). This accords with the national forest policy immediately after the 1939–1945 war to create a strategic reserve of timber.

Figure 1 shows all estimated changes in land cover of 100 km^2 and over, and presents a different picture to Table 1. The largest changes that now appear are those between arable land and improved grassland, probably a result of crop rotation; very little of this predominantly agricultural land has been afforested. The new plantations have largely been established on unimproved grassland and on heather moorland. Whilst only 1679 km^2 of heather moorland has been afforested, 1893 km^2 has changed to unimproved grassland (probably as a result of intensive grazing, see Miles (1988)) or semi-improved grassland. Figure 1 indicates the dynamic aspect of land-use change, especially on the more intensively used agricultural land.

However, the data in Table 1 can also be used to investigate the balance between land cover that is natural or semi-natural and that which is greatly modified by the human population (Fig. 2). In the former category are broadleaved and coniferous woodland, scrub, bracken, heather moorland, mires and wet ground, unimproved and semi-improved grassland, rock, cliff, and both standing and running natural water. In the latter category are plantations, mixed woodlands and parklands, improved grassland, arable land, quarries, built land, and standing man-made and running canalized water. The increase in the modified features in most parts of Scotland is noticeable, though the variance is great with the predominance of these features in the Central Belt of Scotland and along the east coast.

These data indicate the pattern of change in the post-war years, but change has continued. The data from the 1940s to 1970s (Anon, 1993), reflecting a change over an average 26-year period, and the data from the 1970s to 1980s (E. C. Mackey, pers. comm.), reflecting the further change over an average 16-year period, for Central Region, are compared in Table 2. In this part of Scotland afforestation has increased, wetland

Table 1. Data from the National Countryside Monitoring Scheme in Scotland, for the average 26-year period from the 1940s to the 1970s. Scottish Natural Heritage (Tudor et al., 1994)

Land cover or linear feature (km^2 or km)	Net change (km^2 or km)	95% confidence interval (km^2 or km)	% Change
Broadleaved woodland	−213	62	−14.0
Broadleaved plantation	8	20	+20.2
Coniferous woodland	−96	76	−51.4
Coniferous plantation	2135	574	+226.0
Mixed woodland	−49	62	−8.4
Young plantation	2584	765	+525.2
Parkland	−11	24	−13.5
Scrub[a]	40	–	+6.1
Unimproved grassland	−1194	675	−9.5
Semi-improved grassland	454	295	+12.8
Improved grassland	248	596	+3.0
Arable	−148	526	−1.8
Heather moorland	−2741	810	−17.8
Blanket mire	−1064	452	−9.3
Lowland raised mire	−43	29	−23.1
Standing natural water	−45	46	−25.3
Standing man-made water	87	87	+54.4
Developed land[b]	559	–	+32.7
Hedgerow	−15 946	3229	−37.2
Unsurfaced track	7896	2785	+25.2

[a]Includes both tall and low scrub categories; with pooling, no confidence interval can be given.
[b]Includes built transport corridor, recreation and quarry categories; no confidence interval can be given.

sites have diminished, the length of hedgerows has declined, and the amounts of developed land and tracks have continued to increase. Table 2 indicates something of the dynamic nature of land cover, with the extent of improved grassland increasing to the 1970s and then decreasing, whilst the area of arable land has shifted in the opposite direction.

Land-cover estimates (Tables 1 and 2) reflect the patterns of land use – forestry, agriculture, industrial and urban use, etc. These land uses are themselves determined by the social structures and needs of the local community (Munton et al., 1992). However, the overall potential for competing forms of land use has an environmental determinant (Usher, 1992), for example the ability to grow certain crop species. But which crop species is grown is an option that is economically driven. It is the

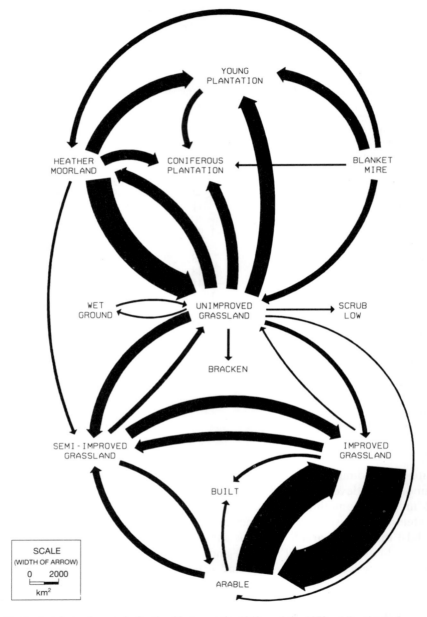

Fig. 1. Land-use changes in Scotland between the 1940s and the 1970s. All estimated changes in excess of 100 km² are shown; the width of the arrows is proportional to the amount of the change, as indicated by the scale.

15. Land-use Patterns and Arthropod Communities

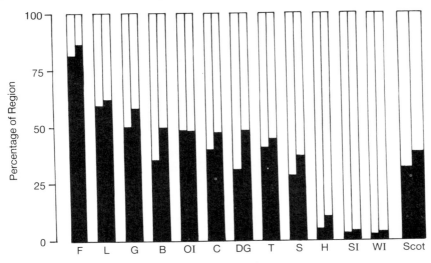

Fig. 2. The area distribution of natural, semi-natural and modified land cover types in the Regions of Scotland. The natural and semi-natural land cover types (open boxes) include broadleaved and coniferous woodland, scrub, bracken, heather moorland, wetlands, standing and natural running waters, unimproved and semi-improved grassland, rock and cliff. The modified/altered land cover types (filled boxes) include broadleaved, coniferous and young plantations, mixed woodland, parkland, standing man-made and canalized waters, improved grassland, arable, quarry, recreation, built land and transport corridors. The Regions of Scotland are: B, Borders; C, Central; DG, Dumfries and Galloway; F, Fife; G, Grampian; H, Highland; L, Lothian; OI, Orkney Islands; S, Strathclyde; SI, Shetland Islands; T, Tayside; WI, Western Islands (Outer Hebrides); Scot, Scotland as a whole. They are arranged in approximately increasing order of natural and semi-natural habitat.

historical, economic and social factors of today's human population that determine the patterns of land use that are seen now, and the patterns of change that will be observed during the next decade or more. The historical deforestation of Scotland (Anderson 1967) has led to the need for reforestation on a scale that is greater than in most other northern temperate countries. Land-use change into the future cannot therefore be independent of change in the past.

Another important source of data on land use change is the Countryside Survey 1990 (Bunce et al., 1992), which will have three major outputs, namely:

- an inventory of the major land cover categories and estimates of their ecological characteristics;

Table 2. A comparison of land cover data from the 1940s to 1970s and from the 1970s to late 1980s of the NCMS, Central Region of Scotland from Anon (1993) and Scottish Natural Heritage (unpublished data)

Land cover or linear feature (km² or km)	Net change 1940s–1970s (km² or km)	% Change	Net change 1970s–1980s (km² or km)	% Change
Broadleaved woodland	−3.6	−5.4	−0.1	−0.2
Broadleaved plantation	10.6	+163.4[a]	−0.1	−0.6
Coniferous woodland	−0.2	−65.3[a]	0.02	+17.3[a]
Coniferous plantation	41.3	+113.8	28.9	+37.8[a]
Mixed woodland	−5.7	−20.1	5.8	+25.8
Young plantation	15.4	+190.1[a]	59.0	+243.7
Parkland	−4.0	−76.5[a]	1.8	+148.9[a]
Scrub	7.3	+22.5	2.2	+5.6
Unimproved grassland	11.2	+1.5	−105.6	−13.6[a]
Semi-improved grassland	−53.9	−30.4	1.2	+0.9
Improved grassland	45.4	+10.7	−27.0	−5.8
Arable	−37.2	−9.0[a]	51.8	+13.8[a]
Heather moorland	−45.3	−12.5	−30.8	−9.6
Blanket mire	−4.1	−6.7	−0.2	−0.4
Lowland raised mire	−3.7	−11.5	−8.8	−30.8
Standing natural water	−0.1	−0.9[a]	−0.2	−2.3
Standing man-made water	0.9	+2.9[a]	−1.2	−3.7
Developed land[b]	43.8	+32.9[a]	26.5	+15.0[a]
Hedgerow	−2156.6	−41.1[a]	−904.3	−29.4[a]
Unsurfaced track	256.8	+17.7	511.1	+30.2[a]

[a] These are more reliable data since the standard error is less than half the estimated net change.
[b] Includes built transport corridor, recreation and quarry categories.

15. Land-use Patterns and Arthropod Communities

- a summary of changes over the last three decades by comparison with the data of earlier surveys; and
- baseline data for future monitoring.

Although neither study is specifically concerned with arthropods, the land-cover data, and the changes reflecting land-use practices, will be important in interpreting changes in arthropod distributions and abundances, and in predicting changes that are likely to take place in the future.

B. The Effects on Arthropod Assemblages

As the land cover responds to the pattern of land use, considerable change can be expected in the composition of the arthropod assemblages. Using the NCMS data (Table 1) to predict what changes might occur, D. Horsfield (pers. comm.) has commented that the 226% increase in coniferous plantations is likely to lead to a large increase in a relatively few species that are associated with the foliage of spruce (*Picea*) trees, including Hemiptera (mainly psyllids and aphids), Lepidoptera, Hymenoptera (mainly sawflies) and Diptera (particularly aphid-feeding hoverflies), as well as a few Coleoptera (e.g. bark beetles and weevils). As the litter layer develops, this is likely to favour many of the soil invertebrates, especially mites and springtails. However, these increases will be balanced by losses of insects characteristic of the former habitats. If unimproved grassland had been planted, there may have been considerable losses of Orthoptera, Hemiptera, Coleoptera, Diptera and Lepidoptera. If blanket mire had been afforested, the main groups to be reduced are likely to be Odonata, Diptera, Coleoptera and spiders, many of which are likely to be local or rare species with northern distributions. If heather moorland had been afforested, the losses are likely to include Hemiptera, Coleoptera, Lepidoptera, Hymenoptera and spiders. The mosaic of different ages of the heather moorland would be lost and this would affect species of both the earlier and later successional phases of the burning cycle used in heather moor management (Usher and Thompson, 1993).

These are gross changes which indicate little of ecological significance; the aphid species of grassland may be replaced by the aphid species of spruce forest. It is changes in species complement which are important. Two examples will be used – the reduction in the linear extent of hedgerows and the establishment of woodland on arable land.

1. Hedgerows

Hedgerows, a linear feature of the countryside, are both being removed for conversion to arable land and falling into disuse as relicts (i.e. as lines of trees or scattered bushes). The data in Table 1 indicate a loss of the order of 600 km per annum in Scotland between the 1940s and 1970s. The decline in the length of hedges seems to be general, having occurred throughout Great Britain (Barr et al., 1992).

Jones (1992) studied the woodlice and millipedes of hedgerows in Oxfordshire with an aim of understanding how hedgerow fragmentation would affect these invertebrate groups. The two groups are not species rich (21 species, see Table 3). In arable farmland they are largely confined to the hedgerow habitat and do not have a strongly seasonal appearance. Jones found that the five hedgerows without millipedes were smaller (a mean of 330 m^2 compared with 692 m^2, $F_{1,29} = 6.45$, $P < 0.05$) and less base rich (pH of 6.2 compared with 7.5, $F_{1,29} = 8.27$, $P < 0.01$). As many hedgerows have a similar width, this is indicative that longer hedges are more likely to support millipede populations. Furthermore, Jones found that the connectance (the number of hedgerow connections) was important in determining the assemblage of woodlice and millipedes (*Cylindroiulus punctatus*, *Brachydesmus superus* and *Oniscus asellus* are indicative of well-connected hedges).

Jones' (1992) study demonstrated that the removal of hedgerows is not the only factor to influence these invertebrate assemblages. There are very subtle influences at work indicating that it is the pattern of the countryside – the number of times a hedge connects with other hedges, whether a hedge abuts onto a wood, or the density of hedgerows and woods in the area – that influences the species composition of an individual hedge. This has clear implications for the management of hedgerows (Jones et al., 1991) in agricultural environments; 10 km of interconnected hedgerows may prove to be more species rich than 10 km of fragmented hedgerows.

2. Woodland in Arable Farmland

With agricultural surpluses in the northern temperate zone, and with an increasing demand for wood products, there has been an increased interest in converting arable land into woodlands. A perceived benefit of such farm woodlands is their wildlife enhancement (Insley, 1988) since it was felt that they would be richer in plants, vertebrates and invertebrates than the arable ecosystems that they would replace. Several studies of established farm woodlands in the Vale of York have endeavoured to

Table 3. The 21 species of millipedes and woodlice collected from 31 hedgerows in Oxfordshire. Data from Jones (1992)

Species	Number of individuals in all hedgerows	Number of hedgerows where species was recorded
Millipedes		
Archiboreoiulus pallidus (Brake-Birks)	1	1
Blaniulus guttulatus (F.)	3	1
Brachydesmus superus (Latzel)	78	14
Cylindroiulus caeruleocinctus (Wood)	1	1
Cylindroiulus latestriatus (Curtis)	1	1
Cylindroiulus punctatus (Leach)	54	11
Glomeris marginata (Villers)	125	11
Nanogona polydesmoides (Leach)	4	1
Ophiodesmus albonanus (Latzel)	1	1
Ophyiulus pilosus (Newport)	9	4
Polydesmus angustus (Latzel)	12	7
Polydesmus gallicus (Latzel)	15	4
Polydesmus inconstans (Latzel)	10	1
Polyxenus lagurus (L.)	1	1
Tachypodoiulus niger (Leach)	163	17
Woodlice		
Armadillidium vulgare (Latreille)	32	11
Oniscus asellus (L.)	130	24
Philoscia muscorum (Scopoli)	788	31
Porcellio scaber (Latreille)	865	30
Trichoniscus pusillus (Brandt)	151	17
Trichoniscus pygmaeus (Sars)	1	1

quantify this wildlife gain, if it exists. Certainly, there appear to be richer communities of Cryptostigmata within farm woods than in the surrounding arable land (Sgardelis and Usher, 1994). Figure 3 demonstrates the sharp discontinuity in both species richness and number of individuals at the woodland margin. This discontinuity could be caused by the frequent use of agrochemicals in the field, by the repeated disturbance of the soil structure in the field, or by the greater quantity of organic matter in the woodland.

For surface dwelling arthropods the results may be different (Bedford and Usher, 1994). Carabid beetles in two woods provide evidence that the field, near the woodland margin, has the greatest species richness (Fig. 4), with species such as *Calathus piceus* and *Abax parallelepipedus* characteristic of the woodland and *Calathus melanocephalus*, *Trechus quadristriatus*, *Bembidion lampros*, *B. tetracolum* and *Pterostichus*

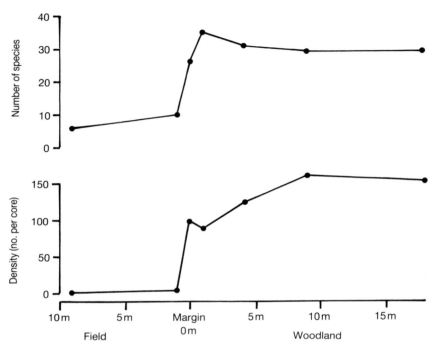

Fig. 3. The Cryptostigmata (moss mites) collected in the surface 10 cm of soil along a transect from 9 m into an arable field to 18 m into a farm woodland. The 0 m point indicates the boundary between wood and field. The upper graph shows the species richness and the lower graph the number of individuals per core. There were too few Cryptostigmata to analyse data from samples 40 m into the field. From Sgardelis and Usher (1994).

melanarius characteristic of the field and field-edge habitats. Similarly for the spiders there is evidence that the woodland itself is less species rich than the margin (Fig. 4), with *Lepthyphantes flavipes*, *Diplocephalus picinus* and *Monocephalus fuscipes* characteristic of the woodland, and *Erigone* spp., *Bathyphantes* spp. and *Pardosa* spp. characteristic of the field and field margin.

These studies of spiders, ground beetles and mites indicate that large changes are likely to occur in the arthropod assemblages where arable land is afforested, though they indicate neither the speed of change nor the factors driving the change. Using an island biogeographic approach, Keiller and Usher (1995) analysed the moth species that occurred in 18 farm woodlands (Fig. 5). The three factors that were considered to be important – area of wood, shape of wood, habitat remnants (e.g.

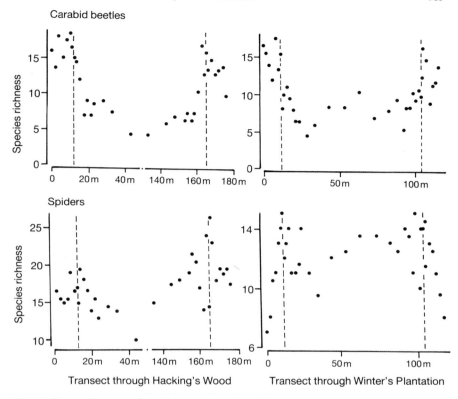

Fig. 4. Scatter diagrams of the richness of species of Carabidae (upper diagrams) and spiders (lower diagrams) along transects of pitfall traps running through Hacking's Wood (left diagrams) and Winter's Plantation (right diagrams) into arable fields on each side of the wood. The dashed lines indicate the boundaries between the woods and the fields. From Bedford and Usher (1994).

hedgerows and copses) in the vicinity – are all associated with the pattern of land use.

From these studies it is possible to start generalizing about the change of arthropod assemblages when land use changes. First, it is likely that there will be large changes in species composition, though the time scale for these changes is currently unknown. Second, the extent of that change will depend on the nature of the land-use change, be it total or patchy, and whether or not small patches of the previous land cover remain. Third, different types of land cover tend to be associated with different species richnesses, as demonstrated by Peng *et al.* (1993) for agroforestry and neighbouring arable systems. The importance of studying species,

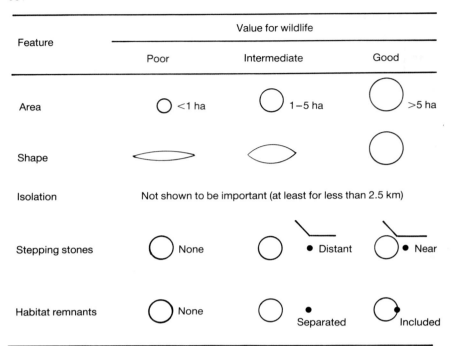

Fig. 5. A diagrammatic representation of the conservation value of designed farm woodlands based on a criterion of the diversity of Macrolepidoptera. Habitat remnants are shown as small black circles and hedgerows by straight lines. From Keiller and Usher (1995).

rather than families or orders, is shown by a study of moths on different land classes in Britain (Bunce *et al.*, 1993). Whereas the overall abundance of moths may not vary greatly, the species composition or the proportional abundance of the species may vary considerably (Table 4). By studying the insects of various land-cover types, and by examining the transboundary differences, it is increasingly possible to predict the induced changes in the arthropod assemblages resulting from land-use changes. These major changes – heather moorland to forest, unimproved grassland to arable fields, reduction of hedgerows or natural running waters – have drastic effects that can largely be predicted; it is the more subtle changes within a land-cover type that are more difficult to predict.

III. SUBTLE CHANGES WITHIN LAND USES

The most common changes that effect invertebrate assemblages are those induced by slight change of land use. These may result from increased

Table 4. The average catches of the nine most common species of moths in traps, grouped in relation to four land classes. Data from Bunce et al. (1993)

Species	Lowland cereal	Lowland grassland	Marginal upland	Upland
Luperina testacea (flounced rustic)	37	18	0	0
Agrotis exclamationis (heart and dart)	33	48	6	0
Spilosoma luteum (buff ermine)	10	20	0	0
Diarsia rubi (small square spot)	6	20	4	18
Spilosoma lubricipeda (white ermine)	16	31	15	6
Orthosia gothica (Hebrew character)	31	42	124	100
Xanthorhoe montanata (silver ground carpet)	19	21	69	96
Hydraecia micacea (rosy rustic)	6	5	105	25
Cerapteryx graminis (antler)	2	2	94	57

stocking rates of sheep or cattle, greater abstraction of water for urban, industrial or agricultural use, a move towards more intensive production of a forest, changes in the use of pesticides for arable crops, greater recreational use resulting in trampling of the vegetation, or from changes in the crop species. Whenever questions relating to these more subtle changes in land use and their effects on invertebrates are asked, the answer tends to be "the overwhelming message is the lack of data which are (a) comprehensive, i.e. wide geographical coverage; and (b) integrated, i.e. direct links from land use to land cover to invertebrates" (O. W. Heal, pers. comm.). There are, however, signs that integrated research projects are commencing, even if the more comprehensive nature of the research is still to be planned and realized. Two examples are considered here.

A. Grassland Invertebrates: the Effects of Grazing Mammals

A research project, which commenced in 1993, is investigating the effects of grazing density on upland invertebrates in Snowdonia. This Countryside Council for Wales project clearly comes into Heal's "integrated" category, with its four main aims being to assess the effects on invertebrate populations of the length of time since vegetation was last grazed, to estimate the relative contribution of above-ground and below-ground plant production to this process, to establish methods of long-term monitoring of invertebrates, and to compare the availability of invertebrate food, in grazed and ungrazed areas, to upland birds and other predators. The geographical coverage is, unfortunately, minimal, and such research could profitably be repeated in other upland areas.

Although the effects of grazing pressure on sward structure are widely understood, there are few studies of the invertebrates. One notable exception was the study by King and Hutchinson (1976), King et al. (1976) and Hutchinson and King (1980) on the effects of sheep stocking density on a number of insect groups. All groups decreased in both abundance and biomass, and usually in species richness, as stocking levels were increased, except for the Formicoidea and larvae of Scarabaeidae. It remains to be proven whether the results of this study at 1000 m altitude on Australian pasture are valid in north temperate situations. However, it does indicate that studies of invertebrates need to have wide taxonomic coverage, not least because different species, families or orders will react in different ways. Gibson et al.'s (1992) study of the spiders of grazed grasslands indicated that the geometry of the environment determined which species, or families, would be favoured by different levels of grazing.

There is a much more extensive literature of the effects of grassland management on the occurrence and abundance of butterflies (reviewed for Europe by Thomas, 1991). For *Lysandra bellargus* (adonis blue), Thomas (1983) estimated that about one-third of the extinct colonies had disappeared because the downland habitat had been ploughed or otherwise improved agriculturally; the remaining two-thirds had gone extinct because changing economic conditions meant that it was no longer economic to have ungulates grazing the downland sites. Similarly, Thomas (1991) showed that *Hesperia comma* (silver-spotted skipper) is dependent upon the grazing of its downland habitat, and that if the grazing is too intensive the potential larval food plants are too short for adult females to lay eggs on them. However, Bourn and Thomas (1993) have shown that *Aricia agestis* (brown argus) requires lush, sheltered growth of its larval food plant, with higher organic nitrogen content, to stimulate oviposition.

The study by Völkl et al. (1993), comparing mown, sheep-grazed and abandoned calcareous grasslands, also contributes to understanding how a number of insect species react, indicating once again that studies need to be at the species level, emphasizing the taxonomic comprehensiveness of any research study. Management targeted at one species will almost certainly reduce the population size of, or even eliminate, several other species. In terms, then, of the management of grasslands we need studies that are:

- geographically comprehensive, i.e. that they have been repeated at a number of sites so that there is an assurance that the results are not site specific;
- taxonomically comprehensive, i.e. that more than one species, genus, family or even order of invertebrate has been studied; and
- integrated, i.e. that the links between land use, land cover and invertebrates have been established.

It is only on the basis of such research that one can gain confidence that the effects of land-use change on arthropod assemblages will be understood, but without temporal replication it still leaves unanswered the question of how climate change will interact with the process.

B. Woodland Invertebrates: the Effects of Forest Management

Several aspects of woodland management have a direct influence on invertebrate assemblages. Of importance to the Lepidoptera is the presence of sunny rides with many nectar-producing plants. Of importance to some species of Coleoptera, Hymenoptera and Diptera is the presence of dead wood. In areas where non-native species are being grown the design of the woodlands can be important for enhancing both the abundance and diversity of insects. Although there are many aspects to forest management, only the decline in coppicing, the loss of dead wood in the managed forest, and the management of forest rides and tracks, will be considered here.

First, there has been a declining market for coppice. This economic factor has meant that there has been a substantial decrease in the area of broadleaved woodland managed under a coppice regime. For example, Peterken (1981) documented the decline in coppice woodland on ancient woodland sites in Rockingham Forest, Northants, from 8442 ha in 1650 to 1389 ha in 1972, but he indicated that at least 90% of the 1972 area was no longer cut as coppice. The decline has affected those species that are particularly associated with the early stages of the re-establishment of a forest cover, such as *Mellicta athalia* (heath fritillary butterfly) (Fuller and Warren, 1990). Usher and Jefferson (1991) and Pollard and Yates (1993) quoted the example of the re-introduction of coppice management at Gait Barrow National Nature Reserve (see Fig. 6). With only one site being coppiced, it is impossible to be certain that *Argynnis adippe* (high brown fritillary) has responded to the altered management of the forest, but two coppice cycles and other circumstantial evidence indicates that this is the most likely reason for the large increases in the population during the first 3–4 years following coppicing. Although most studies of coppiced woodland have been related to butterflies, it is likely that many other species of arthropods will be influenced by the creation of sunnier and more open conditions.

Second, the managed forest generally contains significantly less dead wood than the unmanaged forest. Speight (1989) defined saproxylic invertebrates as those that are "dependent, during some part of their life cycle, upon the dead or dying wood of moribund or dead trees (standing or fallen), or upon wood-inhabiting fungi, or upon the presence of other saproxylics". A consideration of the habitats occupied by saproxylic species indicates that many are unlikely to be found, at least in any abundance, in plantation forests. Working on a European scale, Speight listed 138 species of Coleoptera (mainly in families Buprestidae, Cerambycidae and Elateridae), 64 species of Diptera (mainly in family Syrphidae), 11 species of Hemiptera and 13 species of Hymenoptera that he considered could be used as indicative of forests that have international importance for nature conservation. The study is particularly useful in raising awareness about a group of invertebrates that are threatened by a change of forest management. For example, change from native broad-

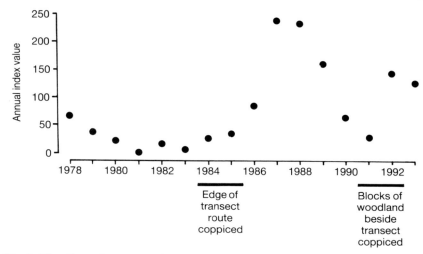

Fig. 6. The effects of recommencing coppice management in Gait Barrow NNR on *Argynnis adippe* (high brown fritillary). Taken from Usher and Jefferson (1991) with an update to 1993 by R. Petley-Jones (pers. comm.). Previous coppicing had taken place in about 1970.

leaved trees to exotic conifers has the inevitable result that invertebrates feeding on only one species of dead wood, such as the xylomyiid fly *Solva interrupta* whose larvae feed only under the bark of moribund aspen, totally lose their habitat. However, for many species, the decline in the amount of dead wood through the cutting of trees at younger ages, or the removal of wood for fuel, means that a basic resource is scarcer, that it can only support smaller populations, and that these populations are likely to become increasingly fragmented. Such a fragmented distribution is demonstrated by *Callicera rufa*, a hoverfly that breeds in rot-holes in old *Pinus sylvestris* (Scots pine) trees (Rotheray and MacGowan, 1990).

Accounts of saproxylic invertebrates in Europe need to be considered in the light of national conservation programmes. Kirby and Drake (1993) have set the British fauna into its European context and have identified the particular habitats that are most under threat in Britain. One particularly pertinent question relates to the perceived need for stringent forest hygiene measures to be applied to all tree crops. In the British situation it seems that the risk of insect damage to either trees or timber is virtually nil when ancient broadleaved trees or deadwood are retained. There is, however, a very real risk to many coniferous plantations, especially pine and spruce, of insect attack, posing problems for the conservation of ancient Caledonian pine forest in Scotland. Kirby and

Drake (1993) concluded with a "Conservation Guide for the Entomological Investigation of Deadwood Habitats". It is perhaps indicative of the scarcity of some of the habitats of the saproxylic fauna that the guide recommends that a small hammer and nails are carried so that dead bark that has been examined can be nailed back in place!

Third, the environment of a woodland area can be modified by the management of the rides and tracks. The majority of research is again related to butterflies, but it is possible that these can be used as indicator species for other groups (e.g. Banwell and Crawford, 1992). A study of *Leptidea sinapis* (wood white butterfly) has shown that it requires a relatively shaded ride (Warren, 1985). Using the results of studies of 12 rides, Warren and Fuller (1990) demonstrated the changes in *L. sinapis* populations as coniferous plantations develop. In five cases the young stands have sunny rides (less than 20% shade) initially, but as they cast more shade the butterflies increased their usage of the rides. In four heavily shaded rides (60–90% shade) there were very few butterflies. These studies indicate an optimal shading of 20–50%, with the butterflies rapidly decreasing their use of the rides as the shade increased from 50 to 60%. However, what are optimal conditions for one species may not be for another; Greatorex-Davies *et al.* (1993) presented data for 39 butterfly species in seven ancient woodlands. They recognized shade-tolerant and shade-intolerant species. As conifers planted into ancient woodlands grow, the shade-tolerant species will maintain a presence, though the numbers of the shade-intolerant species will decline and the rarer ones may not all survive. It is therefore likely that some species will become locally extinct.

Whilst the majority of these studies have been performed in the south of England, Ravenscroft (1994) investigated the occurrence and abundance of *Carterocephalus palaemon* (chequered skipper) in Scotland. This species occurs mainly in the open deciduous woodlands associated with damp areas where the larval food plant, *Molinia caerulea*, grows, but the females require a nectaring source before egg laying commences. Again the method of forest management is critical for the continued existence of this species.

Although so many of these studies relate to butterflies, there are some studies of other groups that can be used to suggest how forests should be structured. The Forestry Commission recommended both ecological work and landscaping of the forest roads so that they become more favourable places for wildlife (Carter and Anderson, 1987). Buse and Good (1993) investigated the relationship between staphylinid beetles and coniferous forest design and management. Young (1992) discussed how even-aged coniferous stands can be made richer for their insect species. Carter and

Anderson (1987) suggested bays of short vegetation, to encourage flowering plants and nectaring insects, along forest roads. Buse and Good (1993) suggested more forest edge, and hence they recommended small habitat islands within the forest. Young (1992) recommended the mixture of a few native tree species, especially along the edges and by the rides. Both Speight (1989) and Kirby and Drake (1993) were advocating the retention of dead wood and other habitat types for the saproxylic fauna. The overwhelming message coming from these studies and recommendations is that a diversity of habitats within an even-aged forest plantation is important if a diverse arthropod assemblage is desired.

IV. DISCUSSION

Land-use change is driven by ecological, socioeconomic and political factors (Whitby, 1992). The causes of land-use change are often economic pressures, and the type of change is governed partly by the social structures of human communities but more importantly by the ecological constraints that determine the narrow range of options that is available (and these options will change in the face of global climate change). As land use changes the land cover, huge changes can be expected in the species complement, diversity, numbers and biomass of arthropods. Even small changes in land use, hardly influencing the land cover, can have large effects on many arthropod species. This discussion will concentrate on six topics that have relevance to the changing assemblages of arthropods.

First, **fragmentation** is a feature of many extensive forms of land use whereby small remnants of natural or semi-natural vegetation are left in a landscape modified by agricultural or silvicultural monocultures (Saunders et al., 1987). By and large this is seen to reduce biological diversity of the invertebrate species characteristic of the fragments (e.g. Webb, 1989), although the "edge effects" due to species characteristic of the neighbouring habitat types may increase apparent diversity (e.g. Webb and Hopkins, 1984), at least in the short term. The crucial question is how speedily the process of "relaxation", the extinction of species on the patch so that it equilibrates to a sustainable number, will occur. Although there is an increasing number of studies of island biogeography for insects on habitat fragments, the dynamics of such systems are still far from being understood. Some studies are starting to tackle this issue by establishing islands, for example within arable land (Thomas et al., 1991, 1992), and analysing how species composition changes and how insects move around the landscape of patch mosaics.

Second, what are the **genetic effects** on the invertebrates? It is surprising how little is known about the population genetics of wild insect species in the northern temperate zone and the effects of habitat management or change on the genetic variability of these populations. As part of their Species Recovery Programme, English Nature are funding research into this topic (I. F. G. McLean, pers. comm.), focusing the work primarily on rare butterfly species but planning to extend it to other Lepidoptera and Orthoptera.

Third, can any prediction be made about **pests**? Some of the scarcer forest insects have become pests – increased to large population size and eating parts of a commercially valuable tree species – when faced with monocultures. With possible climate change, more species have the potential to impact on economic activities in this way; a review by Watt et al. (1990) clearly links fecundity of *Panolis flammea* (pine beauty moth) with accumulated summer temperature. On the other hand, physiological adaptations may uncouple hosts and pests; Watt et al. (1990) analysed *Operophthera brumata* (winter moth) whose larvae feed on the bursting buds of a large number of species. If climate warming is a gradual process, then the moths are likely to evolve adaptations to the earlier bud flush so that egg hatching remains correctly synchronized; if climatic variability becomes greater, then this synchrony may be lost, leading to a potential decline in this species. As outlined in Anon (1991), climate change could have huge implications for the regional flora and fauna, with insects being amongst the first species to respond to change. However, it remains an area for active research to predict what tomorrow's pest species are likely to be.

Fourth, there are **transboundary effects**. These are likely to be most marked in freshwater systems that drain land with changing forms of land use. The largest effect is due to changing acidity of the water drainage from the land, as demonstrated by Rundle et al. (1992), with the consequential mobility of aluminium as freshwaters become strongly acid. It is, therefore, in relation to coniferous afforestation that most studies have been targeted. Weatherley et al. (1993) again emphasized the importance of water acidity, but also their data "indicated that different management techniques can generate a range of habitat types necessary to support different (aquatic) invertebrate taxa". Besides these effects of pH change, there are also eutrophication effects (e.g. Maitland et al., 1990) resulting from the use of fertilizers. However, most research has been targeted at a single habitat without recognizing such transboundary effects. It is only research at the landscape scale that will elucidate ecological processes that are occurring at these large scales.

15. Land-use Patterns and Arthropod Communities 393

Fifth, how does one **conserve** arthropods in the face of so much change? Should programmes for species recovery be established (e.g. English Nature, 1992), targeting conservation activities on a few rare and threatened species? Among the invertebrates on which the English Nature programme has focused are *Armandia cirrhosa* (lagoon sandworm), *Dolomedes plantarius* (fen raft spiders), *Gryllus campestris* (field cricket), *Decticus verrucivorus* (wart-biter cricket), *Cicadetta montana* (New Forest cicada), *Graphoderus zanatus* (a water beetle), *Assocetia caliginosa* (reddish buff moth), *Thetidia smaragdaria* (Essex emerald moth), *Lycaena dispar* (large copper butterfly) and *Maculinea arion* (large blue butterfly). Ten species is a small number compared with the size of the invertebrate fauna of England, but it represents a beginning for active conservation of some of the rarer species. Their population sizes are now low, fragmented, and it becomes important to understand what genetic effects have taken place. Can sufficient habitat be retained, created or managed for such species, and should translocation, either within England or between England and continental Europe, take place so as to increase the genetic variability of these populations? How will these small populations change under a changing climate? Such questions need some sort of risk assessment, both in terms of the demography of the populations themselves (e.g. Burgman *et al.*, 1993) and in terms of the pollution effects on the habitats in which they live (e.g. Bengtsson and Torstensson, 1988).

Finally, this leads to the subject of **prediction**. Can one predict how land-use change and the consequential change in land cover will determine which species of arthropods will become locally extinct and which will survive? What will be their changes in abundance? Which other species will colonize the area and how might they interact with the extant species? The work of Luff *et al.* (1992), looking at the probability of occurrence of selected invertebrates in relation to a classification of land-cover classes, is a step in this direction. The application of ecological theory to the strategies of individual species, although better developed for plants (Bunce *et al.*, 1993), is important. Such studies pose difficult research problems, especially when the autecology of the individual invertebrate species is often poorly understood. However, they are essential if a more integrated approach to viewing large-scale ecological processes is to provide benefits to managers of natural resources, be they foresters or agriculturalists faced with pest-management problems, fisheries managers faced with understanding the food requirements of their stocks, or conservationists endeavouring to protect local, regional, national or international biodiversity.

Acknowledgements

I thank the many people who have made information available to me in preparing this review. In particular, I thank Ed Mackey, Dr Philip Boon, Dr David Horsfield and Dr Neil Ravenscroft (Scottish Natural Heritage), Professor Gareth Wyn Jones (Countryside Council for Wales), Dr Ian McLean (English Nature), Professor Bill Heal (Institute of Terrestrial Ecology) and Dr Peter Saunders (Department of the Environment). I also thank Ed Mackey, Dr Neil Ravenscroft and Dr Des Thompson for commenting on a draft of this chapter.

REFERENCES

Anderson, M. L. (1967). "A History of Scottish Forestry". Nelson, London.
Anon (1991). "The Potential Effects of Climate Change in the United Kingdom. First Report of the United Kingdom Climate Change Impacts Review Group". HMSO, London.
Anon (1993). "National Countryside Monitoring Scheme. Scotland: Central". Scottish Natural Heritage, Edinburgh.
Banwell, J. L. and Crawford, T. J. (1992). "Key Indicators for British Wildlife: Butterfly Monitoring Scheme". Unpublished report, University of York.
Barr, C. J., Bunce, R. G. H., Cummins, R. P., French, D. D. and Howard, D. C. (1992). Hedgerow changes in Great Britain. In "Institute of Terrestrial Ecology: 1991–1992 Report", pp. 21–24. Natural Environment Research Council, Swindon.
Bedford, S. E. and Usher, M. B. (1994). Distribution of arthropod species across the margins of farm woodlands. *Agriculture, Ecosystems and Environment*, in press.
Bengtsson, G. and Torstensson, L. (1988). "Soil Biological Variables in Environmental Hazard Assessment: Concepts for a Research Programme". *National Swedish Environmental Protection Board Report* **3499**.
Bourn, N. A. D. and Thomas, J. A. (1993). The ecology and conservation of the brown argus butterfly *Aricia agestis* in Britain. *Biol. Cons.* **63**, 67–74.
Bunce, R. G. H., Barr, C. J. and Fuller, R. M. (1992). Integration of methods for detecting land use change, with special reference to Countryside Survey 1990. In "Land Use Change: the Causes and Consequences" (M. C. Whitby, ed.), pp. 69–78. HMSO, London.
Bunce, R. G. H., Howard, D. C., Hallam, C. J., Barr, C. J. and Benefield, C. B. (1993). "Ecological Consequences of Land Use Change". Department of the Environment, London.
Burgman, M. A., Ferson, S. and Akçakaya, H. R. (1993). "Risk Assessment in Conservation Biology". Chapman & Hall, London.
Buse, A. and Good, J. E. G. (1993). The effects of conifer forest design and management on abundance and diversity of rove beetles (Coleoptera: Staphylinidae): implications for conservation. *Biol. Cons.* **64**, 67–76.
Carter, C. I. and Anderson, M. A. (1987). "Enhancement of Lowland Forests Ridesides and Roadsides to Benefit Wild Plants and Butterflies". *For. Commission Res. Inf. Note* **126**.
English Nature (1992). "Species Recovery Programme" (A folder containing A4 information sheets on species included in the programme). English Nature, Peterborough.

Fuller, R. J. and Warren, M. S. (1990). "Coppiced Woodlands: their Management for Wildlife". Nature Conservancy Council, Peterborough.

Gibson, C. W. D., Hambler, C. and Brown, V. K. (1992). Changes in spider (Araneae) assemblages in relation to succession and grazing management. *J. Appl. Ecol.*, **29**, 132–142.

Greatorex-Davies, J. N., Sparks, T. H., Hall, L. M. and Marrs, R. H. (1993). The influence of shade on butterflies in rides of coniferised lowland woods in southern England and implications for conservation management. *Biol. Cons.* **63**, 31–41.

Hutchinson, K. J. and King, K. L. (1980). The effects of sheep stocking level on invertebrate abundance, biomass and energy utilization in a temperate, sown grassland. *J. Appl. Ecol.* **17**, 369–387.

Insley, H. (1988). "Farm Woodland Planning". *For. Commission Bull.* **80**.

Jones, S. H. (1992). "The Landscape Ecology of Hedgerows with Particular Reference to Island Biogeography". D Phil thesis, University of York.

Jones, S., Fry, R. and Lonsdale, D. (1991). Hedgerows and arable field margins. *In* "Habitat Conservation for Insects – a Neglected Green Issue" (R. Fry and D. Lonsdale, eds), pp. 116–132. Amateur Entomologists' Society, Middlesex.

Keiller, S. W. J. and Usher, M. B. (1995). The Macrolepidoptera of farm woodlands: determinants of diversity and community structure. *Cons. Biol.*, in press.

King, K. L. and Hutchinson, K. J. (1976). The effects of sheep stocking intensity on the abundance and distribution of mesofauna in pastures. *J. Appl. Ecol.* **13**, 41–55.

King, K. L., Hutchinson, K. J. and Greenslade, P. (1976). The effects of sheep numbers on associations of Collembola in sown pastures. *J. Appl. Ecol.* **13**, 731–739.

Kirby, K. J. and Drake, C. M. (1993). "Dead Wood Matters: the Ecology and Conservation of Saproxylic Invertebrates in Britain". *English Nature Science* **7**.

Luff, M. C., Eyre, M. D., Cherrill, A. J., Foster, G. N. and Pilkington, J. G. (1992). *In* "Land Use Change: the Causes and Consequences" (M. C. Whitby, ed.), pp. 102–110. HMSO, London.

Maitland, P. S., Newson, M. D. and Best, G. A. (1990). "The Impact of Afforestation and Forestry Practice on Freshwater Habitats". *Focus on Nat. Conserv.* **23**.

Miles, J. (1988). Vegetation and soil change in the uplands. *In* "Ecological Change in the Uplands" (M. B. Usher and D. B. A. Thompson, eds), pp. 57–70. Blackwell Scientific, Oxford.

Munton, R. J. C., Lowe, P. and Marsden, T. (1992). Forces driving land use change: the social, economic and political context. *In* "Land Use Change: the Causes and Consequences" (M. C. Whitby, ed.), pp. 15–27. HMSO, London.

Parry, M. (1990). The potential impact on agriculture of the greenhouse effect. *Land Use Policy*, **April 1990**, 109–124.

Parry, M. L., Porter, J. H. and Carter, T. R. (1990). Agriculture: climate change and its implications. *Trends Ecol. Evol.* **5**, 318–322.

Peng, R. K., Incoll, L. D., Sutton, S. L., Wright, C. and Chadwick, A. (1993). Diversity of airborne arthropods in a silvoarable agroforestry system. *J. Appl. Ecol.* **30**, 551–562.

Peterken, G. (1981). "Woodland Conservation and Management". Chapman & Hall, London.

Pollard, E. and Yates, T. J. (1993). "Monitoring Butterflies for Ecology and Conservation: the British Butterfly Monitoring Scheme". Chapman & Hall, London.

Ravenscroft, N. O. M. (1994). The conservation of the chequered skipper butterfly *Carterocephalus palaemon* (Pallas) in Scotland, in press.

Rotheray, G. E. and MacGowan, I. (1990). Re-evaluation of the status of *Callicera rufa* Schummel (Diptera: Syrphidae) in the British Isles. *The Entomol.* **109**, 35–42.

Rundle, S. D., Lloyd, E. C. and Ormerod, S. J. (1992). The effects of riparian management and physicochemistry on macroinvertebrate feeding guilds and community structure in upland British streams. *Aquatic Cons.* **2**, 309–324.

Saunders, D. A., Arnold, G. W., Burbridge, A. A. and Hopkins, A. J. M. (eds) (1987). "Nature Conservation: the Role of Remnants of Native Vegetation". Surrey Beatty, Chipping Norton, New South Wales.

Sgardelis, S. P. and Usher, M. B. (1994). Responses of soil Cryptostigmata across the boundary between a farm woodland and an arable field. *Pedobiologia* **38**, 36–49.

Speight, M. C. D. (1989). "Saproxylic Invertebrates and their Conservation". *Council of Europe, Nature and Environment Series* **42**.

Thomas, J. A. (1983). The ecology and conservation of *Lysandra bellargus* (Lepidoptera: Lycaenidae) in Britain. *J. Appl. Ecol.* **20**, 59–83.

Thomas, J. A. (1991). Rare species conservation: case studies of European butterflies. *In* "The Scientific Management of Temperate Communities for Conservation" (I. F. Spellerberg, F. B. Goldsmith and M. G. Morris, eds), pp. 149–197. Blackwell Scientific, Oxford.

Thomas, M. B., Wratten, S. D. and Sotherton, N. W. (1991). Creation of "island" habitats in farmland to manipulate populations of beneficial arthropods: predator densities and emigration. *J. Appl. Ecol.* **28**, 906–917.

Thomas, M. B., Wratten, S. D. and Sotherton, N. W. (1992). Creation of "island" habitats in farmland to manipulate populations of beneficial arthropods: predator densities and species composition. *J. Appl. Ecol.* **29**, 524–531.

Tudor, G. J., Mackey, E. C. and Underwood, F. M. (1995). "The National Countryside Monitoring Scheme: the Changing Face of Scotland: 1940s–1970s". Scottish Natural Heritage, Battleby, Perthshire.

Usher, M. B. (1992). Land use change and the environment: cause or effect? *In* "Land Use Change: the Causes and Consequences" (M. C. Whitby, ed.), pp. 28–36. HMSO, London.

Usher, M. B. and Jefferson, R. G. (1991). Creating new and successional habitats for arthropods. *In* "The Conservation of Insects and their Habitats" (N. M. Collins and J. A. Thomas, eds), pp. 263–291. Academic Press, London.

Usher, M. B. and Thompson, D. B. A. (1993). Variations in the upland heathlands of Great Britain: conservation importance. *Biol. Cons.* **66**, 69–81.

Völkl, W., Zwölfer, H., Romstöck-Völkl, M. and Schmelzer, C. (1993). Habitat management in calcareous grasslands: effects on the insect community developing in flower heads of *Cynarea. J. Appl. Ecol.* **30**, 307–315.

Warren, M. S. (1985). The influence of shade on butterfly numbers in woodland rides, with special reference to the wood white *Leptidea sinapis. Biol. Cons.* **33**, 147–164.

Warren, M. S. and Fuller, R. J. (1990). "Woodland Rides and Glades: their Management for Wildlife". Nature Conservancy Council, Peterborough.

Watt, A. D., Ward, L. K. and Eversham, B. C. (1990). *In* "The Greenhouse Effect and Terrestrial Ecosystems in the UK" (M. G. R. Cannell and M. D. Hooper, eds), pp. 32–37. HMSO, London.

Weatherby, N. S., Lloyd, E. C., Rundle, S. D. and Ormerod, S. J. (1993). Management of conifer plantations for the conservation of stream macroinvertebrates. *Biol. Cons.* **63**, 171–176.

Webb, N. R. (1989). Studies on the invertebrate fauna of fragmented heathland in Dorset, U.K., and the implications for conservation. *Biol. Cons.* **47**, 153–165.

Webb, N. R. and Hopkins, P. J. (1984). Invertebrate diversity on fragmented *Calluna* heathland. *J. Appl. Ecol.*, **21**, 921–933.

Whitby, M. C. (ed.) (1992). "Land Use Change: The Causes and Consequences". HMSO, London.

Young, M. R. (1992). Conserving insect communities in mixed woodlands. *In* "The Ecology of Mixed-species Stands of Trees" (M. G. R. Cannell, D. C. Malcolm and P. A. Robertson, eds), pp. 277–296. Blackwell, Oxford.

16

Insects as Indicators of Land-use Change: A European Perspective, Focusing on Moths and Ground Beetles

M. L. LUFF AND I. P. WOIWOD

I. Introduction	400
II. Ground Beetles and Macrolepidoptera as Indicator Groups	401
III. Types of Land-use Change	403
IV. Ground Beetles and Land-use Type	403
V. Lepidoptera and Land-use Type	404
VI. "Intensity" of Land Use	409
VII. Monitoring the Effects of Land-use Change	410
VIII. Habitat Scale and Pattern	413
IX. Predicting the Effects of Land-use Change	415
X. Conclusions	417
References	417

Abstract

The use of ground beetles (Carabidae) and macrolepidoptera (butterflies and moths) as indicators of land-use change is reviewed, with particular reference to European species. The suitability of these insects as indicators is discussed, in relation to their diversity, taxonomic stability, ease of monitoring, seasonality and feeding habits.

The relationships between ground beetles and type of land use depend on the vegetation cover type, and on species' responses to environmental factors such as soil moisture and altitude. If the distribution of cover types is known, species' distributions can be modelled. Macrolepidoptera diversity is related to land classes on both the national and local scale, as evidenced by data from the Rothamsted light trap network. Hedges, bushes and woodland produce the greatest diversity. Within any cover type, management affects ground beetle assemblages both directly, and by altering the physical and biotic environment. Woodland management has considerable effect on Lepidoptera by affecting the amount of light available. Monitoring has shown long-term changes in carabid assemblages, related to increasing arable land use; the most seriously affected species are those with poor dispersal powers. An example of predicting the effects of land use change on carabids in the Tyne catchment (north-east England) is given. The importance of habitat scale and pattern to the persistence of both carabids and macrolepidoptera is stressed. It is concluded that the fragmentation of habitats into small isolated patches presents a potentially severe threat to poorly dispersing species.

I. INTRODUCTION

The habitat of insects must, by its definition, supply the needs of each insect species throughout its lifetime. These needs will comprise, at the very least, food and suitable climatic conditions, and may also include shelter from disturbance and natural enemies. Land use is a major determinant of these needs for all terrestrial insects, and possibly also for many aquatic species. In this chapter we review existing work and consider some likely impacts of land-use changes in the UK on two taxonomic groups of insects, Carabidae (ground beetles) and macrolepidoptera (butterflies and moths). It must be emphasized that the importance of insects as land-use indicators is by no means restricted to these taxa, nor to the UK. Overall insect biodiversity in the tropics is dramatically affected by land use and management both in agriculture and forestry (Wolda, 1983; Holloway, 1991; Holloway and Stork, 1991; Sutton and Collins, 1991). The taxa of insects used as indicators of land-use change in tropical or warm temperate climates include ants (Andersen, 1987), lepidopteran larvae (Janzen, 1988) and adults (Pinheiro and Ortiz, 1992). We hope, however, that the geographically and taxonomically more limited examples in this chapter will serve to outline principles of the relationships between insects and land use, that can be applied more widely.

The first part of the chapter considers briefly the factors that are known to determine the habitats for Carabidae and macrolepidoptera, and the types of change to which they will respond. The effect of land-use change on insects can be studied at three levels; on individual species, on the composition of species in a habitat (an assemblage or community) or on simplified measures of the overall structure of that assemblage, such as species richness, diversity or biomass. We give examples of all of these. The importance of scale of land use and habitat pattern is considered. Large-scale and long-term monitoring studies have been carried out on both groups of insects (e.g. Taylor, 1986; Turin and den Boer, 1988) and can be used to assess the impacts of land use and its change. These and other studies can be used to develop predictive models of the impact of future land-use changes on these insects; an example of one such model, as developed in the NERC/ESRC Land Use Programme (O'Callaghan, 1992) is given.

II. GROUND BEETLES AND MACROLEPIDOPTERA AS INDICATOR GROUPS

As pointed out by Woiwod and Thomas (1993), all researchers think that their particular study group is especially suitable as an indicator. Criteria for insects as indicators have been put forward by Holloway (1983) and Foster (1987) among others; particular points stressed by these authors are:

(1) Diversity. The group must have enough species to represent, and be characteristic of, a wide range of habitats. However, there should not be too many species, or the practicalities of identification become too great. Carabidae, with 349 species in Britain and Ireland (Luff, 1994) falls within the range (300–500) recommended by Foster (1987) whereas the macrolepidoptera with about 950 species in Britain (Emmett, 1991) is considerably higher. Despite this, regular large-scale monitoring has been shown to be feasible (Taylor, 1986; Woiwod and Harrington, 1994).

(2) Taxonomic stability. Ideally any group to be used should not only be readily identifiable, but should also be taxonomically stable, so that the nomenclature is not seriously confused. For the British ground beetles and macrolepidoptera this is ensured by a long history of study by both amateur and professional entomologists over the last 150 years. That is not to say that insects cannot be used to monitor changes in environment without such a taxonomic background. Hutcheson (1990) was able to use beetles identified only to family level to characterize insect communities in New Zealand, and west African foliage arthropods could be characterized by family and feeding guild (Basset et al., 1992). More generalized indices such as body size, biomass and species richness based on morpho-species (Operational Taxonomic Units, Basset et al., 1992) can also be used

often without reference to the actual named species present. This will continue to be particularly the situation with tropical faunas (e.g. Morse et al., 1988; Holloway et al., 1990). The relationships between body size, biomass and diversity are discussed by Stork (1988).

(3) Monitoring techniques. If insects are to be used as indicators of change, a simple and replicable sampling technique is needed. For ground beetles, this has long been the role of the ubiquitous pitfall trap. Although the relationship between trap catch and beetle population density is unclear (Adis, 1979; Baars, 1979), this method does at least indicate presence/absence of particular species, and if whole season catches are used, these give a species-specific indication of population size (Baars, 1979) that can be used to compare between habitats and seasons. Nocturnal macrolepidoptera can be sampled routinely using light-trapping methods. These have similar advantages and disadvantages to pitfall trapping, as samples of individual species provided only relative measures of density which cannot be used directly for comparison between species although they can be used to indicate annual population changes within species (Taylor, 1986; Woiwod and Harrington, 1994). For day-flying macrolepidoptera (mainly butterflies), a successful fixed transect method has been developed as the basis for a national monitoring scheme in Britain (Pollard et al., 1986; Pollard and Yates, 1993).

(4) Seasonality. An indicator group should be capable of being caught throughout much of the year, so that changes taking place at different times can be detected, and sampling time of the insects is not too critical. This is true for ground beetles where the species used are of moderate to large size, and relatively long-lived, but for macrolepidoptera which are usually only active as adults for short periods a more regular sampling throughout the annual cycle is required in order to cover the full range of species.

(5) Phytophagy. Foster (1987) considered that host-specific phytophages are not useful indicators, as it would be easier to assess the plants directly. However, it is not always easy to monitor plants directly in tropical forest situations (Holloway, 1983), and Woiwod and Thomas (1993) give examples of how Lepidopteran population size changes by a factor of from three to nearly 30 times more than that of their food plants over periods of 3–7 years following habitat change. Thus, even when monitoring plant populations may be easier, it gives a slower and therefore less sensitive indication of habitat change than do the populations of insects feeding on those plants. In effect, phytophagous insects magnify vegetation and microclimatic effects which may be quite subtle and even undetectable directly except from detailed long-term plant studies (e.g. see Erhardt and Thomas, 1991).

In contrast to Lepidoptera, carabid beetles are mostly carnivorous, with some being omnivores or generalist plant feeders. They respond, however, to changes in their physical environment, and there is considerable evidence, some of which is presented later, that soil moisture in particular is a major determinant of the beetle assemblage in any particular habitat. Together, ground beetles and larger Lepidoptera therefore provide potential indication of land-use changes through both the vegetation composition, its structure, and the underlying soil parameters.

III. TYPES OF LAND-USE CHANGE

Monitoring of land-use change in Britain has benefited from an objective classification of land classes developed by the Institute of Terrestrial Ecology (ITE) (Bunce et al., 1981). This system uses a multivariate analysis of cartographic features in order to characterize each of 32 land classes. The actual land use in each land class can then be assessed by survey, stratified according to land class (Bunce et al., 1992).

If the impact of land-use change is to be predicted, we need to distinguish between two types of change (Rushton, 1992). The first is in "type" of land use, where the vegetation cover type changes from one type to another, for example from upland grazing to forestry, or intensive arable to set aside. The second change is in "intensity" of land use, where management changes within a vegetation cover type. This would include, for example, changes in grazing management on pasture, or of agrochemical usage on an arable crop. The following sections give some examples of the impact of such changes on our selected insect groups.

IV. GROUND BEETLES AND LAND-USE TYPE

Analyses of presence/absence species lists from whole countries or regions within a country (Hengeveld and Hogeweg, 1979; Luff et al., 1989; Turin et al., 1991) have shown that particular land-cover types have characteristic ground beetle faunas. The main determinants of the fauna seem to be soil moisture, distance from actual running or standing water, altitude and vegetation cover. Many species with relatively restricted soil moisture and altitude requirements appear to occur in many land-cover types. Examples of the responses of individual species of *Pterostichus* to some of these parameters are given in Rushton et al. (1991).

Depending on the exclusiveness of a species' requirements, it may occur widely in many land-cover types, or be restricted to a few (Fig. 1).

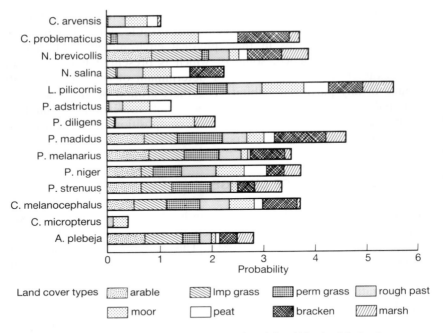

Fig. 1. Probabilities of occurrence of selected species of Carabidae in eight land-cover types. From Luff et al. (1992).

If the proportions of the land-cover types in each land class are known, then the probabilities of occurrence of each species in that land class can be calculated and its distribution mapped. Figure 2 shows the calculated distributions in Redesdale, Northumberland, of two carabid species with contrasting cover-type preferences.

As an example of the effects of change in land use, Luff and Rushton (1988) show the effects on ground beetles of upland improvement, following various procedures. Soil disturbance has a severe effect on large species such as *Carabus*, where highly mobile species such as *Loricera pilicornis* are relatively numerous in improved sites (Fig. 3). The effects at the assemblage level are clear from ordinations (Rushton et al., 1989), with the physical disturbance of the soil during improvement being more important than the subsequent application of pesticides.

V. LEPIDOPTERA AND LAND-USE TYPE

In contrast to ground beetles, for which pitfall trapping gives primarily presence/absence data in any habitat, and the number of species in any

16. Insects as Indicators of Land-use Change

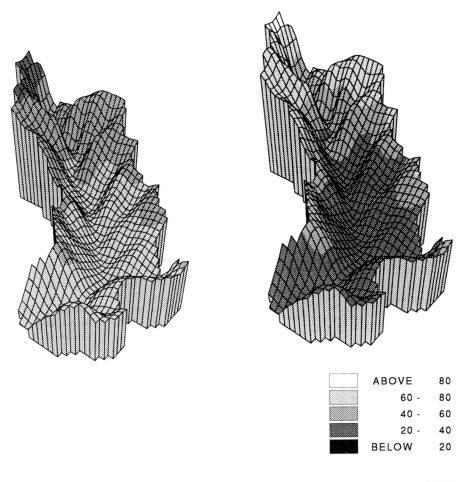

	ABOVE	80
	60 -	80
	40 -	60
	20 -	40
	BELOW	20

Fig. 2. Calculated distributions of *Nebria brevicollis* (left) and *Carabus problematicus* (right) in the Rede catchment, Northumberland. From Luff *et al.* (1992).

one site seldom exceeds 20, macro-moths sampled by the Rothamsted light-trap network are more numerous both in species and in individual abundance, and population trends can be monitored (Woiwod and Dancy, 1987; Crawford, 1991). The pattern of density distribution for any particular species is extremely complex and not yet well understood in most cases (Woiwod and Thomas, 1993). However, species diversity has been shown to be a measurable summary of the whole moth community structure which is influenced by environmental change, and so can be

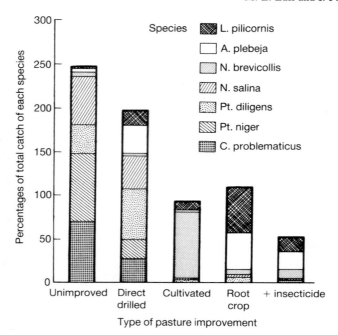

Fig. 3. Relative abundances of some Carabidae following upland pasture improvement. Data from Luff and Rushton (1988).

used to look at the relationship with land use (Taylor, 1978; Taylor et al., 1978).

The parameter α from the log-series distribution has been shown to be the best diversity parameter for a variety of reasons (Taylor et al., 1976) and has now been calculated from over 350 light-trap sites throughout Britain (Fig. 4). Although there are latitudinal and altitudinal effects presumably related to temperature (Turner et al., 1987), the pattern of diversity is closely related to the ITE land-class distribution. This is perhaps not surprising as the land classes themselves are based on cartographic features which are also largely determined by latitude and altitude.

At the individual species level, many species of moths are significantly related to the land class system, and hence to the land cover types comprising each land class (Woiwod and Thomas, 1993). The prime determinant of the land class DECORANA first axis score is altitude, so that the contrasting density distributions of species with differing altitude responses is clear.

16. Insects as Indicators of Land-use Change 407

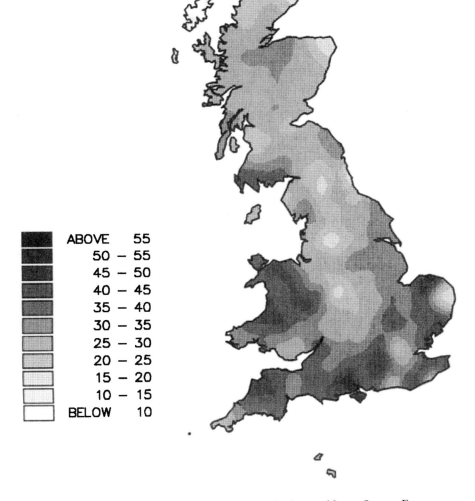

Fig. 4. Moth diversity (α) from 359 light traps of the Rothamsted Insect Survey. From Woiwod and Thomas (1993).

In order to assess the effects of land use within this general picture, a smaller scale of study is needed. This has been provided recently by the network of 26 light traps run on the Rothamsted Estate since 1990. The annual diversity (α) of moths from these samples (Fig. 5) has been related to the local land use within a 50-m and 100-m radius round each trap

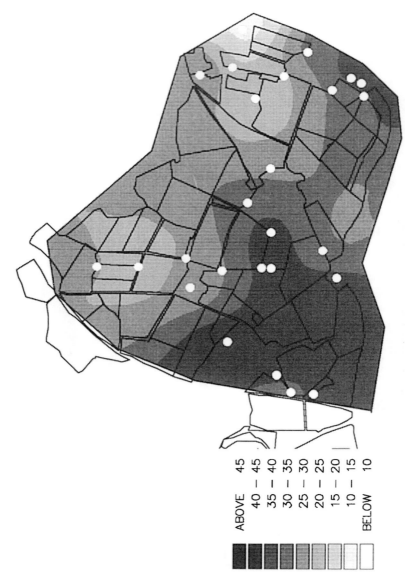

Fig. 5. Moth diversity (α) from 26 sites on the Rothamsted estate in 1990. From Woiwod and Thomas (1993).

(Woiwod and Thomas, 1993). The land use in the smaller radius circle accounted for more of the between-trap variability.

Hedges, bushes and woodland were found to produce the highest diversity, urban habitats and permanent grass the lowest. The same general conclusions were reached by Taylor *et al.* (1978) using data from the national network but because the larger scale analysis included extra variability due to climate and soil type the explanatory power of the relationship was lower (Woiwod and Thomas, 1993). A similar but slightly more complex relationship exists between local land use and total numbers of moths, an estimate of biomass. Standardized transect counts of butterflies at 670 sites throughout Great Britain have also shown a similar land use, species richness pattern (Thomas, 1983).

VI. "INTENSITY" OF LAND USE

Although, as outlined above, the type of land use at any location will clearly affect the diversity and species composition of both ground beetles and moths, many land-use changes involve variations in "intensity" within a cover type rather than from one cover type to another. For ground beetles, these have been studied both at the individual species level, and by multivariate analyses of the whole fauna trapped within particular vegetation types or crops under differing management regimes.

Within intensively managed grasslands in northern England and southern Scotland, the carabid fauna was determined primarily by altitude and soil-based parameters such as water content, bulk density and amount of organic matter (Eyre *et al.*, 1990). There was a significant effect of pesticide application, but this was relatively minor.

In a more uniform arable environment, the Boxworth experiment considered the effects of intensity of pesticide applications on intensive cereals (Greig-Smith, 1989). Three levels of treatment were: (a) full insurance – routine prophylactic application of insecticides, fungicides and herbicides; (b) supervised – pesticides only applied if pest, disease or weed levels exceeded predetermined thresholds; (c) integrated – using thresholds as above, but with choice of varieties and husbandry to reduce pesticide inputs. The treatments had an effect both on individual species of ground beetles (Burn, 1992) and at the community level (Fig. 6). Within a range of crops, the crop type seems to be more important than pesticide usage in determining the carabid assemblage (Booij and Noorlander, 1988), as year to year effects were more important than the management system. Overall, however, organic farming systems seem to support more carabids than conventional ones (Luff, 1987; Booij and Noorlander, 1992) due to a combination of increased prey availability, reduced disturbance and greater weed cover.

The responses of individual carabid species to grassland management intensity have been analysed by ordinating and scoring the various management characteristics (Rushton et al., 1990a). The probabilities of occurrence of various carabids were then related to this management variable by logistic regression, showing responses to intensity of management that were negative (large, non-flying species), positive (smaller, highly mobile species), or which showed an optimum, intermediate management intensity.

Management of scrub and woodland habitats also influences the ground beetle fauna; scrub management on the Castor Hanglands NNR affected the ground beetle communities through the extent of bare ground and grass tussocks in the resulting habitat (Rushton et al., 1990b). Within managed forests, there are clear changes in the carabid assemblage, following the cycle of cutting and replanting (Day and Carthy, 1988) or depending on the stand composition (Butterfield and Benitez-Malvido, 1992). Any changes in these management practices will thus influence the beetle community present. Any disturbance to the established forest generally increases the carabid diversity in the resultant cleared or scarified areas (Parry and Rodger, 1986), although such increase may be short-lived.

One of the major important features of woodland for Lepidoptera is the amount of light reaching the herbaceous vegetation as suitable light levels are essential for butterfly oviposition (Warren, 1985; McKay, 1991). This emphasizes the importance of management practices such as coppicing (Warren and Thomas, 1992) and management of woodland rides (Greatorex-Davies et al., 1993). Combined insect responses to both light and/or leaf quality in woodlands have been shown by Greatorex-Davies et al. (1994) for Heteroptera and Coleoptera in woodland rides in southern England, by Basset (1992) for arthropods in Australian rainforest and by Pinheiro and Ortiz (1992) for fruit-feeding butterflies in Brazil.

VII. MONITORING THE EFFECTS OF LAND-USE CHANGE

Previous sections have given examples of how both type and intensity of land use can influence the particular groups of insects studied. However, if these insects are to be used to monitor land-use change, we need to know the scale of natural variation in their diversity, species composition

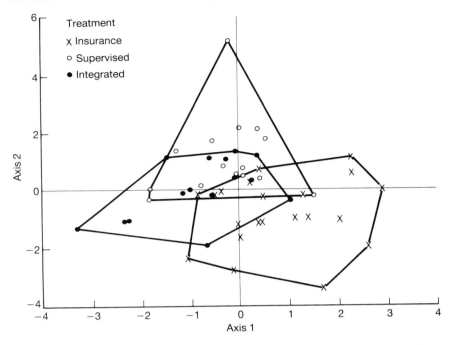

Fig. 6. DECORANA ordination of field/year combinations from the Boxworth experiment, based on their polyphagous predators; polygons enclose the data from each type of treatment.

and abundance. It is here that the few long-term studies of insects at particular sites become important, as they serve to show the extent of natural variation in these biological features (Woiwod, 1991).

Changes in the diversity (α) of moths caught in three sites at Rothamsted (Fig. 7) show that whilst there are considerable (and apparently uncorrelated) fluctuations from year to year at each site, sampling over a number of years shows consistent trends, such as the greater diversity in woodland than in the fields. It also suggests that the field diversity has dropped since 1950, when it was at the present woodland level.

The assemblage of ground beetles in a weedy arable plot in the Tyne valley has been monitored since 1981 (details in Luff, 1990). There has been a decreasing trend in species richness, with associated change in α (as a measure of diversity) (Fig. 8). This has taken place without obvious local changes in land use, and provides some measure of apparent

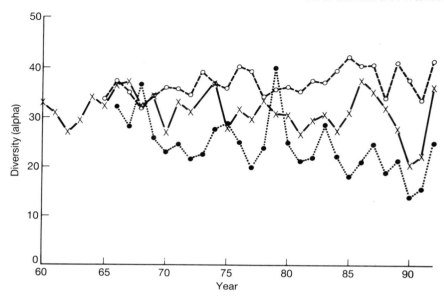

Fig. 7. Diversity (α) of macrolepidoptera from three light traps at Rothamsted since 1960: ×, Barnfield; ●, Allotment; ○, Geescroft. Redrawn from Woiwod (1991).

"noise" against which any effects of changes in land use would have to be measured.

Longer-term studies have been carried out in The Netherlands, where the distribution of carabids has been analysed using data from 1880 to the present (Hengeveld, 1985; Turin and den Boer, 1988; Desender *et al.*, 1994). Hengeveld (1985) has argued that the overall change (and to some extent increase) in the species' distributions this century is caused by climatic factors, and that effects of human influences seem to be of minor importance. The hypothesis tested by Turin and den Boer (1988) was that increasing habitat fragmentation and isolation leads to a decline in the poorly dispersing species, which according to the founding hypothesis of den Boer (1977) will become extinct in small, isolated habitat fragments. The carabids were divided into four groups according to their dispersal power and tolerance of intensive agriculture. The well-dispersing, tolerant species have increased markedly, especially in the latter part of this century (Fig. 9); there has been some increase in the well-dispersing but agriculturally intolerant species, but no clear trend in those poorly dispersing species that are found in discrete, non-agricultural habitats. Thus land-use change effects seem to be superimposed on the overall

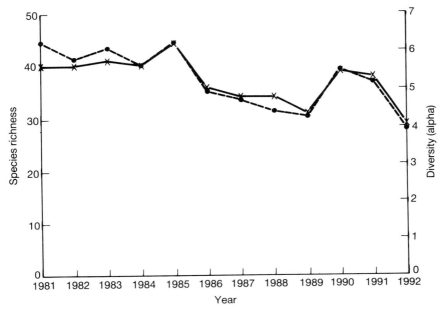

Fig. 8. Diversity (α, ●) and species richness (×) of Carabidae from a weedy arable field near Newcastle since 1981.

long-term pattern of change. Desender et al. (1994) claim that in Belgium there has been a marked change in ground beetle communities since about 1950, associated with habitat loss and agricultural intensification.

VIII. HABITAT SCALE AND PATTERN

Several workers have stressed that effects of land use on insects will vary according to the spatial scale being considered. Hengeveld (1987) and Woiwod and Thomas (1993) give examples from Carabidae and macro-lepidoptera, respectively. At the landscape scale, habitat fragmentation has already been referred to, but other features to be considered are linear habitats such as hedges and field margins.

The effects of habitat fragmentation on heathland ground beetles are particularly well documented. De Vries and den Boer (1990) analysed the poorly dispersing *Agonum ericeti* Panz. in Drenthe (The Netherlands). This species cannot bridge distances greater than about 200 m between reproduction sites. In larger areas of heath (> 50 ha) the species has

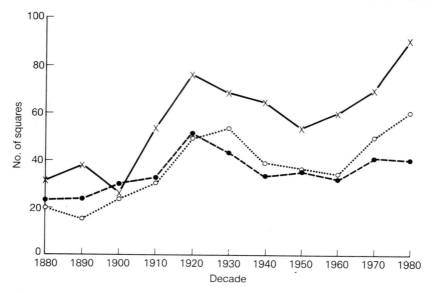

Fig. 9. Numbers of Netherlands 10-km squares from which Carabidae of three ecological groups have been recorded since 1880 (corrected for sampling intensity): +, agriculture tolerant; △, dispersers; ●, non-dispersers. Data from Turin and den Boer (1988).

persisted, whereas in small patches, especially if these have long been isolated, it is often now extinct. More generally, de Vries (1994) has shown that heath carabid species with poor powers of dispersal will decline in habitats of less than about 70 ha, whereas well-dispersing species can persist in habitats as small as 8 ha. The effects of heath fragmentation on indices of invertebrate diversity were also considered by Hopkins and Webb (1984), and Webb (1989a,b). Again, species with poor powers of dispersal tended to occur only on the largest heathlands.

Another example of the combined importance of habitat size and distance between habitat patches is that of *Hesperia comma* L., the silver spotted skipper butterfly (Thomas *et al.*, 1992). The species will only lay eggs on *Festuca ovina* in the earliest stages of vegetation succession (Thomas *et al.*, 1986). Census of all such sites on the North and South Downs showed that the percentage occupancy of suitable habitats was greatest in larger patches, and in those closest to other such suitable habitats. Extinction from otherwise suitable sites was caused by changes in grazing pressure, that is in intensity of land use.

The role of linear habitats has been extensively considered. Hedges can act either as habitats or as barriers to movement for butterflies (Woiwod and Thomas, 1993). Although the hedge bottom may support carabid populations by virtue of the vegetation cover present (Pollard, 1968; Asteraki *et al.*, 1992), it does not support a woodland beetle fauna, but functions rather as a herbaceous field margin. However Burel (1992) has shown that the overall carabid assemblage is related to the previous pattern and history of hedgerows in the landscape. Field margins and headlands are known to support a diverse insect fauna (Dennis and Fry, 1992), and selective management of these can increase the diversity of butterflies (Dover *et al.*, 1990) as well as that of carabids and other insects (Hassall *et al.*, 1992). A recent development is the artificial and deliberate introduction of such strips into arable fields in order to increase the populations of polyphagous predators such as Carabidae (Thomas, 1990; Thomas *et al.*, 1991).

In the same way, intercropping arable with production hedges can increase the populations of natural enemies close to the crop (Peng *et al.*, 1993).

IX. PREDICTING THE EFFECTS OF LAND-USE CHANGE

As outlined above, land-use change can conveniently be divided into changes between and within land-cover types. The effects of changes between cover types in northern England were modelled by Luff *et al.* (1992), using a hierarchical matrix model (Rushton, 1992). Changes in land cover in Cumbria (Nature Conservancy Council, 1987) were used to predict changes in the probability of occurrence of 14 species in five ITE land classes over a 30-year period, using nationwide data on the occurrence of cover types in each land class. All species were predicted to decrease overall; this decrease was greatest in the large, poorly dispersing species such as those of the genus *Carabus*, and least in highly active species favouring more intensive management, such as *Amara plebeja* Gyll. and *Pterostichus melanarius* Illig.

To be more realistic, this model needs to take into account the local (rather than national) distribution of cover types in each land class (Rushton *et al.*, 1994). This gives predictions that are not only more accurate spatially, but which also can be reasonably validated using new data from sites that were not used when deriving the original species/cover type matrix. Comparison of observed and predicted probabilities of occurrence showed a good general agreement, but with some noticeable deviations. The individual predicted and observed probabilities of occurrence differed, however, in only 14 out of the 112 individual species/land class combinations.

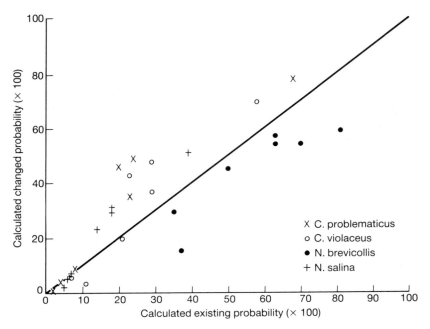

Fig. 10. Calculated existing and predicted probabilities of occurrence of selected Carabidae, following "de-intensification" of agriculture in the Tyne catchment. The diagonal line indicates no change. Data from Rushton *et al.* (1994).

The model was used by Rushton *et al.* (1994) to predict the effects of possible "de-intensification" of upland agriculture in the Tyne catchment. Arable, temporary and permanent pasture cover types were assumed to change to rough grazing; the proportion of change ranged from 100% in upland land classes to 25% in the lowland classes. Figure 10 shows the predicted decrease in *Nebria brevicollis* (F.), a species of the intensively managed grasslands, and increases in *N. salina* Fairm & Lab., as well as of *Carabus* species.

This model still takes no account, however, of the actual landscape pattern, so that the effects of habitat size and separation, considered in the previous section, are not yet allowed for. This may require the use of satellite remote-sensed data on actual distribution of cover types within the landscape, integrated with ground survey of the actual habitats present (see Bunce *et al.*, 1992).

X. CONCLUSIONS

It is evident, from the examples presented here, that the groups of insects that we have selected are indeed good indicators of many types of land-use change. Between them they respond to features of soil structure, microclimate, vegetation structure and species composition. Long-term monitoring work suggests that there are both short- and long-term fluctuations in insect diversity, and in the distribution and abundance of individual species. We are still some way, however, from fully separating the effects of climate, land use and natural population variability.

Models are being developed in order to assess the likely ecological impact of land-use change, using insects among other indicators of such change, and can reasonably predict changes in the overall probabilities of occurrence of particular species in each cover type or habitat. These are not yet at a level of sophistication, however, to predict accurately the spatial patterns of individual species within a landscape.

There is evidence from several sources that a key threat to insects in the changing British landscape is the fragmentation into small, isolated patches of habitats frequented by specialist species with poor dispersal powers. Even in areas of large habitat such as grassland, land-use changes are tending towards a reduction in species diversity, largely through adverse effects on the larger and possibly less mobile species of insect.

References

Adis, J. (1979). Problems of interpreting arthropod sampling with pitfall traps. *Zool. Anz.* **202**, 177–184.

Andersen, A. N. (1987) Ant community organisation and environmental assessment. *In* "The Role of Invertebrates in Conservation and Biological Survey" (J. D. Majer, ed.), pp. 43–52. Western Australian Department of Conservation and Land Management, Perth.

Asteraski, E. J., Hanks, C. B. and Clements, R. O. (1992). The impact of the chemical removal of the hedge-base flora on the community structure of carabid beetles (Col., Carabidae) and spiders (Araneae) of the field and hedge bottom. *J. Appl. Entomol.* **113**, 398–406.

Baars, M. A. (1979). Catches in pitfall traps in relation to mean densities of carabid beetles. *Oecologia* **41**, 25–46.

Basset, Y. (1992). Influence of leaf traits on the spatial distribution of arboreal arthropods within an overstory rainforest tree. *Ecol. Entomol.* **17**, 8–16.

Basset, Y., Aberlenc, H.-P. and Delvare, G. (1992). Abundance and stratification of foliage arthropods in a lowland rain forest of Cameroon. *Ecol. Entomol.* **17**, 310–318.

Boer, P. J. den (1977). Dispersal power and survival. Carabids in a cultivated countryside. *Misc. Papers Landbouwhogeschool, Wageningen* **14**, 1–190.

Booij, C. H. J. and Noorlander, J. (1988). Effects of pesticide use and farm management on carabids in arable crops. "Environmental effects of pesticides". *British Crop Prot. Council Monogr.* **40**, 119–126.

Booij, C. H. J. and Noorlander, J. (1992). Farming systems and insect predators. *Agric. Ecosys. Env.* **40**, 125–135.

Bunce, R. G. H., Barr, C. J. and Whittaker, H. A. (1981). An integrated system of land classification. Natural Environment Research Council. *Institute of Terrestrial Ecology Ann. Rep.* **1980**, 28–33. Natural Environment Research Council.

Bunce, R. G. H., Barr, C. J. and Fuller, R. M. (1992). Integration of methods for detecting land use change, with special reference to Countryside Survey 1990. *In* "Land Use Change: the Causes and Consequences" (M. C. Whitby, ed.), pp. 69–78. HMSO, London.

Burel, F. (1992). Effect of landscape structure and dynamics on species diversity in hedgerow networks. *Landscape Ecol.* **6**, 161–174.

Burn, A. J. (1992). Interactions between cereal pests and their predators and parasites. *In* "Pesticides, Cereal Farming and the Environment. The Boxworth Project" (P. Greig-Smith, G. Frampton and T. Hardy, eds), pp. 110–131. HMSO, London.

Butterfield, J. and Benitez-Malvido, J. (1992). Effect of mixed-species tree planting on the distribution of soil invertebrates. *In* "The Ecology of Mixed-Species Stands of Trees" (M. G. R. Cannell, D. C. Malcolm and P. A. Robertson, eds), pp. 255–265. *British Ecological Society Special Publication* **11**. Blackwell, Oxford.

Crawford, T. J. (1991). The calculation of index numbers from wildlife monitoring data. *In* "Monitoring for Conservation and Ecology" (B. Goldsmith, ed.), pp. 255–248. Chapman & Hall, London.

Day, K. R. and Carthy, J. (1988). Changes in carabid beetle communities accompanying a rotation of Sitka spruce. *Agric. Ecosys. Env.* **24**, 407–415.

Dennis, P. and Fry, G. L. A. (1992). Field margins: can they enhance natural enemy population densities and general arthropod diversity on farmland? *Agric. Ecosys. Env.* **40**, 95–115.

Desender, K., Dufrêne, M. and Maelfait, J. P. (1994). Long term dynamics of carabid beetles in Belgium: a preliminary analysis of the influence of changing climate and land use by means of a database covering more than a century. *In* "Carabid Beetles: Ecology and Evolution" (K. Desender, M. Dufrêne, M. Loreau, M. L. Luff and J.-P. Maelfait, eds), pp. 247–252. Kluwer, Dordrecht.

Dover, J. Sotherton, N. W. and Gobbett, K. (1990). Reduced pesticide inputs on cereal field margins: the effects on butterfly abundance. *Ecol. Entomol.* **15**, 17–24.

Emmet, A. M. (1991). Chart showing the life history and habits of the British Lepidoptera. *In* "The Moths and Butterflies of Great Britain and Ireland" (A. M. Emmet and J. Heath, eds), pp. 61–301. Harley Books, Colchester.

Erhardt, A. and Thomas, J. A. (1991). Lepidoptera as indicators of change in the semi-natural grasslands of lowland and upland Europe. *In* "The Conservation of Insects and their Habitats" (N. M. Collins and J. A. Thomas, eds), pp. 213–236. *15th Symposium of the Royal Entomological Society*. Academic Press, London.

Eyre, M. D., Luff, M. L. and Rushton, S. P. R. (1990). The ground beetle (Coleoptera, Carabidae) fauna of intensively managed agricultural grasslands in northern England and southern Scotland. *Pedobiologia* **34**, 11–18.

Foster, G. N. (1987). The use of Coleoptera records in assessing the conservation status of wetlands. *In* "The Use of Invertebrates in Site Assessment for Conservation" (M. L. Luff, ed.), pp. 8–18. University of Newcastle, Newcastle upon Tyne.

Greatorex-Davies, J. N., Sparks, T. H., Hall, M. L. and Marrs, R. H. (1993). The influence of shade on butterflies in rides of coniferised lowland woods in southern England and implications for conservation management. *Biol. Cons.* **63**, 31–41.

Greatorex-Davies, J. N., Sparks, T. H. and Hall, M. L. (1994). The response of Heteroptera and Coleoptera species to shade and aspect in rides of coniferised lowland woods in southern England. *Biol. Cons.* **67**, 255–273.
Greig-Smith, P. W. (1989). The Boxworth project – environmental effects of cereal pesticides. *J. Roy. Agric. Soc. Engl.* **150**, 171–187.
Hassall, M., Hawthorne, A., Maudsley, M., White, P. and Cardwell, C. (1992). Effects of headland management on invertebrate communities in cereal fields. *Agric. Ecosys. Envir.* **40**, 155–178.
Hengeveld, R. (1985). Dynamics of Dutch beetle species during the twentieth century (Coleoptera, Carabidae). *J. Biogeogr.* **12**, 389–411.
Hengeveld, R. (1987). Scales of variation: their distinction and ecological importance. *Ann. Zool. Fennici* **24**, 195–202.
Hengeveld, R. and Hogeweg, P. (1979). Cluster analysis of the distribution patterns of Dutch carabid species (Col.). *In* "Multivariate Methods in Ecological Work" (L. Orloci, C. R. Rao and W. M. Stiteler, eds), pp. 65–86. International Co-operative Publishing House, Maryland.
Holloway, J. D. (1983). Insect surveys – an approach to environmental monitoring. *Atti XII Congr. Naz. Ital. Entomol. Roma (1980)* **1**, 231–261.
Holloway, J. (1991). Biodiversity and tropical agriculture: a biogeographic view. *Outlook on Agriculture* **20**, 9–13.
Holloway, J. D. and Stork, N. E. (1991). The dimensions of biodiversity: the use of invertebrates as indicators of human impact. *In* "The Biodiversity of Microorganisms and Invertebrates" (D. L. Hawksworth, ed.), pp. 37–62. CAB International, Wallingford.
Holloway, J. D., Robinson, G. S. and Tuck, K. R. (1990). Zonation in the Lepidoptera of northern Sulawesi. *In* "Insects and the Rain Forests of South East Asia (Wallacea)" (W. J. Knight and J. D. Holloway, eds), pp. 153–166. The Royal Entomological Society, London.
Hutcheson, J. (1990). Characterization of terrestrial insect communities using quantified, Malaise-trapped Coleoptera. *Ecol. Entomol.* **15**, 143–151.
Hopkins, P. J. and Webb, N. R. (1984). The composition of the beetle and spider faunas on fragmented heathlands. *J. Appl. Ecol.* **21**, 935–946.
Janzen, D. H. (1988). Ecological characterisation of a Costa Rican dry forest caterpillar fauna. *Biotropica* **20**, 120–135.
Luff, M. L. (1987). Biology of polyphagous ground beetles in agriculture. *Agric. Zool. Rev.* **2**, 237–278.
Luff, M. L. (1990). Spatial and temporal stability of carabid communities in a grass/arable mosaic. *In* "The Role of Ground Beetles in Ecological and Environmental Studies" (N. Stork, ed.), pp. 191–200. Intercept, Andover.
Luff, M. L. (1994) "Provisional Atlas of the Carabidae (ground beetles) of Britain and Ireland". Institute of Terrestrial Ecology, Huntingdon, in press.
Luff, M. L. and Rushton, S. P. (1988). The effects of pasture improvement on the ground beetle and spider fauna of upland grasslands. *Aspects of Applied Biology* **17**, 67–74.
Luff, M. L., Eyre, M. D. and Rushton, S. P. R. (1989). Classification and ordination of the habitats of ground beetles (Coleoptera, Carabidae) in north-east England. *J. Biogeog.* **16**, 121–130.
Luff, M. L., Eyre, M. D., Cherrill, A. J., Foster, G. N. and Pilkington, J. G. (1992). Use of assemblages of invertebrate animals in a land use change model. *In* "Land Use Change: the Causes and Consequences" (M. C. Whitby, ed.), pp. 102–110. HMSO, London.
McKay, H. V. (1991). Egg-laying requirements of woodland butterflies: brimstones (*Gonepteryx rhamni*) and alder buckthorn (*Frangula alnus*). *J. Appl. Ecol.* **28**, 731–743.

Morse, D. R., Stork, N. E. and Lawton, J. H. (1988). Species number, species abundance and body length relationships of arboreal beetles in Bornean lowland rain forest trees. *Ecol. Entomol.* **13**, 25–37.

Nature Conservancy Council (1987). "Changes in the Cumbrian countryside". *Research and Survey in Nature Conservation* **6**. Nature Conservancy Council, Peterborough.

O'Callaghan, J. R. (1992). In "Land Use Change: the Causes and Consequences" (M. C. Whitby, ed.), pp. 79–87. HMSO, London.

Parry, W. H. and Rodger, D. (1986). The effect of soil scarification on the ground beetle fauna of a Caledonian pine forest. *Scot. Forestry* **40**, 1–9.

Peng, R. K., Incoll, L. D., Sutton, S. L., Wright, C. and Chadwick, A. (1993). Diversity of airborne arthropods in a silvoarable agroforestry system. *J. Appl. Ecol.* **30**, 551–562.

Pinheiro, C. E. G. and Ortiz, J. V. C. (1992). Communities of fruit-feeding butterflies along a vegetation gradient in central Brazil. *J. Biogeogr.* **19**, 505–511.

Pollard, E. (1968). Hedges III. The effect of removal of the bottom flora of a hawthorn hedgerow on the Carabidae of the hedge bottom. *J. Appl. Ecol.* **5**, 125–139.

Pollard, E. and Yates, T. (1993). "Monitoring Butterflies for Ecology and Conservation". Chapman & Hall, London.

Pollard, E., Hall, M. L. and Bibby, T. J. (1986). "Monitoring the Abundance of Butterflies, 1976–1985". *Research and Survey in Nature Conservation* **2**. Nature Conservancy Council, Peterborough.

Rushton, S. P. (1992). A preliminary model for investigating the ecological consequences of land use change within the framework of the ITE land classification. In "Land Use Change: the Causes and Consequences" (M. C. Whitby, ed.), pp. 111–117. HMSO, London.

Rushton, S. P., Luff, M. L. and Eyre, M. D. (1989). Effects of pasture improvement and management on the ground beetle and spider communities of upland grasslands. *J. Appl. Ecol.* **26**, 489–503.

Rushton, S. P., Eyre, M. D. and Luff, M. L. (1990a). The effects of management on the occurrence of some carabid species in grassland. In "The Role of Ground Beetles in Ecological and Environmental Studies" (N. Stork, ed.), pp. 209–216. Intercept, Andover.

Rushton, S. P., Eyre, M. D. and Luff, M. L. (1990b). The effects of scrub management on the ground beetles of oolitic limestone grassland at Castor Hanglands National Nature Reserve, Cambridgeshire, UK. *Biol. Cons.* **51**, 97–111.

Rushton, S. P., Luff, M. L. and Eyre, M. D. (1991). The habitat characteristics of *Pterostichus* species in grasslands. *Ecol. Entomol.* **16**, 91–104.

Rushton, S. P., Wadsworth, R. A., Cherrill, A. J. Eyre, M. D. and Luff, M. L. (1994). Modelling the consequences of land use change on the distribution of Carabidae. In "Carabid Beetles: Ecology and Evolution" (K. Desender, M. Dufrêne, M. Loreau, M. L. Luff and J.-P. Maelfait, eds), pp. 353–360. Kluwer, Dordrecht.

Stork, N. E. (1988). Insect diversity: facts, fiction and speculation. *Biol. J. Linn. Soc.* **35**, 321–337.

Sutton, S. L. and Collins, N. M. (1991). Insects and tropical forest conservation. In "The Conservation of Insects and their Habitats" (N. M. Collins and J. A. Thomas, eds), pp. 405–424. Academic Press, London.

Taylor, L. R. (1978). Bates, Williams, Hutchinson – a variety of diversities. In "Diversity of Insect Faunas" (L. A. Mound and N. Waloff, eds), pp. 1–18. Blackwell, Oxford.

Taylor, L. R. (1986). Synoptic dynamics, migration and the Rothamsted insect survey. *J. Anim. Ecol.* **55**, 1–38.

Taylor, L. R., Kempton, R. A. and Woiwod, I. P. (1976). Diversity statistics and the log-series model. *J. Anim. Ecol.* **45**, 255–272.

Taylor, L. R., French, R. A. and Woiwod, I. P. (1978). The Rothamsted insect survey and the urbanisation of land in Britain. *In* "Perspectives in Urban Entomology" (G. W. Frankie and C. S. Koehler, eds), pp. 31–65. Academic Press, New York.

Thomas, C. D., Thomas, J. A. and Warren, M. S. (1992). Distribution of occupied and vacant butterfly habitats in fragmented landscapes. *Oecologia* **92**, 563–567.

Thomas, J. A. (1983). A "WATCH" census of common British butterflies. *J. Biol. Educ.* **17**, 333–338.

Thomas, J. A., Thomas, C. D., Simcox, D. J. and Clarke, R. T. (1986). The ecology and declining status of the silver-spotted skipper butterfly (*Hesperia comma*) in Britain. *J. Appl. Ecol.* **23**, 365–380.

Thomas, M. B. (1990). The role of man-made grassy habitats in enhancing carabid populations in arable land. *In* "The Role of Ground Beetles in Ecological and Environmental Studies" (N. Stork, ed.), pp. 77–85. Intercept. Andover.

Thomas, M. B., Wratten, S. D. and Sotherton, N. W. (1991). Creation of "island" habitats in farmland to manipulate populations of beneficial arthropods: predator densities and emigration. *J. Appl. Ecol.* **28**, 906–917.

Turin, H. and den Boer, P. J. (1988). Changes in the distribution of carabid beetles in The Netherlands since 1880. II. Isolation of habitats and long-term trends in the occurrence of carabid species with different powers of dispersal (Coleoptera, Carabidae). *Biol. Cons.* **44**, 179–200.

Turin, H., Alders, K., den Boer, P. J., van Essen, S., Heijerman, T., Laane, W. and Penterman, T. (1991). Ecological characterization of carabid species (Coleoptera, Carabidae) in the Netherlands from thirty years of pitfall sampling. *Tijdsch. voor Entomol.* **134**, 279–304.

Turner, J. R. G., Gatehouse, G. M. and Corey, C. A. (1987). Does solar energy control organic diversity? Butterflies, moths and the British climate. *Oikos* **48**, 195–205.

Vries, H. H. de (1994). Size of habitat and presence of ground beetle species. *In* "Carabid Beetles: Ecology and Evolution" (K. Desender, M. Dufrêne, M. Loreau, M. L. Luff and J.-P. Maelfait, eds), pp. 253–259. Kluwer, Dordrecht.

Vries, H. H. de and den Boer, P. J. (1990). Survival of populations of *Agonum ericeti* Panz. (Col., Carabidae) in relation to fragmentation of habitats. *Neth. J. Zool.* **40**, 484–498.

Warren, M. S. (1985). The influence of shade on butterfly numbers in woodland rides, with special reference to the wood white, *Leptidea sinapis*. *Biol. Cons.* **33**, 147–164.

Warren, M. S. and Thomas, J. A. (1992). Butterfly responses to coppicing. *In* "Ecology and Management of Coppice Woodlands" (G. P. Buckley, ed.), pp. 249–270. Chapman & Hall, London.

Webb, N. R. (1989a). Studies on the invertebrate fauna of fragmented heathland in Dorset, UK, and the implications for conservation. *Biol Cons.* **47**, 153–165.

Webb, N. R. (1989b). The invertebrates of heather and heathland. *Bot. J. Linn. Soc.* **101**, 307–312.

Woiwod, I. P. (1991). The ecological importance of long-term synoptic monitoring. *In* "The Ecology of Temperate Cereal Fields" (L. G. Firbank, N. Carter, J. F. Darbyshire and G. R. Potts, eds), pp. 275–304. Blackwell, Oxford.

Woiwod, I. P. and Dancy, K. J. (1987). Synoptic monitoring for migrant insect pests in Great Britain and Western Europe. VII. Annual population fluctuations of macrolepidoptera over Great Britain for 17 years. *Rothamsted Experimental Station Report* **1986** (Part 2) 235–262.

Woiwod, I. P. and Harrington, R. (1994). Flying in the face of change – the Rothamsted Insect Survey. *In* "Long-term Research in Agricultural and Ecological Sciences" (R. A. Leigh and A. E. Johnston, eds), CAB International, Wallingford, Oxon, in press.

Woiwod, I. P. and Thomas, J. A. (1993). The ecology of butterflies and moths at the landscape scale. *In* "Landscape Ecology in Britain" (R. Haines-Young and R. G. H. Bunce, eds), pp. 76–92. *Working Paper* **21**. Department of Geography, University of Nottingham.

Wolda, H. (1983). Spatial and temporal variation in abundance in tropical animals. *In* "Tropical Rain Forest: Ecology and Management" (S. L. Sutton, T. C. Whitmore and A. C. Chadwick, eds), pp. 93–105. Blackwell, Oxford.

Part V.
Short Communications

17. The Response of Chironomidae (Diptera) Faunas to Climate Change
    S. J. BROOKS
18. Global Warming, Population Dynamics and Community Structure in a Model Insect Assemblage
    A. J. DAVIS, L. S. JENKINSON, J. H. LAWTON, B. SHORROCKS AND S. WOOD
19. Gaseous Air Pollutants – Can We Identify Critical Loads for Insects
    G. R. PORT, K. BARRETT, E. OKELLO AND A. DAVISON
20. Effects of Changing Land Use on Eucalypt Dieback in Australia in Relation to Insect Phytophagy and Tree Re-establishment
    R. A. FARROW AND R. B. FLOYD
21. Modelling the Population of *Hydrotaea irritans* Using a Cohort-based System
    J. D. AUSTIN AND J. E. HILLERTON
22. Effects of Ivermectin Residues in Cattle Dung on Dung Insect Communities Under Extensive Farming Conditions in South Africa
    C. H. SCHOLTZ AND K. KRÜGER
23. Monitoring the Response of Tropical Insects to Change in the Environment: Trouble with Termites
    P. EGGLETON AND D. E. BIGNELL
24. Potential Use of Suction Trap Collections of Aphids as Indicators of Plant Biodiversity
    S. E. HALBERT, M. D. JENNINGS, C. B. COGAN, S. S. QUISENBERRY AND J. B. JOHNSON
25. Shifts in the Flight Periods of British Aphids: a Response to Climate Warming?
    R. A. FLEMING AND G. M. TATCHELL
26. Potential Changes in Spatial Distribution of Outbreaks of Forest Defoliation Under Climate Change
    D. W. WILLIAMS AND A. M. LIEBHOLD

17

The Response of Chironomidae (Diptera) Faunas to Climate Change

S. J. BROOKS

I. Introduction	425
II. Chironomidae as Environmental Indicators	426
III. Sampling Methods	427
IV. Taphonomy	427
V. Results	427
References	429

Abstract

The value of Chironomidae (Diptera) as indicators of environmental change is reviewed with particular reference to their response to long-term fluctuations in temperature. The methods of sampling chironomid larval head capsules from lake sediment cores are described together with the conditions of preservation of the head capsules. Preliminary results from analysis of a lake sediment core from the late-glacial period (10 000–12 000 yr BP) are briefly described and demonstrate the sensitivity of the chironomid fauna to climate change.

I. INTRODUCTION

Chironomidae have long been known to be good indicators of water quality (see Chapter 12) and are particularly sensitive to eutrophication (e.g. Gamm, 1927). They have also been used to define the trophic status of lakes (Brundin, 1956) and, more recently, have been used as indicators of acidification (Wiederholm and Eriksson, 1977), salinity (Paterson and Walker, 1974) and heavy metal pollution (Kansanen and Jaakkola, 1985).

With growing interest in the effects of global warming, chironomids are now under scrutiny for their potential as indicators of climate change.

II. CHIRONOMIDAE AS ENVIRONMENTAL INDICATORS

Chironomidae have several attributes that make them good indicators of environmental change:

(1) Their larvae occur in almost all aquatic biotopes.
(2) Chironomid larvae are abundant in non-marine aquatic ecosystems. Among benthic macroinvertebrates only oligochaetes are more abundant but oligochaetes are poorly preserved in lake sediments.
(3) The heavily chitinized head capsules of chironomid larvae are well preserved in lake sediments and are usually identifiable to generic, if not species, level.
(4) The chironomid fauna is usually species rich. This makes the family sensitive to environmental change, since different species will respond differently to a variety of environmental factors.
(5) Chironomids are *in situ* indicators. Some proxy indicators, particularly pollen, may originate a considerable distance from the sampling site.
(6) Because of their rapid generation time and the mobility of the adults, Chironomidae are able to respond quickly to environmental change.

Recent work has demonstrated (Walker *et al.*, 1991; Wilson *et al.*, 1993) that Canadian chironomid faunas change across a latitudinal gradient and the factor that best accounted for this change was summer surface water temperature. Data based on the composition of the chironomid taxa in a lake sediment core sample have enabled the reconstruction of past temperature profiles.

There are a number of ways in which climatic change may affect the chironomid fauna. Chironomini tend to dominate in warm, eutrophic waters. They are able to tolerate low oxygen concentrations due to the presence of haemoglobin in their body fluids. In cold lakes, warm-adapted species are unable to complete their life cycles. Lack of food in cold, oligotrophic lakes may also be a limiting factor which tends to favour the smaller orthoclad, diamesine or tanytarsine species. Littoral species also tend to be excluded from cold lakes. Changes in the substrate and aquatic or terrestrial macrophytes brought about by climatic perturbations may also influence the species composition of the chironomid fauna.

17. The Response of Chironomidae (Diptera) Faunas to Climate Change

III. SAMPLING METHODS

In order to assess the environmental impact of climate change, sample sites should be as remote from human interference as possible. Consequently, suitable lakes are to be found in high montane, alpine or subarctic locations. The sediment core is usually taken from the deepest part of the lake where it is likely to have the longest undisturbed record. In addition, such a sample will include profundal species. Unlike littoral species, these are least likely to be affected by short-term seasonal fluctuations and are most likely to reflect long-term climatic changes. Having taken the sample, the sediment core is extruded and divided into 2, 5 or 10-mm sections, depending on the degree of resolution required. About 2 g of sediment usually provides sufficient material for analysis. The sediment is first deflocculated in warm KOH, then sieved through a 90-μm mesh. Finally, the chironomid larval head capsules are removed individually and slide mounted for identification.

IV. TAPHONOMY

The head capsules preserved in the lake sediments are derived from the larval exuviae. All Chironomidae have four larval instars. However, the first and second instars often have insufficient chitin to be preserved. Sometimes, the fourth larval exuviae may remain attached to the pupal exuviae after the adult has emerged. Preservation of these head capsules in the lake sediment is unlikely as the pupal exuviae are usually blown to the edge of the lake. Larval head capsules can be preserved in excellent condition but often the antennae, mandibles and associated mouthparts are missing. Head capsules of Orthocladiinae are often split in half longitudinally. Studies have shown a good correlation between life and death chironomid assemblages (Walker *et al.*, 1984).

V. RESULTS

To illustrate how Chironomidae react to climate change, some preliminary results of an analysis I am conducting on sediments derived from Lake Kråkenes in Norway are shown (Fig. 1). These demonstrate how Chironomini genera are present during the warm Allerød (12 000 yr BP) and Holocene (10 000 yr BP) eras but are absent during the cold Younger Dryas (11 000 yr BP). In contrast, the cold-adapted Diamesinae, together with the Orthocladiinae genera, *Heterotrissocladius* and *Paracladius*, and

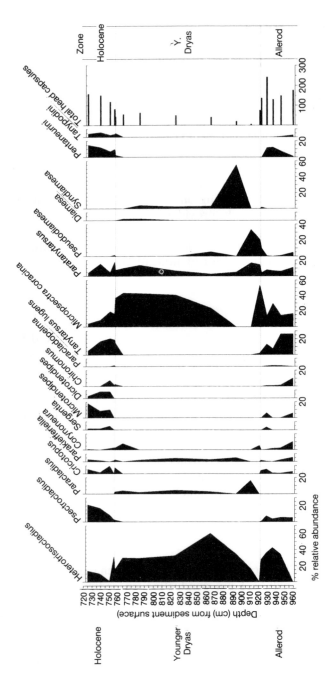

Fig. 1. The relative abundances of Chironomidae in the sediment core from Kråkenes Lake, Norway (10 000–12 000 BP).

the Tanytarsini taxa, *Micropsectra coracina* and *Paratanytarsus*, dominate in the Younger Dryas. The *Heterotrissocladius* peak towards the end of the Allerød suggests a cold phase at the end of this period that may correspond to the Gerzensee Oscillation (see Levesque *et al.*, 1993). The short *Heterotrissocladius* peak at the beginning of the Holocene may reflect an influx of cold glacial melt waters following a rise in air temperature.

Acknowledgements

I am grateful to Dr Hilary Birks, University of Bergen for help and encouragement. The work on Lake Kråkenes sediments was part funded by the Norwegian Science Foundation

REFERENCES

Brundin, L. (1956). Die bodenfaunistischen Seetypen und ihre Anwendbarkeit auf die Südhalbkugel. Zugleich eine Theorie der produktionsbiologischen Bedeutung der Glazialen Erosion. *Inst. Freshwater Res., Drottningholm* **37**, 186–235.

Gamms, H. (1927). Die Geschichte der Lunzer seen, Moore und Walder. *Int. Revue ges. Hydrobiol. Hydrogr.* **18**, 304–387.

Kansenen, P. H. and Jaakkola, T. (1985) Assessment of pollution history from recent sediments in Lake Vanajavesi, southern Finland. I. Selection of representative profiles, their dating and chemistratigraphy. *Ann. Zool. Fenn.* **22**, 13–55.

Levesque, A. J., Mayle, F. E., Walker, I. R. and Cwynar, L. C. (1993). A previously unrecognized late-glacial cold event in eastern North America. *Nature* **361**, 623–626.

Paterson, C. G. and Walker, K. F. (1974). Recent history of *Tanytarsus barbitarsus* Freeman (Diptera: Chironomidae) in the sediments of a shallow, saline lake. *Aust. J. Mar. Freshw. Res.* **25**, 315–325.

Walker, I. R., Fernando, C. H. and Paterson, C. G. (1984). The chironomid fauna of four shallow humic lakes and bog pools in Atlantic Canada and their representation by subfossil assemblages in the surficial sediments. *Hydrobiology* **112**, 61–67.

Walker, I. R., Smol, J. P., Engstrom, D. R. and Birks, H. J. B. (1991). An assessment of Chironomidae as quantitative indicators of past climatic change. *Can. J. Fish. Aquat. Sci.* **48**, 975–987.

Wiederholm, T. and Erikson, L. (1977). Benthos of an acid lake. *Oikos* **29**, 261–267.

Wilson, S. E., Walker, I. R., Mott, R. J. and Smol, J. P. (1993). Climatic and limnological changes associated with the Younger Dryas in Atlantic Canada. *Climate Dynamics* **8**, 177–187.

18

Global Warming, Population Dynamics and Community Structure in a Model Insect Assemblage

A. J. DAVIS, L. S. JENKINSON, J. H. LAWTON,
B. SHORROCKS AND S. WOOD

I. Introduction	432
II. Rationale	432
III. Methods	433
IV. Results	434
References	438

Abstract

The effects of global warming on insect communities are unlikely to be simple and may involve catastrophic changes in community composition because of temperature-sensitive competitive and trophic interactions. Such changes cannot be investigated in single species laboratory studies or in the complex conditions of the field (see Chapter 1). We therefore describe a method for investigating the effects of global warming on communities using a simplified, replicatable, laboratory ecosystem of three *Drosophila* species and their parasitoids. Each ecosystem consists of linked cages in a temperature cline of 10–25°C, current conditions, or 15–30°C, after global warming. The linkage between cages allows adult insects to migrate between temperatures.

We present initial results demonstrating temperature-sensitive competitive interactions between *Drosophila* species. We show that each species reaches its highest population density at different temperatures in single-species ecosystems.

I. INTRODUCTION

The title chosen for this symposium indicates the current concern about environmental change and its effects. To meet this concern we not only need effective methods of investigation but also methods that allow us to offer the public, and the politicians, sensible predictions. In the research described here we aim to develop an appropriate methodology, and predictive models, for the responses of insects and insect communities to environmental change. Our primary interest is in the effects of global warming but we believe the rationale and methodology are applicable to studying the effects of other environmental changes.

In this chapter we discuss the rationale of our approach, explain the method we have adopted and conclude by reporting our initial results.

II. RATIONALE

The presentations at this symposium demonstrate that the effects of environmental change on insects are complex. Even single factors acting on single insect species produce complexity and, to date, almost all the research on predicting the responses of insects to climate change has focused on single species. However, we know that species exist, not on their own, but in dynamic relationships, horizontally with competitors and vertically with pathogens, predators and parasites. These interactions, particularly competition, are known to be temperature sensitive (e.g. Moore, 1952) and, consequently, global warming may not simply move existing species assemblages to higher altitudes but may alter community composition. Communities may flip into completely different states (Lawton, Chapter 1, this volume). Changes in community composition imply cascading further changes since we know that the removal or decline of species in one trophic level has amplified consequences on other trophic levels (Paine, 1966; Lawton and Hassell, 1984). These changes are of more than local importance since warm-adapted insects dispersing from population centres could swamp cool-adapted species already pushed into suboptimum habitats by global warming.

The outcome of these interactions cannot be predicted from studies on single species or on interactions in isolated cages nor can we even use such methods to study these phenomena, yet it is vital that we do study them. The alternative of investigating all possible interactions in the field would be an enormous task and is probably impractical, even assuming large-scale concentration on a paradigmatic species (Lawton, Chapter 1, this volume).

18. Global-warming, Population Dynamics and Community Structure

There is, fortunately, an intermediate alternative. We can construct simplified ecosystems in the laboratory where physical factors can be controlled and population parameters can be easily measured. We can then identify what parameters govern their behaviour. In addition these systems can be disturbed from equilibrium and their subsequent readjustments investigated.

The beauty of this approach is that we can replicate our study and thus obtain predictive power impossible with field studies. The crucial parameters we identify can then be built into models which we can use to investigate the outcome of particular changes.

III. METHODS

In the *Drosophila* Unit at Leeds we have set up five series of eight linked population cages with a pair of cages in each of four incubators (Fig. 1)

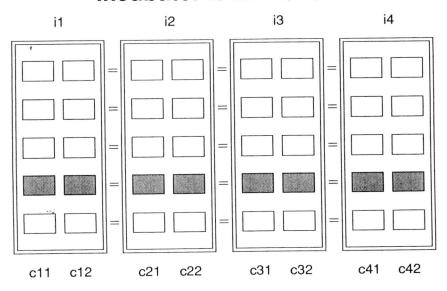

Fig. 1. The arrangement of perspex cages, each $160 \times 120 \times 330$ mm (c1.1–c4.2), in one set of four linked incubators (i1–i4) showing, shaded, an experimental cage-series of eight linked cages.

and duplicated this arrangement to give 10 cage-series. The temperature in each incubator, and thus within the cages, can be controlled to within 1°C. We are aware that changing the temperature changes many other variables, including humidity and food quality, but temperature is the driving variable in our system, much as it is in the field.

Our experimental organisms are *Drosophila*, essentially grazers on micro-organisms, and some of their parasitoids. *Drosophila* are easily maintained, have short generation times and a wide range of different parasitoids (Carton *et al.*, 1986). These parasitoids have marked effects on wild *Drosophila* populations (Driessen *et al.*, 1990) and may alter the outcome of species interactions (Boulétreau *et al.*, 1991). We are using *Drosophila* for these entirely practical reasons not because it is important to know how they will respond to global warming nor because we want to reassemble *Drosophila* communities under different temperatures.

The behaviour of the system is first assessed with one species in each cage-series and all four incubators in each set at 20°C. Then we produce a temperature cline by setting incubators 1–4 at, successively, 10, 15, 20 and 25°C. This spread represents about 20° of latitude in the mid-temperate region; for example, from Spain to northern England.

The populations in each cage are assessed by standardized partial counts or mark–release–recapture (MRR) and the movement between incubators measured by MRR, direct observation or video. The cages with the highest population counts indicate the temperature optima of the species. Thereafter the clines can be shifted upwards by 4–5°C, to run from 15 to 30°C, mimicking the temperature rise expected with global warming (Bennetts, Chapter 3, this volume).

The whole process will then be repeated with two species in each cline to reveal whether temperature optima are shifted along the cline by interspecific interactions. The second trophic level, the parasitoids, will then be added. By juggling with the species and strain of parasitoid we will be able to exert similar or different mortalities on each host species.

IV. RESULTS

Because we want to investigate the effects of interactions between species in this system we need to demonstrate that interactions existed. Using *Drosophila melanogaster, D. simulans* and *D. subobscura* we set up pairs of species in cages at each of the four temperatures used in the initial clines. At the two lower temperatures *D. subobscura* succeeds against either *D. simulans* or *D. melanogaster* but is clearly excluded by these species at 20 and 25°C. *Drosophila melanogaster*, however, succeeds

18. Global-warming, Population Dynamics and Community Structure 435

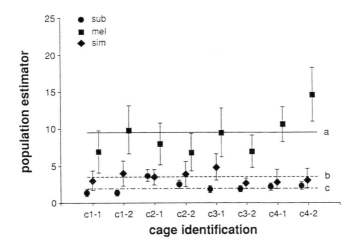

Fig. 2. *Drosophila melanogaster* ($n = 18$), *D. simulans* ($n = 24$) and *D. subobscura* ($n = 24$) populations in cage-series at 20°C, with fitted horizontal lines for each species (respectively, a, b and c). Error bars = ±1 SE.

against *D. simulans* at all four temperatures, contrary to earlier work (Moore, 1952). There are, therefore, temperature-sensitive interactions between these species.

The same three species were established in the cage-series. Three replicates were set up for each species and each replicate was run for several weeks. Populations were estimated in cage series and clines by weekly counts of the flies in a standard grid on each cage. This estimator is closely related to the real number of flies (population = $320 + 9.88 \times$ estimate, $r = 0.814$, $P < 0.001$). In cage-series at 20°C species populations can be described by horizontal lines (Fig. 2) indicating that populations are not influenced by the equipment. The populations of *D. melanogaster* are higher than those of *D. simulans* which in turn are higher than those of *D. subobscura*; an arrangement that accords both with our results (Table 1) and with the literature.

In clines, the pattern is very different with clear effects of temperature on populations of *D. melanogaster* (Fig. 3), *D. simulans* (Fig. 4) and *D. subobscura* (Fig. 5). However, each species is affected differently since

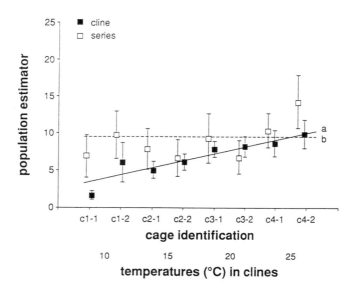

Fig. 3. *Drosophila melanogaster* populations in cage-series at 20°C ($n = 18$, fitted curve b) and in temperature clines ($n = 24$, fitted curve a). Error bars = ±1 SE.

Table 1. The outcome of competition between pairs of *Drosophila* species at four different temperatures

Temperature (°C)	Competing pairs of *Drosophila* species		
	mel/sim	mel/sub	sim/sub
10	mel	sub	sub
15	mel	sub	sub
20	mel	mel	sim
25	mel	mel	sim

18. Global-warming, Population Dynamics and Community Structure   437

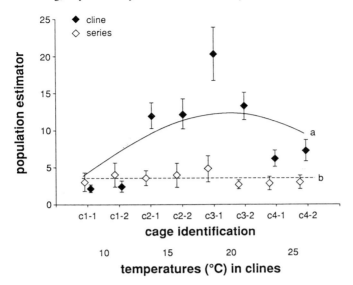

Fig. 4. *Drosophila simulans* populations in cage series at 20°C ($n = 24$, fitted curve b) and in temperature clines ($n = 24$, fitted curve a). Error bars = ±1 SE.

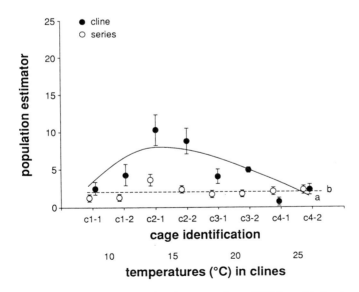

Fig. 5. *Drosophila subobscura* populations in cage series at 20°C ($n = 24$, fitted curve b) and in temperature clines ($n = 18$, fitted curve a). Error bars = ±1 SE.

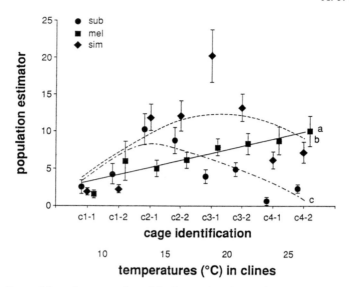

Fig. 6. Drosophila melanogaster ($n = 24$), *D. simulans* ($n = 24$) and *D. subobscura* ($n = 18$) populations in temperature clines with fitted curves for each species (respectively a, b and c). Error bars = ± 1 SE.

each species has a different temperature optimum with the highest populations of *D. melanogaster* are at 25°C, of *D. simulans* at 20°C and of *D. subobscura* at 15°C. It will be interesting to see whether these optima persist when the species are run together in the same clines.

These initial results encourage us to believe that our approach provides a means of investigating the complexities of community responses to global warming and of gaining information from which generalized predictions might sensibly be made.

Acknowledgements

This project is funded under the AFRC Global Environmental Response Programme contract PG24/605 awarded in 1993 to Professors B. Shorrocks and J. H. Lawton.

REFERENCES

Boulétreau, M., Fouillet, P. and Allemand, R. (1991). Parasitoids affect competitive outcomes between the sibling species, *Drosophila melanogaster* and *Drosophila simulans*. *Redia* **74**, 171–177.

Carton, Y., Boulétreau, M., van Alphen, J. J. M. and van Lenteren, J.C. (1986). The *Drosophila* parasitic wasps. *In* "The Genetics and Biology of Drosophila" (M. Ashburner, H. L. Carson and J. N. Thompson, eds),Vol 3e, Chapter 39, pp. 347–394. Academic Press, London.

Driessen, G., Hemerik, L., van Alphen, J. J. M. (1990). *Drosophila* species, breeding in stinkhorn (*Phallus impudicus* Pers.) and their larval parasitoids. *Neth. J. Zool.* **40**, 409–427.

Lawton, J. H. and Hassell, M. P. (1984). Interspecific competition in insects. *In* "Ecological Entomology" (C. B. Huffaker and R. L. Rabb, eds), pp. 451–495. John Wiley, Chichester.

Moore, J. A. (1952). Competition between *Drosophila melanogaster* and *Drosophila simulans*. I. Population cage experiments.*Evolution* **6**, 407–420.

Paine, R. T. (1966). Food web complexity and species diversity. *Am. Nat.* **122**, 240–285.

19

Gaseous Air Pollutants – Can We Identify Critical Loads for Insects?

G. R. PORT, K. BARRETT, E. OKELLO AND A. DAVISON

I. Introduction 442
II. Possible Effects of Deposited Nitrogen on Insects 442
III. Approach 444
IV. Results 446
    A. Attack Rate 446
    B. Growth Rate 446
    C. Adult Weight 446
    D. Survival 450
    E. Population Growth 450
V. Discussion 450
    References 452

Abstract

Ammonia and oxides of nitrogen are significant air pollutants which, when deposited, may act as plant fertilizers. This chapter considers the value of the critical load concept for assessing impact of deposited nitrogenous pollutants on insects.

The most immediate impact is on herbivorous insects when nitrogen alters the nutritional quality of the foodplant by altering plant chemistry. Whilst some insect herbivores show no change in performance, possibly due to compensatory feeding, others show a change in performance as the nutritional quality of the food changes. Where performance of the insect improves it is due to one or more of increased growth rate, shorter time to maturity, greater adult weight, increased fecundity and improved survival. All of these may contribute to population growth. Furthermore, population density may increase as a consequence of increased attack on plants with more nitrogen.

Experiments involving different rates of either NH_4 or NO_3 nitrogen being added to beech. (*Fagus sylvatica* L.) and birch (*Betula pendula* Roth. and *Betula pubescens* Ehrh.) are described. Insect herbivores on the trees show changes in some of the performance parameters described above. Of particular interest was the trend for a regular increase in performance with increasing NH_4 nitrogen. In contrast high rates of NO_3 nitrogen did not correlate with the greatest performance by some herbivores. The positive response to even the lowest rate of nitrogen addition (equivalent to $20 \text{ kg ha}^{-1} \text{ year}^{-1}$) suggests that, for some insect herbivore species at least, the critical loads will be very small.

I. INTRODUCTION

Emissions of NO and NO_2 or NH_3, from combustion processes or animal wastes respectively, are increasing and causing environmental concern because they are major sources of acidification. They can also act as plant fertilizers. Current rates of deposition of total nitrogen in the UK are estimated to vary from less than $0.2 \text{ kg ha}^{-1} \text{year}^{-1}$ to over 80 in some locations (UKPORG, 1990; Anon., 1993). (By comparison, the winter wheat crop in the UK has about $180 \text{ kg ha}^{-1} \text{ year}^{-1}$ applied as fertilizer.) The UN Economic Commission for Europe and the UK Department of the Environment are using the *critical load* concept as the basis for the control of such emissions (Nilsson and Grennfelt, 1988). The critical load is defined very broadly as the threshold of pollutant input to a receptor, above which there are effects (Bull, 1991; Davison, 1993). Terrestrial communities, soils and water bodies, differ in their sensitivity to acidification and nitrogen fertilization so critical loads have to be assessed and mapped for each type of receptor.

Most research on critical loads has been done for soils, freshwaters and vegetation (Hornung and Skeffington, 1993). Here we consider the value of the critical load concept for assessing impact of deposited nitrogenous pollutants on insects. We have limited our discussion to nitrogen because of its importance and significance in affecting the performance of herbivorous insects (see Brown, Chapter 10, this volume).

II. POSSIBLE EFFECTS OF DEPOSITED NITROGEN ON INSECTS

Deposited nitrogen may affect the insect fauna in several ways. The most direct effect is when the nitrogen alters the nutritional quality of the foodplant by altering plant chemistry and this we shall consider below.

19. Gaseous Air Pollutants

More indirectly, deposited nitrogen may ultimately alter the species composition of vegetation which would preclude some herbivorous insects, but favour others. A more subtle and indirect effect is when deposited nitrogen affects the growth pattern of the vegetation so that vegetation structure (architecture) is altered. Again this will disadvantage some herbivorous insects whilst favouring others (Jansson *et al.*, 1991).

There has been much interest in the effects of changing nitrogen content of the food on insects (McNeill and Southwood, 1978; Mattson, 1980) and there have been numerous studies where this has been one of the factors under consideration. Some of this work relates to *natural variation* in the nitrogen content of the food. However, we are more concerned with the effects of adding nitrogen to the foodplant. Yet, even where the experiments have involved nitrogen addition, many of the results are difficult to relate to possible effects of deposited nitrogen for the following reasons:

(1) First, nitrogen is deposited as both NO_x and NH_y, but proportions vary in different parts of Britain. For example, NO_x is likely to predominate in urban centres, but in many rural areas NH_y is probably the most important form (Sutton *et al.*, 1993). Previous studies have sometimes simply compared the effects of manipulating nitrogen content of the food without identifying the nitrogen source for the plant. Others have used treatments of ammonium nitrate (NH_4NO_3) to modify nitrogen levels. In both cases it is impossible to identify specific effects of NO_x or NH_y, or to determine the importance of differences in the ratio.

(2) Second, although there have been many fertilizer studies, most have used application rates which are too great to represent nitrogen deposition from the atmosphere, which does not usually exceed 80 kg ha^{-1} $year^{-1}$. In many cases other nutrients were also added, such as phosphate, and the additional effects of this are unknown.

(3) Third, nitrogen deposition from the atmosphere is continuous although not necessarily at a constant rate. In most fertilizer experiments, a single or a few applications are used rather than frequent applications that come closer to mimicking deposition. There are likely to be very different effects on the physiology of plants receiving a single, large addition of nitrogen compared to regular additions.

Notwithstanding these difficulties, the responses of herbivorous insects to increasing nitrogen available to the foodplant can be summarized as follows:

(1) At very low levels of nitrogen the insects do not survive. They either starve to death or grow very slowly and succumb to pathogens or other mortality factors (Smith and Northcott, 1951; Ohmart et al., 1985). These low levels of nitrogen are usually only found in artificial conditions where the insect can be maintained on a synthetic diet.
(2) Above a minimum threshold level of nitrogen, insects can survive and reproduce on the foodplant.
(3) As the nitrogen content of the food increases above the threshold, then insects either show a relatively constant performance or their performance improves (often regarded as an increase in fitness) (Scriber, 1984).
(4) Those insects that show relatively constant performance probably adjust their consumption rate so that they compensate for the changing quality of food (compensatory feeding) (Slansky and Feeny, 1977; Auerbach and Strong, 1981; Ohmart et al., 1985; Simpson and Simpson, 1990).
(5) Where performance improves, as the nitrogen content of the food increases, this may be due to changes in one or more of the following parameters:

   (a) Increased growth rate which results in either
   (b) shorter time to maturity (with reduced exposure to mortality factors (Loader and Damman, 1991; Thomas and Hodkinson, 1991)), and/or
   (c) greater adult weight (Myers and Post, 1981). Which is usually linked to
   (d) increased fecundity.
   (e) Improved survival (possibly linked to the shorter time to reach maturity).

   All of these will contribute to population growth.
(6) In addition to these proximate effects on the success of individual insects, population growth may also occur as a consequence of increased attack on plants with more nitrogen (Silvanima and Strong, 1991).

III. APPROACH

To elucidate the effects of deposited nitrogen, an experiment has been started, involving manipulation of both the rates and the species (NH_4 or NO_3) of nitrogen being added. The plant species selected for the study were both trees: beech (*Fagus sylvatica* L.) and birch (*Betula pendula*

Roth. and *Betula pubescens* Ehrh.). These species are hosts to a number of herbivore species and our aim is to relate the performance of selected herbivores to changes in plant chemistry caused by the treatments. We hope to identify chemical components which are affected by the nitrogen additions and which correlate with invertebrate performance. The geographic variation in these components can then be studied with the aim of identifying areas of Britain where there might be effects of deposited nitrogen.

We are still engaged in the detailed chemical analysis of material collected from the experiment and our results will be published in full elsewhere. However, some of the data on insect performance (below) show some interesting effects of the treatments and illustrate some of the problems in attempting to identify critical loads.

The experimental treatments are set out in Table 1. Two-year-old, bare-rooted, sapling trees were transplanted in early March into a low-nutrient, silty-sandy soil within large tubs. There were 12 replicates of each treatment for beech and birch, set in randomized blocks. The trees were left in the open thus exposing them to natural colonization by insects from surrounding vegetation.

The nutrient additions were applied in solution to the soil surface. No attempt was made to apply solutions to the foliage as it was considered this would result in unacceptable variation in loading between trees within a treatment. Furthermore, the aim was to produce trees with a range of nutrient contents, rather than to mimic precisely particular deposition rates. Most treatments (w) were applied weekly, starting in late March, and scaled to achieve the desired annual rate of addition. One treatment (s) was applied once annually in March.

Of several insects that occurred on the trees, two have been studied in

Table 1. Rates (equivalent to kg ha^{-1} year^{-1}), forms and frequencies of supplementary nitrogen applied in solution to beech and birch saplings. In the experiment discussed here, all plants received supplementary phosphorus and potassium to avoid limitation by these elements

Ammonium as NH_4SO_4	Nitrate as $NaNO_4$
0 weekly	0 weekly
20 weekly	20 weekly
40 weekly	40 weekly
80 weekly	80 weekly
80 single (applied in March)	80 single (applied in March)

some detail. These were the beech weevil *Rhynchaenus fagi* L. which as a larva mines inside the beech leaves and is a chewing herbivore, and the aphid *Euceraphis betulae* Koch which is a phloem feeder on birch.

IV. RESULTS

A. Attack Rate

Increasing rates of ammonium or nitrate addition to beech increase the attack rate by the beech weevil *R. fagi*. A single application of fertilizer in March produces a similar response to weekly applications at the same annual rate (Fig. 1). In calculating attack rate the numbers of mines on each tree were counted, but trees with no mines were excluded from the estimates. This was to separate the effects of chance colonization from effects of treatments of the oviposition behaviour of the weevil. There was no significant difference ($P > 0.05$) in the number of saplings which were not infested in each treatment. This suggests that the treatments did not affect host colonization, but did affect oviposition behaviour on the trees. Regression analysis of the data (excluding the single application of fertilizer) shows that addition of ammonium has a strong, but not quite significant effect on attack rate ($P = 0.059$, $r^2 = 0.829$), whereas the effect of nitrate addition is significant ($P = 0.011$, $r^2 = 0.966$).

B. Growth Rate

The aphid *E. betulae* on birch showed higher mean relative growth rate (MRGR) (see Brown, Chapter 10, this volume, for definition) as the rate of ammonium additions increased (Fig. 2(a)). Regression analysis (as in "A" above) indicated a significant effect of ammonium addition on MRGR ($P = 0.012$, $r^2 = 0.963$). The response to nitrate addition suggests that greatest growth is achieved when plants receive weekly additions equivalent to 20 kg ha^{-1} year^{-1}, above this rate growth declines (Fig. 2(b)). The linear regression was not significant ($P = 0.547$, $r^2 = 0.000$).

C. Adult Weight

R. fagi adult weight shows a positive trend with increasing additions of ammonium. The trend with nitrate is not uniform (Fig. 3). However, in

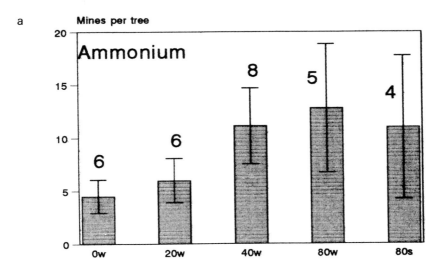

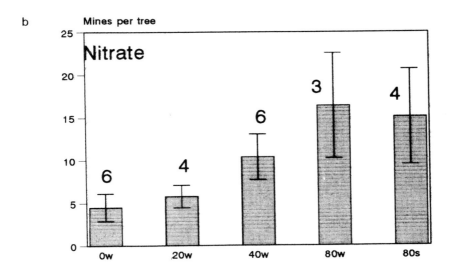

Fig. 1. Attack rate (mean mines per tree ± SE) by the beech weevil, *R. fagi*, on beech trees receiving different rates and frequencies of supplementary nitrogen as ammonium (a) or nitrate (b). Figures above the bars represent numbers of trees contributing to the mean (trees which were not attacked were excluded from the calculation of attack rate).

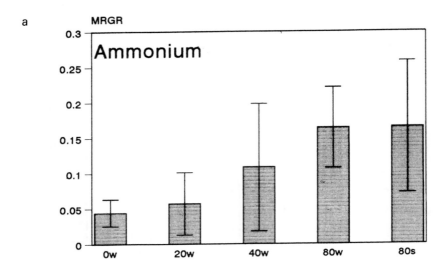

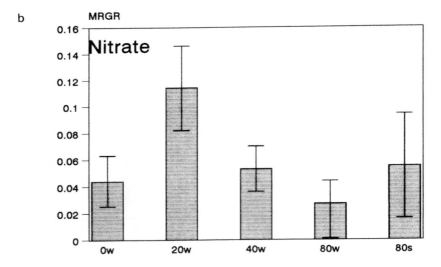

Fig. 2. Mean relative growth rate (mean ± SE, $n = 3$) of birch aphid, *E. betulae*, on birch trees receiving different rates and frequencies of supplementary nitrogen as ammonium (a) or nitrate (b).

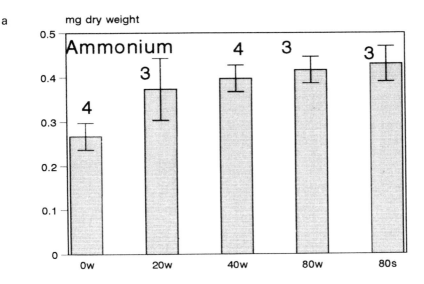
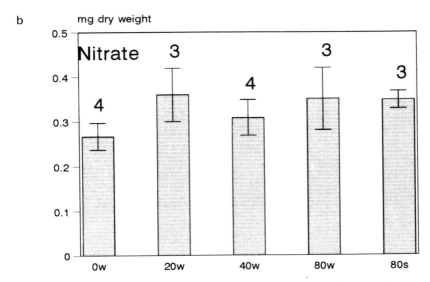

Fig. 3. Mean adult weight (mg dry wt ± SE) of the beech weevil, *R. fagi*, emerging from pupae collected from mines on beech trees receiving different rates and frequencies of supplementary nitrogen as ammonium (a) or nitrate (b). Figures above the bars represent numbers of trees contributing to the mean (trees with less than two mines were excluded from the calculation of mean adult weight).

neither case is the linear regression significant. Similar results were obtained with *E. betulae*.

D. Survival

There were no consistent patterns in the survival of *R. fagi* between eggs laid and mature larvae, with survival averaging 80%. No data were obtained for *E. betulae* because of the difficulty of identifying individual aphids in uncontrolled environments.

E. Population Growth

Once *R. fagi* adults emerge from the pupa they are free to disperse away from the larval food plant and do not reproduce until the following year. Thus it was not possible to monitor any effects on population size caused by feeding on plants in the different treatments. By contrast, the aphid *E. betulae* completes many generations a year on birch. As might be expected from the growth rate studies on this species, the population on plants in the different treatments did vary. With increasing rates of ammonium the population also increased, the largest population on many sampling dates being that on plants receiving 80 kg ha^{-1} year^{-1} as a single application (Fig. 4(a)). In the figure, the regression of population size against different rates of ammonium applied weekly is significant ($P = 0.05$, $r^2 = 0.851$). With increasing rates of nitrate the largest population was usually on those plants receiving 20 kg ha^{-1} year^{-1}. At higher rates of addition the population size declined (Fig. 4(b)). In the figure, the regression of population size against different rates of nitrate applied weekly is not significant.

V. DISCUSSION

The preliminary data from the experiment suggest that the species of nitrogen pollution (N_{ox} or N_{hy}) may have a substantial effect on the response of some herbivorous insects. High rates of NO_4 addition did not correlate with greater growth of individuals or populations of *E. betulae* on birch. Thus, in evaluating nitrogen inputs into a system it will be necessary to distinguish between N_{ox} and N_{hy}.

Even the lowest rate of nitrogen addition (equivalent to 20 kg ha^{-1}

19. Gaseous Air Pollutants

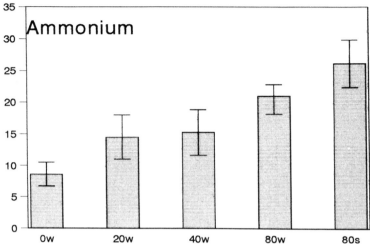

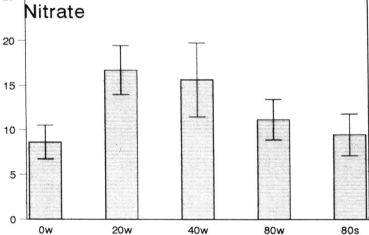

Fig. 4. Population size of birch aphid, *E. betulae*, (mean aphids on top 15 leaves ± SE, assessed on 25 June 1993, $n = 12$) on birch trees receiving different rates, forms and frequencies of supplementary nitrogen.

year^{-1} of NO_3 or NH_4) produced a change in the performance parameters of the two insect species considered here. This suggests that, for some insect herbivore species at least, the critical loads (taking a strict definition) will be very small. However, whether small improvements in performance are ultimately reflected in increased population size will depend on the complex suite of regulatory factors, both biotic and abiotic, influencing a population. The objectively measured *critical loads* will be used to derive *target loads* (Bull, 1991), which are considered to be achievable targets for pollution control. If the aim is to avoid perturbation of natural populations, it may be that in estimating target loads, the buffering effects of natural population regulation may be taken into account. However, where natural regulation is weaker, as with some multivoltine insects such as aphids (r strategists), the target load may of necessity be very close to the critical load.

Acknowledgements

This work is funded by the UK Department of Environment.

REFERENCES

Anon. (1993). "Air Pollution and Tree Health in the United Kingdom". HMSO, London.

Auerbach, M. J. and Strong, D. R. (1981). Nutritional ecology of *Heliconia* herbivores: Experiments with plant fertilization and alternative hosts. *Ecol. Monogr.* **51**, 63–83.

Bull, K. R. (1991). The critical loads/levels approach to gaseous pollutant emission control. *Env. Poll.* **69**, 105–123.

Davison, A. W. (1993). Patterns of air pollution: critical loads and abatement strategies. In "Managing the Human Impact on the Natural Environment" (M. Newson, ed.), pp. 109–127. Belhaven Press, London.

Hornung, M. and Skeffington, R. A. (eds) (1993). "Critical Loads: Concepts and Applications". HMSO, London.

Jansson, R. K., Leibee, G. L., Sanchez, C. A. and Lecrone, S. H. (1991). Effects of nitrogen and foliar biomass on population parameters of cabbage insects. *Entomologia Exp. Appl.* **61**, 7–16.

Loader, C. and Damman, H. (1991). Nitrogen content of food plants and vulnerability of *Pieris rapae* to natural enemies. *Ecology* **72**, 1586–1590.

McNeill, S. and Southwood, T. R. E. (1978). The role of nitrogen in the development of insect/plant relationship. In "Biochemical Aspects of Plant and Animal Coevolution" (J. B. Harborne, ed.), pp 77–98. Academic Press, London.

Mattson, W. J. (1980). Herbivory in relation to plant nitrogen content. *Ann. Rev. Ecol. Syst.* **11**, 119–161.

Myers, J. H. and Post, B. J. (1981). Plant nitrogen and fluctuations in insect populations: A test with the Cinnabar Moth – Tansy Ragwort system. *Oecologia* **48**, 151–156.

Nilsson, J. and Grennfelt, P. (1988). "Critical Loads for Sulphur and Nitrogen". *UN-ECE/Nordic Council Workshop Report, Sokloster, Sweden, March 1988*. Nordic Council of Ministers, Copenhagen.
Ohmart, C. P., Stewart, L. G. and Thomas, J. R. (1985). Effects of food quality, particularly nitrogen concentrations, of *Eucalyptus blakelyi* foliage on the growth of *Paropsis atomaria* larvae (Coleoptera: Chrysomelidae). *Oecologia* **65**, 543–549.
Scriber, J. M. (1984). Host-plant suitability. *In* "Chemical Ecology of Insects" (W. Bell and R. Carde, eds), pp. 159–202. Chapman & Hall, London.
Simpson, S. J. and Simpson, C. L. (1990). The mechanisms of compensation by phytophagous insects. *In* "Insect–Plant Interactions 2" (E. A. Bernays, ed.), pp. 111–160. CRC Press, Florida.
Silvanima, J. V. C. and Strong, D. R. (1991). Is host-plant quality responsible for the population pulses of salt-marsh planthoppers (Homoptera: Delphacidae) in northwestern Florida? *Ecol. Entomol.* **16**, 221–232.
Slansky, F. and Feeny, P. (1977). Stabilization of the rate of nitrogen accumulation by larvae of the cabbage butterfly on wild and cultivated food plants. *Ecol. Monogr.* **47**, 209–228.
Smith, D. S. and Northcott, F. E. (1951). The effects on the grasshopper, *Melanopus mexicanus mexicanus* (Sauss.) (Orthoptera: Acrididae), of varying the nitrogen content in its food plant. *Can. J. Zool.* **29**, 297–304.
Sutton, M. A., Pitcairn, C. E. R. and Fowler, D. (1993). The exchange of ammonia between the atmosphere and plant communities. *Adv. Ecol. Res.* **24**, 301–393.
Thomas, A. T. and Hodkinson, I. D. (1991). Nitrogen, water stress and feeding efficiency of lepidopteran herbivores. *J. Appl. Ecol.* **28**, 703–720.
UKPORG (1990). "Oxides of Nitrogen in the United Kingdom". *The Second Report of the United Kingdom Photochemical Oxidants Review Group*. Department of the Environment, London.

20

Effects of Changing Land Use on Eucalypt Dieback in Australia in Relation to Insect Phytophagy and Tree Re-establishment

ROGER A. FARROW AND ROBERT B. FLOYD

I. Introduction 456
II. Land Degradation and the Need for Trees 456
III. Eucalypt Dieback, its Causes and Remedies 456
IV. Selecting Insect Resistance in Trees for Re-establishment on Farmland 457
V. Variations in Resistance to Insect Attack 458
References 459

Abstract

Dieback of eucalypts in Australian farmland has invariably followed tree clearing and pasture improvement. Dieback has been correlated with a complex of factors associated with changing land use and weather but the one common factor to most rural dieback is insect phytophagy. Programmes to re-establish trees in dieback-affected areas, in order to arrest land degradation and restore agricultural productivity, also face the problem of insect phytophagy. During outbreaks of phytophagous insects on eucalypts, variations in susceptibility to insect attack become apparent between species and between individuals within species. This presents an opportunity of propagating species and genotypes of eucalypts with enhanced resistance to insect feeding for use in areas prone to insect attack, depending on the heritability of such traits.

I. INTRODUCTION

The spread of agriculture across Australia over the past 150 years has been accompanied by widespread tree clearing and loss of understorey to create a characteristic parkland setting of scattered mature eucalypts. Much of this change has been concentrated in woodland ecosystems and in south-east Australia, where more than 90% of woodland has been converted to agriculture (Anon, 1990). Ploughing of native pastures to establish crops and exotic pastures has removed most eucalypt lignotubers, while domestic stock, vermin and introduced pasture plants prevent seedling establishment, resulting in an unsustainable parkland agro-ecosystem.

II. LAND DEGRADATION AND THE NEED FOR TREES

The most obvious sign of land degradation is the loss of topsoil caused by wind and water erosion (Marshall, 1990). A more insidious effect is a rising water table causing extended periods of waterlogging during rainy periods. In some areas, saline water is brought to the surface resulting in dry land salination (Morris and Jenkin, 1990). Loss of shade and shelter from trees also reduces animal production (Lynch and Donnelly, 1980). The role of trees in preventing soil erosion, lowering water tables, providing shelter and developing sustainable agricultural systems is increasingly recognized and there are now major incentives to re-establish native trees in farmland (Bourke and Youle, 1990).

III. EUCALYPT DIEBACK, ITS CAUSES AND REMEDIES

Although senescence contributes to eucalypt decline in parklands, the major cause is dieback induced by the effects of the same environmental changes that have accompanied the creation of a productive pastoral industry in former grassy woodlands. Improved pastures support increased densities of scarabaeid beetle larvae and the foliage-feeding adults are concentrated onto fewer individual eucalypt trees (Mackay, 1978). Increased soil fertility increases the nitrogen content of foliage which may increase the numbers of other defoliating and sap-feeding insects (Landsberg and Wylie, 1983). Other changes in the environment, such as waterlogging, may make trees more susceptible to insect attack through changes in foliage quality (White, 1986), while cycles of defoliation and refoliation also cause feedbacks that improve foliage quality and favour phytophagous insects (Landsberg and Wylie, 1988). It is also suggested that the number of insect parasitoids of insect herbivores is lower in the parkland environment compared with the original woodland community, due to the lack of nectar sources provided by native flowering shrubs (Davidson, 1981) and this is alleged to cause the numbers of phytophagous insects to increase. Research on the biology of thynnid wasps parasitizing Christmas beetles (Ridsdill-Smith, 1968) suggests that they do not have the reproductive capacity and host-seeking ability to prevent outbreaks of their host. Climatic fluctuations are also associated with outbreaks of phytophagous insects on eucalypts (Carne *et al.*, 1980), resulting in episodes of sustained defoliation and dieback on top of any existing chronic defoliation.

There is little that can be done to restore the vitality of older trees suffering insect dieback, as this would require major changes in land use to reduce soil fertility and restore the original native perennial grasses. Fluctuations in weather will continue to have a major influence on outbreaks of phytophagous insects and cannot be modified. It is unclear to what extent increased botanical diversity affects natural enemy populations and whether this would influence the abundance of phytophagous insects or prevent outbreaks. In any case such modifications are not a particularly practical option in intensively managed farmland. However, a more important consideration is the need to ensure the survival of new plantings on farmland to replace trees lost through dieback. The progeny of trees affected by insect feeding is equally susceptible to attack. Our studies indicate that more than 50% annual growth increment may be lost to insect herbivory in plantations in improved pastures with some trees dying completely (Floyd and Farrow, 1994), even though many species of natural enemy are observed attacking insect herbivores in such systems.

IV. SELECTING INSECT RESISTANCE IN TREES FOR RE-ESTABLISHMENT ON FARMLAND

It is widely accepted that species and genotypes tolerant to salt and waterlogging are required for some farmland sites (Cremer, 1990), but there has been some reluctance to extend this approach to other attributes such as insect resistance. The collection of seed from local surviving trees (the local provenance) tends to ignore the fact that these survivors may no longer be adapted to the farmland environment and furthermore their seed may be inbred because of isolation (Pederick, 1987). In CSIRO trials, non-local species of eucalypt have generally survived and grown better than species native to the locality (Floyd and Farrow, 1994).

V. VARIATIONS IN RESISTANCE TO INSECT ATTACK

Variations in resistance to insect attack have been observed at the interspecific level (Pryor, 1952), at the intraspecific level between populations (Farrow et al., 1994), within populations (Journet, 1980; Floyd et al., 1994) and even within individual trees as somatic mutations (Edwards et al., 1990). The distribution and persistence of resistant and susceptible phenotypes in wild populations suggests that such variations are heritable. During insect outbreaks, resistant genotypes produce much more seed than susceptible forms (Farrow, unpublished), presumably because energy reserves go towards reproduction rather than to the replacement of lost foliage. There is little opportunity for more resistant genotypes to evolve in nature because of negligible seedling survival in grazed situations.

Eucalypts allocate a relatively high proportion of carbon to the production of terpenes, phenols and tannins, which are assumed to act as chemical defences against herbivores. The terpenes may comprise 50–100 different interrelated compounds (Boland et al., 1991). The low nutrient status and thickness of cuticular waxes may also confer innate resistance to insect feeding. Eucalypts are outbreeding species which preserve high genetic heterogeneity for a range of characters, including the terpenoids and cuticular waxes. Each species can exhibit different proportions of terpenoid compounds as stable polymorphs which take the form of biotypes or chemotypes within populations and biological races or chemical races between populations (Boland et al., 1991). Similar polymorphic variation can occur in physical traits such as leaf waxiness (Floyd and Farrow, 1994).

Variations in susceptibility to insect attack have been related to the proportions of different terpenoids between species, between populations within species, between individuals in the same population and within individual trees as somatic mutations (Edwards et al., 1990, 1993; Floyd and Farrow, 1994) but not to total oils or phenolics (Fox and Macauley, 1977; Morrow and Fox, 1980).

As early as 1952, Pryor stated that trees planted on farms would only survive if they had resistance to insect attack and concern has also been expressed about the potential susceptibility of monoclonal plantations to insect attack. It is now possible to select insect-resistant genotypes for farmland areas subject to insect-induced dieback. Planting insect-resistant genotypes of local species duplicates and accelerates the processes of natural selection that would occur in natural populations if recruitment was successful. Insect resistant trees can be produced by clonal propagation, seed from controlled pollination, or seed from orchards derived from seed collected from open-pollinated resistant trees after all insect-susceptible progeny have been removed. These solutions to insect problems can also be applied to plantation systems whose primary aim is wood production. Resistance mechanisms may vary between insect species and the mechanisms for a sap-feeding insect may be quite different from those of a leaf-chewer (Floyd and Farrow, 1994) so that the breeding strategy would have to be directed at the prevailing pest of a particular region. This may provide cross resistance to other pests belonging to similar functional groups.

The insect community of farmland trees tends to be dominated by a few abundant species as a result of agricultural practice. The introduction of eucalypts with elevated resistance to particular herbivores will alter this community and give rise to new equilibria, although these are never likely to duplicate those existing in the healthy, remnant woodland communities that still exist on some farms. It is these communities that must be better managed, rather than plantations or agroforests, to preserve insect biodiversity.

References

Anon (1990). "Atlas of Australian Resources – Vegetation". Aust. Gov. Pub. Serv., Canberra.
Boland, D. J., Brophy, J. J. and House, A. P. N. (eds) (1991). "Eucalyptus Leaf Oils". Inkata, Melbourne.
Bourke, S. and Youle, R. (1990). The revegetation of Victoria: the first stages. A description of the revegetation movement in Victoria. In "Australian Ecosystems: 200 years of Utilisation, Degradation and Reconstruction" (D. A. Saunders, A. J. M. Hopkins and R. A. How, eds). Proc. Ecol. Soc. Aust. 16, 135–140.
Carne, P. B., McInnes, R. S. and Green, J. P. (1980). Seasonal fluctuations in the abundance of two leaf-eating insects. In "Eucalypt Dieback in Forests and Woodlands" (K. M. Old, G. A. Kile and C. P. Ohmart, eds), pp. 121–126. CSIRO, Melbourne.
Cremer, K. W. (ed.) (1990). "Trees for Rural Australia". Inkata Press, Melbourne.
Davidson, R. L. (1981). Correcting past mistakes – loss of habitat for predators and parasites of pasture pests. Proc. 3rd Australasian Conf. Grassl. Invert. Ecol. 3, 199–206.
Edwards, P. B., Wanjura, W. J. and Brown, W. V. and Dearn, J. M. (1990). Mosaic resistance in plants. Nature 347, 434–435.
Edwards, P. B., Wanjura, W. J. and Brown, W. V. (1993). Selective herbivory by Christmas beetles in response to intraspecific variation in Eucalyptus terpenoids. Oecologia 95, 551–557.
Farrow, R. A., Floyd, R. B. and Neumann, F. G. (1994). Inter-provenance variation in

resistance of blue gum (*Eucalyptus globulus*) juvenile foliage to autumn gum moth (*Mnesampela privata* Guenée) (Lepidoptera: Geometridae). *Aust. For.*, **57**, 65–68.

Floyd, R. B. and Farrow, R. A. (1994). The potential role of natural insect resistance in the integrated pest management of eucalypt plantations in Australia. *In* "Proc. Symp. Biotechnological and Environmental Approaches to Forest Pest Management, 28–30 April 1993" (S. Halos, ed.). Diliman, Philippines.

Floyd, R. B., Farrow, R. A. and Neumann, F. G. (1994). Inter- and intra-provenance variation in resistance of red gum foliage to insect feeding. *Aust. For.*, **57**, 45–48.

Fox, L. R. and Macauley, B. J. (1977). Insect grazing on *Eucalyptus* in response to variations in leaf tannins and nitrogen. *Oecologia* **29**, 145–162.

Journet, A. R. P. (1980). Intraspecific variation in food plant favourability to phytophagous insects: psyllids on *Eucalyptus blakelyi*. *Ecol. Entomol.* **5**, 249–261.

Landsberg, J. and Wylie, F. R. (1983). Water stress, leaf nutrients and defoliation: a model of dieback of rural eucalypts. *Aust. J. Ecol.* **8**, 27–41.

Landsberg, J. and Wylie, F. R. (1988). Dieback of rural trees in Australia. *Geojournal* **17**, 231–237.

Lynch, J. J. and Donnelly, J. B. (1980). Changes in pasture and animal production resulting from the use of windbreaks. *Aust. J. Agric. Res.* **31**, 967–979.

Mackay, S. M. (1978). Dying eucalypts of the New England Tablelands. *Forests and Timber* **14**, 18–20.

Marshall, C. J. (1990). Control of erosion. *In* "Trees for Rural Australia" (K. W. Cremer, ed.), pp. 369–376. Inkata Press, Melbourne.

Morris, J. D. and Jenkin, J. J. (1990). Trees in salinity control. *In* "Trees for Rural Australia" (K. W. Cremer, ed.), pp. 357–366. Inkata Press, Melbourne.

Morrow, P. A. and Fox, L. R. (1980). Effects of variation in *Eucalyptus* essential oil yield on insect growth and grazing damage. *Oecologia* **45**, 209–219.

Pederick, L. A. (1987). "Reducing the Effects of Inbreeding in Eucalypts". *Forest Management and Research Branch Leaflet* **4**. Department of Conservation and Natural Resources, Melbourne.

Pryor, L. D. (1952). Variation in resistance to leaf-eating insects in some eucalypts. *Proc. Linn. Soc. N.S.W.* **27**, 364–369.

Ridsdill-Smith, T. J. (1968). "A Wasp Parasite of Scarabaeidae". MSc thesis, University of New England, Armidale.

White, T. C. R. (1986). Weather, Eucalyptus dieback in New England, and a general hypothesis of the cause of dieback. *Pac. Sci.* **40**, 69–89.

21

Modelling the Population of *Hydrotaea irritans* Using a Cohort-Based System

J. D. AUSTIN AND J. E. HILLERTON

Abstract

This chapter describes the application of a cohort-based mathematical simulation modelling system for the population of the fly *Hydrotaea irritans*, which is thought to be responsible for the transmission of summer mastitis. Comparisons are made with field collected data, and the results of running the model with an increased temperature scenario are discussed.

The distribution and abundance of Muscid flies associated with farm animals are closely related to climate and prevailing weather (Hammer, 1941). Diseases known to be transmitted by such flies in Europe, including summer mastitis by *Hydrotaea irritans* and infectious bovine keratoconjunctivitis and nematode infections transmitted by *Musca autumnalis*, as well as production limitation caused by fly worry, coincide with peaks of fly abundance (Hillerton, 1987). Generally the flies are more spatially widespread than their economic and health impact, which is highly seasonal. The influence of climate on the fly populations is poorly understood and hence the possible consequences of climatic variation for fly-transmitted diseases are unknown. This chapter describes the influence of climate on the population biology of the fly *Hydrotaea irritans*, which is ubiquitous to temperate Europe. Its population structure is projected by a mathematical simulation model using field and literature data. The model predictions are tested with field observations from seven sites in the UK, France and northern Spain.

A cohort-based model (INSIVE) has been constructed to predict seasonality and population abundance. Basic data on the development of

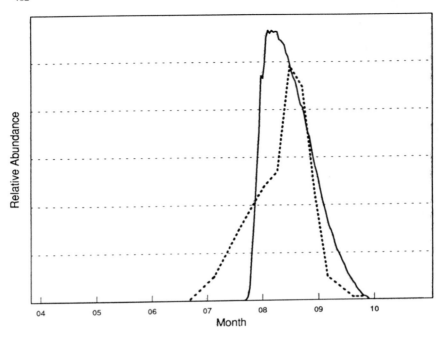

Fig. 1. INSIVE prediction of adult activity of *Hydrotaea irritans* (———), for 25 years using long-term average soil temperatures, and one year's field data from Berkshire in 1992 (----).

the fly have been taken from the literature (Kirkwood, 1976; Robinson, 1979). The population predictions are based on the development of cohorts, derived from the number of eggs laid in each successive day in the breeding season. Each cohort is aged and development estimated by a day-degree algorithm using daily temperature based on monthly averages from an Oxford weather station over 42 years (1949–1991). The predicted emergence window for adults has been adjusted to fit UK field observations by selecting a threshold temperature for development of 13.7°C. Running the population model for a period of 25 years produces a stable output and predicts a synchronized emergence and a unimodal activity curve (Fig. 1). This agrees with field data and the description of a univoltine population (Robinson, 1979; Liebisch, 1987).

Increasing the baseline temperature data set by 1°C to simulate slight climatic warming produces a predicted population with a more gradual adult emergence and a bimodal activity curve indicating a second generation could be possible (Fig. 2). Field data from the UK suggest that

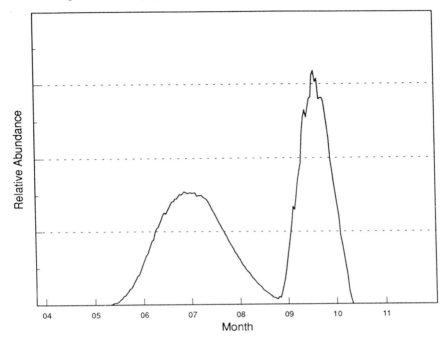

Fig. 2. INSIVE prediction of adult activity of *Hydrotaea irritans* (———) when the underlying temperature during the development phase was 1°C higher than the long-term average soil temperature.

this may have occurred in 1974 (Luff and Robinson, 1977) and again in 1984 (Ball *et al.*, 1985). Our own very recent observations from northern Spain suggest a second generation may occur there. The initial speculation is that, given a small amount of climatic warming in temperate Europe the seasonality of this fly could extend in the UK and its adult activity and abundance change with implications for disease transmission and the period for which farm animals are at risk.

The model will now be refined by the inclusion of further physiological data and testing in comparison with field observations. The basic model will then be extendable to other muscids and possibly *Culicoides* sp. which are important as vectors of disease in temperate Europe.

Acknowledgement

This work is supported by the Ministry of Agriculture, Fisheries and Food.

REFERENCES

Ball, S. G., Port, G. R. and Luff, M. L. (1985). Aspects of the reproductive biology of some cattle visiting Muscidae (Diptera) in North-East England. *Vet. Parasit.* **18**, 183–196.

Hammer, O. (1941). Biological and ecological investigations on flies associated with pasturing cattle and their excrement. *Vidensk. Medd. Dansk. Naturh. Foren.* **105**, 141–393.

Hillerton, J. E. (1987). Summer mastitis; vector transmission or not? *Parasit. Today* **3**, 121–123.

Kirkwood, A. C. (1976). Ovarian and larval development of the sheep headfly, *Hydrotaea irritans* (Fallen) (Diptera, Muscidae). *Bull. Entomol. Res.* **66**, 757–763.

Liebisch, A. (1987). Vector biology of flies on grazing cattle in Germany. In "Summer Mastitis" (G. Thomas, H. J. Over, U. Vecht and P. Nansen, eds), pp. 109–115. Martinus Nijhoff, Dordrecht.

Luff, M. L. and Robinson, J. (1977). Computer-drawn three-dimensional surfaces for biological illustration. *Med. Biol. Illust.* **27**, 41–46.

Robinson, J. (1979). The sheep headfly, *Hydrotaea irritans*: biology of immature stages in the soil. *Bull Entomol. Res.* **69**, 589–598.

22

Effects of Ivermectin Residues in Cattle Dung on Dung Insect Communities Under Extensive Farming Conditions in South Africa

C. H. SCHOLTZ AND KERSTIN KRÜGER

I. Introduction 466
II. Methods 467
  A. Allocation 467
  B. Treatment of Cattle 467
  C. Sampling Procedure 467
  D. Statistical Analyses 468
III. Results 468
IV. Discussion 468
  References 470

Abstract

In the first study of its kind under subtropical conditions, experiments were carried out to determine the effect of ivermectin usage on dung insect communities under extensive farming conditions. Two groups of 20 beef-breed cows were treated with a single injectable dose of ivermectin at the manufacturer's specifications. Two similar groups of cows were used as controls. The cows ranged free in 80-ha camps. The insect communities in the camps were sampled prior to treatment then monitored for 3 months after treatment. Dung insect diversity was lower 1 month after treatment in camps with treated cattle than in the control camps, but 2 months thereafter there were no discernible differences between them. We conclude that, although ivermectin residues apparently depress insect populations initially, the populations recover within 2 months thereafter with the result that long-term effects on dung fauna may not be as severe as some previous studies have claimed.

I. INTRODUCTION

Ivermectin is one of the avermectin parasiticides – that is, a derivative of the compound avermectin B_1 (abamectin) that is used in livestock to control nematodes and endo- and ectoparasitic arthropods. It is administered to the livestock in various forms including oral drenches, subcutaneous injections, topical solutions and by sustained-release boluses (Campbell, 1989). A characteristic of the compound is that, regardless of the method of administration, most of the given dose is ultimately excreted by the treated animals in the faeces. The concentration and the length of time the drug residues are present within the faeces depend largely on the size of the dose, method of administration and duration of treatment. The drug's action is not confined to parasitic nematodes and arthropods, because livestock faeces, particularly those of cattle, attract and harbour large numbers of insects and other invertebrates. The potency of ivermectin, its presence in livestock faeces, and the use of the faeces by many invertebrates, are undisputed facts (Strong, 1993). The question is, does the excreted drug exert any significant effect on the invertebrate dung fauna? Some studies indicate that the drug residues in dung drastically affect some of the insects that develop in cattle dung (Strong, 1993).

In southern Africa there are about 780 species of dung beetles of the scarabaeid subfamily Scarabaeinae and about 60 species of dung-dwelling Aphodiinae. Additionally there are several hundred species of dung-frequenting staphylinid, histerid and hydrophilid beetles, most of which are predators (Doube, 1991). The beetles show considerable specialization with respect to soil type and vegetation cover, diel and seasonal activity, the age and type of dung used, and their foraging and reproductive strategies (see Doube, 1991).

There is some anecdotal evidence that certain dung beetle species' distribution and abundance have changed dramatically as a result of agricultural practices (Scholtz and Chown, 1993).

South Africa has a large cattle industry, with about 8 million cattle that generate about 11% of the country's agricultural income. Most of the beef cattle are farmed extensively. As a rule mature cattle are not treated with anthelmintics but calves may be treated at weaning (usually 7–8 months old).

Because of the concern that ivermectin usage could cause serious environmental effects, as claimed by various authors (see Strong, 1993), we set out to try to measure this effect on dung insect communities under normal extensive farming conditions in South Africa. The richness of the dung fauna in this region and its environmental importance contributed to the concern that pasture contamination with undegraded bovine dung pats, such as was observed in Australia (see Waterhouse, 1974), could arise if the dung insects were greatly disturbed. The approach we followed was somewhat different to that used in previous studies since we attempted to assess the putative depressive effects on individual species and on the communities and to determine whether the species and communities recovered (either as a result of sublethal effects of the ivermectin residues in the dung, emergence from dung deposited before treatment, or replacement by immigration from surrounding areas) in a reasonable time. This chapter reports on some aspects of that study.

II. METHODS

The trials were done at Parys (26.54°S 27.37°E) on the Highveld of central South Africa; the area is a highland (about 1350 m above sea level) temperate grassland. The history of the cattle was well known. The trials ran from December 1992 to March 1993, thus the austral summer and peak beetle activity period. Cattle were present on surrounding farms.

A. Allocation

Prior to treating the cattle with ivermectin a herd of 80 cows, some with calves, were divided randomly into four equal-sized groups and confined in two pairs of adjacent paddocks of 80–100 ha. The two pairs of paddocks were situated about 200 m apart and abutted other paddocks with untreated cattle.

B. Treatment of Cattle

The cattle in two groups were injected with a single dose of ivermectin, according to the manufacturer's specifications; those in the other two groups served as controls.

C. Sampling Procedures

On the day prior to treatment, and at monthly intervals for 3 months thereafter, 10 artificial 1-kg pats, made from mixed dung from cattle from another, untreated herd, were placed at 50-pace intervals in a transect across each paddock. One day after placing out the pats, 50 mm of underlying soil were removed and placed individually in Berlese-type extractors. The procedure after treatment was the same as that before treating the cattle except that in addition to the artificial pats placed in the paddocks 10 fresh, natural dung pats were also labelled in each paddock. Extracted insects were identified (or sorted to morphospecies) and enumerated. Some of the larger ball-rolling or dung-burying scarabaeines may have been undersampled by this sampling method. Temperature, relative humidity and rainfall were monitored continuously for the duration of the trial.

D. Statistical Analyses

Since diversity is accepted as a measure of wellbeing of the system we calculated and compared (Shannon) diversity indices and rank/abundance of the dung insects of the treatment and control dung samples.

III. RESULTS

About 50 000 specimens representing about 74 beetle (49 scarabaeines, 13 aphodiines, 10 staphylinids, 2 hydrophilids) and five fly species were collected. Insect diversity and rank/abundance in paddocks with treated cattle decreased relative to the controls 1 month after treatment but returned to similar levels to those of the controls within 2 months (Figs 1, 2).

IV. DISCUSSION

Although the cattle were maintained under normal farm conditions we created a worst-case scenario by treating all the cattle in the treatment paddocks (as opposed to the management norm of only treating weaners), so if there were major effects one would expect there to be an observable population depression. One month after treatment the dung insect populations were lower in treated paddocks than in the controls but 2 months thereafter there were no discernible differences between them.

What is of concern is not so much the localized loss of a species or a habitat because this readily occurs at the local level naturally for a variety of reasons, but the unnatural loss of the ability of species to recover, evolve or respond to various perturbations (Fairweather, 1993). There is

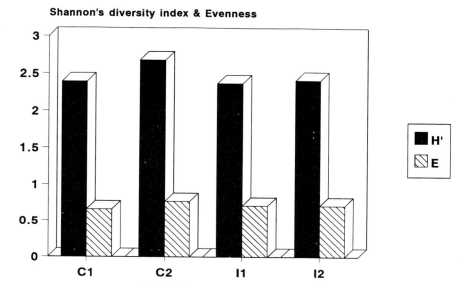

Fig. 1. Shannon's diversity index (H') and evenness (E) of dung insects in artificial control pats in four paddocks prior to treatment with ivermectin; two later served as control (C1 and C2) paddocks while the cattle in the other two (I1 and I2) were treated with ivermectin.

little evidence to suggest that adult insects are killed by exposure to ivermectin residues but the treatment has mainly larvicidal effects and may depress fecundity (Roncalli, 1989) of those insects exposed to the dung while it is toxic. Although we were unable to determine the exact cause of the initial population depression, this was clearly compensated for within 2 months, so that the nett population effects and environmental consequences would have been ameliorated. Since only parts of the population would be affected and mortality would fall unevenly in different sexes and different age classes the population would be "rescued" (Brown and Kodric-Brown, 1977) by immigration from local subpopulations. In spite of apparently high species numbers in dung, dung systems support relatively few kinds of trophic pathways (Schoenly et al., 1991) and these loosely connected food webs are more resistant to species removals than are highly connected food webs (Pimm, 1991).

What we were unable to determine from the preliminary results of this experiment was whether any keystone species were removed whose loss may cause effects to cascade throughout the community. Some of the

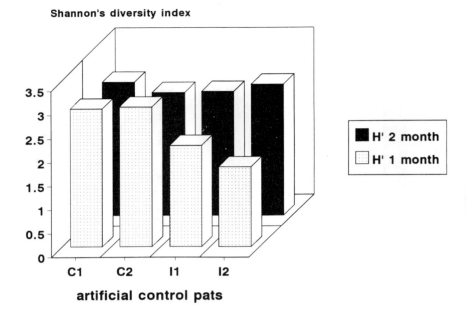

Fig. 2. Comparison of Shannon's diversity index (H') of dung insects in artificial control pats 1 month and 2 months after treatment of cattle with a single standard injection of ivermectin; control paddocks (C1 and C2) and paddocks with treated cattle (I1 and I2).

large rollers or buryers amongst the scarabaeines, which may have been undersampled, may be more important than their numbers suggest.

Our results indicate that dung insect populations apparently recover from whatever depressive effects ivermectin treatment has on them. There is growing evidence that other management practices such as dipping or spraying against ectoparasites may have significant effects on dung insects (Bianchin et al., 1992). Thus the use of various antiparasitics in combination could cause deleterious effects on dung insects. Consequently, general farm management practices may be largely responsible for the perceived changes in composition of dung insect species.

REFERENCES

Bianchin, I., Honer, M. R., Gomez, A. and Koller, W. W. (1992). Efeito de alguns carrapaticidas/inseticidas sobre *Onthophagus gazella*. *EMBRAPA Comun. Tec.* **45**, 1–7.
Brown, J. H. and Kodric-Brown, A. (1977). Turnover rates in insular biogeography: effect of immigration on extinction. *Ecology* **58**, 445–449.

Campbell, W. C. (ed.), (1989). "Ivermectin and Abamectin". Springer, New York.
Doube, B. M. (1991). Dung beetles of Southern Africa. In "Dung Beetle Ecology" (I. Hanski and Y. Cambefort, eds), pp. 133–155. Princeton University Press, Princeton.
Fairweather, P. G. (1993). Links between ecology and ecophilosophy, ethics and the requirements of environmental management. *Aust. J. Ecol.* **18**, 3–19.
Pimm, S. L. (1991). "The Balance of Nature: Ecological Issues in the Conservation of Species and Communities". The University of Chicago Press, Chicago.
Roncalli, R. A. (1989). Environmental aspects of the use of ivermectin and abamectin in livestock: effects on cattle dung fauna. In "Ivermectin and Abamectin" (W. C. Campbell, ed), pp. 173–181. Springer, New York.
Schoenly, K., Beaver, R. A. and Heumier, T. A. (1991). On the trophic relations of insects: a food-web approach. *Am. Nat.* **137**(5), 597–638.
Scholtz, C. H. and Chown, S. L. (1993). Insect conservation and extensive agriculture: the savanna of southern Africa. In "Perspectives in Insect Conservation" (K. J. Gaston, T. New, and M. J. Samways, eds), pp. 75–95. Intercept, Andover.
Strong, L. (1993). Overview: the impact of avermectins on pastureland ecology. *Vet. Parasit.* **48**, 3–17.
Waterhouse, D. F. (1974). The biological control of dung. *Sci. Am.* **230**, 100–109.

23

Monitoring the Response of Tropical Insects to Changes in the Environment: Troubles with Termites

P. EGGLETON AND D. E. BIGNELL

I. Introduction	474
II. Troubles with Termites	476
A. Biogeographical Variation Between Sites	476
B. Sampling	477
C. Seasonal Variation	486
D. Termite Behaviour	489
E. Work-force and Resources	489
F. Identification and Taxonomic Identity	490
III. The Data So Far	491
IV. Conclusion: General Consequences	492
References	494

Abstract

Termite sampling is discussed in the context of changes in land use. The difficulties inherent in such sampling are considered, with an emphasis on the relative strengths and weaknesses of qualitative and quantitative methods. It is concluded that most methods lead to an underestimate of termite abundance and biomass, especially mound-sampling alone, as mounds may account for less than 10% of termite abundance in forest systems. A stratified sampling regime is outlined, which provides a more comprehensive and accurate protocol for the quantitative sampling of termites. Provisional results from a study using this regime are presented and compared with other studies using less comprehensive methods. Differing sampling protocols, however, make comparisons between studies difficult. What comparable results there are suggest that forest clearance leads to greatly reduced levels of termite species richness, abundance and biomass, but that regenerating forest and native tree plantations may have as high or higher species richness, abundance and diversity as primary forest. As a general point, it is concluded that there is a pressing need to standardize sampling methods and integrate data from different insect groups within the same study areas.

I. INTRODUCTION

Levels of overall biodiversity are thought to be highest in tropical forest ecosystems (Marshall, 1992). However, these ecosystems are among the most threatened by changes in land use (Myers, 1989; Sayer and Whitmore, 1991; Harcourt, 1992). Conservationists and other biologists are concerned both to monitor and conserve this diversity for a number of reasons. These include the preservation of potentially valuable organisms and genes, the maintenance of ecosystem stability, and the protection of environments from the consequences of forest clearance on global climate change (Myers, 1989, Groombridge, 1992). In addition, there is a strong moral case against accelerating rates of extinction beyond natural background levels (Sayer and Whitmore, 1991; Lawton and May, 1994).

In any discussion of tropical forest ecosystems insects inevitably take centre stage. Estimates of insect diversity in such systems are many and varied (Erwin, 1982, 1983; Stork, 1988), but all authors would agree that insects are the most diverse and abundant group. However, despite recent interest in tropical deforestation and associated extinction rates, there are relatively few studies dealing with the responses of tropical insects to environmental change (Sutton and Collins, 1991). This may partly be due to the relative remoteness of tropical biomes, but sampling tropical insects is also a difficult task, with many problems not necessarily encountered at higher latitudes. In this article we use sampling of termites to illustrate and discuss the difficulties that arise with insect groups in the tropics. We argue that both the complexity of tropical systems and the scale of our ignorance begs for a more integrated and standardized approach, and substantiate our case with a brief critical appraisal of the existing data on termite abundance and diversity. We conclude that the scale of sampling and processing effort required to gather data on tropical insects has not been fully appreciated.

Termites are extremely important components of tropical ecosystems. They have a premier role as decomposers of organic material (and through this contribute significantly to carbon fluxes), they are extremely important conditioners of soil, and are among the most destructive of all

23. Troubles with Termites

Table 1. Abundance and species richness data for the Mbalmayo plots, with estimates of abundance contributions of wood, soil and mound populations. Percentages refer to % of total abundance for each category in each site. j, jackknife estimate of overall species richness. From Eggleton et al. (1995)

Treatment	No./m²			Quant sp. no.	Transect sp. no.
	Wood	Soil	Mounds		
Complete clearance	15	2432	0	11	16 (j = 21)
	0.6%	99.4%	0%		
Weeded fallow	25	1483	0	8	24 (j = 36)
	1.7%	98.3%	0%		
Young plantation	3127	3049	531	31	53 (j = 75)
	46.6%	45.4%	8.0%		
Old secondary	1452	8453	524	31	53 (j = 82)
	13.9%	81.1%	5.0%		
Near primary	168	6385	401	25	46 (j = 64)
	2.4%	91.8%	5.8%		

insect pests. They are also very abundant (up to 10 000 individuals per square metre, see Table 1) and make up perhaps 10% of animal biomass in the tropics (Lee and Wood, 1971; Wilson, 1993).

Brown (1991) emphasizes the importance of termites as an indicator group for biogeographical and ecological analysis and monitoring. On a scale of 24 he gives termites 20 (only surpassed by Heliconiine/Ithomiine butterflies and ants, which he gives 21), stressing a number of termite attributes: their taxonomic and ecological diversification, their relatively sedentary habit, their taxonomic tractability, the fact that individuals are present throughout the year, their functional importance in ecosystems, and their apparently predictable response to disturbance. Termites' sedentary habits mean that they can be sampled directly to give absolute values of abundance and biomass, unlike other more mobile insect groups where relative sampling methods (such as Malaise and flight interception traps) have to be employed.

We are at present undertaking an extensive study of termite diversity at the Mbalmayo Forest Reserve, southern Cameroon as part of the UK's Natural Environment Research Council's Terrestrial Initiative in Global Environment Research (TIGER, see acknowledgements).

At Mbalmayo long-term trials of a number of land-use systems are being carried out under the management of the Republic of Cameroon Office Nationale de Développement des Fôrets. We are part of a consortium investigating the responses of termite assemblages to disturbances accompanying forest clearance and afforestation. Within this project, we feel that we have been able to sample termites, both qualitatively and quantitatively, with a thoroughness not commonly achieved in previous studies of natural tropical ecosystems. We refer extensively to the preliminary results of this study and suggest that our data represent a tentative baseline for comparing other, less comprehensive, studies.

II. TROUBLES WITH TERMITES

In this section we deal with the problems that make ecological termite studies difficult to undertake and the resulting data difficult to interpret.

A. Biogeographical Variation Between Sites

Evolution and geological history ensure that no two sites anywhere within the world have exactly the same complement of species. This is especially true for termites, which disperse poorly and so have high rates of species turnover across relatively short geographical distances, for example only 36% of termite species found in southern Cameroon forests are also found in the forested regions of neighbouring Nigeria (Eggleton et al., unpublished; Johnson et al., 1982). At a continental scale this problem is amplified such that, even at higher taxonomic levels, there is very little similarity between faunas; for example, southern Cameroon and west Amazonia, two of the richest areas in the world for termites, have only 3.4% of genera in common.

Biogeographical taxonomic differences may radically alter ecological patterns. One functional group of termites, the Macrotermitinae (unique amongst termites as fungus-growers) are absent from the Neotropics and Australia. Energetic estimates for termites in a southern Guinea savannah show that the Macrotermitinae account for over two-thirds of the energy flow through the termite assemblage (Wood and Sands, 1978). Although similar estimates do not exist for any Neotropical system, the absence of Macrotermitinae can be predicted to produce major functional differences.

In Australia, both the Macrotermitinae and the soil-feeding Apicotermitinae are absent. The consequences of this are not fully understood, but the small number of humus feeders (Noirot, 1992) and the relatively few species found in rain forest areas (Gay, 1970) suggest a pattern of speciation and niche-exploitation quite different from that of other biogeographical regions (see Eggleton et al., 1994). In Australia and in the tropical Americas, the foraging role of the Macrotermitinae is performed by other groups (notably by species of the Nasutitermitinae), but none of these is associated with a symbiotic fungus and a different functional structure of termite assemblages is likely, especially in savannah-like systems.

These differences can also be seen at smaller scales. In Mbalmayo there are no *Macrotermes* species. These species forage actively in the leaf litter, and are often dominant in other African forest and savannah. Areas relatively close by with *Macrotermes* (forests in Congo and Central Cameroon) appear to have a lower standing crop of leaf litter than Mbalmayo (C. Rouland, pers. comm.).

These biogeographical differences complicate studies that attempt to compare termite assemblages in different habitats across biogeographical regions. However, we cannot even begin to estimate the strength of this effect until we have solved the problems of how to gather data on termite assemblages which are strictly comparable site against site.

B. Sampling

1. The root of the problem: termite spatial distribution

Termites are not distributed randomly within habitats. Being colonial insects, they concentrate around colony centres of different sizes, and in turn these colony centres are scattered unevenly across microhabitats (but see Salick and Tho, 1984, for a system where termites may be randomly distributed). Figure 1 shows schematically the complex local distribution of termites in the Mbalmayo Forest Reserve. We have split the termite species into 12 separate microhabitat groups, in order to emphasize the putative complexity of the system. Although these microhabitats represent real biological entities that it would be desirable to sample, in fact it is very difficult to sample each termite colony in each different microhabitat separately. This is because of the difficulty of defining the exact limits of a colony (i.e. both the colony centre and the colony's ramifying foraging tunnels). In reality sampling is normally practicable in four areas: (a) termites in the soil (either foraging or in subterranean nests); (b) termites in mounds (as primary or secondary occupants); (c) termites in the canopy (in nests, foraging on dead branches still on trees, in suspended soil); and (d) termites in wood (either foraging or nesting in the logs on the floor of the forests); (for a breakdown of these areas within the Mbalmayo forest reserve, see Table 3). Previous termite studies have mostly concentrated on one of these sampling areas or have employed relative (baiting) or qualitative approaches (Table 2).

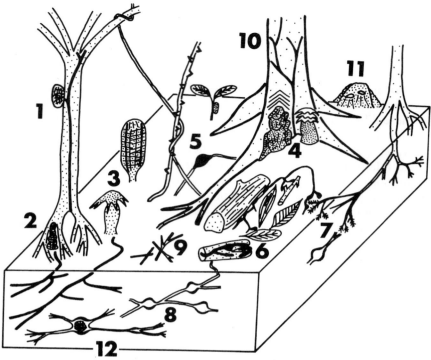

Fig. 1. The principal microhabitats and probable functional groupings of termites in the humid forest zone of southern Cameroon (diagram schematic and not to scale). 1. Termites nesting arboreally and foraging widely (normally within runways) within the canopy and on the ground (also includes termites nesting and feeding on dead wood that remains attached to trees in the canopy, i.e. Kalotermitidae); typically active wood-feeders. 2. Termites constructing epigeal mounds associated with stilt roots; typically soil feeding Termitinae foraging widely in the soil profile. 3. Termites constructing free-standing epigeal mounds not obviously associated with trees; typically soil feeding Termitinae foraging widely in the soil profile. 4. Termites in large epigeal mounds, frequently multiple constructions containing large colonies with many secondary inhabitants, associated with buttress roots; typically soil-feeding Termitinae foraging widely in the soil profile, but secondary species may include Apicotermitinae, soil and wood feeding Nasutitermitinae and Macrotermitinae. 5. Purse nests constructed with soil but attached to low vegetation and connected to the soil by runways on plant stems; typically soil-feeding soldierless Apicotermitinae, some possibly feeding at the root/soil interface. 6. Termites associated with decaying wood and other organic matter; a very diverse assemblage including wood-feeding Termitidae such as some Termitinae and Macrotermitinae that colonize whole logs and branches, and many colonies of Apicotermitinae apparently confined to the wood/soil interface. 7. Termites associated with root hairs; a speculative group possibly including some Apicotermitinae and some species of *Microtermes*. 8. Macrotermitinae forming polycalic subterranean networks, often at considerable depth and foraging widely at the surface of the ground. 9. Very fine twigs and dead plant stems may be hollowed out by small termites, especially *Microtermes* (see text); 10. Wood-feeding termites foraging in the high canopy but nesting underground, or termites establishing large colonies in heartwood; 11. Large but generally well-spaced hard carton mounds of wood-feeding Termitinae. 12. Entirely subterranean termites with diffuse or concentrated nests; typically soil-feeding Apicotermitinae (and ?Termitinae) foraging widely in the soil profile.

2. Qualitative sampling

Termites have often been sampled casually, without an explicit sampling protocol. If the sampling effort is large enough then this can give good estimates for species richness. Collins' (1977) estimate of 43 species for casual sampling from the Edea rain forest in southern Cameroon is close to Eggleton *et al.*'s (1995) qualitative transect estimate of 46 for a similar (slightly drier) forest 170 km away. Constantino (1992) attempted to standardize the collecting effort between his sites by intensive collecting in 50×50 m plots, but there was no explicitly defined quantitative sampling regime.

Eggleton *et al.* (1995) suggest a different approach. Working at five sites within the Mbalmayo Forest Reserve they sampled extensively along a 100×2 m transect, splitting the transect into equal area quadrats for analysis purposes. This enabled statistical approaches to be employed (i.e. species/sampling effort curves, and first order jackknife estimates of total species richness, see tables 1 and 2), which enhance confidence in the data despite its non-quantitative nature. De Souza and Brown (1994) employed a very similar transect approach to examine the effects of forest fragmentation on termite assemblages in Central Amazonia.

The advantage of these qualitative methods is that it is easier to collect all the species than it is to estimate each species' abundance. This is especially true of termite species that nest in trees but forage in the leaf litter; estimates of colony size are very difficult, but confirming that the species is present is very easy. The great limitation is that only species richness can be measured.

3. Baits

Termites will come to a number of different cellulose-based baits (e.g. toilet rolls, wood blocks, etc.), and this has been used for semi-quantitative sampling. Baiting is a good way of gaining rough species richness estimates and some relative estimates of abundance and biomass. This is often used when absolute quantitative sampling is impracticable.

Baiting, however, has major drawbacks. There is a bias in the samples obtained towards species feeding primarily on cellulosic resources (excluding soil-feeding species) or those actively foraging at the soil surface (excluding species that are permanently subterranean or arboreal). Baiting is also more likely to select for foraging castes, excluding others (reproductives, soldiers) that facilitate identification and taxonomic description. Baiting also declines in effectiveness as baits are consumed or dry out.

Table 2 Summary of termite assemblage parameters from previous studies. C, casual sampling; SQ, semi-quantitative sampling; T, transect sampling; M, mound sampling; SC, soil cores; SM, soil monoliths; SP, soil pits; S(B), soil sample with termites extracted by Berlese–Tullberg funnel; W, wood samples; j, jackknife estimates. In species number section "all" or "one", "two" etc. refer to studies where total species richness is not given or where one or a few specific species are involved. Baiting studies excluded. For explanation of these sampling methods see text. Figures for abundance and biomass are means. Table modified and augmented from Wood and Sands (1978)

Biome	Sampling protocol	Number of species	No. m^{-2}	Biomass (g m^{-2})	Author
Tropical savannah					
N. Guinea savannah, W. Africa	M	19	110–2860		Sands (1965a,b)
Grass savannah, central Africa	M	Single	612–701	1.3–1.9	Hebrant, in Bouillon (1970)
Savannah woodland, southern Africa	M	Single	139–1711		Dangerfield (1990b)
Savannah woodland, north Australia	SC	All	2000		Lee and Wood (1971)
"Savannah"	SC	Single	70		Bouillon et al. (1962), Bouillon (1964)
Secondary southern Guinea savannah, W. Africa	SC	22	2966	3.6	Wood et al. (1977)
Savannah Woodland, southern Africa	SM	All	100	0.8	Dangerfield (1990b)
Disturbed S. Woodland, southern Africa	SM	All	49	1.5	Dangerfield (1990b)
S. Guinea savannah, W. Africa	SC, M	23	4402	11.1	Wood et al. (1977, unpublished)
Savannah woodland, West Africa	SC, M	Four	879		Ohiagu (1979)
Secondary S. woodland, West Africa	SC, M	One	860		Ohiagu (1979)
"Derived" savannah, West Africa	SP, M	36	861	1.7	Josens (1972)
Tropical forest					
Rainforest, West Africa	C	43			Collins (1977)
Rainforest, Amazonia	C	25–61			Constantino (1992)
Swamp forest, Amazonia	C, SQ	11			Constantino (1992)
Semi-deciduous forest, West Africa	T	46, j=64			Eggleton et al. (1995)
Secondary semi-deciduous forest, West Africa	T	53, j=82			Eggleton et al. (1995)
Rainforest, Central Amazonia	T	17–28			De Souza and Brown (1994)
Rainforest fragments, Central Amazonia	T	4–7			De Souza and Brown (1994)
Rainforest, South America	M	Single	87–104	0.1	Wiegert (1970)
Rainforest, Malaysia	M	Four	1330	3.4	Matsumoto (1976)
Dry deciduous forest, N.E. Thailand	SP	All	100	0.1	Watanabe et al. (1984)
Mixed Dipterocarp, West Malaysia	SC	All	779–1603	1.1–1.8	Collins (1980a)

Habitat	Code	n	Value	Reference
Lower Montane forest, West Malaysia	SC	All	38	Collins (1980a)
Upper Montane forest, West Malaysia	SC	All	99–295	Collins (1980a)
Mixed Dipterocarp, West Malaysia	SC	All	1148	Collins et al. (1984)
Alluvial forest, West Malaysia	SC	All	254	Collins et al. (1984)
Kerangas forest, West Malaysia	SC	All	1408	Collins et al. (1984)
Rainforest, West Indies	SC	All	4450	Strickland (1944)
Rainforest, Central Amazonia	S (B)	All	3–453	Bandeira (1979)
Riparian forest, West Africa	SC	33	2646	Wood et al. (1982)
Semi-deciduous forest, West Africa	SC	31	3163	Wood et al. (1982)
Semi-deciduous forest, West Africa	SC	All	3163	Wood and Johnson (unpublished)
Riverine forest, Central Africa	SC, M	All	1000	Maldague (1964)
Mixed Dipterocarp, West Malaysia	SP, M	57	3160–3810	Abe (1979)
Semi-deciduous forest, West Africa	SP, M, W	30	6954	Eggleton et al. (unpublished)
Secondary/forestry systems				
Tree plantation (1 year), West Africa	T	16, j=21		Eggleton et al. (1994a)
Tree plantation (4 year), West Africa	T	53, j=75		Eggleton et al. (1994a)
Secondary forest, Central Amazonia	S (B)	All	708–1109	Bandeira (1979)
Secondary semi-deciduous forest, West Africa	SP, M, W	36	10 429	Eggleton et al. (unpublished)
Tree plantation (1 year), West Africa	SP, M, W	11	2447	Eggleton et al. (unpublished)
Tree plantation (2 year), West Africa	SP, M, W		119	Eggleton et al. (unpublished)
Tree plantation (5 year), West Africa	SP, M, W	28	6707	Eggleton et al. (unpublished)
Agricultural systems				
Weeded bush fallow, West Africa	T	24, j=36	17–621	Eggleton et al. (1994a)
Pasture, Central Amazonia	S(B)	All	7	Bandeira (1979)
Cleared and burned forest, N.E. Thailand	SP	All		Watanabe et al. (1984)
Grazed pasture, West Africa	SC	20	2010	Wood et al. (1977)
Maize (first year), West Africa	SC	8	1553	Wood et al. (1977)
Maize (8–24 years), West Africa	SC	4	6825	Wood et al. (1977)
Sugar cane plantation, West Africa	SC	8	4800	Wood et al. (1982)
Maize, Southern Africa	SM	All	0	Dangerfield (1990b)
Grass fallow, Southern Africa	SM	All	51	Dangerfield (1990b)
Eucalyptus stand, Southern Africa	SM	All	19	Dangerfield (1990b)
Weeded bush fallow, West Africa	SP, M, W	8	1508	Eggleton et al. (unpublished)

Values column (right side): 0.1; 0–0.7; 1.8; 0.5; 3.6; ; ; 6.9; 8; 8; 11; 8.7–10.1; ; ; ; ; ; ; ; ; ; <0.1; 2.8; 1.7; 18.9; 5.4; 0; 0.1; 0.8

Because of these difficulties we will not discuss baiting methods in detail, although there are a number of studies which have used them effectively for purposes other than the estimation of assemblage parameters (e.g. rates of wood decomposition due to termites, Abe, 1980; reproductive isolating mechanisms, Wood, 1981; foraging activity, Pearce, 1990, Pearce et al., 1990; the effects of habitat fragmentation, De Souza, 1993). Termiticide application using baits forms an important part of the control of termite pests (e.g. Duncan et al., 1990).

4. Mounds

Colonies are sometimes used as sampling units rather than individuals, because the colony is the unit of selection in social insects (see, for instance, Wilson, 1992). However, such studies tend to be limited to surveys of epigeal mounds (e.g. Watanabe et al., 1984; Redford, 1984; Noble et al., 1989). Of the 110 species collected in the Mbalmayo forest reserve by systematic sampling protocols only eight (7%) build obvious epigeal mounds, six (5%) build arboreal nests, and one (<1%) builds purse nests on shrubby vegetation, leaving 96 (88%) with subterranean nests, nesting in wood, or as secondary occupants (Table 3). These figures imply that a colony-level approach sampling only epigeal mounds will lead to serious species richness underestimates, especially if secondary occupancy is not taken into account (see below). However, approaches like Abe and Matsumoto's (1979) who dug into the soil to identify large subterranean mounds may be more accurate.

There have been several studies that sample mound populations to obtain estimates of assemblage species richness, abundance and biomass (see Table 2). Some studies have used a transect approach (Dangerfield, 1990a), others have used radioisotopes to obtain mark–recapture estimates of population sizes (Easey and Holt, 1989). Studies have often concentrated on one or a few obvious (apparently dominant) species (e.g. Ohiagu, 1979, see also Table 2). However, as with species richness, mounds may not hold a significant proportion of termite abundance or biomass. Even mound-building species may have a large number of their individuals outside the nest, foraging, at any one time. Ohiagu (1979) showed that more than half of the populations of four *Trinervitermes* species were in the soil rather than in the mounds. At an assemblage level this bias may be much worse; in Mbalmayo, less than 10% of the sampled termites were in mounds (Table 1).

Mounds, however, can be very important sources of diversity, due to the frequent presence of secondary termite inhabitants within them. Although secondary occupants usually form small colonies and may not be highly abundant, mounds can contain many species of them (e.g.

Table 3. Distribution of termites in different habitats at the Mbalmayo site, southern Cameroon. Data combined for all plots. No. of species is the number that are found primarily in that group of microhabitats, secondary occupants not included. Nearly all the species can be collected from soil samples, although they may not nest or forage there. Plot areas 20 × 30 m for quantitative estimates and 100 × 2 m for transect estimates (see text)

Microhabitat	Type	No. of species	Genera with colony centres in microhabitat
1	Arboreal	6 (+s)	*Nasutitermes*, *Microcerotermes*
2, 3, 4, 11	Epigeal	8 (+s)	*Cubitermes*, *Thoracotermes*, *Procubitermes*, ?*Noditermes*, *Cephalotermes*, *Termes* (many secondaries)
5	Purse	1? (+s)	*Astalotermes* (and secondaries)
6, 9, 10	Wood	c.30	*Nasutitermes*, *Microcerotermes*, *Termes*, *Coptotermes*, *Schedorhinotermes*, most Macrotermitinae
7	Roots	?	?some Apicotermitinae
8, 12	Soil	>100	Most Apicotermitinae, some Termitinae

+s, plus secondary inhabitants.

Redford, 1984). Our own work confirms this – 40 of the 110 species recorded from the Mbalmayo Forest Reserve have been found in other termites' mounds or nests (although some are also found elsewhere).

5. Soil

Many termite species forage in the soil to some extent (see Table 3) and a large number are restricted to soil with both the colony centre and foraging galleries concealed under the ground. However, soil sampling is not easy (Sands, 1972). Approaches differ in the number and size of soil samples taken. These range from a few very large pits (up to 10 m × 20 m × 30 cm (depth), (Abe and Matsumoto, 1979), through a moderate number of medium-sized soil monoliths (25 cm × 25 cm × 30 cm, Dangerfield, 1990b) or pits (50 cm × 50 cm × 50 cm, Watanabe and Ruaysoongnern, 1984; 20 cm × 20 cm × 50 cm, Eggleton et al. (1995) to a large number of small soil cores (10 cm diameter × 100 cm deep, Wood et al., 1982). Constraints on soil sample size (e.g. labour and transportation) are discussed below.

There are problems with any size of soil sample. Digging a few large pits risks either under- or overestimating termite abundance due to patchiness. On the other hand, most small soil cores will have no termites in them, which can lead to statistical problems and can dent the enthusiasm of locally recruited field assistants. In addition, soil cores can be pushed deeper than pits can be dug, but there are diminishing returns below about 50–100 cm (Wood et al., 1982), (except in those systems with *Odontotermes* and *Macrotermes* species dominant, where there may be significant termite numbers at 2–3 m). One clear advantage of small cores is the relative volumetric constancy of the samples and their rapid execution, minimizing the possibility that termites will escape the core before it is extracted. Soil pits are difficult to dig to a consistent size over a long period of sampling. The notional 20 cm × 20 cm × 50 cm pits employed at Mbalmayo (Fig. 3) have an excavation volume of 20 litres, but a 1-cm error in all three principal dimensions would generate an error of over 2000 cm^3 (i.e. greater than 10% of the original total volume).

Soil samples give more accurate estimates of overall termite abundance than mound samples (see Table 1). However, they probably underestimate species richness as they may not sample species intimately associated with mounds and dead wood.

6. Arboreal termites

Arboreal nests are extremely difficult to sample, as most of them occur above 2 m. Fogging is not a suitable technique for dislodging termites, as even if they are affected by the insecticide spray they remain inside their nests or foraging tunnels. Scaling or felling of trees is usually impractical. Some termites (e.g. *Coptotermes*) live wholly within the dead (but still affixed) branches of trees or within rotted heartwood, leaving no external signs of their presence.

7. Wood

Sampling has rarely been extended to wood found lying on the forest floor, and where it has sampling has been qualitative (Abe, 1979) or semi-quantitative (Collins, 1983) rather than quantitative. The Mbalmayo studies suggest that this may represent a large underestimation in certain habitats (e.g. the young plantation plot in Table 1).

Wood-feeding termites are often very difficult to dislodge from gallery complexes formed within items of fallen dead wood. Species that form colony centres in logs and branches can insinuate themselves in huge numbers. Failure to sample larger items of wood may underestimate termite abundance.

8. Towards a standard sampling protocol?

In the Mbalmayo studies we have tried to overcome some of the sampling problems discussed above. We have used a stratified sampling regime (described below and see Figs 1 and 2), designed to take into account the patchy distribution and trophic specialization of the termite assemblages, and to combine both quantitative and qualitative protocols. In addition, repeat samples have been taken in the second year of the project.

The sampling programme differed slightly between the first and second years. In 1992 a sampling area of 30 m × 20 m was marked within each hectare plot and where necessary cleared along the perimeter lines by machete to permit access. Ten quadrats, each of 2 m × 2 m, were marked within each sampling area after the selection of co-ordinates by a random process. All dead wood within each quadrat was removed to the laboratory and dissected fibre by fibre. In the centre of the quadrat a 20 cm × 20 cm × 50 cm pit was dug and again the sample was taken back to the laboratory and hand-sorted. All removed termites were put into 70% alcohol for later processing at the Natural History Museum, London. Qualitative sampling from a belt transect of 100 m × 2 m was undertaken 3 months later (see Eggleton *et al.*, 1995).

Using the qualitative transects as a species-richness baseline, it was clear that we were underestimating species richness using this technique (i.e. quantitative sampling picked up less than half of the estimated total species richness in the plots (Table 1)). This appeared to be due to the undersampling of termites (especially soil-feeding species) that live at the wood/soil interface and in microsites such as soil between the buttress roots of trees. Therefore, in the second quantitative sampling we followed the same regime but additionally removed the top 5 cm from the pit and sorted it separately (a so-called soil scrape). We also took another nine 20 cm × 20 cm × 5 cm scrapes from within each quadrat so that we have strictly comparable pit samples as well as additional scrape samples. Though not yet analysed fully, initial results suggest that the addition of scrape samples will give a closer approximation to the species richness levels estimated by the transect approach.

After all other forms of sampling had been completed, all termite mounds and nests in the 20 cm × 30 m plots were removed for the extraction and enumeration of insects within them. Termites so sampled were identified to species. In the case of very large mounds a combined homogenization and subsampling procedure was employed to estimate total numbers.

Figures 2 and 3 show schematically the sampling protocols in the 2 years and Table 1 presents a summary of our preliminary results. Although this protocol is open to a number of criticisms (especially in regard of the number of pits, depth of pits and lack of sampling or arboreal nests) we believe it to be one of the first attempts to sample termites at all spatial scales. We hope that it will form the basis of a standardized sampling regime.

A notable feature of the data is the small proportion of the total termite population found in recognizable mounds and nests, which strikingly illustrates the potential inaccuracies of mound-centred sampling protocols (see above). Termite densities in mature forest (estimated as 10 466 m^{-2} for a secondary site and 7130 m^{-2} for near primary woodland, Table 1) are amongst the highest ever recorded (see Table 2). This is partly in accordance with biogeographical prediction (Eggleton *et al.*, (1994) but also illustrates the advantages of a thorough sampling programme.

C. Seasonal Variation

Tropical systems are often thought to be aseasonal, or to be broadly predictable in their seasonality. In fact, there is increasing evidence that seasonality is a very important factor in tropical insect studies (Wolda, 1988).

Seasonality can affect termite abundances (Bandeira (1979) and Ohiagu (1979) found it to be higher during the wet season), the proportion of

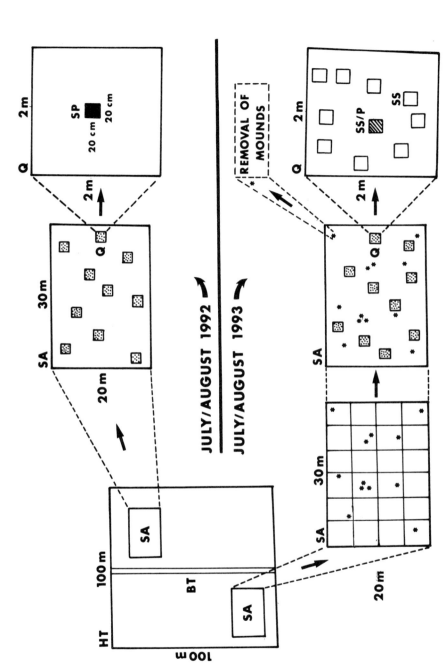

Fig. 2. Overall scheme of quantitative sampling in each 1-ha plot (HT). SA, sampling area; Q, quadrat; SP, soil pit; SS, soil pit; SS/P, soil scrape over pit; *, mounds, purse nests and other termite structures. See text and Fig. 3 for details.

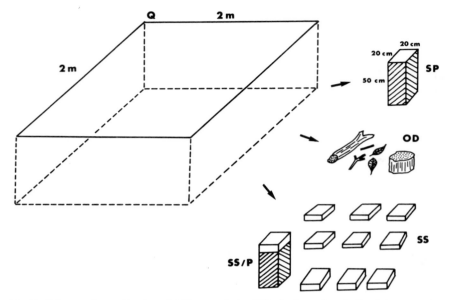

Fig. 3. Scheme of sampling from 2 × 2 m quadrats. OD, dead wood and other lying organic debris; SP, soil pit; SS, soil scrape, SS/P, soil scrape extended to 50 cm to sample deeper layers. Note, the combined volume of the 10 scrapes is the same as one complete soil pit, but those termites associated with tree buttresses and fine roots (which apparently contribute significantly to overall species diversity) are more efficiently sampled. See text for further details.

individuals in nests and mounds and the vertical stratification of foraging individuals in the soil (Ohiagu, 1979; Wood *et al.*, 1982.) These data have serious implications for short studies without replication that are made only at one particular time of year.

Seasonal difference in climatic conditions may also influence termites' response to disturbance. Although pristine forests may be buffered to a certain extent from seasonal changes, in cleared or disturbed forest systems the surviving species are likely to be those that are able to cope with severe differences between wet and dry seasons.

Seasonality often constrains when it is practicable to sample. The dry season may be the easiest time to take soil samples (but coring of very dry soils is difficult), while in the wet roads may become impassable, and habitats extremely difficult to work in (in heavy rain soil pits may fill with water and excavated material is difficult to handsort). However, dry season sampling may underestimate peak termite abundance as overall population sizes may be lower. Indeed, termite biomass may be highest in the wet season when the fully mature alates are released from colony centres.

D. Termite Behaviour

Termites have behaviour patterns that cause sampling problems. Throughout the day there are different levels of a dynamic equilibrium between nest-inhabiting and foraging individuals (reviewed in Wood, 1978). This means that the time of day when sampling is carried out may influence the result; most tropical field work begins at first light and sampling is generally completed by midday, leaving the later hours for specimen extraction and sorting.

Sampling itself may disturb the termites. Some termites may escape from soil pits or cores before they can be extracted (e.g. fast moving Macrotermitinae), although this is unlikely to be a serious problem for soil-feeding termites (mostly very sluggish). Some sampling methods are specifically designed to prevent termites escaping from soil samples (i.e. Dangerfield's (1990b) monoliths).

Termites may also abandon mounds if the area around them is disturbed. A large *Thoracotermes macrothorax* mound on one of the Mbalmayo sites was abandoned by its occupants within a few days of the site being partially weeded for sampling purposes (Eggleton *et al.*, unpublished). The extent to which these behaviours bias termite sampling has yet to be quantified, but light weeding (usually by machete) to permit access is a normal preliminary to sampling in forest sites which are overgrown with underbrush (e.g. plantations, secondary woodland) and the possibility that such site preparation is in itself a serious disturbance is clearly worrying.

Termite behaviour may also affect how easily termites can be extracted from samples. Many termites (especially soil-feeders) tend to retreat into pockets of wetter substrate as samples dry out. This makes it hard to extract them using conventional Berlese–Tullberg methods that drive invertebrates out of substrates by drying. Studies that use Berlese–Tullberg funnels to estimate abundances of a range of invertebrate groups may underestimate termite abundance (e.g. Bandeira, 1979; Collins *et al.*, 1984; Stork and Brendell, 1993). Termites generally have to be extracted by hand (see below), although Maldague (1964) pioneered a flotation system which might be suitable for small-scale sampling.

E. Work-force and Resources

Hand sampling termites is a labour-intensive process. Table 4 gives a breakdown of sampling efforts for different parts of the Mbalmayo project, along with the effort required to sample larger areas and more taxa. Fourteen weeks of intensive sampling have taken 36 worker-months

Table 4. Table showing sampling and processing times for the Mbalmayo termite work so far and speculative times for sampling a whole hectare both for termites (first three rows) and all insects (last row). For a hectare we have assumed a total of 10 000 spp. (based on Hammond's (1990) estimates). Sampling time estimates are based on the TIGER experience of termite sampling. For processing we have assumed an average of 4 h per species (based on data for termites, Eggleton et al., unpublished, and beetles, Mawdsley, unpublished) and an 8-h working day, 330 days a year

Sample size/type	% of area	Sampling time	Processing time
100-m transect	2	20 days	30 days
Stratified sampling of 20 × 30 m area (10 pits etc.)	7	1 month	1 month
Stratified sampling of whole hectare (170 pits etc.)	7	17 months	17 months
Stratified sampling of whole hectare (all insects)	10	5.6 years	15 years

of effort. Very few studies are able to bring such a level of resources into the field. For example, termite studies by postgraduate students working on their own preclude the use of large numbers of soil pits or intensive wood sampling. Many such studies are reliant on relative sampling methods (e.g. De Souza, 1993; De Souza and Brown, 1994).

F. Identification and Taxonomic Identity

Although sampling and extracting termites from their substrates is a laborious task, identification of the termites in the lab is generally even more time consuming (see Table 4).

Termites have an advantage over many insect groups in that it is possible to name a good proportion of them (in the Mbalmayo study we have been able to name, at least tentatively, 70% of the species). Putting names on specimens is not a taxonomic luxury – it is the only way to make species-composition comparisons between sites possible. However sorting to *named* species is a harder task than sorting to recognizable taxonomic units (RTUs, Rees, 1983), as comparisons with existing specimens and literature descriptions are required. In addition, there are numerous termite genera that need complete revisions and where species level identifications are very difficult.

If results are to remain comparable between sites it is essential that new species discovered in ecological studies be described. This may be a major task (i.e. there are three new genera and 32 new species awaiting description from the Mbalmayo work) and is often unlikely to be possible given the presently low level of expertise in termite taxonomy (Wilson, 1993).

III. THE DATA SO FAR

Termites have been the focus of a relatively large number of studies that attempt to estimate assemblage parameters in different ecosystems. We summarize the most important of these in Table 1. If we look just at West Africa, we find that estimates of termite abundance for savannahs (49–4402 m^{-2}) overlap with that of forests (1000–7130 m^{-2}), with that of secondary forests/plantations (119–10 466 m^{-2}) and that of agricultural systems (1507–6825 m^{-2}). As each study uses a different sampling regime it is difficult to judge whether these overlaps are real or not. Data is only comparable within individual studies that have looked at more than one site and have used consistent sampling regimes.

Termite abundance, biomass and species richness are generally greatly reduced when forest is cleared (Collins, 1980b; Wood et al., 1982; Eggleton et al., 1995; Eggleton et al., unpublished). However, the data are sometimes equivocal. Wood et al. (1977) recorded very high abundances from 8 to 24-year-old maize plots in Nigeria, and Bandeira (1979) records higher abundances in pasture than in rain forest in Central Amazonia. However, at least in the case of Wood et al. (1977), overall species richness was much lower in the 8–24-year-old maize plots than in primary woodland (Table 2).

There appears to be a biogeographical element that influences the effects of forest clearance. Where savannah and forests are contiguous abundances may not be severely depressed in areas cleared for cultivation (Wood et al., 1977, 1982). Savannah species may colonize the cleared areas (Watanabe et al., 1984) and so buffer the effects of clearance. In other forests relatively far from savannah regions or where savannah-evolved species have never become established (Eggleton et al., 1995), clearance may have a greater effect on termite populations. In Mbalmayo, a 1-year-old plantation that had been completely cleared by bulldozing showed a massive drop in abundance and species richness from the mature forest that had preceded it, although at this stage some forest termites were still present in the plot. Repeated sampling in the second year showed an even greater drop, with the implication that termite populations would be almost completely eliminated until sufficient canopy or underbrush cover was re-established to sustain a forest-typical fauna (Table 2). Much lower species richness and abundance values were also recorded from a treeless weeded fallow plot than for a primary control plot.

In contrast, regenerating forests may have higher species richness and abundance (Table 2, Bandeira, 1979; Eggleton et al., 1995; Eggleton et al., unpublished). The preliminary Mbalmayo results show that the highest abundance and species richness is in a mature secondary site (Table 1), while a 4-year-old plantation and a near primary site have a very similar abundance and species richness. There also appears to be qualitative differences in species richness and taxonomic composition between the secondary and primary plots (Eggleton et al., 1994a). These differences may be due to the higher levels of dead wood present in the secondary sites (J. H. Lawton, pers. comm.) and thus available, directly or indirectly, to termites.

IV. CONCLUSIONS: GENERAL CONSEQUENCES

Termites are more labour intensive to sample than most insect groups. Other soil macro-invertebrates can usually by sampled using smaller soil cores and Berlese–Tullberg type extraction methods. More mobile insects are not easy to sample quantitatively, although canopy fogging is a very useful approach. For most studies, qualitative and relative sampling protocols will have to be used.

Stratified approaches that attempt to sample all the insects in a particular area (say a hectare) are impracticable (see Table 4). Just sampling and processing a representative sample of termites from a 1-ha site is estimated to be a 34 worker-month task, and termites are likely to represent less than 1% of the species richness of an area (even though they may be over 10% of the biomass). Our rough estimate of 20 worker-years (Table 4) for a complete 1-ha analysis of all insects is probably an underestimate.

The problem of taxonomic identity is far more serious for most other insect groups than for termites. Beetles collected from Sulawesi (Hammond, 1990) and Brunei Darussalam (Mawdsley, unpublished data) that are being processed at the Natural History Museum, London, have nearly all had to be sorted to RTUs within families rather than identified to species. Comparisons of the two faunas except in the broadest taxonomic and ecological sense is thus impossible. More efficient ways of describing species must be found (possibly computerized), such that RTU determinations can quickly be converted into species assignments. Only then will larger-scale biogeographical comparisons of faunas be possible.

Our studies on termites have made us aware of the urgent need both to standardize and integrate insect studies (see also discussion in Sutton and Collins, 1991). An agreed set of sampling protocols needs to be developed at suitable scales, so that data from different areas can be compared. Without such standardization many studies will be doomed to stand in isolation, and so add relatively little to our understanding of general patterns of insect responses to land-use changes.

Perhaps as important is the choice of particular sites for studies. Site choice should address particular land uses that are, or may become, general features of tropical ecosystems. For example, in the Mbalmayo Forest Reserve agroforestry plots are being set up where agricultural crops are grown under a tree plantation canopy. We intend to extend our termite sampling into those areas, as we believe that agroforestry is an important component of African land-use regimes (Nwoboshi, 1982).

Once sites are chosen, a number of taxonomic groups can be studied in parallel within those sites. Although it would be preferable if groups used in such studies could be those that are recognized as "indicator groups" (i.e. Brown, 1991; Sutton and Collins, 1991), this is often not possible, as the choice of study groups is often constrained by whatever expertise is available at the time.

The Mbalmayo site is an especially good place to attempt such integrated studies, as it is easily accessible, has a complex mosaic of different land uses (i.e. degrees of human-induced disturbance), and there is good infrastructure support from the governmental organizations already based there. So far studies have been made of nematodes, beetles and birds within the five sites that have been sampled for termites. We hope that further groups (e.g. ants, Watt and Stork, pers. comm.) will be studied as the project progresses. The results of such integrated studies will begin to piece together the different taxonomic components of biological responses to land-use changes.

Acknowledgements

We thank Nigel Stork and Richard Harrington for inviting us to write this chapter. We also thank the other members of the TIGER 1.5 consortium (John Lawton, Bill Sands, Tom Wood and Mike Hodda) who have contributed to the work described in this chapter. In addition, we thank Helaina Black, Brian Waite, Robert Belshaw, Eileen Wright, David Jones, Nigel Stork, Peter Hammond, Gavin Gillman, Dieudonne Nguele, Andy Roby, Nick Bignell, Corinne Rouland, Alan Watt, Gerry Lawson, Zak Tchoundjeu, Stan Claasen and two anonymous referees who have helped in various ways both in Cameroon and the UK.

Nick Mawdsley kindly provided some data from his Brunei work and discussed some of the ideas in this manuscript with us. We acknowledge the financial support provided by the Natural Environment Research Council through its Terrestrial Initiative in Global Environment Research Programme, award no. GST/02/625.

References

Abe, T. (1979). Studies on the distribution and ecological role of the termites in a lowland rain forest of West Malaysia (2) food and feeding habits of termites in the Pasoh forest reserve. *Jap. J. Ecol.* **29**, 121–135.

Abe, T. (1980). Studies on the distribution and ecological role of the termites in a lowland rain forest of West Malaysia (4) The role of termites in the process of wood decomposition in Pasoh Forest Reserve. *Rev. Ecol. Biol. Sol.* **17**, 23–40.

Abe, T. and Matsumoto, T. (1979). Studies on the distribution and ecological role of the termites in a lowland rain forest of West Malaysia (3) distribution and abundance of termites in Pasoh forest reserve. *Jap. J. Ecol.* **29**, 337–351.

Abe, T. and Watanabe, H. (1983). Soil macrofauna in a tropical rainforest and its adjacent cassava plantation in Okinawa – with special reference to the activity of termites. *Physiol. Ecol. Japan* **20**, 101–104.

Bandeira, A. G. (1979). Ecologia des cupins (Insecta: Isoptera) da Amazônia Central: efeitos do desmatamento sobre as populacoes. *Acta Amazonica* **9**, 481–499.

Bouillon, A. (1964). Etude de la composition des sociétés dans trois espèces d'*Apicotermes* Holmgren (Isoptera, Termitinae). In "Etudes sur les Termites Africains" (A. Bouillon, ed.), pp. 181–196. Masson et Cie, Paris.

Bouillon, A. (1970). Termites of the Ethiopian region. In "Biology of Termites" (K. Krishna and F. M. Weesner, eds), Vol. 2, pp. 153–280. Academic Press, London.

Bouillon, A., Lekie, R. and Mathot, G. (1962). Etudes sur les termites africains. 1. Distribution spatiale et essai sur l'origine et la dispersion des espèces du genre *Apicotermes* (Termitinae). *Studia Universitatis Lovanium Faculté des Sciences* **15**, 1–35.

Brown, K. S. (1991). Conservation of Neotropical environments: insects as indicators. In "The Conservation of Insects and their Habitats" (N. M. Collins and J. A. Thomas, eds), pp. 349–404. Academic Press, London.

Collins, N. M. (1977). Oxford expedition to the Edea-Marienberg Forest Reserve, United Republic of Cameroon. *Bull. Oxf. Univ. Explor. Club, New Series* **3**, 5–15.

Collins, N. M. (1980a). The distribution of soil macrofauna on the West Ridge of Gunung (Mount) Mulu, Sarawak. *Oecologia* **44**, 263–275.

Collins, N. M. (1980b). The effect of logging on termite (Isoptera) diversity and decomposition processes in lowland Dipterocarp forests. In "Tropical Ecology and Development" (J. I. Furtado, ed.), pp. 113–121. International Society of Tropical Ecology, Kuala Lumpur.

Collins, N. M. (1983). Termite populations and their role in litter removal in Malaysian rain forests. In "Tropical Rain Forest: Ecology and Management" (S. L. Sutton, T. C. Whitmore and A. C. Chadwick, eds), pp. 311–325. Blackwell Scientific Publications, Oxford.

Collins, N. M., Anderson, J. M. and Vallack, H. W. (1984). Studies on the soil invertebrates of lowlands and montane rain forests in the Gunung National Park. *Sarawak Mus. J.* **51**, 19–33.

Constantino, R. (1992). Abundance and diversity of termites (Insecta: Isoptera) in two sites of primary rain forest in Brazilian Amazonia. *Biotropica* **24**, 420–430.

Dangerfield, J. M. (1990a). The distribution and abundance of *Cubitermes sankurensis* (Wassmann) (Isoptera; Termitidae) within a Miombo woodland site in Zimbabwe. *Afr. J. Ecol.* **28**, 15–20.

Dangerfield, J. M. (1990b). Abundance, biomass and diversity of soil macrofauna in savanna woodland and associated managed habitats. *Pedobiologia* **34**, 141–150.

Duncan, F. D., Nel, A., Batzofin, S. H. and Hewitt, P. H. (1990). A mathematical approach to rating food acceptance of the harvester termite, *Hodotermes mossambicus* (Isoptera: Hodotermitidae) and the evaluation of baits for its control. *Bull. Entomol. Res.* **80**, 277–287.

Easey, J. F. and Holt, J. A. (1989). Population estimation of some mound-building termites (Isoptera: Termitidae) using radioisotope methods. *Material Organismen*, **24**, 81–91.

Eggleton, P., Bignell, D. E., Sands, W. A., Waite, B., Wood, T. G. and Lawton, J. H. (1995). The species richness of termites under differing levels of forest disturbance in the Mbalmayo Forest Reserve, Southern Cameroon. *J. Trop. Ecol.* **11**, 1–14.

Eggleton, P., Williams, P. H. and Gaston, K. J. (1994). Explaining global termite diversity: productivity or history? *Biod. Cons.*, **3**, 318–330.

Erwin, T. L. (1982) Tropical forests: their richness in Coleoptera and other arthropod species. *Coleopt. Bull.* **36**, 74–75.

Erwin, T. L. (1983). Beetles and other insects of tropical forest canopies at Manaus, Brazil, sampled by insecticide fogging. *In* "Tropical Rain Forest: Ecology and Management" (S. L. Sutton, T. C. Whitmore and A. C. Chadwick, eds), pp. 59–75. Blackwell Scientific Publications, Oxford.

Gay, F. J. (1970). Isoptera (termites). *In* "The Insects of Australia" (D. F. Waterhouse, ed.), pp. 275–293. Melbourne University Press, Melbourne.

Groombridge, B. (ed.) (1992). "Global Biodiversity". Chapman & Hall, London.

Hammond, P. H. (1990). Insect abundance and diversity in the Dumoga-Bone National Park, N. Sulawesi, with special reference to the beetle fauna of lowland rain forest in the teh Toraut region. *In* "Insects and the Rain Forests of South East Asia (Wallacea)" (W. J. Knight and J. D. Holloway, eds), pp. 197–254. Royal Entomological Society of London, London.

Harcourt, C. (1992). Tropical moist forests. *In* "Global Biodiversity" (B. Groombridge, ed.), pp. 256–275. Chapman & Hall, London.

Holloway, J. D., Kirk-Spriggs, A. H. and Khen, C. V. (1992). The response of some rain forest insect groups to logging and conversion to plantation. *Phil. Trans. R. Soc., B* **335**, 425–436.

Johnson, R. A., Lamb, R. W., Sands, W. A., Shittu, R. M., Williams, R. M. C. and Wood, T. G. (1982). A check list of Nigerian termites (Isoptera) with brief notes on their biology and distribution. *Niger. Fld* **45**, 50–64.

Josens, G. (1972). Etudes biologiques et écologique des termites (Isoptera) de la svane de Lamto (Côte d'Ivoire). *Annl Soc. R. Zool. Belg.* **103**, 169–176.

Lawton, J. H. and May, R. M. (1994). "Estimating Extinction Rates". Oxford University Press, Oxford.

Lee, K. E. and Wood, T. G. (1971). "Termites and Soil". Academic Press, London.

Maldague, M. E. (1964). Importance des populations de termites dans les sols equatorial. *Transactions of the 8th International Congress of Soil Science, Bucharest, 1964*, **3**, 743–751.

Marshall, A. G. (1992). The Royal Society's south-east Asian rain forest programme: an introduction. *Phil. Trans. R. Soc., B* **335**, 327–330.

Matsumoto, T. (1976). The role of termites in an equatorial rain forest ecosystem of West Malaysia. 1. Population density, biomass, carbon, nitrogen and calorific content and respiration rate. *Oecologia* **22**, 153–178.

Matsumoto, T. and Abe, T. (1979). Studies on the distribution and ecological role of the termites in a lowland rain forest of West Malaysia (II) Leaf litter consumption on the forest floor. *Oecologia* **38**, 261–274.

Myers, N. (1989). "Deforestation Rates in Tropical Forests and their Climatic Implications". Friends of the Earth, London.

Noble, J. C., Diggle, P. J. and Whitford, W. G. (1989). The spatial distributions of termite pavements and hummock feeding sites in a semi-arid woodland in eastern Australia. *Acta Oecol. Gener.* **10**, 355–376.

Noirot, C. H. (1992). From wood- to humus-feeding: an important trend in termite evolution. *In* "Biology and Evolution of Social Insects" (J. Billen, ed.), pp. 107–119. Leuven University Press, Leuven.

Nwoboshi, L. C. (1982). "Tropical Silviculture, Principles and Techniques". University Press, Ibadan.

Ohiagu, C. E. (1979). Nest and soil populations of *Trinervitermes* spp. with particular reference to *T. geminatus* (Wasmann), Isoptera) in Southern Guinea Savanna near Mokwa, Nigeria. *Oecologia* **40**, 167–178.

Pearce, M. J. (1990). A new trap for collecting termites and assessing their foraging activity. *Trop. Pest Manag.* **36**, 310–311.

Pearce, M. J., Cowie, R. H., Pack, A. S. and Reavey, D. (1990). Intraspecific agression, colony identity and foraging distances in Sudanese *Microtermes* spp. (Isoptera: Termitidae: Macrotermitinae). *Ecol. Entomol.* **15**, 71–77.

Redford, K. H. (1984). The termitaria of *Cornitermes cumulans* (Isoptera: Termitidae) and their role in determining a potential keystone species. *Biotropica* **16**, 112–119.

Rees, C. J. C. (1983). Microclimate and the flying Hemiptera fauna of a primary lowland rain forest in Sulawesi. *In* "Tropical Rain Forest: Ecology and Management" (S. L. Sutton, T. C. Whitmore and A. C. Chadwick, eds), pp. 121–136. Blackwell Scientific Publications, Oxford.

Salick, J. and Tho, Y. Y. (1984). An analysis of termite fauanae in Malayan rainforests. *J. Appl. Ecol.* **21**, 547–561.

Sands, W. A. (1965a). Mound population movements and fluctuation in *Trinervitermes ebenerianus* Sjöstedt (Isoptera, Termitidae, Nasutermitinae). *Insectes Sociaux* **12**, 49–58.

Sands, W. A. (1965b). Termite distribution within man-made habitats in West Africa, with special reference to species segregation in the genus *Trinervitermes* (Isoptera, Termitidae, Nasutitermitinae). *J. Anim. Ecol.* **34**, 557–571.

Sands, W. A. (1972). Problems in attempting to sample tropical subterranean termite populations. *Ekol. Pol.* **20**, 23–31.

Sayer, J. A. and Whitmore, T. C. (1991). Tropical moist forests: destruction and species extinction. *Biol. Cons.* **55**, 199–213.

De Souza, O. F. F. (1993). "The Effect of Cerrado Fragmentation on Termites (Isoptera)". PhD thesis, Imperial College, London.

De Souza, O. F. F. and Brown, V. K. (1994). Effects of habitat fragmentation on Amazonian termite communities. *J. Trop. Ecol.*, **10**, 197–206.

Stork, N. E. (1988). Insect diversity: facts, fiction and speculation. *Biol. J. Linn. Soc.* **35**, 321–337.

Stork, N. E. and Brendell, M. J. D. (1993). Arthropod abundance in lowland rain forest of Seram. *In* "The Natural History of Seram" (I. D. Edwards, A. A. Macdonald and J. Proctor, eds), pp. 115–130. Intercept, Andover, UK.

Strickland, A. H. (1944). The arthropod fauna of some tropical soils. *Trop. Agric.* **21**, 107–114.

Sutton, S. L. and Collins, N. M. (1991). Insects and tropical forest conservation. *In* "The

Conservation of Insects and their Habitats" (N. M. Collins and J. A. Thomas, eds), pp. 405–424. Academic Press, London.

Watanabe, H. and Ruaysoongnern, S. (1984). Effects of shifting cultivation on soil macrofauna in Northeastern Thailand. *Mem. Coll. Agric. Kyoto Univ.* **125**, 35–43.

Watanabe, H., Takeda, H. and Ruaysoongnern, S. (1984). Termites of Northeastern Thailand with special reference to changes in species composition due to shifting cultivation. *Mem. Coll. Agric. Kyoto Univ.* **125**, 45–57.

Wiegert, R. G. (1970). Energetics of the nest-building termite *Nasutitermes costalis* (Holm.) in a Puerto Rican forest. *In* "A Tropical Rain Forest. A Study of Irridation and Ecology at El Verde, Puerto Rica" (H. T. Odum, ed.), Vol. 1, pp. 57–64. Division of Technical Information, United States Atomic Energy Commission.

Wilson, E. O. (1992). The effects of complex social life on evolution and biodiversity. *Oikos* **63**, 13–18.

Wilson, E. O. (1993). "The Diversity of Life". Harvard University Press, Harvard, MA.

Wolda, H. (1988). Insect seasonality: why? *A. Rev. Ecol. Sys.* **19**, 1–18.

Wood, T. G. (1978). Food and feeding habits of termites. *In* "Production Ecology of Ants and Termites" (M. V. Brian, ed.), pp. 55–80. Cambridge University Press, Cambridge.

Wood, T. G. (1981). Reproductive isolating mechanisms among species of *Microtermes* (Isoptera: Termitidae) in the Southern Guinea Savanna near Mokwa, Nigeria. *In* "Biosystematics of Social Insects" (P. E. Howse and J.-L. Clément, eds), pp. 309–325. Academic Press, New York.

Wood, T. G. and Sands, W. A. (1978). The role of termites in ecosystems. *In* "Production Ecology of Ants and Termites" (M. V. Brian, ed.), pp. 245–292. Cambridge University Press, Cambridge.

Wood, T. G., Johnson, R. A. and Ohiagu, C. E. (1977). Populations of termites (Isoptera) in natural and agricultural ecosystems in southern Guinea savanna near Mokwa, Nigeria. *Geo-Eco-Trop* **1**, 139–148.

Wood, T. G., Johnson, R. A., Bacchus, S., Shittu, M. O. and Anderson, J. M. (1982). Abundance and distribution of termites (Isoptera) in a Riparian forest in the Southern Guinea savanna vegetation zone of Nigeria. *Biotropica* **14**, 25–39.

Natural Environment Research Council through its Terrestrial Initiative in Global Environment Research Programme, award no. GST/02/625.

24

Potential Use of Suction Trap Collections of Aphids as Indicators of Plant Biodiversity

S. E. HALBERT, M. D. JENNINGS, C. B. COGAN,
S. S. QUISENBERRY AND J. B. JOHNSON

> I. Introduction 499
> II. Idaho Suction Trap Collections Reflect Surrounding
>   Vegetation 501
>   References 503

Abstract

A case is put for the use of aphids as indicators of habitat biodiversity. Using Sagebush (*Artemsia* spp) and its associated host-specific aphids as an example, a significant correlation is shown between the number of plant and aphid species at nine suction trap locations.

I. INTRODUCTION

The United States has an extensive system of National Parks and preserves under various jurisdictions; however, reasons for preservation are numerous and not necessarily ecological. The modern concept of Gap Analysis (Scott *et al.*, 1993) combines knowledge of existing vegetation, faunal distribution and geostatistics analyses to locate regions of maximum biodiversity. If such habitats could be selected and targeted for conservation, protection of more species per unit area of preserved land than is currently achieved would result. The goal of Gap Analysis is to ensure that all habitat types are represented in appropriate preserves.

In order to assess habitat biodiversity, indicator groups of organisms must be found that reflect biodiversity. Currently, vegetation, vertebrates and butterflies are used to assess biodiversity. We suggest that aphid samples from suction traps may also be valuable in assessing biodiversity in North American temperate regions.

Aphids fulfil several of the criteria listed by Noss (1990) for good indicators of habitat biodiversity. First, aphids have the ability to reproduce rapidly and thus respond to changing vegetation. Second, aphids are widely distributed across north temperate regions. They are sufficiently abundant and diverse to provide a robust indicator of biodiversity. Third, the suction trap surveys in Europe (Taylor, 1984) and western North America (Allison and Pike, 1988; Halbert et al., 1990; Pike et al., 1990) comprise an existing system for synoptic measurement of aphid biodiversity. Survey systems do not exist for other groups of insects. Fourth, flight densities of some species of aphids, e.g. cereal aphids, closely track ecologically significant parameters such as host plant phenology (Halbert et al., 1990).

Additionally, aphids fulfil several more criteria for suitable indicator taxa as outlined by Pearson and Cassola (1992). Most aphids have very limited host ranges and are thus sensitive indicators of habitat changes. Most aphid flight activity is local (Halbert et al., 1990; Loxdale et al., 1993), so their distribution reflects that of the species comprising local vegetation. For example, suction trap collections of *Myzus lythri* (Schrank) were used to predict two unreported locations for *Lythrum salicaria* L. (purple loosestrife), listed as a noxious weed in the state of Idaho, and weed populations were later verified (Halbert and Voegtlin, 1994). Finally, aphids certainly have economic importance that justifies dedication of resources for research. Suction trap surveys of aphids are routinely used for both ecological studies and agricultural pest monitoring (Halbert et al., 1990; Taylor, 1984).

The Idaho suction trap survey system, in operation since 1985, has generated distribution data on over 300 species or species groups of aphids. On the basis of comparisons of collections with crop development, other phenological indicators and known host plant ranges, we estimate that suction traps reflect flight activity within a 50 km radius of the trapping site, provided that no major changes in elevation occur within that distance (Halbert et al., 1992a). This is in general agreement with European estimates (Dedryver et al., 1991).

The disadvantage to using aphids as indicators of habitat biodiversity in North America is that taxonomically, North American aphids are poorly known compared with other insect groups such as butterflies and tiger beetles that are currently used as indicators (Pearson and Cassola, 1992). Several large North American aphid genera have not been revised taxonomically for a number of years, and winged forms that are collected in suction trap samples are often undescribed or simply not included in keys to species.

II. IDAHO SUCTION TRAP COLLECTIONS REFLECT SURROUNDING VEGETATION

The state of Idaho is located in the northwestern United States. It is approximately 765 km from north to south (117° W longitude) and 505 km from east to west (42° N latitude). The state encompasses a wide variety of climatic and ecological zones, including deserts, forests, montane vegetation and agricultural land. Elevation in Idaho ranges from about 440 m to 3800 m. Idaho's varied topography and climates make it an ideal place to investigate biodiversity indicators.

In spite of the taxonomic limitations detailed above, data indicate that aphid collections are species rich in areas that have diversified plant communities. The two areas with the highest numbers of aphid species are Bonners Ferry in northern Idaho, and Parma in southwest Idaho. Bonners Ferry is in a valley surrounded by diverse types of montane and forest vegetation. The area contains relict populations retained within the northern Idaho glacial refugium (Daubenmire, 1952) and appears to include vegetation more typical of the eastern United States as well. Parma is surrounded by diverse desert vegetation and agricultural land that supports at least 60 different crops. The percentage of native aphid species in the Parma collections is lower than at Bonners Ferry. Aberdeen, which is surrounded by agricultural land producing only five major crops, is relatively low both in average number of aphid species collected and in percentage of native species in the collections.

A group of plant species with a diverse, specific and distinctive aphid fauna was selected to test our hypothesis that aphid species could be used as indicators of habitat biodiversity. Sagebrush (*Artemisia* spp.) is a very common native plant in Idaho with approximately twelve species or subspecies occurring in the State. Thirty to 40 aphid species in nine different genera occur almost exclusively on sagebrush. They are a well-defined group and easily distinguishable from aphids that infest other plants. From the Gap Analysis mapping project, we were able to list the species of sagebrush occurring within 30 km of nine suction trap locations. The locations were chosen for geographic diversity and continuous long-term operation (Fig. 1). A scatter plot shows the number of sagebrush species or subspecies and the number of aphid species specific to sagebrush that were collected at each of the trap locations (Fig. 2). These data indicate a moderately strong relationship ($r = 0.75$, $n = 9$,

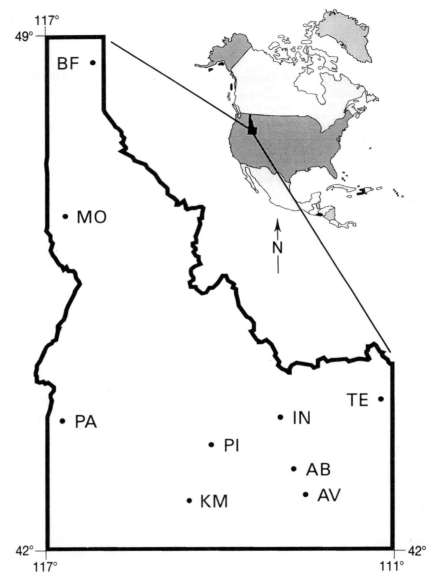

Fig. 1. Map showing Idaho suction trap locations used in analysis of aphid and sagebrush biodiversity. Abbreviations indicate suction trap locations: BF, Bonners Ferry; MO, Moscow; PA, Parma; KM, Kimberly; AB, Aberdeen; PI, Picabo; IN, Idaho National Energy Laboratory Reservation; AV, Arbon Valley, TE, Tetonia.

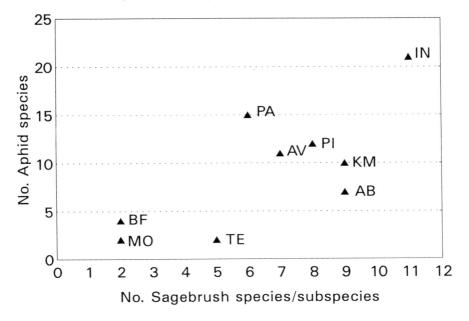

Fig. 2. Scatter plot showing sagebrush species or subspecies and sagebrush specific aphid species for nine suction trap locations in Idaho. Abbreviations as in Fig. 1.

$P = 0.02$) between sagebrush and aphid species biodiversity. Sagebrush vegetation is most common and diverse in the southern Idaho desert, and very little occurs in northern Idaho. Accordingly, diversity of sagebrush and associated aphid species are highest in southern Idaho. Aphid collections reflect host plant biodiversity.

Acknowledgements

We thank Peggy Bain for producing the graphics for this chapter.

REFERENCES

Allison, D. A. and Pike, K. S. (1988). An inexpensive suction trap and its use in an aphid monitoring network. *J. Agri. Entomol.* **5**, 103–107.

Daubenmire, R. (1952). Plant geography of Idaho. In 'Flora of Idaho' (R. J. Davis, ed.), pp 1–17. Brigham Young University Press, Provo, UT.

Dedryver, C. A., Gamon, A. and Gillet, H. (1991). Plurilocal assessment of the risk of primary infection of young cereals by BYDV in autumn in the west of France. *Acta Phytopathol. et Entomol. Hungarica* **26**, 51–57.

Halbert, S. E., Connelly, B. J., and Sandvol, L. (1990) Suction trapping of aphids in western North America. *Acta Phytopathol. et Entomol. Hungarica* **25**, 411–422.

Halbert, S. E., Elberson, L. and Johnson, J. B. (1992a). Suction trapping of Russian wheat aphid: What do the numbers mean? *In* "Proceedings of the Fifth Russian Wheat Aphid Conference, Fort Worth, TX, 26–28 January, 1992" (W. P. Morrison, ed.), pp. 282–297. Texas A&M University, Lubbock, TX.

Halbert, S. E. and Voegtlin, D. J. (1994). *Weed Technology* **8** (in press).

Loxdale, H. D., Hardie, J., Halbert, S. E., Foottit, R., Kidd, N. A. and Carter, C. I. (1993). The relative importance of short- and long-range movement of flying aphids. *Biological Reviews* **68**, 291–311.

Noss, R. F. (1990). Indicators for monitoring biodiversity: A hierarchical approach. *Conservation Biology* **4**, 355–364.

Pearson, D. L. and Cassola, F. (1992). *Conservation Biology* **6**, 376–391.

Pike, K. S., Allison, D. W., Low, G., Bishop, G. W., Halbert, S. E. and Johnston, R. L. (1990). Cereal aphid vectors: A western regional (USA) monitoring system. *In* "World Perspectives on Barley Yellow Dwarf" (P. A. Burnett, ed.), pp. 282–285. CIMMYT, Mexico D.F., Mexico.

Scott, J. M., Davis, F., Csuti, B., Noss, R., Butterfield, B., Groves, C., Anderson, H., Caicco, S., D'Erchia, F., Edwards Jr, T. C., Ulliman, J. and Wright, R. G. (1993) Gap analysis: A geographical approach to protection of biological diversity. Wildlife Monographs No. 123. Washington Wildlife Society, Bethesda, MD.

Taylor, L. R. (1984). An international standard for the synoptic monitoring and dynamic mapping of migrant pest aphid populations. *In* "The Movement and Dispersal of Agriculturally Important Biotic Agents" (D. R. MacKenzie, C. S. Barfield, G. C. Kennedy and R. D. Berger with D. J. Taranto, eds), pp. 337–418. Claitor's Publishing Division, Baton Rouge, LA.

25

Shifts in the Flight Periods of British Aphids: a Response to Climate Warming?

R. A. FLEMING AND G. M. TATCHELL

I. Introduction	505
II. Methods	506
III. Results	507
IV. Discussion	508
References	508

Abstract

Evidence is given for an advance in the flight phenology of three to six days for five aphid species in Britain over the past twenty five years. The mean temperature has risen by 0.4°C over the same period, suggesting that aphids may be good biological indicators of climate warming.

I. INTRODUCTION

The importance of climatic factors in determining the phenology (temporal distribution) of the developmental stages in insect life cycles has long been recognized. Because developmental rates are positively correlated with heat accumulation (e.g. Candolle, 1855), one might generally expect insect developmental rates to have increased in response to the recent global warming (Houghton *et al.*, 1990) and phenologies to have advanced. We have examined the extensive records of the Rothamsted Insect Survey (RIS) suction trap network in the UK (Taylor, 1986) for evidence of such trends in the spring flight periods of selected aphid species.

II. METHODS

Aphids typically have a marked seasonal cycle in the production of alate forms and the timing of their flights. These flights can involve movement between different host plant species and represent a fundamental aspect of the aphid life cycle (Dixon, 1985). The RIS suction traps provide continuous standardized samples of aphids flying at a height of 12.2 m above ground (Macaulay et al., 1988). Since 1964, when the first trap was established at Rothamsted, a network has been developed throughout Britain (Taylor, 1986). Samples are collected daily from the traps during the main period of aphid flight from mid-April to early November, and weekly for the rest of the year. All aphids in each sample are identified to species and counted.

This chapter concerns five aphid species (*Drepanosiphum platanoidis* (Schr.), *Elatobium abietinum* (Walk.), *Microlophium carnosum* (Theo.), *Periphyllus testudinaceus* (Fern.), and *Phorodon humuli* (Schr.)) of the roughly 350 recorded by the RIS. These five are common and widely distributed species with well-defined flight periods and relatively well-studied biologies. The eight trap sites selected were fairly evenly spaced from Wye ($51.2°$ N latitude) to Dundee ($56.5°$ N latitude) along the eastern side of the UK (Fleming and Tatchell, 1994, fig. 1).

For species with multiple flight periods, the spring migration was studied and the end of the spring flight at each site was identified by visual inspection after plotting the mean number of aphids trapped on each date against Julian date (i.e. the number of days since January first). For each species, this "cutoff date" was associated permanently with the site and did not vary from year to year. (In the long run, this "cutoff date" could be expected to advance with the rest of the spring flight phenology, so the use of a fixed date is conservative as it will tend to produce underestimates of a phenological advance. However, using advancing "cutoff dates" runs the risk of overestimating the magnitude and frequency of phenological advances in the spring flight.) There was often some overlap of flight periods, but it rarely presented difficulties for the analysis (Fleming and Tatchell, 1994).

To describe each flight phenology, the Julian date by which 5, 10, 25, 50 and 75% of the total flight had occurred was determined. These centiles were calculated for each species-site combination with at least eight spring flights for which 20 or more individuals were recorded. The Julian dates corresponding to each selected centile were then regressed against the year of the flight, producing 170 different simple linear regressions.

The consistency of any trends in a species' flight phenology at different

25. Shifts in the Flight Periods of British Aphids

Table 1. Results of the 34 simple linear regressions and five multiple linear regressions for each centile date against year. The total of 34 includes all species at all sites providing data points for at least 8 years. The second and third rows give the number of these simple linear regressions which had negative slopes, and the probability of the corresponding one-tailed sign test, respectively. The fourth and fifth rows give the estimated means and standard errors (in days) from the multiple linear regressions indicating the extent of the estimated phenological advances among the five species

Centile	5	10	25	50	75
Number negative	28	30	28	26	22
P-value	0.0001	<0.0001	0.0001	0.0009	0.045
Mean (days)	5.94	6.41	4.47	3.93	3.50
Standard error	1.18	1.72	2.01	2.01	1.83

latitudes, and over different population densities, was also investigated. The data from different sites were pooled for each centile date–species combination. The Julian dates corresponding to each selected centile date–species combination were then regressed against year of the flight, latitude of the site, and the logarithm of the population density in multiple linear regressions. With the linear effects of site latitude and population density thus removed, the coefficient for flight year provided an estimate of the species' general rate of phenological shift in days per year. To estimate the extent of a species' phenological shift, the magnitude of the coefficient for flight year was multiplied by the range of the observations in years.

III. RESULTS

Of the 170 regressions of Julian date (for each selected centile) against the year of the flight (for each species–site combination), the few with statistically significant slopes could be explained by chance alone. However, regardless of the significance of individual slopes, most of the 34 slopes estimated for each centile were negative (Table 1). For instance, 30 of the 34 slopes were negative for the tenth centile dates of the flight phenology distributions. The probability of this occurring by chance alone is less than 1 in 10 000. The multiple linear regression results indicate the magnitude of this phenological shift. They suggest that the flight phenologies for the five species studied have advanced by an average of about 3–6 days over the last 25 years (Table 1).

IV. DISCUSSION

These results indicate that the flights of the species examined have tended to occur increasingly early during the period of study. During the same period, the climate has become warmer. For instance, data reported by Jones and Wigley (1990) show that, since 1964, temperature and year are positively correlated ($r = 0.62$, $P = 0.0007$), and the mean temperature in the northern hemisphere has increased by about 0.4°C.

The lack of statistically significant slopes found in the individual regressions of centile dates against year may be the result of the small magnitude of this temperature trend relative to the year-to-year variability in climate and to the "noise" introduced by various biological and ecological factors which also influence phenology. By the same token, the small but widespread response among species and sites may indicate that aphid flight phenology is sensitive to, and an ideal biological indicator of, climate warming.

REFERENCES

Candolle, A. P. de. (1855). "Geographique Botanique". Raisonée, Paris.
Dixon, A. F. G. (1985). "Aphid Ecology". Blackie, Glasgow.
Fleming, R. A. and Tatchell, G. M. (1994). Long term trends in aphid flight phenology consistent with global warming: methods and some preliminary results. In "Individuals, Populations, and Patterns in Ecology" (S. R. Leather, A. D. Watt, N. J. Mills and K. F. A. Walters, eds) pp. 63–71. Intercept, Andover.
Houghton, J. T., Jenkins, G. J. and Ephraums, J. J. (eds) (1990). "Climate Change: The IPCC Scientific Assessment", Cambridge University Press, Cambridge.
Jones, P. D. and Wigley, T. M. L. (1990). Global warming trends. *Sci. Am.* **263**, 66–73.
Macaulay, E. D. M., Tatchell, G. M. and Taylor, L. R. (1988). The Rothamsted Insect Survey "12 metre" suction trap. *Bull. Entomol. Res.* **78**, 121–129.
Taylor, L. R. (1986). Synoptic dynamics, migration and the Rothamsted Insect Survey. *J. Anim. Ecol.* **55**, 1–38.

26

Potential Changes in Spatial Distribution of Outbreaks of Forest Defoliators under Climate Change

DAVID W. WILLIAMS AND ANDREW M. LIEBHOLD

Abstract

Changes in the geographical ranges and spatial extent of pest outbreaks are likely consequences of greenhouse warming. We investigated potential changes in spatial distribution of outbreaks of two species of Lepidoptera, *Choristoneura occidentalis* and *Lymantria dispar*, in Oregon and Pennsylvania, USA, using maps of historical defoliation and climate in a geographic information system. Relationships between defoliation status and the climatic variables were developed using discriminant function analysis. We investigated five climatic change scenarios: an increase of 2°C, a 2°C increase with a small increase and decrease in precipitation, and equilibrium projections of temperature and precipitation by two general circulation models (GCMs). Projections of spatial change were qualitatively similar for both species in the first three scenarios. With an increase in temperature alone, the projected defoliated area decreased relative to ambient conditions. With increases in temperature and precipitation, the defoliated area increased. Conversely, the defoliated area diminished to negligible size when temperature increased and precipitation decreased. For *C. occidentalis*, one GCM predicted a large increase in the defoliated area and the other about the same as under ambient conditions. In contrast, no defoliation by *L. dispar* was predicted under either GCM scenario. The results are discussed in terms of potential changes in forest composition.

A general prediction for spatial change of insect populations under greenhouse warming is that their ranges will move poleward and toward higher elevation (Porter *et al.*, 1991). The frequency and spatial extent of insect outbreaks also may increase because of enhanced overwinter survival and a lengthened growing season. We investigated the potential changes in spatial distribution of outbreaks of two species of Lepidoptera, the western spruce budworm, *Choristoneura occidentalis* Freeman, and the gypsy moth, *Lymantria dispar* (L.), in the States of Oregon and Pennsylvania, respectively. We did so by mapping current outbreak areas and climate variables using a Geographic Information System, relating defoliation to climate statistically, and then extrapolating potential new defoliation areas under several climate change scenarios.

Maps of the frequency of aerially detectable defoliation (i.e. more than 30% of foliage lost) in 2×2 km grid cells were assembled using historical aerial sketch map data collected in Oregon and Pennsylvania from 1947 to 1979 and from 1969 to 1989 respectively. Mean monthly temperature maxima and minima and precipitation in each cell were interpolated from historical (30-year) weather station data using multiple regressions that included latitude, longitude, and elevation as independent variables.

A relationship between defoliation status and the climatic variables was estimated using a linear discriminant function (Manly, 1986). Individual grid cells were classified as defoliated or not defoliated dependent upon the value of the discriminant function, which is a linear combination of the climatic variables. A stepwise procedure was used to select only the most significant of the 36 variables for inclusion in the function. For the budworm, the analysis yielded a function of 28 variables and had a squared canonical correlation of 0.31. The function was highly significant (likelihood ratio = 0.69, $P = 0.0$, $n = 61\ 809$) and classified 82.2% of the grid cells correctly. The analysis for gypsy moth yielded a function of just five variables (mean daily minimum temperature in December and mean monthly precipitation in January, May, September, and December) with a squared canonical correlation of 0.07. This function also was highly significant (likelihood ratio = 0.93, $P = 0.0$, $n = 24\ 952$), but classified only 60.0% of cells correctly.

To extrapolate climate change scenarios, changes in temperature and precipitation were added to the climatic variables in the discriminant function, a new value for each grid cell was computed, and each cell was reclassified depending upon the value. Five climate change scenarios were considered: an increase of 2°C over the entire State, an increase of 2°C and an increase of 0.5 mm precipitation per day, an increase of 2°C and a decrease of 0.5 mm, and temperature and precipitation changes as predicted by two general circulation models (GCMs), the Goddard Institute for Space Studies (GISS) (Hansen *et al.*, 1983) and Geophysical Fluids Dynamics Laboratory (GFDL) (Manabe and Wetherald, 1986)

Table 1. Percentages of areas of Oregon and Pennsylvania projected as defoliated by western spruce budworm and gypsy moth, respectively, under climate change scenarios

Scenario	Percentage area defoliated	
	Oregon	Pennsylvania
Ambient temperature and precipitation	24.7	50.2
Increase 2°C	7.2	9.5
Increase 2°C, increase 0.5 mm day^{-1}	63.0	79.7
Increase 2°C, decrease 0.5 mm day^{-1}	0.1	0.0
GFDL model	88.3	0.0
GISS model	23.8	0.0

models. The GCMs projected equilibrium changes after 100 years at doubled carbon dioxide levels for each State.

Projections of spatial change were qualitatively similar for both pest species for the first three scenarios (Table 1). With an increase in temperature alone, the projected defoliated area decreased relative to the base scenario under ambient conditions and was concentrated at higher elevation (Fig. 1(a)–(b)). With an increase in both temperature and precipitation, the defoliated area increased (Fig. 1(c)). Conversely, the defoliated area diminished to negligible size when increased temperature was accompanied by decreased precipitation (Fig. 1(d)). For the budworm in Oregon, the GFDL model predicted a large increase in the defoliated area, and the GISS predicted an area about the same as under ambient conditions (Fig. 1(e)–(f)). In contrast, no defoliation by gypsy moth was predicted for either GCM scenario in Pennsylvania (Table 1). The difference resulted in part from the more extreme climate projections (i.e. higher temperatures with less precipitation) for Pennsylvania than for Oregon.

In addition to insect abundance, a susceptible forest type is an implicit component of defoliation. Thus, the predicted changes in patterns of defoliation may reflect changes in forest composition as a result of climate change. For example, a recent study of potential geographical shifts of tree species under the GISS and GFDL scenarios suggested that yellow birch and sugar maple, two potential although not preferred hosts of gypsy moth, ultimately may shift their ranges northward and almost completely out of Pennsylvania (Davis and Zabinski, 1992).

Unfortunately, similar information is not available for oak species, the preferred hosts. However, if their ranges undergo similar shifts, such

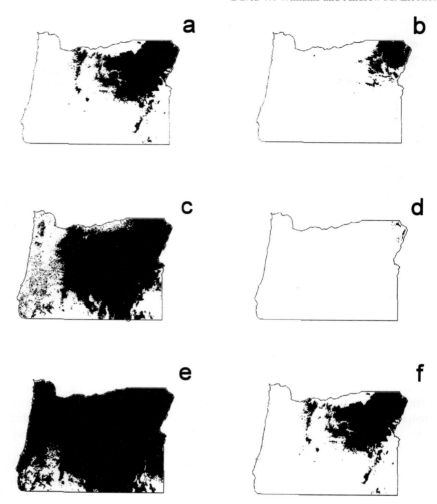

Fig. 1. Potential outbreak areas of western spruce budworm in Oregon under climate change scenarios: (a) ambient temperature and precipitation; (b) 2°C increase; (c) 2°C increase with 0.5 mm day^{-1} increase; (d) 2°C increase with 0.5 mm day^{-1} decrease; (e) GFDL model; (f) GISS model.

movement may provide one explanation for the absence of gypsy moth defoliation in the State predicted by the analysis.

REFERENCES

Davis, M. B. and Zabinski, C. (1992). Changes in geographical range resulting from greenhouse warming: effects on biodiversity in forests. *In* "Global Warming and Biological Diversity" (R. L. Peters and T. E. Lovejoy, eds), pp. 297–308. Yale University Press, New Haven.

Hansen, J., Russell, G., Rind, D., Stone, P., Lacis, A., Lebedeff, S., Ruedy, R. and Travis, L. (1983). Efficient three-dimensional global models for climate studies: Models I and II. *Monthly Weather Rev.* **3**, 609–662.

Manabe, S. and Wetherald, R. T. (1986). Reduction in summer soil wetness induced by an increase in atmospheric carbon dioxide. *Science* **232**, 626–628.

Manly, B. F. J. (1986). "Multivariate Statistical Methods. A Primer". Chapman & Hall, London.

Porter, J. H., Parry, M. L. and Carter, T. R. (1991). The potential effects of climatic change on agricultural insect pests. *Agric. For. Meteorol.* **57**, 221–240.

Index

Numbers in italics refer to figures and tables

A

Abax parallelepipedus 381
Ablabesmyia longistyla and *A. monilis* 286
absolute extinction rates 326
abundance
    and climate changes 13–14, 149, 183
        prediction 60, 61, 63, 65, 68–9, 72, 81
    and land-use changes 344
    and pollution 202–7, 284
    of termites 477–8
    of tsetse flies 183
    see also geographical distribution
Acacia saligna 313
Acanthocnemidae 302
Acer spp
    A. monspesulanum 35
    A. saccharum 11, 199, 201, 203–4, 209, 210
acidification 340, 392, 425
    see also quality *under* water
Acidota quadrata 39
Acridoidea 301
Acyrthosiphon spp
    A. pisum 131, 132, 137
        and pollution 230, 231, 234, 238, 239
    A. svalbardicum 126, 144, 145
Aedes aegypti and *A. albopictus* 159–60, 172
aerosols, effect of 52–3
Africa
    climate changes 99, 299
        disease and insect vectors 159, 160, 162, 163–4, 165, 166, 169–70, 171–2
        prediction of distribution 66, 74
    land-use changes 298, 401
        environmental disturbances 303–9, 311–15
        species richness 301–2
        termites 475–84, 489–91, 493
        pollution: ivermectin residues in cattle dung 465–71
        tsetse flies *see* remote sensing

Agonum ericeti 413
agriculture
    climate changes and prediction of distributions 75, 79, 85–6, 88
    see also environment *below*
    environment for crop insect pests and climate changes 92–121
        direct effects 95–7
        increased carbon dioxide 94
        likely adjustments 97–9
        see also European Corn Borer
    land-use changes
        and arthropod communities 372, 373, 375, 377, 379, 385
        and extinctions 336–8, 341–3
        indicators 402, 406, 409, 411, 412, 415–16
        and trees (including agroforestry) 380–4, 385, 457–9, 481, 491, 493
        see also crops and arable farming; livestock
Agrotis exclamationis 385
Aiolopus thalassinus 263
air pollution *see* carbon dioxide; gaseous air pollutants
Alaska: climate changes 31, 37, 39
alfalfa 234
allelic diversity reduced 14–15
Allerod 427–9
Amara plebeja 404, 406, 415
Amazonia 353, 476, 479, 491
American trypanosomiasis 160, 164
ammonium pollution 441, 446–51
Andricus quercuscalicis 20–1
anholocyclic pests *see* aphids
animals 66, 306
    as disease hosts 162–3, 164
    and host crops 84–5
    see also livestock
Anopheles spp 160
    A. arabiensis 170
    A. culicifacies 169

Anopheles spp–contd.
   A. dirus 171
   A. gambiae 166
   A. minimus 171
   A. stephensi 172
Anotylus gibbulus 39–40
Antarctica 54, 299, 312
ants 313, 335, 360, 386, 400
AOGCMs (Atmosphere-Ocean GCMs) 50
aphids
   Aphis spp
      A. craccivora 137
      A. fabae 137, 230–1, 234–40, 242–3
      A. pomi 226
      A. rumicis 234
   and climate changes 19, 41, 97, 125–53
      critical loads 441–53
      essential biology 128–30
      shifts in flight periods 505–8
      see also aphids *under* temperature
   as indicators of plant biodiversity 499–505
   and land-use changes 334, 379
   and pollution 256
      gaseous air pollutants 221, 226, 227, 228–32, 234–44
   and temperature *see under* temperature
Aphodiinae 466, 468
   Aphodius bonvouloiri, A. holdereri and *A. jacobsoni* 41
Apicotermitinae 476, 478, 483
Aplothorax burchelli 328
Apoidea 301
aquatic areas *see* water
arable farming *see* crops and arable farming *and also* environment *under* agriculture
Aradus cinnamomeus 260, 262
Archiboreoiulus pallidus 381
Arctic insects 312
   aphids *see Acyrthosiphon* spp
   fossils in Britain 36–41
Argynnis adippe 334, 388, 389
Argyrocupha malagrida
   A. m. cedrusmontana 301
   A. m. malagrida 302, 311
   A. m. maryae 301
Aricia agestis 387
Armadillidium vulgare 381
Armandia cirrhosa 393
Artemisia spp (sagebrush) 209, 501–3

   A. tridentata 199, 201, 210, 211, 214
   A. vulgaris 100
arthropod communities *see under* land-use changes
Asaphidion cyanicorne 44–5
Asclepias curassavica 233
Asia
   climate changes 34, 37, 39, 79, 99
   disease and insect vectors 160, 162, 171
   Quaternary 41, 42
   land-use changes 313
      extinctions 327, 328–9, 353–4, 360
   pollution 244, 275
   termites 480–1, 492
Aslauga australis 315
aspen *see under Populus*
Assocetia caliginosa 393
Astalotermes 483
Atlantic Ocean 54, 57
Atmosphere-Ocean GCMs 50
Australia
   climate changes and distribution prediction 61, 62, 66, 79, 82, 83–4
   land-use changes 298, 328, 386, 410
      environmental disturbances 303, 305, 308, 311, 314
      eucalypt dieback 455–60
      species richness 301–2
   livestock 467
   pollution 265
   termites 476–7, 480, 481
Austria 21

B

Bangladesh 354
bark
   beetles 35, 226, 379
   bugs 260, 262
barley *see Hordeum vulgare*
Bathyphantes spp 382
beans and pollution
   and elevated atmospheric carbon dioxide 198, 200, 202, 203, 208–9, 211
   gaseous air pollutants 229, 230, 231, 233, 234, 235, 239–40, 243
beech *see Fagus*
beetles *see* Coleoptera
Belgium 21, 104, 413

Index

Bembidion spp
  B. grisvardi 43–4
  B. ibericum 43–4
  B. lampros 381
  B. octomaculatum 42–3
  B. tetracolum 381
Bemisia tabaci 199, 200, 203, 207
Betula pendula and *B. pubescens* 209, 442, 444–5, 446, 448, 450, 451
BIOCLIM 65, 66, 75, 76, 179
biodiversity 7
  aphids as indicators of 499–505
  conservation 499–505
  evolution of southern 299–301
  reduced 14–15
  see also richness *under* species
biological control 113–14, 117, 310
Biological Monitoring Working Party 273–4
birch *see Betula*
birds 162–3
biting midge (*Culicoides*) 86
blackflies 160, 165
Blaniulus guttulatus 381
bog, marsh and mire and land-use changes 338, 340
  and arthropod communities 371, 375, 379
Boophilus microplus 82–3
Boreaphilus henningianus 40
Brachydesmus superus 380, 381
Brassica oleracea 136
Brazil
  land-use changes 301, 327, 353, 360, 410
  termites 476, 479, 491
Brevicoryne brassicae 131, 137, 231
Britain
  climate changes 56–7, 104
  and agricultural pests 98, 99–100
  and aphids 441–53, 505–8
    see also climate changes *under* aphids
  global warming experiments 431–9
  and prediction of distribution 61, 62, 79, 81, 85
  Quaternary 31, 34, 35, 36–46
  spread of species 17–19, 21
  fly population modelled 461–3
  land-use changes
    extinctions 322, 329, 331–4, 336, 337–51, 353, 354–9
    see also indicators *under* land-use changes

pollution
  Chironomidae as indicators of water quality 271, 273–4, 275, 277, 280–7, 289
  gaseous air pollutants 220, 222, 223, 225, 227, 229, 238, 240, 243–4, 442
  global warming and population dynamics 431–9
Brunei 492
Buprestidae 388
butterflies *see* Lepidoptera
BYDV (barley yellow dwarf virus) 140–3

C

CAB International 86
Caccobius schreberi 35
cadmium 251, 253, 257, 259, 262, 264
Calacanthia tribomi 42
Calathus melanocephalus and *C. piceus* 381, *C. micropterus* 404
calcium 257, 287
Caledia captiva 66
Callicera rufa 389
Cameroon (Mbalmaya Forest Reserve) and termites 475–9, 482–4, 489–93
Campanulotes defectus 358
Canada
  climate changes 41, 136, 426
  pollution 277, 278, 279
  see also North America
Carabidae (ground beetles) 343, 381–3
  Carabus spp 415
    C. arvensis 404
    C. maeander 36
    C. problematicus 404, 405, 406, 416
    C. violaceus 416
  extinctions 341, 343
  as indicators *see* indicators *under* land-use changes
carbon-based compounds 209–11
carbon dioxide 5, 6, 8, 10
  and aphids 127
  elevated atmospheric 197–216
    approaches to study of 198–201
    insect performance and abundance 202–7
    plant chemistry 207–13
    temperature 213–15

Carbon dioxide–*contd.*
  and Hadley Centre transient climate experiment *Plate 1*, 50, 52–3, 55
  increase and agriculture 94
  and prediction of distributions in changed climate 61, 77
Caribbean *see* Latin America
Carterocephalus palaemon 334, 390
cattle *see* livestock
Cenozoic 31
Central America see Latin America and Caribbean
Cephalotermes 483
Cerambycidae (longhorn beetles) 335, 343, 388
Cerapteryx graminis 385
Ceratomegilla ulkie 41
Cercopidae 336
Cercyonis sthenele sthenele 336
cereals 244
  aphids on 126, 147
  rice paddies and disease 162, 171
  see also Hordeum; maize; rice; *Triticum*
CFCs 224
Chaetocnema ectypa 200, 203, 207
Chagas' disease 160, 164
Chaitophorus populicola 234
changes *see* environmental changes
chemical(s)
  defences of trees 458
  see also drugs; fertilizers; pesticides
Chiloxanthus stellatus 42
China
  climate changes 34, 41, 70
  disease and insect vectors 160, 162, 171
  land-use changes 313, 354
Chironomidae
  Chironomus spp 272, 278, 289, 428
    C. anthracinus 272, 276, 287, 288
    C. plumosus 275, 286, 288
    C. riparius (*C. thummi*) 272, 273, 275, 286, 288
  and climate changes 32, 86
    response to 425–9
  and pollution *see* quality *under* water
chlorofluorocarbons 224
Chlorolestes apricans Plate 9, 306, 315
Choristoneura occidentalis 510–13
Chrysolina septentrionalis 42
cicadas 393

Cicadetta montana 393
Cinara pilicornis 228, 230, 232, 234–5, 241, 244
Circellium bacchus 306
circulation model limitations 57
CIS (Commonwealth of Independent States) *see* Russia and CIS
climate changes 3, 4–6, 12, 14, 29–193, 311–12, 457
  and agriculture *see* environment *under* agriculture
  and aphids *see* climate changes *under* aphids
  and forest defoliators 510–14
  and land-use changes 372, 392, 400
  predictions *see* distributions *under* future
  southern hemisphere 298–9
  time scales of 6–7
  see also disease and insect vectors; Hadley Centre; Quaternary; remote sensing; temperature
CLIMEX model 60, 61–2, 67–74, 75–7, 86–7
  applications 78, 80–1, 83, 86
  geographical distributions 69–74
  seasonal phenology and relative abundance 68–9
  and tsetse flies in Africa 179–80
climograms and prediction of distributions in changed climate 65, 75–6, 86
closed populations *see* fragmentation
cold *see* temperature
Coleoptera
  and climate changes
    prediction of distributions 70–3, 81, 83
    Quaternary 32, 33, 35, 40–3, 46
  and land-use changes 410, 456–7
    arthropod communities 379, 386, 388, 393
    extinctions 322, 327, 328, 335, 337–44, 348–51, 359
    southern hemisphere 302, 306, 310, 312
  and pollution 226
    elevated atmospheric carbon dioxide 198, 200, 203, 207
    gaseous air 226, 229, 233, 240
    ivermectin residues in cattle dung 466–7, 468
    metals 255, 257, 263
  see also Carabidae; Cerambycidae; scarabaeids

Collembola 252, 254, 259, 260–1, 263, 301
Colophon spp 312
Colorado beetle *see Leptinotarsa decemlineata*
Columbicola extinctus 358
Colymbetes dolabratus 38–9
Commonwealth of Independent States *see* Russia and CIS
community
  effects of metal pollution 259
  extinction 351–3
  competition between species and global warming 431–9
Congo 477
conservation 65, 314, 322, 393, 456
  biodiversity 499–505
  forests 308, 474, 492
constancy, species 20–1, 30–1, 34–5
copper 251–2, 254–60, 262, 264
coppicing 388, 389, 410
Coptotermes 483, 485
corridors between isolated areas needed 46
Corynoneura spp 428
cotton *see Gossypium hirsutum*
cottonwood *see Populus*
Countryside Council for Wales 386
Countryside Survey 377, 379
Crataegus monogyna 226
Cretaceous 329–30
crickets 301, 393
Cricotopus spp 428
  C. bicinctus 275
  C. intersectus 286, 287, 288
critical concentration of metals 258
critical loads for insects *see under* gaseous air pollutants
crops and arable farming
  and climate changes 85–6, 88, 97
  and land-use changes
    arthropod communities 373, 375, 377, 379
    elevated carbon dioxide 198–203, 208–9, 211, 214
    gaseous air pollutants 228, 229, 231, 233–4, 238–41, 243
    indicators 402, 406, 409, 411, 412, 415–16
    southern hemisphere 306–7, 315
    woodland in 380–4, 385, 415–16
  see also agriculture; pests

Cryptochironomus spp 279
Cryptostigmata 381–2
Cubitermes 483
Culicidae
  Culex spp 160, 162
    C. quinquefasciatus 165, 172
  see also mosquitoes
Culicoides spp 463
  C. brevitarsis 86
Curculionidae *see* weevils
Curimopsis cyclolepidia 41
Cylindroiulus spp
  C. caeruleocinctus and *C. latestriatus* 381
  C. punctatus 380, 381
cynipid gallwasp 20–1

D

Dacus tryoni 66
dams 165
  see also irrigation
damselflies *Plates 9–10*, 306–7, 313–15
Danaus plexippus 233, 243
Danthonia 314
day-length 83–4, 118
DECORANA 406, 411
Decticus verrucivorus 393
  integrated with inductive *see* CLIMEX
deficiency of metals 254–5
deforestation 171–2, 198
  and land-use changes 307–9, 324–5, 338, 353, 377, 491
  see also trees
deformities induced by pollution 279–80
Delia radicum 85
Deloneura immoculata 314
dengue 159–62
Denmark 240, 276
density dependence 183
  see also abundance
Dermaptera 328
Diacheila spp
  D. artica 38
  D. polita 37, 38, 39
Diamesa spp 427–8
diapause 118, 126
Diarsia rubi 385
Dicrotendipes spp 428
Diplocephalus picinus 382

Diptera
  and climate changes 32
  and land-use changes 328, 379, 388, 389
  and pollution 252, 258
    see also quality under water
  see also Chironomidae
discriminant analysis Plates 3–8, 66, 86, 185–7
disease
  of barley 140–3
  and insect vectors 157–74
    and land-use changes 171–2
    and mites 163
    and ticks 81, 160, 163
    see also disease under flies and temperature
  livestock 182, 184, 255, 461–4
dispersal see migration
distribution see geographical distribution
Diuraphis noxia 97, 133, 504
diversity 401, 406–7, 412–13
dock and dock leaf beetle 233–4
Dolomedes plantarius 393
Drepanosiphum spp
  D. acerinum 131–2
  D. platanoidis 130, 131–2, 506–8
Drosophila spp 21
  D. melanogaster 434–6, 438
  D. simulans 434–5, 437–8
  D. subobscura 434–5, 437–8
  and global warming 431–9
  and metals 263–4
drought-induced stress 96, 113
drugs: ivermectin residues in cattle dung 465–71
Dryas (era) 428–9
Dryas octopetala 144
Dryococelus australis 310
dung
  beetles see scarabaeids
  ivermectin residues in 465–71

E

earwigs 328
Ecchlorolestes spp
  E. nylephtha 307–8, 313
  E. peringueyi Plate 10, 307–8, 313
Ecoclimatic index 70

Ectopistes migratorius 358
Effective Temperature Sum 100–2, 103
Eichhornia crassipes 311
EI (Ecoclimatic index) 70
Elateridae 388
Elatobium spp
  E. abietinum 214, 230, 232, 234, 506–8
Euceraphis spp
  E. betulae 446, 448, 450, 451
elements
  natural selection of 254–8
  periodic table of 252
  see also metals
elephantiasis 160, 164–5
elevated atmospheric carbon dioxide see under carbon dioxide
Eltham copper 311
Empoasca spp 200, 203, 207
EMSYS (European Mapping System) 102, 106, 109
encephalitis, Japanese 160, 162–3
endemism 39, 42, 301–2, 351–4
Endochironomus spp
  E. impar/dispar 272, 286, 287, 288
  E. tendens 286, 288
England see Britain
English Nature 392, 393
environmental changes, response of insects to 3–24
  dynamics of phytopahgous populations 7–12
  and tropical insects see termites
  see also climate changes; land-use changes; pollution
environmental pressures see under southern hemisphere
Ephemeroptera 273, 328
Epilachna varivestris 229, 233, 240
Epinotia tedella 220
Erigone spp 382
erosion 305, 342–3, 456
ETS (Effective Temperature Sum) 100–2, 103, 107–8
eucalypts 308, 309
  dieback and land-use changes 455–60
  Eucalyptus grandis 209
Euoniticellus spp 306
Europe
  aphids 148, 500
  climate changes 50, 56–7, 148

Index

disease 160, 162, 171, 461–4
global patterns of transient change 56–8
prediction of distribution 73, 79–80, 83
Quaternary 34, 35, 36–46
spread of species 17, 19–21
see also environment *under* agriculture
land-use changes 314, 387, 389, 393
extinctions 328, 331, 336, 346
indicators 399, 412–14
pollution
Chironomidae as indicators of water quality 272, 276–7, 280
gaseous air pollutants 219, 220, 227–8, 240, 243, 244, 442
metals 260, 262–3
see also Britain
European Corn Borer 99–118
described 99–100
impact of climate changes 106–12
implications of climate changes 111–17
potential distribution 102–3, 107–10
projected climate changes 105–6
refining approach 117–18
see also mapping European Corn Borer *and also* environment *under* agriculture
European Mapping System *see* EMSYS
eutrophication 392, 425
evolution
and extinction 329–30
and fire 303–4
and metals 254, 260
rapid, lack of *see* species constancy
of southern biodiversity 299–301
see also genetic effects
exclusion, competitive 7
exotic invasions in southern hemisphere 310–11
experiments
climate 431–9
see also Hadley Centre
multi-factorial 5, 10
pollution *see under* gaseous air pollutants
external threats and extinctions 335–41
extinctions 321–66, 412
and biogeography 351–4
co-extinction of parasites 358
external threats 335–41
parameters 329–35
present knowledge 326–9

rates 323–5, 351–8
relative rates 326, 354–8, 359
see also vulnerability

F

Fagus sylvatica 226, 442, 444, 446–7, 449–50
fen 338, 340
fertilizers 208
see also critical loads *under* gaseous air pollutants
Festuca ovina 414
Fiji 328
filariasis, lymphatic 160, 164–5
Finland, pollution in
gaseous air pollutants study 227–8
metals 260, 262–3
fire 303–5, 315
fleas 358
flies 85, 389, 468
and disease 163, 164, 461–4
see also trypanosomiasis
population modelled 461–4
see also fruit flies; tsetse flies
flight activity of pests
aphids and temperature 126, 135–6, 143, 147, 505–8
European Corn Borer 104, 110, 114
see also migration
forestry and reforestation 400, 410
and arthropod communities 371–2, 374–9, 387–91
Forestry Commission 390
see also deforestation; trees/forest
Formicoidea *see* ants
fossils 330
see also Quaternary
fragmentation of habitats
and barriers to spread of species 10, 19, 46
and land-use changes 391, 412
burnt areas 305
extinctions 331–3, 335, 336, 359
forests 308, 313
game areas 305
France
climate changes 103, 114, 116, 146, 148
fly population modelled 461–3
spread of insects 21, 103

Frankliniella spp 200, 207
fruit flies 66
  see also Drosophila
fuel burning *see* pollution
fumigation studies *see* chamber studies
future and predictions
  climate changes 6–7, 10
    and agriculture 105–6
    Hadley Centre transient climate experiment 50, 53, 57
    tsetse flies in Africa *Plates 6* and *7*, 177, 178–9, 186, 189–90
    see also distributions *below*
  distributions in changed climate 59–90
    comparison of approaches 74–8
    impacts *see* impacts of climate changes
    see also estimating potential *under* geographical distribution
  gaseous air pollutants research 243–5
  land-use changes 393
    effects of 415–16
    extinctions 353–4
    southern hemisphere 313–15

G

gallwasp 20–1
Gap Analysis mapping 500, 501
gaseous air pollutants 219–47
  critical loads for insects 441–53
    adult weight 446, 449–50
    attack rate 446, 447
    deposited nitrogen 442–4
    growth rate 446, 448
    population growth 450, 451
  direct effect on plants 222–5
  evidence of insect/plant/air pollution interactions 225–8
  experimental manipulations 228–42
    closed chamber studies 228–39
    field-based 240–2
    open-top chamber studies 239–40
  predictions and future research 243–5
  summary of current knowledge 242–3
  see also nitrogen dioxide; ozone; sulphur dioxide
gases *see*2 carbon dioxide; gaseous air pollutants
Gastrophysa viridula 233

general circulation models (GCMs) 16, 78
  agriculture and pests 105–6, 111, 118
  forest defoliators 510, 511–12
  Hadley Centre transient climate experiment 50–1, 53–5, 57–8
genetic effects 7, 20–1, 392
  of metal pollution 263–4
  see also evolution
geographical distribution and range 12–18
  and CLIMEX model 69–74
  estimating potential 60, 61–74
    BIOCLIM 65, 66, 75, 76, 179
    climograms 65, 75–6, 86
    multivariate statistics 65–6, 75–6, 86
    physiological data and population dynamics models 66–7
    reliability and accuracy of data sets 74
    see also CLIMEX
  and extinctions 344–5
  fragmented landscapes and barriers to spread of species 19
  limitations 14–15
    see also climate changes; land-use changes; pollution
  local abundance and size of range 13–14
  and migration 12–18
  and Quaternary climate changes 29, 36–45
  rates of spread 16–18
  tsetse flies *see* remote sensing
  see also spread of species
geographical information system 66, 74, 87, 186, 511
Geophysical Fluids Dynamics Laboratory 511–13
Germany
  land-use changes 328
  pollution 220, 276–7
  spread of species 17, 21, 104
GFDL *see* Geophysical Fluids Dynamics
GI (Growth Index) 70, 73, 86
GIS *see* geographical information system
GISS *see* Goddard Institute of Space Studies
glaciation 299, 427–8
  see also Quaternary
global changes and southern hemisphere 311–12
global extinctions 331, 333–5
global patterns of transient change in climate 50–5
  aerosols, effect of 53

Index

magnitude of change 50–2
pattern of *Plate 1*, 53–4
sea level rises 54–5
global warming and greenhouse gases 52–3, 198, 312
  and agricultural pests 105–12, 505–8
  implications of 137–8
  and population dynamics 431–9
  prediction of distributions 78, 81–2
  see also carbon dioxide; climate changes; Hadley Centre; temperature
Glomeris marginata 381
Glossina spp (tsetse flies) 66, 163
  G. morsitans Plates 3–6, 65, 178, 181, 182–4, 187–90
  G. pallidipes Plates 4 and 7–8, 178, 187, 188–90
  G. palpalis 182
  see also remote sensing; trypanosomiasis
Glycine max (soybean)
  and elevated atmospheric carbon dioxide 198, 200, 202, 208–9
  and gaseous air pollutants 229, 233, 239–40
Glyptotendipes pallens 286, 287, 288
Goddard Institute of Space Studies (GISS) 105–16, 511–13
Gondwanaland 299
  relict damselflies *Plates 9–10*, 306–8, 313, 315
Gossypium hirsutum (cotton) 198, 199, 200, 202, 208–9, 214
gradients, gaseous air pollutants along 227–9
grains *see* cereals
Graphoderus zanatus 393
grasshoppers 66
  and land-use changes 299, 304, 309, 314
  and pollution 199, 255, 263
grasslands
  and climate changes 19, 97, 165, 182
  expansion of 97
  and land-use changes 314, 414
    arthropod communities 371, 373–9, 385
    extinctions 337–8, 340
    indicators 406, 409–10, 416
    invertebrates and grazing mammals 386–7
    southern hemisphere 299, 304–5, 314–15
  see also cereals; savannah

grazing animals *see* livestock
Greece 104
greenhouse gases *see* carbon dioxide; global warming
ground beetles *see* Carabidae
GROWEST model 86
growing season and impacts of climate changes 79–82, 95
Growth Index 70, 73, 86, 179
Grylloidea 301
Gryllus campestris 393
Guinea 476
Gymnobelideus leadbeateri 74

H

habitats *see* fragmentation
Hadley Centre transient climate experiment *Plates 1–2*, 50–8, 128
  over Western Europe 56–8
  see also global patterns of transient change
Hadramphus spp
  H. stilborcarpae 310
  H. tuberculatus 335
Haemaphysalis longicornis 79–80
harvesting date changes and climate changes 98–9
Hawaii 328–9
health problems *see* disease
heat *see* global warming; temperature
heathland and moorland 19, 337–8, 340, 413–14
  and arthropod communities 371, 374–9, 385
hedgerows 337–8, 375, 379, 380
  and indicators 399, 409, 414–15
Heliothis virescens 255
Helophorus spp 39
Hemiptera
  and elevated atmospheric carbon dioxide 198, 199, 200, 201, 203, 207
  and land-use changes 379, 388
  and Quaternary climate changes 33, 42, 46
  rates of spread 17
  see also aphids
herbivorous insects *see* insects
Hesperiidae 314, 354
  Hesperia comma 19, 332, 333, 334, 387, 414
Heteroptera 42, 410

Heterotrissocladius spp 427–9
  H. grimshawi 272, 286–7, 288, 289
  H. marcidus 289
hexapod 254
Hippodamia arctica 41–2
histerids 466
Holoboreaphilus nordenskioeldi 39
Holocene 427–9
Homoptera 328, 336
hops (*Humulus*) 100
Hordeum vulgare (barley)
  and climate changes 97
  disease 140–3
  and land-use changes 373
  and pollution 228, 231, 234, 238, 243
host animals 162–3, 164
  see also livestock
host plants *see* plants
hoverflies 379, 389
human activities 46, 159–60
  see also agriculture; extinctions; impacts of environmental pressures; land-use changes; urbanization
humidity *see* moisture; precipitation
Humulus spp 100
Hungary 17, 21
hunting 305–6
Hyadaphis tataricae 504
Hydraecia micacea 385
hydrogen fluoride pollution 226
hydrophilids 39, 466, 468
Hydrotaea irritans 461–4
Hymenoptera 33, 226, 328
  and land-use changes 379, 388

I

Iberian peninsula
  climate changes 39, 41, 56, 57, 104
  fly population modelled 461–3
Idaho suction traps and plant biodiversity 499–505
IGBP *see* International Geosphere-Biosphere Programme
impacts of climate changes 78–85
  scenarios 78
  studies 78–9
  see also nature of impacts
impacts of environmental pressures in southern hemisphere 303–12
  arable farming 306–7
  deforestation 307–9
  exotic invasions 310–11
  fire 303–5, 315
  global changes 311–12

hunting, pastoralism and livestock 305–6, 315
introduced forest patches 309
urbanization 311
water impoundments 309–10
inbreeding 336
India 162
indicators 425
  Chironomidae as 426
  see also quality *under* water
  of plant biodiversity
    aphids as 499–505
    termites as *see* termites
  see also under land-use changes
Indonesia 162, 360, 492
industry 342–3, 375
  see also pollution; urbanisation
insecticides *see under* pesticides
insects *see* climate changes; land-use changes; pollution
INSIVE 461–3
Institute of Terrestrial Ecology 403, 406, 415
'intensity' of land-use 403, 409–10
Intergovernmental Panel on Climate Change 50–1, 54, 55, 93, 105–6, 158
International Geosphere-Biosphere Programme 192
introductions 336
invasions 17–18, 307, 309
  see also migration
IPCC *see* Intergovernmental Panel on Climate Change
Ireland 38, 40, 280, 401
  see also Britain
Iridomyrmex humilis 313
iron 251, 254–8
irrigation and climate changes 70, 98, 113, 170, 171
islands 328–9, 331, 391
  see also Britain; New Zealand; Pacific
isolated areas *see* fragmentation
Isotoma olivacea 260–1
Italy 21, 104, 160
ITE *see* Institute of Terrestrial Ecology
IUCN 325, 327, 356
ivermectin residues in cattle dung 465–71
Ixodes ricinus 79, 81

J

Japan 162, 275
Japanese encephalitis (JE) 160, 162–3
Junonia coenia 199, 201, 202–6

Index

K

Kalotermitidae 478
Keiferia lycopersicilla 233
Kenya: disease and insect vectors *Plates 4–8*, 169–70
  tsetse flies 178, 181, 182, 184–5, 187–9

L

Labidura herculeana 328
ladybirds 41–2
Lake District tarns 271, 280–7, 289
  geology 280–1
  methods 282–4
  results 284–7
  sampling sites 281–2
  see also Chironomidae
Lamingtoniidae 302
land and temperature 50–3
land-use changes 298–420
  and aphids 127
  and arthropod communities (Britain) 371–96
    effects on assemblages 379–84
    forest management 387–91
    grazing mammals 386–7
    hedgerows 380
    patterns of change 373–9
    subtle changes 384, 386–91
    woodland in arable farming 380–4, 385
  and climate changes 372, 392, 400
  disease and insect vectors 171–2
  and eucalypt dieback 455–60
  habitat scale and pattern 413–15
  indicators of (ground beetles and macrolepidoptera) 399–420
  'intensity' of land use 403, 409–10
  monitoring effects of change 410–13
  predicting effects of 415–16
  types of change 403–9
  see also extinctions; southern hemisphere
Land Use Programme (NERC/ESRC) 401
Laos 162
Latin America and Caribbean
  disease and insect vectors 159, 164, 165
  land-use changes 299, 301, 305, 410
  extinctions 327, 353, 360
  termites 476–7, 479, 481, 491
latitude and climate changes 83–4
Laurasia 299
Lauterbornia coracina (*Microspectra coracina*) 276, 428
lead 251, 253, 259–63
leaf beetles 233

leaf hoppers 226
leaf-miners 17
leishmaniasis 160, 164
leisure and recreation 342–3, 377
Lepidochrysops hypoplia and *L. lotana* 314, 315
Lepidoptera
  and climate changes
    agriculture *see* European Corn Borer
    in model system 11
    Quaternary 33
    rates of spread 17–18
  and land-use changes
    arthropod communities 379, 382–5, 387–90, 392–3
    extinctions 327, 328, 331–4, 335, 336, 337–43, 346–8, 353, 354, 359, 360
    forest defoliators 510–14
    indicators *see* indicators *under* land-use changes
    in southern hemisphere 301–2, 309, 311, 314–15
  and pollution
    elevated atmospheric carbon dioxide 198, 199–208, 210, 214
    gaseous air 220, 233, 243
    metals 255, 260
  see also in particular Lymantria dispar
Lepthyphantes flavipes 382
Leptidea sinapis 390
Leptinotarsa decemlineata (Colorado beetle) 70–3, 81, 83
lice 358
life cycle and reproduction
  and climate changes 68–9, 79, 81–2
    aphids and temperature 126, 128–30, 139–44
  and extinctions 336, 347
lima bean *see Phaseolus lunata*
Liriodendron tulipifera 209
livestock
  diseases 182, 184, 255, 461–4
  grassland invertebrates and grazing mammals 386–7
  ivermectin residues in cattle dung 465–71
  land-use changes in southern hemisphere 305–6
  see also animals; ticks
local extinctions 331–3, 334
longhorn beetles (Cerambycidae) 335, 343, 388
Loricera pilicornis 404, 406
Lucilia cuprina 255, 258
Luperina testacea 385
lupins 231
Luxembourg 104

Lycaena dispar 393
lycaenid butterflies 301–2, 311, 314–15, 393
Lycopersicon esculentum (tomato) and pollution 233
  carbon dioxide 198, 199, 201, 203, 208–9
Lymantria dispar 11, 83, 233
  and elevated atmospheric carbon dioxide 201, 202, 203–4, 206, 208, 210
  as forest defoliator 510–12, 514
Lyme disease 160, 163
lymphatic filariasis 160, 164–5
Lysandra bellargus 387
Lythrum salicaria (purple loosestrife) 500, 503

M

macrolepidoptera *see* Lepidoptera
Macropelopia adaucta (*M. goetghebueri*) 272, 286–7, 288, 289
Macropus spp 66
Macrosiphum spp
  M. albifrons 231
  M. euphorbiae 131
Macrotermitinae 476–8, 483
  Macrotermes spp 477, 484
Maculinea arion 335, 393
Madagascar 170, 171
maize 491
  effect of pests on *see* European Corn Borer
Malacosoma disstria 11, 196, 201, 203–4, 208, 210
malaria 158, 160, 165–71
Malaysia 480–1
management practices
  and extinctions 338, 340–3
  see also forestry
manganese 255–7
maple *see* Acer
mapping
  European Corn Borer 100–4
    bioclimatic model 100–2
    European Mapping System 102
    peak flight activity 104
    potential distribution 102–3
  Gap Analysis 500, 501
  tsetse flies in Africa 180, 181–2, 184–7, 188
marsh *see* bog, marsh and mire
masarine wasps 303
mass extinction event 329–30
mastitis 461–4
Match IIndex 61
Mauritius 328

mayfly 273
mean relative growth rate and gaseous air pollutants study 20, 221, 235–40
Megoura viciae 137
Melanopus spp
  M. differentalis 201, 203, 208, 214
  M. sanguinipes 201, 203, 208
Mellicta athalia 333, 388
Mentha piperita 198, 201, 209, 211
mercury 251, 253, 257, 259
metals 227, 252–68, 275, 425
  natural selection of elements 254–8
  pollution 258–64
    bioaccumulation 260–2
    community effects 259
    genetic effects 263–4
    individuals and species, effect on 262–3
metapopulations 331, 336
meteorology 75
  see also climate
methane 224
Metopolophium dirhodum 126
  and gaseous air pollutants 228, 229, 231, 234, 238
  and temperature 131, 147–8
Mexican bean beetle 229, 233
Mexico 164
Microcerotermes 483
microhabitats 323, 477–8
Microlophium carnosum 506–8
Microspectra coracina (*Lauterbornia coracina*) 276, 428–9
Microtendipes spp 286, 428
Microtermes 478
midges *see* Chironomidae
migration/mobility of species 349
  barriers to *see* fragmentation
  and climate changes 7, 12–18, 20, 34, 65
  aphids and temperature 133–4
  and geographic range 12–18
  see also flight activity
millipedes 380–1
Miocene 31
mires *see* bog, marsh and mire
mites 163, 382
mobility *see* migration/mobility
moisture
  and agricultural pests 95–6, 97, 113, 118
  and disease 164–5
  prediction of distributions in changed climate 78, 83
  soil *Plate 2*, 57
  see also precipitation; water
Molinia caerulea 390
monarch butterfly 233, 243
Mongolia 41

Index

Monocephalus fuscipes 382
morph determination of aphids and
    temperature 126, 136–7, 144
mortality of insects
    and climate changes 65, 133–4
    and pollution 444
mosquitoes 264
    and disease 158, 159–60, 161, 162, 165–72
moss 41, 382
moths *see* Lepidoptera
movement *see* migration
MRGR *see* mean relative growth rate
mugwort 100
multi-impact studies 214
multivariate statistics prediction of
    distributions in changed climate 65–6,
    75–6, 86
multivoltine species and impacts of climate
    changes 81–2
Musca autumnalis 461
Myrmica sabuleti 335
Myzocallis kuricola 137
Myzus spp
    M. lythri 500
    M. persicae 126
        and pollution 231, 256
        and temperature 131, 133–4, 136, 139,
        143, 146–9

N

Nanogonap olydesmoides 381
Nasutitermes 483
Nasutitermitinae 476, 478
National Biodiversity Action Plans 353
National Countryside Monitoring Scheme
    371, 374–5, 378–9
natural selection of elements 254–8
Nature Conservancy Council 415
nature of impacts of climate changes 79–85
    change in severity of limiting factors 82–3
    extreme events, evidence of 83
    host crops and animals 84–5
    interaction with other insect species 85
    latitude and climate changes 83–4
    length of growing season and spatial
        consequences 79–82
nature reserves 499–500
    isolated *see* fragmentation
NCMS *see* National Countryside Monitoring
    Scheme
NDVI (normalized difference vegetation
    index) 188–9

Nebria spp
    N. brevicollis 262, 404, 405, 406, 416
    N. salina 404, 406, 416
nematodes and disease 164–5
Nepal 170
Netherlands
    land-use changes
        and extinctions 334, 346, 347
        indicators 412, 413–14
    spread of species 21, 104
new crops and over-wintering of pests 97
New Zealand
    climate changes 79
    land-use changes 298, 401
        environmental disturbances 307–8, 310
        extinctions 328, 335
    pollution 276
nickel 260, 265
Nicrophorus americanus 344
Nigeria 183, 476, 482
Nilaparvata lugens 17
nitrogen dioxide and nitrogen pollution 5,
    96, 220, 224, 228, 242
    and carbon dioxide 197, 207–9, 212–13
    direct effect on plants 223
    experiments 229, 230–2, 237–9
    future research 243–4
    see also critical loads *under* gaseous air
        pollutants
Noditermes 483
normalized difference vegetation index 188–9
North America
    aphids as indicators of plant biodiversity
        499–505
    climate changes 426
        agriculture 97, 102, 112
        disease and insect vectors 160, 162, 164
        and ECB 99
        and forest defoliators 510–14
        prediction of distribution 65, 71–3
        Quaternary 36, 41
        spread of species 17
    land-use changes 308, 313
        extinctions 328, 336, 344
    pollution
        Chironomidae as indicators of water
            quality 273, 276–80
        gaseous air pollutants study 239–40,
            243, 244
    see also Canada; United States

northern hemisphere
  climate changes 50–1, 54
  see also aphids; Asia; Europe; North America
Norway 289, 427–8
Norway spruce see under Picea
Nymphalis polychloros 334

O

oak see Quercus
oceans see sea
Ochthebius dilatatus and *O. figueroi* 39
Odonata 33, 301, 306, 379
Odontotermes spp 484
Olophrum boreale 39
Omaliinae 39
Onchocerca volvulus 165
onchocerciasis 160, 165
Oniscus ascellus 380, 381
Oniticellus fulvus 35
Onthophagus fracticornis, *O. furcatus* and *O. massai* 35
Onychiurus armatus 260–1
Onychophora 308
Operophtera brumata 392
Ophidesmus albonanus 381
Ophyiulus pilosus 381
Orthocladinae 275, 289, 427–8
  Orthocladius consobrinus 286, 288
Orthoptera
  and carbon dioxide pollution 198, 201, 203, 208, 214
  and land-use changes 308, 328, 379, 392
Orthosia gothica 385
Ostrinia nubilalis see European corn borer
overgrazing 305
overlapping ranges 13
over-wintering of pests 97, 162
Oxytelus see *Anotylus*
ozone pollution 220, 226, 242–3, 312
  direct effect on plants 223–5
  experiments 228, 231–6
  future research 244

P

Pacific areas
  climate changes 159, 170
  land-use changes and extinctions 328–9, 353–4
  see also Asia; Australia; New Zealand
Pagastiella orophila 286, 287, 288, 289
palaeoecological indicators of water quality, Chironomidae as indicators of 277–9
palaeontology see fossils
Palpomyia sp 274–5
Pangaea 299
Panolis flammea 392
Papua New Guinea 170
Paracladius spp 427–8
Paracladopelma spp 428
Parakiefferiella spp 428
Paralucia pyrodiscus lucida 311
parasites 457
  extinction 358
  parasiticides in dung 465–71
  and Quaternary climate changes 33
  used in biological control 117
  vectors of disease see under disease
Paratanytarsus 428–9
Pardosa spp 382
parthenogenetic insects see aphids
passenger pigeon 358
pastoralism see livestock
pea aphid see *Acyrthosiphon*
peas 231, 234
peat cutting 342–3
Pectinophora gossypiella 200, 203, 205–7
Pentaneurini spp 428
peppermint see *Mentha piperita*
performance and carbon dioxide pollution 202–7
Periphyllus testudinaceus 506–8
Permian 329–30
pesticides 409
  insecticides 99, 116–17, 259, 406
  parasiticides in dung 465–71
pests, crop
  and climate changes 17, 65, 79, 86
    temperature 126, 131, 133, 135–7, 140, 147–8
  see also environment under agriculture
  and land-use changes 392
  and pollution 231, 237, 241
  see also aphids; crops
Petrova resinella 260
Phaenopsectra flavipes 272, 286, 287, 288
Phaseolus spp
  P. lunata 198, 200, 203, 208–9, 211

Index

P. vulgaris 229, 239, 240, 243
Phasmoptera 328
phenology
    of aphids and temperature 144–9
    European Corn Borer 110–12
Philaenus spumarius 336
Philippines 162, 354
Philoscia muscorum 381
Phorodon humuli 506–8
phosphorus 277
Phragmites 19
Phthiraptera 358
Phyllonorycter leucographella and *P. platani* 17
phytophagous populations 402
    dynamics of 7–12
    see also insects; plants
Picea spp (spruce) 17, 228, 379
    aphid *see* Elatobium
    budworm 510–13
    P. abies (Norway spruce) 220, 226, 228, 241
    P. sitchensis (Sitka spruce) 212, 214, 230, 232, 234–5
Pieris rapae 17
pigeon, passenger 358
pigs 162–3
pine bark bugs 260, 262
pine resin gall moth 260
Pinus spp 209
    P. ponderosa 226
    P. radiata 309
    P. sylvestris (Scots pine) 17, 389
    and pollution 226, 227–8, 234
Plagiodera versicolora 233
Plantago lanceolata 199, 201, 202, 209, 210, 211
planthoppers 17
planting date changes and climate changes 98–9, 143–4
plants 7–12, 17, 84–5
    biodiversity, aphids as indicators of 499–505
    chemistry and pollution 207–13
    and pollution *see* gaseous air pollutants
    see also crops; grasslands; heathland; trees
Plasmodium spp
    P. falciparum 166, 168, 171
    P. vivax 166
Plecoptera 273, 328

Pleistocene *see* glaciation
Plejebus argus 332, 333
pollution 5, 127, 197–291, 425
    and disease 165
    and extinctions 342–3
    see also carbon dioxide; gaseous air pollutants; metals *and* quality *under* water
Polydesmus angustus, *P. gallicus* and *P. inconstans* 381
Polypedilum spp 286
    P. laetum 272
Polyxenus lagurus 381
population
    dynamics 7–12, 444
        see also environmental changes
    flies, model of 461–4
    growth and gaseous air pollutants 450, 451
    metapopulations 331, 336
    see also geographical distribution
Populus spp (cottonwood)
    P. deltoides 10, 233–4
    P. tremuloides (aspen) 11
    and carbon dioxide pollution 199, 201, 202–4, 206, 208–9, 210, 211
Porcellio scaber 381
possum (*Gymnobelideus*) 74
potatoes 142–3
potential crop changes and climate changes 96–7
potential distribution of European Corn Borer 102–3, 107–10
precedence and geographical distribution 177, 178–9, 182
precipitation 8
    and agricultural pests 96, 117, 118, 127
    and disease 162
    Hadley Centre transient climate experiment *Plate 2*, 56–7
    prediction of distributions in changed climate 78, 83
    see also moisture; water
prediction *see* future and prediction
Procladius spp 279
    P. flavifrons 286, 287
    P. sagittalis 272, 286, 287, 288
Procubitermes 483
Protanypus morio 288
Prunus padus 140
Psectrocladius spp 428

Psectrocladius spp–*contd.*
  P. obvius 286, 287, 288
  P. platypus 286, 287
Pseudochironomus prasinatus 286
Pseudodiamesa spp 428
Pseudoplusia includens 200, 202, 203, 206, 208
psyllids 291, 379
Pterolocera spp 314
Pterostichus spp 403
  P. adstrictus, *P. madidus* and *P. strenuus* 404
  P. diligens and *P. niger* 404, 406
  P. melanarius 381–2, 404, 415
Pycnoglypta lurida 39
Pyralidae 99

Q

Quaternary climate changes 12, 20, 29–47, 299
  cycle of glacials and interglacials 34–45
    changes in geographical ranges 36–45
    last glacial 36
    last interglacial 35
  entomology 31–3
Quercus spp
  Q. alba 233
  Q. cerris 20
  Q. rubra 11, 199, 201, 202, 203–4, 206, 208–9, 210, 211

R

rainbow trout 306–7
rainfall *see* precipitation
rainforest *see* Amazonia; termites
range *see* geographical distribution
Rattus rattus 310
recreation 342–3, 377
Red Data Book 313
  and extinctions 39–43, 322, 333, 337, 344
    ecological factors 346–8
    phylogenetic status 348–51
Reduviid bugs 160, 163–4
reed 19
reforestation *see* forestry
regression analysis 180

regulation of metals 255–8
relapsing fever 160
relative extinction rate 326, 354–8, 359
relative growth rate 199
remote sensing and changing distribution of tsetse flies in Africa 177–93
  analysing distributions 179–81
  biological and statistical models *Plates 3–5*, 177, 182–90
  mapping 180, 181–2, 184–7
  predictions and future 190–1
  see also Glossina spp
reproduction *see* genetic effects; life cycle and reproduction
RER *see* relative extinction rate
reservoir hosts 164
RGR *see* relative growth rate 199
Rhagoletis mendax 66
Rhinorphipidae 302
Rhipicephalus appendiculatus 66, 74, 163
Rhopalosiphum padi 126
  and gaseous air pollutants 228, 231, 234, 238, 240
  and temperature 131, 134–6, 140–2, 147, 149
Rhynchaenus fagi 239, 446, 447, 449–50
Rhyniella praecursor 254
rice 17
  paddies and disease 162, 171
rinderpest 182, 184
riparian sites 337–8, 340
RIS *see* Rothamsted Insect Survey
rivers
  and disease 165
  and land-use changes 306, 309–10, 311, 343
  water quality 273–6
  see also water
roads and tracks
  and land-use changes 375, 377, 379, 390–1
  pollution 222, 226
Rothamsted Insect Survey 227
  and aphids 125, 131–2, 140–1, 144, 148–9, 505–6
  and land-use change indicators 399, 405, 407–8, 411–12
Rumex obtusifolius 233
Russia and CIS: climate changes 36, 37, 39–40, 103, 170
Rwanda 170, 172

Index 531

S

sagebrush *see Artemisia*
St Helena 328
salination of soil 456, 457
Salmo gairdneri (rainbow trout) 306–7
sand dunes 338, 340
sandflies 160, 164
saproxylic insects 388, 389
savannah
  southern hemisphere 277, 305, 314–15
  and termites 480, 491
  and tsetse flies 182
sawflies 226, 227, 379
scale insects 226
Scandinavia
  climate changes 43, 99–100, 145
  pollution 277–8, 289
scarabaeids (dung beetles)
  and climate changes 35, 40–1
  and land-use changes 306, 386, 456, 466–7, 468
Schedorhinotermes 483
Schizaphis graminum 131
Schizolachnus pineti 228, 230, 232, 234
Scirtothrips aurantii 307
Scolytus koenigi 35
Scotland
  climate changes 506
    aphids 142, 148
    Quaternary 37–8, 40, 44–5
  land-use changes 371, 374–8, 389, 409
  pollution 280
  Scottish Natural Heritage 375, 378
Scots pine *see under Pinus*
seasonality 76
  and disease 162, 163
  and indicators 402
  phenology and CLIMEX model 68–9
  and pollution 287
  variation and termites 486, 488
  winter *see* climate changes *under* aphids
sea and temperature 50–3, 54–5
Sergentia spp 428
Siberia 36
Sicily 35
Simplocaria metallica 41
Siphonaptera 358
Sitka spruce *see under Picea*
Sitobion avenae 126

and gaseous air pollutants 231, 237, 241
and temperature 126, 131, 133, 135–7, 140, 147–8
Slovenia 21
social insects *see* ants; termites
sodium 287, 289
Sogatella furcifera 17
soil
  disturbance 404
  erosion 305, 342–3, 456
  fertility 97
  moisture *Plate 2*, 46
  pollution 255
  salination and waterlogging 456, 457
  and termites 484
Solva interrupta 389
Sorghum spp 244
  S. vulgare 226
South Africa
  climate changes 299
  ivermectin residues in cattle dung 465–71
  land-use changes 298–9, 353
    environmental disturbances 303–5, 306–8, 309, 311, 312, 313–15
    species richness 301–2
  relict species *Plates 9–10*, 306–8, 313, 315
South America *see* Latin America
southern hemisphere
  climate changes 50–1, 54
  environmental pressures 298–319
  endemism 301–2
  evolution of southern biodiversity 299–301
  future 313–15
  land distribution and climatic history 298–9
  species richness 298–9, 301–3, 315
  see also Australia; impacts of environmental pressures; Latin America; New Zealand; South Africa
soybean *see Glycine max*
Spain
  climate changes 39, 41, 104
  fly population modelled 461–3
species
  abundance *see* abundance
  constancy and lack of evolutionary change 20–1, 30–1, 34–5
  diversity *see* biodiversity

species–*contd.*
  richness 413
    and extinctions 323, 350–4
    in southern hemisphere 298–9, 301–3, 315
  Species Recovery programme 392
spiders 379, 382, 383, 386, 393
Spilosoma lubricipeda and *S. luteum* 385
spittlebugs 336
Spodoptera spp
  S. eridania 201, 203, 206, 208
  S. exigua 200, 202, 205, 214
spread of species
  barriers to 19
  rates of 16–18
  see also geographical distribution; migration
spruce see *Picea*
Sri Lanka 169
SSSI status 340
Stackhousia tryonii 265
stag beetles 312
Staphylinidae 39–40, 466, 468
starch 210
stick insect 310
Stictochironomus spp 279
  S. stricticus 286
Stipa 314
stochasticity, environmental 335–6, 352
stonefly 273
stormy weather 57
stress
  on crops 95–6, 113
    see also pests
  indices in CLIMEX model 69–72, 179
sublethal effects of temperature changes 83, 134–5
suction trap data
  and gaseous air pollutants 227
  and plant biodiversity 499–505
  see also Rothamsted
Sudan 165
sugar beet 142–3
sulphur dioxide pollution 5, 220, 242
  direct effect on plants 222
  evidence 226, 227, 228
  experiments 230–1, 237–41
  future research 244
suprapopulation concept 63
Sweden
  climate changes and pests 99–100, 145

pollution 277–8
Switzerland 112
Syncerus caffer 306
Syndiamesa spp 428
Synemon spp 314
Syrphidae 388

T

Tabanus sp 25
Tachypodoiulus niger 381
Taiwan 328
Tanypodinae 289
  Tanypodini spp 428
Tanytarsini
  Lauterbornia coracina 276, 428–9
  Tanytarsus spp 286
    T. lugens 428
Tanzania: disease and insect vectors *Plates 4–8*, 166, 171–2
  tsetse flies 178, 181, 182, 184–5, 187, 187–9
target loads 452
tarns see Lake District tarns
taxonomic stability and indicators 401–2
TBI (Trent Biotic Index) 273
temperature and temperature changes 8, 10
  and abundance 68
  and agriculture see environment *under* agriculture
  and aphids 125–6, 127–8, 130–49
    abundance 149
    development and fecundity 130–2
    individual 130–8
    life-cycle strategies 139–44
    morph determination 126, 136–7, 144
    mortality 133–4
    movement 135–6
    phenology 144–9
    populations 139–48
    sublethal effects 134–5
    warmer climate, implications of 137–8
  and carbon dioxide 213–15
  and Chironomidae 425–9
  and disease and insect vectors 159–71
    dengue 159–62
    Japanese encephalitis 160, 162–3
    leishmaniasis 160, 164
    lymphatic filariasis 160, 164–5
    malaria 158, 160, 165–71

Index

onchocerciasis 160, 165
tick- and mite-borne 81, 160, 163
see also trypanosomiasis
and *Drosophila* assemblage 431–9
and Hadley Centre transient climate experiment *Plate 2*, 56
and pollution 232, 235
and population dynamics 431–9
prediction of distributions in changed climate 68, 69–72, 78, 79, 81–3
and tsetse flies *Plates 5–8*, 178–9
see also global warming; Quaternary
Tenebrio molitor 257
tephritid fly (*Urophora*) 17
Termes 483
termites 360, 473–96
behaviour 489
biogeographical variation between sites 476–7
general consequences 492–3
identification and taxonomic identity 490–1
sampling 477–86, 487
seasonal variation 486, 488
work-force and resources 489–90
Termitidae 478
Termitinae 478, 483
terpenoid compounds 458
Tertiary 299
Tettigoniidae and Tettigonioidea 301
Thailand 160, 162
pollution 244
termites 480, 481
Themeda 314
Thetidia smaragdaria 393
Thoracotermes spp 483
M. macrothorax 489
thrips 307
Thymelicus acteon 333
Thysanoptera 198, 200, 207
Tibet 41
ticks 66, 74, 79–80, 82–3
and disease 81, 160, 163
time scales of climate changes 6–7
see also future
Tinocallis saltans 504
tomato see *Lycopersicon*
Tonga 354
transboundary effects 392
transient climate see Hadley Centre
traps, suction see suction trap data

Trechus quadristatus 381
trees/forest
and agriculture 380–4, 385, 457–9, 481, 491, 493
and climate changes 19, 65
critical loads for insects 441–53
defoliators and 510–14
conservation 308, 474, 492
and land-use changes
in arable farmland 380–4, 385, 415–16
and arthropod communities 371–9, 380–4, 385
dieback and re-establishment 455–60
extinctions 324–5, 337–8, 340–3, 353–4, 359
indicators 399–400, 402, 409–10
management and woodland invertebrates 387–91
southern hemisphere 307–9, 313–14, 455–60
see also deforestation
major species see eucalypts; *Picea*; *Pinus*; *Populus*; *Quercus*
and pollution 11, 255
carbon dioxide 199, 201, 202–4, 206, 208–9, 210, 211, 212, 214
gaseous air pollutants 220, 221, 226, 227–8, 230, 232, 234–5, 241
reforestation see forestry
tropical 324, 344, 353, 359
see also Amazonia; termites
see also deforestation; forestry
Trent Biotic Index 273
Trialeurodes vaporariorum 199, 201, 203, 207
triatomine bugs 164
Trichogramma maidis 117
Trichoniscus pusillus and *T. pygmaeus* 381
Trichoplusia ni 200, 203, 205, 208, 214
Trichoptera 32–3, 328
Trinervitermes 482
Trioza erytreae 309
Triticum aestivum (wheat) 97, 231, 234, 238, 241, 243
tropical insects
fossils in Britain 33, 35, 36
as vectors see *under* disease
see also termites
trout 306–7
trypanosomiasis
African 160, 163–4

trypanosomiasis–*contd.*
  American (Chagas' disease) 160, 164
tsetse flies *see Glossina*; remote sensing; trypanosomiasis
Turkey 104
typhus 163

U

ultraviolet 5, 8, 224
United Kingdom *see* Britain
United Nations Commission for Europe 442
United States 112
  aphids as indicators of plant biodiversity 499–505
  climate changes and forest defoliators 510–14
  disease and insect vectors 162, 164
  land-use changes 313, 328, 336, 344
  pollution 239–40
  see also North America
univoltine species and impacts of climate changes 79–81
urbanization
  and disease transmission 160, 165, 171, 172
  and extinctions 336, 340–3
  and land-use changes 311, 375–6, 409
  and pollution 222, 228–9, 238, 443
Urophora cardui 17
UV-B *see* ultraviolet

V

vectors of disease *see under* disease
vegetation *see* plants
Vespidae 301
Vespula germanica 69–70
Viburnum opulus 239
Vicia faba 230, 231, 235–6, 238, 240, 243
Vietnam 327, 354
virus disease of barley 140–3
viviparity *see* aphids
voltinism 79–82, 345, 347
vulnerability to extinction 328, 341–54
  community 351–3
  ecological correlates 341, 344–6
  ecological factors and RDB status 346–8

  phylogenetic status and RDB status 348–51
  predicting geography of 353–4
  see also extinctions

W

Wales, Countryside Council for 386
  see also Britain
walking aphids and temperature 136
warmer climate *see* global warming; temperature
wasps 69–70, 301, 303, 457
water
  beetles 38–9, 393
  concentration in plants 211
  and disease transmission 160, 162, 165, 170–1
  impoundments in southern hemisphere 309–10
  pollution 342–3
    see also quality *below*
  quality, Chironomidae as indicators of 271–91
    biotic indices and river quality 273–4
    deformities induced by pollution 279–80
    in lakes 271, 276–7, 280–7, 289
    palaeoecological indicators 277–9
    in rivers 273–6
  standing 371, 375, 379, 392
  use by plants 94
  waterlogged soil 456, 457
  weed 310–11
  see also irrigation; Lake District; moisture; precipitation; rivers
WCMC *see* World Conservation Monitoring Centre
weather *see* climate; meteorology
weevils
  and climate changes 441–53
  and land-use changes 310, 335, 341, 343, 379
  and pollution 220
West Indies *see* Latin America
wheat *see Triticum aestivum*
WHO (World Health Organization) 160
willow leaf beetle 233
winter 97, 162
  see also climate changes *under* aphids

Index

woodland *see* trees/forest
woodlice 380–1
World Conservation Monitoring Centre 7, 327, 357
World Health Organization 160
Wuchereria bancrofti 164–5

X

Xanthorhoe montanata 385

xylomid fly 389
Xylotoles costatus 335

Z

Zambia 183
Zavrelia melanura 272, 286, 287, 288
Zimbabwe 74, 181, 182, 187
zinc 251–2, 255–7, 262, 275